Colloidal Gold

Colloidal Gold

Principles, Methods, and Applications

VOLUME 2

Edited by

M. A. Hayat

Department of Biology
Kean College of New Jersey
Union, New Jersey

ACADEMIC PRESS, INC.
Harcourt Brace Jovanovich, Publishers
San Diego New York Berkeley Boston
London Sydney Tokyo Toronto

Cover: *Bacteroides nodus* pili preparation sequentially immunolabeled with heterologous antibody and the 5-nm immunogold probe followed by the homologous antibody and finally the 20-nm immunogold probe; x146,000. Courtesy of J. E. Beesley and M. P. Betts. (*See also p. 245 and accompanying text.*)

This book is printed on acid-free paper. ∞

Copyright © 1989 by Academic Press, Inc.
All Rights Reserved.
No part of this publication may be reproduced or transmitted in any form or by any means, electronic or mechanical, including photocopy, recording, or any information storage and retrieval system, without permission in writing from the publisher.

Academic Press, Inc.
San Diego, California 92101

United Kingdom Edition published by
Academic Press Limited
24–28 Oval Road, London NW1 7DX

Library of Congress Cataloging-in-Publication Data

(Revised for vol. 2)

Colloidal gold.

  Includes bibliographies and index.
  1. Colloidal gold. 2. Microscope and microscopy--
Technique. 3. Proteins--analysis. 4. Immunogold
labeling. I. Hayat, M. A., Date.
QR187.I45C64 1989   574.87'6042   88-7361
ISBN 0-12-333927-8   (v. 1 : alk. paper)
ISBN 0-12-333928-6   (v. 2 : alk. paper)

Printed in the United States of America
89 90 91 92   9 8 7 6 5 4 3 2 1

Contents

List of Contributors	xxi
Preface	xxv
Contents of Other Volumes	xxix

1   Molecular Interactions between Colloidal Gold, Proteins, and Living Cells
P. Baudhuin, P. Van der Smissen, S. Beauvois, and P. J. Courtoy

Introduction	2
Characterization of Colloidal Gold Particle Suspensions	2
Size Distribution	3
Average Atomic Composition of Particles	4
Adsorption of Protein on Particles	5
The Scatchard Analysis	5
Flocculation Curve	7
Factors Affecting Adsorption Parameters	7
Particle Size	7
Molecular Weight of Proteins	7
Effect of pH	9

	Interaction of Protein–Gold with Cells	10
	Interaction of Protein–Gold Complexes with Receptors	10
	In Section and "En Face" Observation	12
	Immunogold Labeling of Cellular Components	13
	References	16
2	**Protein A–Gold: Nonspecific Binding and Cross-Contamination** *Alex D. Hyatt*	
	Introduction	19
	Preparation of Protein A–Colloidal Gold Complexes	20
	Principles and Application of Protein A–Gold Labeling	21
	Single Labeling	21
	Double Labeling	22
	Sources of Error in Protein A–Gold Labeling	25
	Cross-Contamination	26
	Nonspecific Binding	27
	Concluding Remarks	30
	References	31
3	**Role of Tissue Processing in Colloidal Gold Methods** *Geoffrey R. Newman and Jan A. Hobot*	
	Introduction	33
	LR White and Lowicryl K4M	36
	LR White or Lowicryl K4M?	41
	Modern Acrylics and Colloidal Gold	43
	References	43
4	**LR White Embedding Medium for Colloidal Gold Methods** *Geoffrey R. Newman*	
	Introduction	48
	Physicochemical Properties and Advantages of LR White	48
	Routine Processing and Polymerization by Heat	49
	Catalytic Polymerization	54

Contents

Rapid Infiltration and Catalytic Polymerization	56
Tissue Blocks	56
Cell Cultures	59
Infiltration at 22°C and Catalytic Polymerization in the Cold	61
Storing LR White Blocks	65
Sectioning LR White Blocks	66
Semithin Sections	66
Thin Sections	67
Immunolabeling of Sections	68
Semithin Sections	68
Thin Sections	68
On-Grid Labeling	68
Preembedding/On-Grid Labeling	69
Counterstaining of Sections	69
Semithin Sections	69
Thin Sections	69
Concluding Remarks	70
References	71

5 Lowicryls and Low-Temperature Embedding for Colloidal Gold Methods
Jan A. Hobot

Introduction	76
Advantages of Lowicryls	78
Fixation and Choice of Embedding Temperature	79
The Progressive Lowering of Temperature Technique	80
Methods of Achieving Low Temperatures	80
Choice of Dehydration Agent	81
Lowicryls K4M and HM20	82
Composition and Handling of Lowicryls	82
Protocols	83
Progressive Lowering of Temperature Technique	83
Embedding at 0°C (273 K) or 22°C (295 K)	84
Embedding at −20°C (253 K) or −10°C (263 K)	84
Rapid Embedding	85

Polymerization	86
Ultraviolet Light	86
Chemical Catalysts	87
Heat	88
Immunolabeling	88
Lowicryl K4M or HM20?	91
Cryosubstitution	95
Methods for Achieving Substitution Temperatures	98
Methods of Freezing and Tissue Preparation	99
Preparation of Substitution Media	100
Infiltration and Embedding	100
Lowicryls K11M and HM23	101
Composition and Handling	101
Substitution Procedures	102
Lowicryl K11M: Embedding and Polymerization	102
Lowicryl HM23: Embedding and Polymerization	102
Polymerization Methods	103
Immunolabeling	103
Future Developments	105
Freeze-Drying and Low-Temperature Embedding	107
General Comments	109
Storage and Thin Sectioning of Lowicryl Blocks	109
Methods for Immunolabeling of Thin Sections	110
Counterstaining of Thin Sections	111
References	112

6 The Enzyme–Gold Cytochemical Approach: A Review
Moise Bendayan

Introduction	118
Preparation of the Enzyme–Gold Complexes	119
Assays for Assessing the Biological Activity of Enzyme–Gold Complexes	121
Tissue Processing	122
Cytochemical Labeling	122
Procedure	124
Cytochemical Controls	124

Quantitative Evaluations	126
Applications	126
Cytochemistry of Nucleic Acids	127
RNase–Gold Complex	127
DNase–Gold Complex	129
Cytochemistry of Extracellular Matrix Components	133
Elastase–Gold Complex	133
Collagenase–Gold Complex	133
Cytochemistry of Polysaccharides and Glycoconjugates	133
Amylase–Gold Complex	136
Glucosidase–Gold and Galactosidase–Gold Complexes	136
Mannosidase–Gold Complex	138
Neuraminidase–Gold and Hyaluronidase–Gold Complexes	138
Cytochemistry of Fungal and Plant Cell Wall Components	141
Xylanase–Gold Complex	141
Cellulase–Gold Complex	141
Chitinase–Gold Complex	141
Exoglucanase–Gold Complex	141
Pectinase–Gold Complexes	142
Cytochemistry of Phospholipids	142
Phospholipase A_2–Gold Complex	142
Conclusions	144
References	145

7 Preparation and Application of Lipoprotein–Gold Complex
Dean A. Handley and Cynthia M. Arbeeny

Introduction	150
Preparation of Lipoproteins	150
Labeling Procedure	152
Visualization of the Lipoprotein–Gold Complex	154
Application of the HDL–Gold Complex	154
Application of the LDL–Gold Complex	155
Applications of Very Low Density Lipoprotein–Gold Complex	157
Summary	158
References	160

8 Preparation and Application of Albumin–Gold Complex
Sergio Villaschi

Introduction	163
Preparation of Albumin–Gold Complex	164
Preparation and Characterization of Gold Markers	164
Determination of Stabilizing Albumin Concentration and Preparation of Gold Probes	165
Storage of Albumin–Gold Complex	168
Working Protocol	169
Application of Albumin–Gold Complex	170
References	173

9 Label-Fracture Cytochemistry
Frederick W. K. Kan and Pedro Pinto da Silva

Introduction	176
Rationale and Development	177
Methodology	178
Preparation of Cells for Label-Fracture	178
Choice of Marker	178
Fixation	181
Cryoprotection, Mounting, and Freezing of Specimens	181
Fracturing and Cleaning of Replicas	182
Images of Label-Fracture Replicas	183
Applications	187
Molecular Demarcation of Surface Domains of Boar Spermatozoa	187
Mapping of Antigens on Boar Sperm Cell Surface	192
Other Applications	192
Advantages and Limitations	192
New Developments and Outlook	196
Replica-Staining Label-Fracture	196
Fracture-Flip	196
References	198

10 Colloidal Gold Conjugates for Retrograde Neuronal Tracing
Daniel Menétrey and Allan I. Basbaum

Introduction	204
Materials and Methods	204
Preparation of Colloidal Gold	204
Adsorption to Proteins	205
Injection	206
Survival Time	207
Tissue Fixation	207
Tissue Sectioning	207
Tissue Processing	208
Light Microscopy	208
Electron Microscopy	209
Applications	209
Light Microscopy	210
Retrograde Labeling	210
Use in Conjunction with Other Retrograde Tracers	211
Use in Conjunction with Enzymatically (HRP) Localized Tracers	211
Use in Conjunction with Immunohistochemistry	212
Electron Microscopy	215
Single Retrograde Labeling	215
Multiple Retrograde Labeling	220
Use in Combination with Anterograde Tracers	221
Use in Combination with HRP-Based Immunocytochemistry	221
Concluding Remarks and Perspectives	224
References	225

11 Colloidal Gold Labeling of Microtubules in Cleaved Whole Mounts of Cells
Jan A. Traas

Introduction	228
Methodology	229
Specimen Preparation	229
Preparation of Cells	229
Preparation of Grids and Coverslips	230

Fixation	230
Plant Cells in Tissues (Roots of Seedlings and Young Plants)	230
Protoplasts from Plant Cells	231
Animal Cells Grown in Culture	231
The Use of Formaldehyde	231
Cleaving	232
Plant Cells in Tissues (Roots of Seedlings)	232
Protoplasts of Plant Cells	232
Animal Cells in Culture	233
Immunogold Labeling	234
Blocking of Nonspecific Binding	234
Immunolabeling	234
Postfixation and Critical-Point Drying	235
Postfixation	235
Dehydration and Critical-Point Drying	235
Examining the Specimen in the Electron Microscope	236
Controls	236
Negative Staining	236
Immunofluorescence	236
Concluding Remarks	237
References	239

12 Colloidal Gold: Immunonegative Staining Method
Julian E. Beesley

Introduction	243
Methodology	244
Immunolabeling Schedule	246
Variations of the Technique	246
Double-Labeling Experiments	247
Quantitation	248
Assessment	249
Applications	249
Virology	249
Bacteriology	251
Conclusions	252
References	252

13 Immunogold Labeling of Viruses *in Situ*
Sylvia M. Pietschmann, Elda H. S. Hausmann, and Hans R. Gelderblom

Introduction	256
Preembedding Immunoelectron Microscopy versus Postembedding Labeling of Thin Sections for Immunoelectron Microscopy	257
Preembedding Immunoelectron Microscopy of Virus-Infected Cells	258
Preparation of Cells for Preembedding Labeling	258
Coating of Plastic and Glass Surfaces	260
Processing of Cell Monolayers	261
Embedding and Evaluation of Microtest Cultures	264
Immunocryoultramicrotomy	265
Preparation of Cells	265
Incubation and Stabilization of Thin Frozen Sections	273
Use of Semithin Cryosections in Immunocytochemistry	274
Choice of Immunoelectron Microscopic Markers and Labeling Techniques	275
Electron-Dense Markers	275
Tactics of Immunolabeling: Direct versus Indirect Techniques	278
Concluding Remarks	279
References	281

14 Study of Exocytosis with Colloidal Gold and Other Methods
Pieter Buma

Introduction	286
Tissue Specimens	287
Methods for Detecting Exocytosis	287
Rat Tissues	288
Comparison of Different Methods	289
Conventional Fixation	289
TAGO Fixation	289
The TARI Method	289

	Applications and Limitations of Methods for Detecting Exocytosis	292
	Immunocytochemical Staining Procedure	293
	Protein A–Gold Labeling of Exocytotic Release Sites	294
	Comparison of Protein A–Gold Method with Other Immunocytochemical Methods	294
	Applications	296
	References	299

15 Colloidal Gold Labeling of Acrylic Resin-Embedded Plant Tissues
Eliot Mark Herman

Introduction	304
Fixation of Plant Cells	305
Processing and Embedding	305
Lowicryl	306
Lowicryl Embedding of Plant Tissues	306
Ethanol Dehydration and Lowicryl Embedding at Subfreezing Temperature	307
Dehydration with N,N'-Dimethylformamide at Room Temperature and Embedding at $-20°C$	309
Future Prospects of Lowicryl in the Plant Sciences	310
LR White	310
Unosmicated Tissue	311
Osmicated Tissue	312
Chemical Catalyst Polymerization at Subzero Temperature	313
Labeling Sections with Specific Antibodies	313
Sources of Antibodies	313
Protocols for the Indirect Labeling of Acrylic Thin Sections	314
Labeling Lowicryl Sections	315
Labeling Unosmicated LR White Sections	315
Removal of Osmium from LR White Sections and Immunogold Labeling	316
Labeling Osmicated Tissue in LR White Sections without Removal of Osmium	317
Controls; Checking for Nonspecific or Pseudospecific Labeling	317
Control Solutions	318
Control Tissues	319

Contents

 Other Artifacts Common to Plant Tissues 319
  Cell Wall Labeling 319
  Pseudospecific Labeling 320
  Glycoprotein Cross-Reactivity 320

 Summary 320
 References 321

16 Preembedding Immunogold Staining of Cell Surface-Associated Antigens Performed on Suspended Cells and Tissue Sections
Corrado Ferrari, Giuseppe De Panfilis, and Gian Carlo Manara

 Introduction 324
 Immunogold Labeling of Suspended Cells 325
  Blood Cell Preparation 325
  Epidermal Cell Suspension Preparation 325
  Prefixation 325
  Immunogold Labeling of a Single Antigen 326
  Immunogold Labeling of Two Distinct Antigens on the Same Cell Surface 328
  Immunogold Labeling of Three Distinct Antigens on the Same Cell Surface 329
  Immunogold–Silver Staining 332

 Immunogold Labeling of Tissue Sections 333
  Rat Kidney Processing 333
   Localization of MoAb8 in Rat Kidney 333
  Human Skin Processing 333
   Cell Surface Labeling 335

 Concluding Remarks 335
 References 340

17 Colloidal Gold in High-Voltage Electron Microscopy—Ruthenium Red Method and Whole-Cell Mount
Kuniaki Takata and Hiroshi Hirano

 Introduction 346
 Colloidal Gold–Lectin–Ruthenium Red Staining 347
  Labeling Procedure 347
  Sectioning and Microscopic Observation 349

Labeling of Whole-Cell Mounts	351
Labeling Procedure	352
Comparison with Scanning Images	353
References	354

18 Correlative Light and Electron Microscopic Immunocytochemistry on Reembedded Resin Sections with Colloidal Gold
Henderson Mar and Thomas N. Wight

Introduction	358
Technical Requirements	360
Fixation	360
Embedding	360
Sectioning	360
Removal of Resin	361
Immunostaining for Light Microscopy	362
Antibodies	362
Peroxidase Screening	363
Colloidal Gold Immunostaining	363
Microscopic Techniques	364
Electron Microsopy of Immunostained Thick Sections	364
Reembedding Techniques for Electron Microscopy	364
Electron Microscopy of Reembedded 2-μm Sections	365
Electron Microscopy of Deplasticized Adjacent Thin Sections	365
Preparation and Handling of Grids	365
Deplasticization of Thin Sections	366
Colloidal Gold Immunocytochemical Staining	367
Reembedding of Thin Sections	368
General Characteristics of Deplasticized Sections	368
Section Thickness	368
Rate of Resin Removal	369
Preservation of Antigenicity	370
Preservation of Ultrastructure	372
Limitations	375
References	377

19 Streptavidin–Gold Labeling for Ultrastructural *in Situ* Nucleic Acid Hybridization
Robert A. Wolber and Theodore F. Beals

Introduction	380
Principles of Hybridization Technique	381
Fixation	381
Probes	382
Preembedding Ultrastructural Hybridization	382
Cell Pretreatments	383
Hybridization	385
Quantitation	387
Improvements	392
Postembedding Hybridization on Thin Sections	392
Applications	393
References	393

20 Detection of Proteins with Colloidal Gold
Roland Rohringer

Introduction	398
Methodology	399
General, Nonspecific Detection of Proteins on Blots	399
Preparation of Gold Colloids Using White Phosphorus	400
Preparation of Gold Colloids Using Citric Acid	400
Treatment of Blots before Staining with Gold Sols	401
pH and Colloidal Gold Particle Size	401
Staining Protocol	401
Detection of Specific Proteins on Blots	401
General and Specific Protein Staining Performed Sequentially on the Same Blot	404
Negative Gold Stain	406
Quantitative Determination of Protein	407
Applications	408
Selected Examples of the Use of Gold Colloids on Blots	408
References	409

21 Undecagold–Antibody Method
James F. Hainfeld

Introduction	413
History of the Undecagold Cluster	415
Methodology	417
Preparation of Undecagold	417
Purification of the Undecagold Cluster	420
Activation of Gold Cluster	422
Preparation of Fab' Fragments	423
Reaction of Fab' with Au_{11} and Purification	424
Quantitation of Gold Cluster Labeling	424
Electron Microscopy	425
Applications	427
References	428

22 Immunogold Labeling for the Single-Laser FACS Analysis of Triple Antibody-Binding Cells
Thomas H. Tötterman and Roger Festin

Introduction	432
Techniques	433
Cells	433
Antibodies	433
Staining of Cells	434
Flow Cytometry (FCM)	434
Applications	435
Detection of Immunogold Stained Lymphocytes	435
Accuracy of Gold versus Fluorochrome Staining	435
Triple Antibody Staining	436
Comments	438
Advantages	438
Limitations	439
Future Prospects	440
Summary	441
References	441

23 Silver-Enhanced Colloidal Gold for the Detection of Leukocyte Cell Surface Antigens in Dark-Field and Epipolarization Microscopy
M. De Waele

Introduction	444
Immunogold–Silver Staining of Leukocyte Cell Surface Antigens with Monoclonal Antibodies	445
Preparation of the Cell Suspensions	445
Monoclonal Antibodies	445
Colloidal Gold-Labeled Goat Antimouse Antibodies	446
Physical Developer	446
Immunogold–Silver Staining Procedure	447
Examination of the Preparations in Bright-Field, Dark-Field, and Epipolarization Microscopy	448
The Microscope	448
Bright-Field Microscopy	449
Dark-Field Microscopy	449
Epipolarization Microscopy	449
Appearance of the Preparations	451
Standardization of the Immunogold–Silver Staining Procedure	453
Determination of the Silver Enhancement Time	454
Titration of the Colloidal Gold-Labeled Goat Antimouse Antibodies	455
Size of the Gold Probe	455
Titration of the Gold Reagent	455
Titration of the Monoclonal Antibody	458
The Optimal Silver Enhancement Interval	459
Reactivity of the Physical Developer: Influence of Light and Temperature	461
Performance of Two Epipolarization Microscopes	463
Conclusion	464
References	465

Index 469

List of Contributors

Numbers in parentheses indicate the pages on which the authors' contributions begin.

Cynthia M. Arbeeny (149), Department of Medicine, Albert Einstein College of Medicine, Bronx, New York 10461

Allan I. Basbaum (203), Departments of Anatomy and Physiology, University of California, San Francisco, California 94143

P. Baudhuin (1), UCL 75.41 Université de Louvain, Avenue Hippocrate, 75, B-1200 Brussels, Belgium

Theodore F. Beals (379), Department of Pathology, Laboratory Service, Ann Arbor Veterans Administration Medical Center, Department of Pathology, School of Medicine, The University of Michigan, Ann Arbor, Michigan 48105

S. Beauvois (1), Laboratoire de Chimie Physiologique, Université Catholique de Louvain and International Institute of Cellular and Molecular Pathology, B-1200 Brussels, Belgium

Julian E. Beesley (243), Wellcome Research Laboratories, Langley Court, Beckenham, Kent BR3 3BS, England

Moise Bendayan (117), Department of Anatomy, Université de Montréal, Montréal, Québec, Canada H3C 3J7

Pieter Buma (285), Department of Orthopaedy, St. Radboud Hospital, Th. Craanenlaan 1, 6500 HB Nijmegen, The Netherlands

P. J. Courtoy (1), Laboratoire de Chimie Physiologique, Université Catholique de Louvain and International Institute of Cellular and Molecular Pathology, B-1200 Brussels, Belgium

Giuseppe De Panfilis (323), Istituto di Clinica Dermosifilopatica, Universitá di Parma, Viale Gramsci 14, I-43100 Parma, Italy

M. De Waele (443), Laboratory of Hematology and Immunology, AZ-VUB, Laarbeeklaan 101, B-1090 Brussels, Belgium

Corrado Ferrari (323), Istituto di Istologia e Embriologia generale, Universitá di Parma, Viale Gramsci 14, I-43100 Parma, Italy

Roger Festin (431), Clinical Immunology Section, University Hospital, S-751 85 Uppsala, Sweden

Hans R. Gelderblom (255), Robert Koch-Institut des Bundesgesundheitsamtes, Nordufer 20, D-1000 Berlin 65, Federal Republic of Germany

James F. Hainfeld (413), Brookhaven National Laboratory, Biology Department, Upton, New York 11973

Dean A. Handley (149), Mediators and Biomolecular Therapy Section, Monoclonal Antibody Department, Sandoz Research Institute, East Hanover, New Jersey 07936

Elda H. S. Hausmann (255), Robert Koch-Institut des Bundesgesundheitsamtes, Nordufer 20, D-1000 Berlin 65, Federal Republic of Germany

Eliot Mark Herman (303), Plant Molecular Biology Laboratory, U.S. Department of Agriculture, ARS, PSI, PMBL, Room 118, Building 006, BARC-West, Beltsville, Maryland 20705

Hiroshi Hirano (345), Department of Anatomy, Kyorin University School of Medicine, Shinkawa, Mitaka, Tokyo 181, Japan

Jan A. Hobot (33, 75), University of Wales College of Medicine, Electron Microscopy Unit, Heath Park, Cardiff CF4 4XN, Wales

Alex D. Hyatt (19), Australian Animal Health Laboratory, CSIRO, P.O. Bag 24, Geelong, Victoria 3220, Australia

Frederick W. K. Kan (175), Department of Anatomy, Faculty of Medicine, Université de Montréal, C.P. 6128 Succ. A, Montreal, Québec, Canada H3C 3J7

Gian Carlo Manara (323), Istituto di Clinica Dermosifilopatica, Universitá di Parma, Viale Gramsci 14, I-43100 Parma, Italy

Henderson Mar (357), Department of Pathology, School of Medicine, University of Washington, Seattle, Washington 98195

Daniel Menétrey (203), INSERM, U-161, Neurophysiologie Pharmacologique, 2 rue d'Alésia, 75014 Paris, France

Geoffrey R. Newman (33, 47), University of Wales College of Medicine, Electron Microscopy Unit, Heath Park, Cardiff CF4 4XN, Wales

List of Contributors

Sylvia M. Pietschmann (255), Robert Koch-Institut des Bundesgesundheitsamtes, Nordufer 20, D-1000 Berlin 65, Federal Republic of Germany

Pedro Pinto da Silva (175), Membrane Biology Section, Laboratory of Mathematical Biology, National Cancer Institute, Frederick Cancer Research Facility, Frederick, Maryland 21701

Roland Rohringer (397), Agriculture Canada, Research Station de Recherche, 195 Dafoe Road, Winnipeg, Manitoba, Canada R35 2M9

Kuniaki Takata (345), Department of Anatomy, Kyorin University School of Medicine, Shinkawa, Mitaka, Tokyo 181, Japan

Thomas H. Tötterman (431), Clinical Immunology Section, University Hospital, S-751 85 Uppsala, Sweden

Jan A. Traas (227), I.N.R.A., Station d'amélioration des plantes maraichaires, B.P. 94, 84140 Montfavet, France

P. Van der Smissen (1), Laboratoire de Chimie Physiologique, Université Catholique de Louvain and International Institute of Cellular and Molecular Pathology, B-1200 Brussels, Belgium

Sergio Villaschi (163), Division of Anatomic Pathology, Department of Surgery, Second University of Rome "Tor Vergate," Via Orazio Raimondo 1, 00173 Rome, Italy

Thomas N. Wight (357), Department of Pathology, School of Medicine, University of Washington, Seattle, Washington 98195

Robert A Wolber (379), Division of Anatomic Pathology, Vancouver General Hospital, Department of Pathology, University of British Columbia, Vancouver, B.C., Canada V5Z 1M9

Preface

As recently as 1971, Faulk and Taylor introduced colloidal gold as a marker for specific ligands for electron microscopy. Since then the development and application of this method have been phenomenal. Currently, it is the most widely used methodology in the field of immunocytochemistry; it is truly a universal cytochemical method. This rapid development is not surprising considering the useful and important characteristics of the marker. Colloidal gold markers can be prepared with a large number of macromolecules which generally maintain their bioactivity. The macromolecules that have been adsorbed to the gold particles include proteins A and G, immunoglobulins, lectins, toxins, glycoproteins, lipoproteins, dextran, enzymes, streptavidin, hormones, glycoalbumin, transferrin, neomycin, polymyxin, insulin, and bovine serum albumin.

One of the unique properties of the colloidal gold methodology is that gold particles of various dimensions (2–150 nm in diameter) can be prepared easily for multiple labeling; this labeling allows the simultaneous visualization of more than one macromolecule on the same section. Undecagold clusters with a gold atom core of as small as 0.82 nm in diameter can be synthesized for labeling an antibody (Chapter 21, this volume, by J. F. Hainfeld). This wide range of size allows for both low and high resolution studies.

Colloidal gold particles are electron opaque. They display little nonspecific adsorption to the embedding media and can thus be employed with epoxy, methacrylate, and acrylic resins with equal success. These markers can be used for both preembedding and postembedding immunolabeling. Conventional thin sections as well as thin cryosections can be labeled with colloidal gold.

The colloidal gold method is applicable to most microscopical systems including optical microscopy, scanning, transmission, and high voltage electron microscopy, photoelectron microscopy, photon microscopy, fluorescent microscopy, and dark-field and epipolarization microscopy. This marker can be used with both the secondary and backscattered electron modes of the scanning electron microscope. Colloidal gold has been employed in conjunction with label-fracture, replica-staining label-fracture, immunoreplica, and negative staining. Sputter coating with platinum can be used in combination with immunogold labeling to study whole mounts of cytoskeletons for scanning and transmission electron microscopy and for scanning transmission electron microscopy. Correlative light and electron microscopy can also be carried out with immunogold staining.

Enzymes labeled with this marker constitute specific probes for the cytochemical detection of corresponding substrates, thereby facilitating the study of a large variety of molecules. One of the most recent applications of colloidal gold is in the field of *in situ* nucleic acid hybridization; another is in the study of dynamic cellular processes such as endocytosis, membrane fusion, spreading of cells on surfaces, and cellular processing of molecules. Immunogold staining is useful for virus detection through the interaction of antibodies with the surface of virus particles. Immunogold negative staining is an important method in virology. All of the aforementioned and other methods are presented in this three-volume treatise.

Since the method of specimen embedding influences labeling with colloidal gold, a detailed discussion on the use of Lowicryls and LR White is presented in this volume. Procedures to minimize nonspecific background labeling of resin sections with colloidal gold are explained. The progressive lowering of temperature used for embedding in Lowicryls is also detailed.

Colloidal gold also has many nonmicroscopical and noncytochemical applications such as the detection of specific proteins on blots; this application is presented in this volume. In order to understand and interpret the results of any methodology, the principles governing it must be known. In this connection, the limitations of a methodology must also be known. Such information is given in the three volumes.

This volume has developed through the efforts of 36 scientists representing 12 countries. All the contributors are eminent authorities in their respective fields. Indeed, many are the originators of important methods.

Preface

The scientific community is fortunate to have the benefit of a continual expansion and refinement of colloidal gold methodology by those who originated it. I greatly appreciate their participation and promptness in completing the chapters.

M. A. Hayat

Contents of Other Volumes

Volume 1

1. The Development and Application of Colloidal Gold as a Microscopic Probe
   Dean A. Handley
2. Methods for Synthesis of Colloidal Gold
   Dean A. Handley
3. Protein A–Gold and Protein G–Gold Postembedding Immunoelectron Microscopy
   Moïse Bendayan
4. Preparation and Application of Lectin–Gold Complexes
   Nicole Benhamou
5. Colloidal Gold for Multiple Staining
   Jeannine Doerr-Schott
6. Immunogold Labeling of Ultrathin Cryosections
   Paul M. P. van Bergen en Henegouwen
7. Colloidal Gold for Scanning Electron Microscopy
   Marc Horisberger
8. Backscattered Electron Imaging of the Colloidal Gold Marker on Cell Surfaces
   Etienne De Harven and Davide Soligo
9. Silver-Enhanced Colloidal Gold Method
   Lucio Scopsi

10 Silver-Enhanced Colloidal Gold for Light Microscopy
   Gerhard W. Hacker
11 Strategies in Colloidal Gold Labeling of Cell Surfaces and Cytoskeletal Elements of Cultured Cells
   G. Bruce Birrell and O. Hayes Griffith
12 Colloidal Gold–Immunoreplica Method
   M. V. Nermut and A. Nicol
13 Immunogold Double-Diffusion Method
   James Robinson Harris
14 Colloidal Gold-Labeled Agarose–Gelatin Microspherule Method
   Kui-xiong Gao
15 Colloidal Gold for Microbiological Immunoctyochemistry
   Julian E. Beesley
16 Immunogold Labeling of Viruses in Suspension
   Elisabeth Kjeldsberg
17 Plasma Membrane Localization of Proteins with Gold Immunocytochemistry
   Anthony N. van den Pol, Mark Ellisman, and Tom Deerinck
18 Factors Affecting the Staining with Colloidal Gold
   Kinam Park, Haesun Park, and Ralph M. Albrecht

   Index

Volume 3 (in preparation)

1 Choice of Methods
   Eduard Kellenberger and M. A. Hayat
2 Lowicryl Resins
   Werner Villiger
3 Polar or Apolar Lowicryl Resin for Immunolabeling?
   Markus Dürrenberger
4 Comparable Yield of Immunolabel by Resin Sections and Thawed Cryosections
   York-Dieter Stierhof
5 Techniques for the Production of Monoclonal and Polyclonal Antibodies
   K. John Morrow, J. Mleczko, S. Kreuzer, E. Unuvar, and S. King

6   Microinjection of Colloidal Gold Tracers in Nuclear
    Transport Studies
    Steven Dworetzky and Carl Feldherr
7   Location and Identification of Colloidal Gold Particles with
    an Energy Dispersive Analyzer
    Sinikka Eskelinen
8   Immunochemistry of Antibody Binding to Surface
    Immobilized Antigen
    Håkan Nygren and Maria Werthen
9   Simultaneous Demonstration of Antigens on Outer and
    Protoplasmic Surfaces of Plasma Membranes by Replica
    Immunocytochemistry
    G. Rutter and Heinz Hohenberg
10 Detection of Sparse Antigens by Immunogold Method
    Richard W. Anderson and B. Pathak
11 Double Labeling of Antigenic Sites on Cell Surfaces Imaged
    with Backscattered Electrons
    Ellen Namork
12 Preparation and Application of Insulin–Colloidal Gold
    Complex
    Robert M. Smith and Leonard Jarett
13 Application of Colloidal Gold to Diagnostic Pathology
    Guillermo A. Herrera

Index

1

Molecular Interactions between Colloidal Gold, Proteins, and Living Cells

P. BAUDHUIN, P. VAN DER SMISSEN, S. BEAUVOIS, and P. J. COURTOY

Laboratoire de Chimie Physiologique
Université Catholique de Louvain
and
International Institute of Cellular and Molecular Pathology
Brussels, Belgium

INTRODUCTION
CHARACTERIZATION OF COLLOIDAL GOLD
   PARTICLE SUSPENSIONS
  Size Distribution
  Average Atomic Composition of Particles
ADSORPTION OF PROTEIN ON PARTICLES
  The Scatchard Analysis
  Flocculation Curve
FACTORS AFFECTING ADSORPTION PARAMETERS
  Particle Size
  Molecular Weight of Proteins
  Effect of pH

INTERACTION OF PROTEIN–GOLD WITH CELLS
  Interaction of Protein–Gold Complexes with Receptors
  In Section and "En Face" Observation
  Immunogold Labeling of Cellular Components
REFERENCES

INTRODUCTION

The widespread use of colloidal gold particles as markers for electron microscopy can be attributed mainly to four factors: (1) high contrast in the electron microscope allows easy detection of small particles; (2) homogeneity of the particle shape and size, together with the high contrast, avoids confusion with naturally occurring cell structures; (3) nonspecific adsorption properties of colloidal gold particles allow their use as tracers for proteins differing largely in nature and physical characteristics; (4) relatively simple methods can be used to obtain populations of colloidal gold particles of different average sizes, facilitating double labeling. Full advantage of this potential, however, can be taken when the characteristics (at the molecular level) of the gold particle suspensions and the adsorption of proteins on their surface are known. Indeed, such information is necessary in order to achieve quantitative observations at the electron microscope level and to assess their significance at the subcellular level.

This chapter will emphasize the quantitative evaluation of the properties of colloidal gold particle suspensions. It will also describe, in molecular terms, gold–protein complexes and their interactions with living cells and fixed tissues. Experimental data will be presented for illustration purposes; more detailed information on the adsorption of specific proteins has been presented by De Roe *et al.* (1987).

CHARACTERIZATION OF COLLOIDAL GOLD PARTICLE SUSPENSIONS

Our experience is mainly with preparations obtained by the citrate reduction method, originally described by Frens (1973) and Horisberger (1979), as modified by Slot and Geuze (1985) and De Roe *et al.* (1987). Before adsorbing proteins, we routinely characterize the size distribution of particle suspensions and determine concentration and average mass of particles for each preparation of colloidal gold.

Size Distribution

For size distribution analysis, suspensions are sprayed on Formvar-coated grids (Effa Spray Mounter; E. F. Fullam, Schenectady, NY) and examined with the electron microscope. Sizing is performed from micrographs, at a final magnification of 200,000–300,000, as calibrated with respect to a grating replica (2160 lines/mm, E. F. Fullam), photographed, and processed together with the sample. The optomechanical particle size analyzer TGZ3 of Zeiss (Oberkochen, FRG), in which a light disk is fitted on the particle by means of a handwheel, is very convenient, but simpler methods such as the test circles described by Weibel (1979) or more complex computer-based methods may also be used.

As an average, with the usual dispersion (coefficient of variation less than 10% of gold particle diameter), we measure 400 particles per preparation; 100 particles is a minimum for characterizing size. The size distribution histogram is divided in 8 to 10 classes. Size distributions for particle preparations with a nominal diameter of 8 or 15 nm are presented in Fig. 1.1. The clear separation of the two distributions should be noted: it clearly illustrates the possible use of double labeling by controlling the experimental conditions for the reduction of tetrachloroauric solutions.

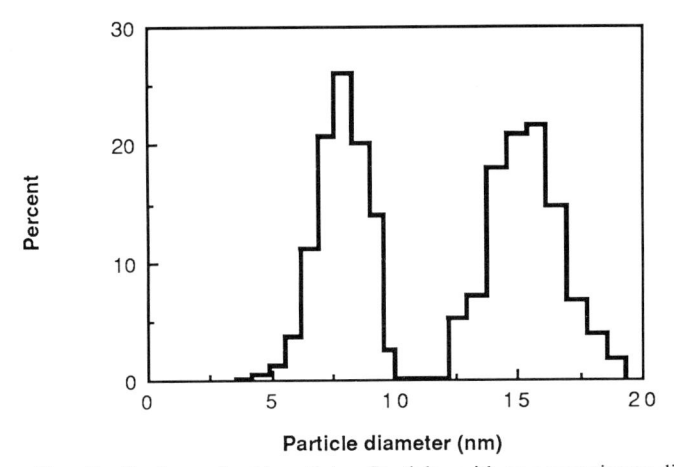

Fig. 1.1. Size distributions of gold particles. Particles with an approximate diameter of ~8 nm were obtained by reduction of 94 ml of 0.259 mM boiling tetrachloroauric acid with 4 ml of 34 mM trisodium citrate and 2 ml of 0.1% tannic acid. Particles with an approximate diameter of ~15 nm were obtained by reduction of 243 ml of 0.30 mM boiling tetrachloroauric acid with 7.5 ml of 38.8 mM trisodium citrate. Both preparations were boiled under reflux for additional 15 min after rapid addition of reductants. Distributions are based on 430 and 370 particles of ~8 nm and 15 nm approximate diameter, respectively.

Our experience indicates that with rapid addition and mixing of reductants, further purification in glycerol density gradients (Slot and Geuze, 1981) can be omitted. Average parameters of the preparations of Fig. 1.1 are listed in Table 1.1. The size distribution yields the true statistical estimate for average diameter, surface, and volume of particles. Although the differences are relatively small owing to the low dispersion of sizes, the average surface or average volume is larger than the surface or the volume of the particle with average diameter. The importance of particle surface measurement will be apparent from the discussion on adsorption properties. As described below, the average volume is used to assess concentration in terms of number of particles per unit volume of suspension.

Average Atomic Composition of Particles

For standardization purposes, gold concentration is best determined by atomic emission spectrometry. However, for simplicity, a correlation should be established between the results obtained with this technique and classical absorption spectrophotometry at a 290-nm wavelength for chloroauric solutions and at 520 nm for colloidal suspensions. The absorbance at 520 nm is not affected by protein adsorption. Furthermore, since the reduction of gold by citrate is complete, the concentration of gold in

TABLE 1.1
Examples of Characterization of Colloidal Gold Suspension[a]

	Approximate Diameter[b]	
Parameter of Particles	8 nm	15 nm
Average diameter, measured (nm)	7.91	15.3
Standard deviation of diameter (nm)	1.05	1.45
Area of average diameter particle (nm^2)	197	737
Volume of average diameter particle (nm^3)	259	1875
Average area (nm^2)	200	744
Average volume (nm^3)	273	1932
Concentration of gold		
(mM)	0.243	0.291
(μg/ml)	47.9	57.3
Average particle mass (g)[c]	4.64×10^{-18}	32.8×10^{-18}
Gold atoms/particle	14,180	100,230
Particle concentration (number/ml)	10.3×10^{12}	1.74×10^{12}

[a] Preparations analyzed are those of Fig. 1.1.
[b] Approximate diameter is defined as the rounded average diameter expected from the procedure used for reduction of gold.
[c] Assuming a density of 17 g/ml.

the colloidal suspension is identical to that of the initial tetrachloroauric solution. We have found that the absorbance of the 0.30 mM chloroauric solution, classically used for preparation of colloidal gold, is 0.878 at 290 nm. For 15-nm particles, a suspension 0.30 mM in gold has an absorbance of 1.153 at 520 nm; the corresponding value for 8-nm particles is 1.125.

The number of gold particles per unit volume (Ackerman *et al.*, 1983) can be estimated from the concentration of gold, taking into account a density of 17 g/ml (Weast, 1979). Average particle mass and the number of gold atoms in a single particle can be computed from the particle concentration and the average particle volume. These two parameters, when multiplied respectively by N, Avogadro's number, and by the atomic weight of gold, can be compared directly to classical molecular weights of biological macromolecules expressed in daltons. Taking the 8- and 15-nm (nominal) diameter particle populations described in Table 1.1, average values of 2.8×10^6 and 19.7×10^6 daltons per particle are obtained.

ADSORPTION OF PROTEIN ON PARTICLES

The Scatchard Analysis

The classical Scatchard analysis (Scatchard, 1949) is applicable to the adsorption of proteins to gold particles. When association of proteins on the particles is studied at various concentrations, typical saturation curves are obtained, which can be linearized according to the equation

$$\frac{B}{F} = \frac{1}{K_d} N_{tot} - \frac{1}{K_d} B$$

where B is the number of protein molecules bound per particle, F the molar concentration of the protein free in solution, K_d the equilibrium constant for dissociation of protein from particles, and N_{tot} the number of available adsorption sites, i.e., the number of molecules per particle at saturation. Hence, the intercept of the line with the abscissa ($B/F = O$) yields the value N_{tot}, while the slope of the line is $-1/K_d$. Figure 1.2 presents an example of such analysis performed on transferrin.

The applicability of Scatchard analysis to the formation of protein–gold complexes indicates a finite number of adsorption sites, leading to saturation. This can be explained best by assuming that a monomolecular shell of protein is formed around the particles. Indeed, direct morphological evidence for this model has been obtained for protein such as low-density lipoprotein (LDL) (Handley *et al.*, 1981b) and ferritin (De Roe *et al.*,

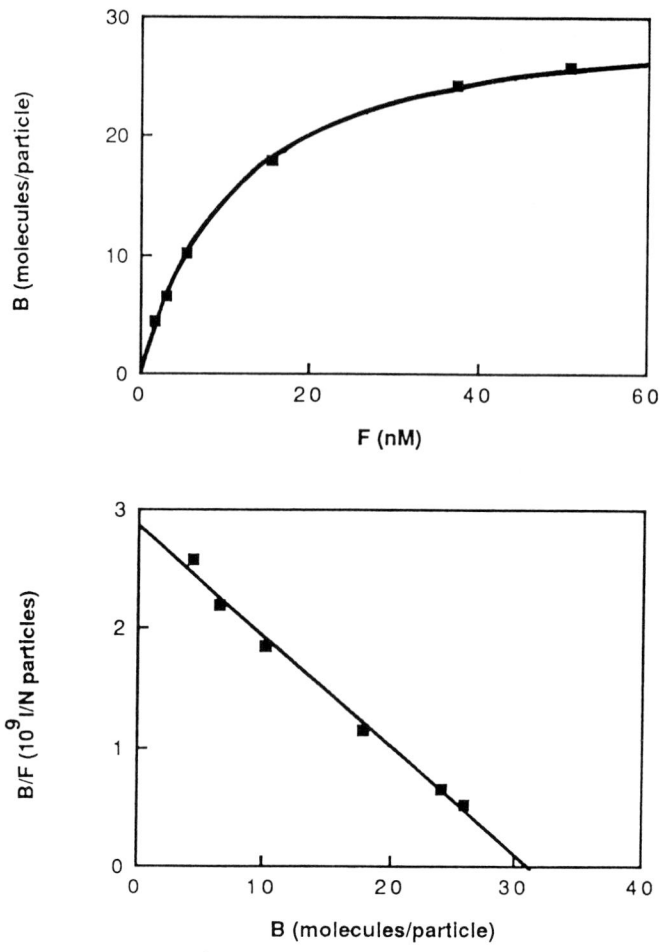

Fig. 1.2. Adsorption of transferrin to 15-nm particles. Suspensions were exposed for 15 min to various concentrations of radiolabeled transferrin dialyzed against 2 mM sodium tetraborate (pH 9) and then stabilized by addition of BSA at a final concentration of 1.7 mg/ml. They were cleared of unbound protein by three centrifugations at $9 \times 10^5 \, g \times$ min (30,000 rpm, 15 min, using a Beckman Preparative Centrifuge). The upper panel illustrates the experimental data. Scatchard analysis (lower panel) yielded an equilibrium constant (K_d) of 9.5 nM and 31 molecules per particle at saturation. The saturation curve was calculated using these values.

1987). The advantage of the analysis is that it fully characterizes the adsorption phenomenon by two independent parameters: the number of sites per gold particle and the dissociation equilibrium constant. As discussed in detail below, a relation between these parameters and molecular weight exists for globular proteins.

Flocculation Curve

A rough approximation of the amount of protein needed for saturating gold particles can also be obtained from a flocculation curve (Fig. 1.3). Flocculation of uncovered, or partially covered, particles is induced by addition of 140 mM NaCl and can be monitored by following the absorbance of the suspension at 580 nm. The concentration minimizing the optical density, and thus avoiding flocculation, indicates saturation.

FACTORS AFFECTING ADSORPTION PARAMETERS

Particle Size

As expected, when a protein such as bovine serum albumin (BSA) is adsorbed to particles of different sizes, the maximum number of molecules which can be associated with a single particle (N_{tot}) is linearly related to the average projection area of the particles. Otherwise, for a given protein, the area occupied by a single molecule was roughly constant irrespective of the particle size.

Molecular Weight of Proteins

From the saturation curves as well as from the effect of particle size, it can be inferred that the area occupied by a protein will be equal to its

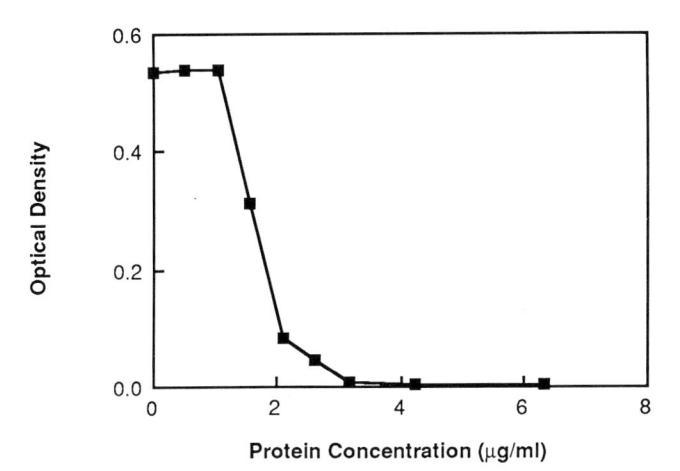

Fig. 1.3. Flocculation curve for protein A. Particles with a diameter of ~8 nm (1 ml of a suspension of 12 × 10^{12} particles/ml) were incubated for 15 min with various concentrations of protein A. Flocculation was induced by addition of NaCl (final concentration 140 mM) and absorbance was measured at 580 nm. (Courtesy of Dr. Ch. Slomianny.)

projection on the gold particle surface. As already mentioned, this model of a monomolecular shell of protein covering the gold particle is confirmed by electron microscopy. In the case of a sphere, the projection area of a protein will be proportional to its volume or mass, raised to a power of 2/3. Hence, for globular proteins, the area occupied by a molecule should be proportional to (molecular weight)$^{2/3}$, and the maximum number of molecules that can be adsorbed on a single particle should be inversely proportional to that quantity. As shown in Fig. 1.4, such relation is indeed observed for several proteins. It should be noted that this implies that the proteins studied have roughly the same density in solution. The fact that galactosylated BSA displays a much lower number of molecules per particle than expected may indicate that the carbohydrate moieties increase the volume of the BSA molecule far in excess of the mass that they represent. Similarly, LDL particles represent a rather special case, since the molecular weight of the lipid part of the LDL particle

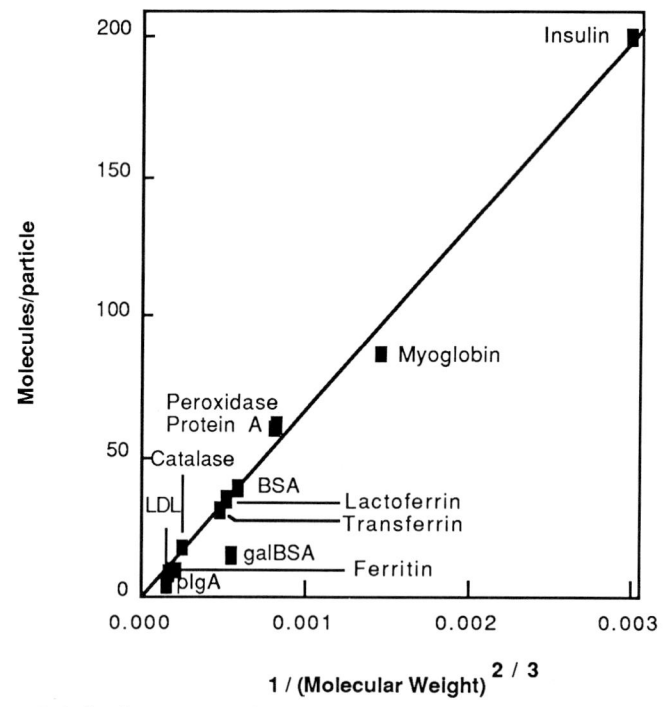

Fig. 1.4. Relation between number of molecules at saturation and molecular weight. Approximate diameter of particles was 15 nm. For globular proteins, a linear relationship is found between the maximum number of molecules bound to a single gold particle and the inverse of (molecular weight)$^{2/3}$. (Redrawn from De Roe et al., 1987.)

TABLE 1.2
Adsorption Parameters of Various Proteins on Gold Particles[a]

Protein	Molecules/Particle (at Saturation)	K_d (nM)
BSA	39	160.3
Catalase	18	36.5
Ferritin	8	81.8
Galactosylated BSA	15	47.1
Horseradish peroxidase	61	11.8
IgA (polymeric)	4	182.5
Insulin	200	3.6
Lactoferrin	35	20.3
LDL	9	73.8
Myoglobin	87	7.6
Protein A	60	10.9
Transferrin	31	10.2

[a] The approximate diameter of the gold particles used is 15 nm.

is not taken into account in Fig. 1.4. We have summarized in Table 1.2 the adsorption parameters measured for the association of a series of proteins to gold particles with a nominal diameter of 15 nm.

An effect of the molecular weight on the equilibrium constant (K_d) is also observed. A correlation, albeit weaker than in Fig. 1.4, exists between the affinity ($1/K_d$) and the 2/3 power of molecular weight. This may be reflecting the possibility that multiplicity of contact points between larger molecules and gold particles increases their affinity.

Effect of pH

An agreement on the effect of pH on the adsorption of proteins to gold particles is lacking. We were unable to observe a relation between the isoelectric point of insulin, myoglobin, BSA, galactosylated BSA, or IgA and the number of molecules associated to a single particle. Moreover, adsorption of myoglobin occurred with the same affinity at pH 5.2, 5.6, and 7.2, although the number of binding sites decreased from 84 to 31 when pH was raised from 5.2 to 7.2.

Since nonstabilized particles flocculate in the presence of added electrolytes, the simplest explanation for the decrease of total binding observed at neutral pH is that the small increase of ionic strength resulting from pH adjustment causes some flocculation of particles. Hence, we do not recommend altering the pH of colloidal suspensions before protein adsorption. In our experience, omitting this step not only simplifies the procedure but also makes further purification by glycerol density gradients unnecessary.

INTERACTION OF PROTEIN–GOLD WITH CELLS

In cell biology, protein–gold complexes can be used as tracers to follow intracellular pathways and, more specifically, to study modalities of cellular uptake and processing of proteins added to the extracellular medium (Handley et al., 1981a; Dickson et al., 1981). Alternatively, they can be applied to fixed cells or tissue sections, as markers of specific binding sites or antigens (Griffiths et al., 1983; Tokuyasu, 1986).

Interaction of Protein–Gold Complexes with Receptors

Adsorption of ligand molecules to gold particles may affect the binding equilibrium with their receptor owing to several factors. We are unable to discuss possible interactions between protein molecules adsorbed to the same particle, since no pertinent data on this aspect are available. It should, however, be stressed that such interactions are likely to occur and that we may thus be introducing an approximation. Indeed, the distance between adjacent molecules may partially preclude simultaneous binding, whereas binding of several adjacent molecules may considerably slow down dissociation of the protein–gold complex from the receptor.

1. Clustering of several ligand molecules on a particle, combined with the mass of the gold particle, results in a high-molecular-weight complex with a much smaller diffusion constant than the free ligand molecule in solution.

2. The adsorption of the ligand to the gold particle may involve the binding site of the biological receptor. The effect will vary according to the valency of the soluble ligand molecules and the surface properties of the ligand, which may promote specific orientations for adsorption. Generally, these factors can be expected to reduce the average affinity of the individual ligand molecules for the biological receptor. Using as an experimental model asialofetuin binding to rat hepatocytes in culture (Figs. 1.5 and 1.6), we have observed an eightfold reduction in the affinity of asialofetuin molecules for the hepatic galactose receptor when asialofetuin is adsorbed to 15-nm gold particles (Beauvois et al., 1987).

3. The number of ligand molecules bound per cell should be higher with ligand–gold complexes than with free soluble ligand molecules. With

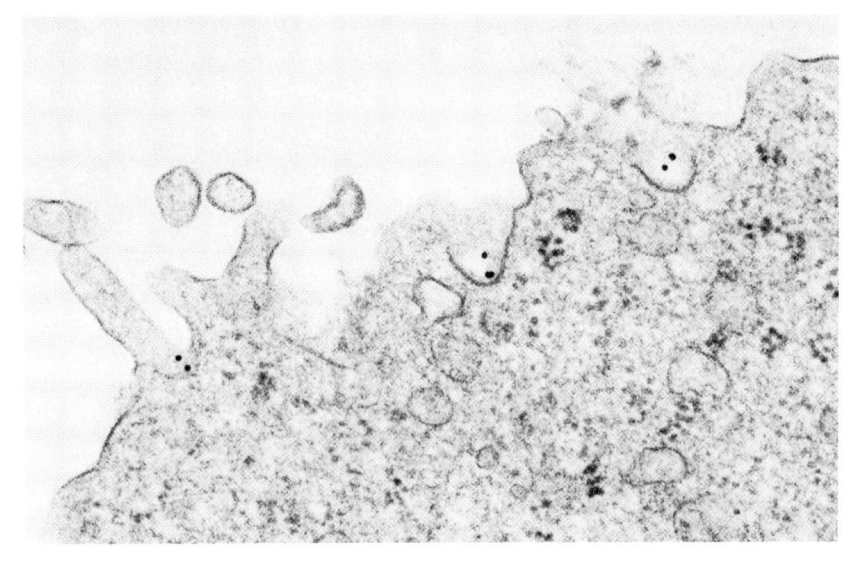

1. Interactions between Gold, Proteins, and Cells

Fig. 1.5. Surface labeling of the galactose-specific receptor by asialofetuin-coated gold particles observed in a section. After binding of ligand–gold complexes for 24 hr and extensive washing, cells were fixed for 1 hr with 1.5% glutaraldehyde in 0.1 M cacodylate buffer (pH 7.4, 22°C) and postfixed with 1% OsO_4 in the same buffer (4°C, 1 hr). After "en bloc" staining with 1% uranyl acetate (4°C, in the dark, for 18 hr), the cells were dehydrated in graded series of ethanol, embedded, sectioned, and poststained with uranyl acetate and lead citrate according to standard procedures. Gold particles are clustered in coated pits; ×52,500.

ligand-coated gold particles, each molecule bound to its biological receptor will also associate to the cell all other molecules adsorbed on the same gold particle. The increase in the number of ligand molecules bound to the cell would then be expected to correspond to the average number of molecules per gold particle. However, using asialofetuin and the hepatic galactose receptor model, we have observed an increase equal to only one-third of the average number of molecules per gold particle. This could reflect the clustering of receptor molecules, which, after immobilization of a gold particle on the membrane, diffuse laterally and bind to adjacent ligand molecules on the same particle.

Owing to the multiplicity of the factors involved, it is thus advisable to check the binding of ligand–gold complexes to the cells which are under study and to compare the parameters of the binding equilibrium with those observed for the ligand occurring free in solution. In fixed material, cross-linking of membrane proteins excludes their lateral diffusion, and multiple binding will occur only to clusters of receptors existing before fixation.

Fig. 1.6. Surface labeling of the galactose-specific receptor by asialofetuin-coated gold particles observed on a whole-mount carbon replica. After binding of ligand–gold complex as in Fig. 1.5, the cells were fixed overnight with 1.5% glutaraldehyde in a 0.1 M cacodylate buffer at 4°C and postfixed with 2% OsO_4 in 0.1 M cacodylate for 4 hr. After dehydration, the cells were slowly air-dried at 4°C and coated with a thin carbon film. These whole-mount carbon replicas were subsequently treated with hydrofluoric acid to dissolve the coverslip; after rinsing in water, the replicas were cleaned for 1 hr with household bleach and examined in the electron microscope. No gold particles are associated with the collagen. The thin offshoots are the microvilli of the adherent hepatocytes; ×52,500.

In Section and "En Face" Observation

Cells in culture represent the experimental model of choice to study the interactions between ligand and receptors, especially in the case of ligand–gold complexes, which can be cleared *in vivo* from plasma either

1. Interactions between Gold, Proteins, and Cells

nonspecifically by reticuloendothelial cells ("colloidopexy") or specifically, after binding to cell surface receptors. Figures 1.5 and 1.6 illustrate the surface labeling which can be observed at 4°C (no internalization occurs at this temperature), using gold particles with an approximate diameter of ~15 nm, coated by asialofetuin. The gold probe was used to study the galactose-specific receptor (Ashwell and Morell, 1974) on the surface of isolated hepatocytes, prepared by the technique described by Wanson *et al.* (1977).

The use of resin sections (Fig. 1.5) allows detailed studies of the structural organization of the biological membranes bearing the receptor and their relationship with specialized membrane domains such as coated pits or the underlying cytoskeleton. However, very few particles can usually be analyzed in a random cell profile, unless the number of receptors is unusually high (e.g., EGF receptors on A431 cells) or the receptors are topologically restricted to a narrow domain, such as the LDL receptors in coated pits of mammalian cells or the flagellar pocket of *Trypanosoma brucei* (Coppens *et al.,* 1988).

To circumvent the rare occurrence of ligand–particle complexes in a section, flat membrane domains can be analyzed by the "en face" observation technique of replicas, as illustrated in Fig. 1.6. The advantage of the replica technique is obvious when interest is focused on the number of receptors per surface unit area and their respective distribution. In whole mounts, a very large area is observed, compared with that which can be observed in thin sections. The disadvantage is loss of structural information. The intensity of labeling of receptors by ligand–gold complexes obtained by the whole-mount technique is comparable to that seen using the scanning electron microscope for the observation of surface antigens (see, for example, de Harven and Soligo, 1986).

Immunogold Labeling of Cellular Components

The study of the heterogeneity of the cellular membrane is an example of an application of the "en bloc" immunogold labeling technique, which will be illustrated by the localization of 5'-nucleotidase in normal rat liver. In this organ, the enzyme is concentrated along bile canaliculi (Chen *et al.,* 1987). Preparations such as those illustrated in Fig. 1.7 were obtained from rat liver fixed by perfusion with 0.5% glutaraldehyde in 0.1 M cacodylate for 10 min. Small blocks of 1 mm^3 were cut and incubated for 1 hr in a 1% lysine solution to inactivate the free aldehydes. They were then incubated for 1 hr in a 1/100 dilution of a polyclonal antiserum against 5'-nucleotidase. After rinsing in 0.1 M cacodylate, the blocks were incubated for 1 hr with a protein A–gold (8-nm) probe for 1 hr (4.6 × 10^{10}

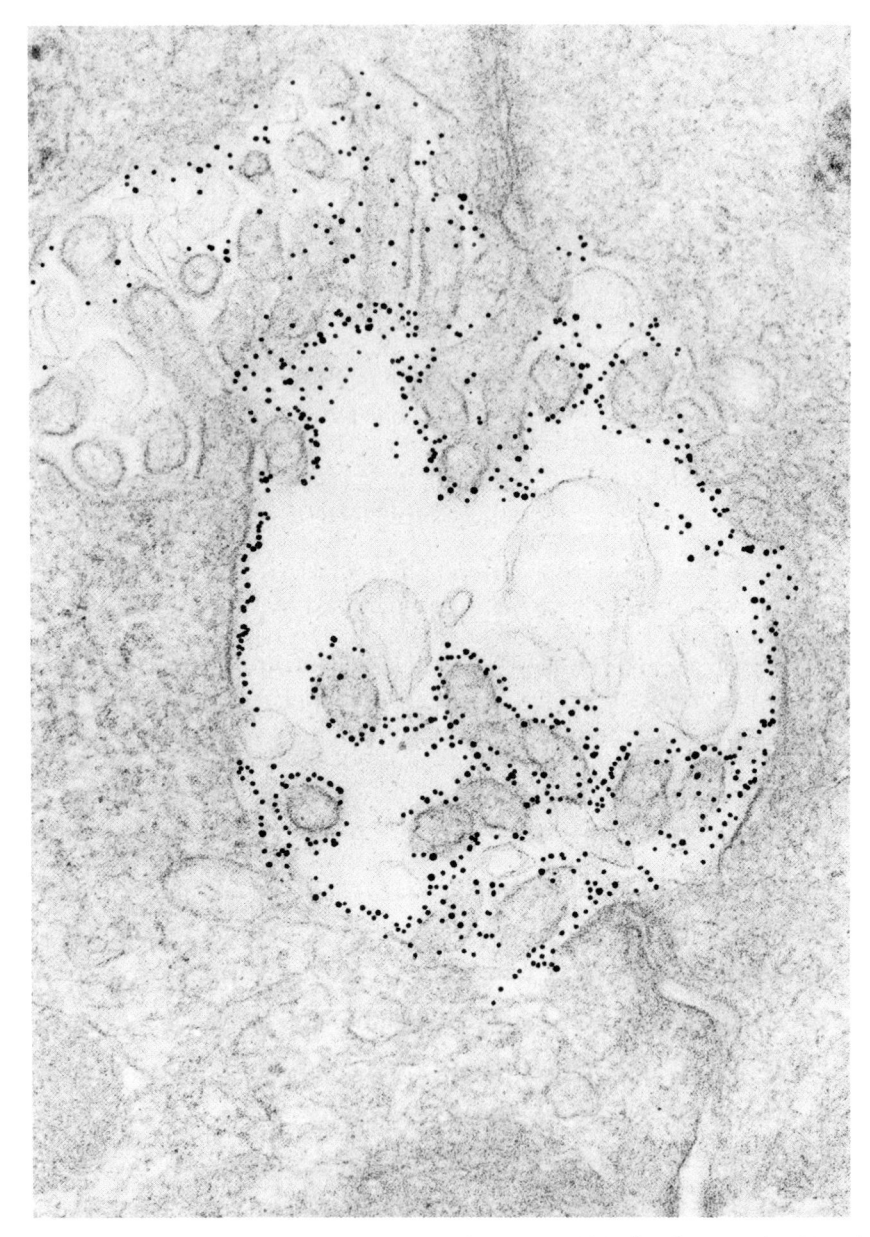

Fig. 1.7. Immunogold labeling. "En bloc" immunogold labeling for 5′-nucleotidase in normal rat liver in the bile canalicular domain, as described in text. Labeling is restricted to the canalicular membrane and microvilli and is absent in intraluminal vesicles; ×70,000.

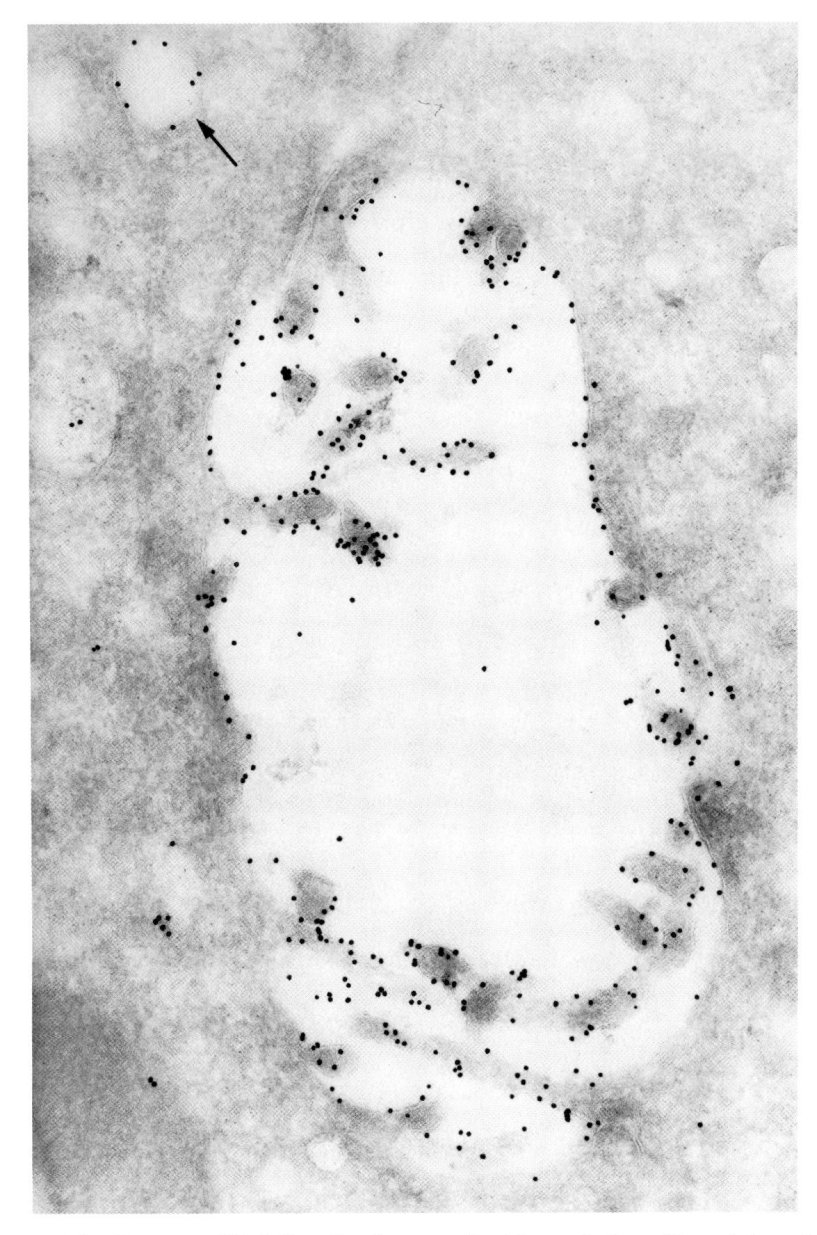

1. Interactions between Gold, Proteins, and Cells 15

Fig. 1.8. Immunogold labeling. Rat liver was fixed by perfusion with a mixture of 5% formaldehyde and 0.1% glutaraldehyde and postfixed in 5% formaldehyde for 1 hr. Thin frozen sections of ~90 nm were incubated with a 1/100 dilution of anti-5′-nucleotidase serum, followed by labeling with protein A–gold complex (4.6 × 10^{10} particles/ml, approximate diameter ~8 nm). In addition to the canalicular membrane, a few pericanalicular vesicles are labeled (arrow); ×50,000.

particles/ml). After washing for several hours in 0.1 M cacodylate, the blocks were fixed again with 1.5% glutaraldehyde to cross-link the antibody–protein A complex, postfixed in OsO_4, dehydrated, and embedded. Very superficial sections were cut and poststained with uranyl acetate and lead citrate according to standard procedures.

This "en bloc" technique is useful if the antibody and the gold complex can freely diffuse to the surface, where the antigen is localized. The staining obtained is more abundant, with a well-preserved morphology, than with the classical thin frozen section methodology, as illustrated in Fig. 1.8.

REFERENCES

Ackerman, G. A., Yang, J., and Wolken, K. W. (1983). Differential surface labeling and internalization of glucagon by peripheral leukocytes. *J. Histochem. Cytochem.* **31**, 433.

Ashwell, G., and Morell, A G. (1974). The role of surface carbohydrates in hepatic recognition and transport of circulating glycoproteins. *Adv. Enzymol.* **41**, 99.

Beauvois, S., Van der Smissen, P., Vael, T., Baudhuin, P., and Courtoy, P. J. (1987). Effects of prolonged incubation at 4°C on asialofetuin and asialofetuin–gold binding to adherent rat hepatocytes. *J. Cell Biol.* **105**, 60a.

Chen, S. D., Widnell, C., Vaerman, J. P., Baudhuin, P., and Courtoy, P. J. (1987). Ultrastructural localization of 5'-nucleotidase in the normal rat liver and after bile duct ligation. *J. Cell Biol.* **105**, 233a.

Coppens, I., Baudhuin, P., Opperdoes, F., and Courtoy, P. J. (1988). Receptors for the host low-density lipoproteins on the hemoflagellate *Trypanosoma brucei*: Purification and involvement in the growth of the parasite. *Proc. Natl. Acad. Sci. U.S.A.* **85**, 6753.

de Harven, E., and Soligo, D. (1986). Scanning electron microscopy of cell surface antigens labelled with colloidal gold. *Am. J. Anat.* **175**, 277.

De Roe, C., Courtoy, P. J., and Baudhuin, P. (1987). A model of protein–colloidal gold interactions. *J. Histochem. Cytochem.* **35**, 1191.

Dickson, R. B., Willingham, M. C., and Pastan, I. (1981). Alpha-2-macroglobulin adsorbed to colloidal gold: A new probe in the study of receptor-mediated endocytosis. *J. Cell Biol.* **89**, 29.

Frens, G. (1973). Controlled nucleation for the regulation of the particle size in monodisperse gold suspensions. *Nature (London), Phys. Sci.* **241**, 20.

Griffiths, G., Simons, K., Warren, G., and Tokuyasu, K. T. (1983). Immunoelectron microscopy using thin, frozen sections: Application to studies of the intracellular transport of Semliki Forest virus spike glycoproteins. *In* "Methods in Enzymology," (S. Fleischer and B. Fleischer, eds.), Vol. 96, p. 466. Academic Press, New York.

Handley, D. A., Arbeeny, C. M., Witte, L. D., and Chien, S. (1981a). Colloidal gold–low density lipoprotein conjugates as membrane receptor probes. *Proc. Natl. Acad. Sci. U.S.A.* **78**, 368.

Handley, D. A., Arbeeny, C. M., Eder, H. A., and Chien, S. (1981b). Hepatic binding and internalization of low-density lipoprotein–gold conjugates in rats treated with 17 alpha-ethinylestradiol, *J. Cell Biol.* **90**, 778.

Horisberger, M. (1979). Evaluation of colloidal gold as a cytochemical marker for transmission and scanning electron microscopy. *Biol. Cell.* **36**, 253.

Scatchard, G. (1949). The attractions of proteins for small molecules and ions. *Ann. N. Y. Acad. Sci.* **51**, 660.

Slot, J. W., and Geuze, H. J. (1981). Sizing of protein A–colloidal gold probes for immunoelectron microscopy. *J. Cell Biol.* **90**, 533.

Slot, J. W., and Geuze, H. J. (1985). A new method for preparing gold probes for multiple-labelling cytochemistry. *Eur. J. Cell Biol.* **38**, 87.

Tokuyasu, K. T. (1986). Application of cryoultramicrotomy to immunochemistry. *J. Microsc. (Oxford)* **143**, 139.

Wanson, J.-C., Drochmans, P., Mosselmans, R., and Ronveaux, M.-F. (1977). Adult rat hepatocytes in primary monolayer culture. Ultrastructural characteristics of intercellular contacts and cell membrane differentiations. *J. Cell Biol.* **74**, 858.

Weast, R. C. (1979). "CRC Handbook of Chemistry and Physics," 59th ed. CRC Press, Boca Raton, Florida.

Weibel, E. R. (1979). "Stereological Methods," Vol. 1. Academic Press, London.

2

Protein A–Gold: Nonspecific Binding and Cross-Contamination

ALEX D. HYATT

Australian Animal Health Laboratory
CSIRO
Geelong, Victoria, Australia

INTRODUCTION
PREPARATION OF PROTEIN A–COLLOIDAL GOLD
    COMPLEXES
PRINCIPLES AND APPLICATION OF PROTEIN A–GOLD
    LABELING
  Single Labeling
  Double Labeling
SOURCES OF ERROR IN PROTEIN A–GOLD LABELING
  Cross-Contamination
  Nonspecific Binding
CONCLUDING REMARKS
REFERENCES

INTRODUCTION

Colloidal gold has been a popular electron-dense marker in immunoelectron microscopy for the past decade. Over this period protein A–gold has been used extensively as a reagent for binding immunoglobulins from many mammalian species. Due to the broad reactivity of the probe and the ease of preparation, protein A–gold remains a popular choice as an immunolabel. The aim of this chapter is to discuss the potential problems associated with protein A–gold as stated in the literature and as recognized in this laboratory. To understand the potential difficulties, it is necessary first to discuss the techniques involved with the preparation of colloidal gold and protein A–colloidal gold conjugates.

PREPARATION OF PROTEIN A–COLLOIDAL GOLD COMPLEXES

Colloidal gold is generally prepared by the reduction of tetrachloroauric acid, $H(AuCl_4) \cdot 4H_2O$, with a range of chemicals including phosphorus (Zsigmondy, 1905; Zsigmondy and Thiessen, 1925), trisodium citrate (Frens, 1973), ascorbic acid (Stathis and Fabrikanos, 1958), sodium borohydride (Tschopp et al., 1982), and trisodium citrate and tannic acid (Slot and Geuze, 1985). The gold colloid is formed from micromolecular units (nuclei) and the increase in particle size is dependent on the rate of formation of nuclei and the rate of crystal growth. Thus, by controlling the reduction of tetrachloroauric acid, gold probes of varying diameters can be produced. Reduction with white phosphorus, ascorbic acid, or trisodium citrate, for example, produces gold particles with average diameters of 5, 12, and 16 nm, respectively (Slot and Geuze, 1981). For further details, see Handley (1989).

An important surface property of colloidal gold is its negative charge. In an aqueous phase, colloidal gold will stay in suspension by electrostatic repulsion. If electrolytes are added, the ion layers of the particles are compressed and so the critical "cohesion" distance is reduced, thereby facilitating flocculation. It has been recognized since 1901 (Zsigmondy, 1901) that gold particles can be stabilized in solution by the addition of proteins. One of the more successful of such proteins is *Staphylococcus aureus* coat protein A. The complexing of protein A (as with other proteins) to colloidal gold involves electrostatic interaction between the negatively charged surface of gold particles and positively charged groups of the proteins. It is generally accepted that the protein is attracted into the action radius of van der Waals–London attractive forces and there firmly binds to the surface of the gold particles.

The important parameters for successful preparation of stable and active protein A–gold complexes are the pH of the colloidal gold suspension (Roth, 1983) and the determination of the minimum amount of protein required for stabilization. Although the isoelectric point for protein A is

5.1, complexes are generally produced at a pH range of 5.9 to 6.2. However, adequate protein binding can be obtained at a pH between 5 and 6 (Slot and Geuze, 1981); the overall bioactivity and stabilization do not appear to be compromised in such complexes. The amount of protein A required for stabilization of colloidal gold is generally determined by the technique described by Zsigmondy and Thiessen (1925). The test results in a color change of the colloidal gold solution from orange-red to blue when there is incomplete stabilization. The minimum amount of protein A required to prevent this effect is taken as an estimation for the stabilization of the gold solution. It is common practice to add 10 to 100% excess protein A to the solution to ensure stabilization (Slot and Geuze, 1981, 1985). The protein A complex is then stored in a buffer (e.g., phosphate-buffered saline, PBS) containing an excess of unrelated protein such as bovine serum albumin (BSA); this further stabilizes the gold complex.

PRINCIPLES AND APPLICATION OF PROTEIN A–GOLD LABELING

The protein A–gold method is a two-stage immunolabeling procedure. The antigenic site within or on the specimen is revealed (indirectly) by the addition of a primary antibody. The excess antibody is removed and protein A–gold added. Protein A has two functional Fc binding regions (Langone, 1982), which interact predominantly at the CH_2 and CH_3 domains of the Fc region of the antibody (Forsgren and Sjöquist, 1966). The binding is strong in most IgG classes of several mammalian species such as human, rabbit, guinea pig, and dog; weaker with IgG classes from horse, cow, and mouse; and weakest in IgG from goat, sheep, rat, and chicken (Forsgren and Sjöquist, 1966; Kronvall et al., 1970, 1974; Biberfeld et al., 1975; Lindmark et al., 1983). Binding to different classes of IgG, human IgA, and IgE and IgM from various species has also been shown (Goudswaard et al., 1978; Langone, 1982; Lindmark et al., 1983). It is obvious from the above that antibodies used in protein A–gold procedures should be generated from species which produce high-affinity antibody–protein A complexes. Furthermore, the antibodies should be of the one class, that is, they should be affinity-purified or monoclonal antibodies.

Single Labeling

Labeling of antigens can occur prior to embedding (preembedding immunocytochemistry) or postembedding of whole or sectioned specimens.

Labeling can therefore occur either on or within tissue sections or whole mounts (for example, whole cells and complete cytoskeletal matrices). Irrespective of the mode of antigenic presentation, the labeling protocol generally follows a set format such as that outlined below. Labeling of structures in Fig. 2.1a,b was produced by the same protocol, the details of which are shown in parentheses. The incubations were performed in plastic petri dishes and solutions changed with micropipettes (Fig. 2.2).

1. Wash (PBS, 3 min).
2. Fixation (0.1% glutaraldehyde in PBS, 2 min; when cytoskeletons are prepared the fixative includes 1% of the non-ionic detergent NP40).
3. Wash (PBS, 3 × 3 min).
4. Blocking step [PBS containing 1% BSA (PBS–BSA), 10 min].
5. Primary antibody (1 : 10 dilution in PBS–BSA, 37°C (310 K), 1 hr).
6. Wash (PBS–BSA, 6 × 3 min).
7. Protein A–gold (1 : 20 in PBS–BSA, 37°C (310 K), 1 hr).
8. Wash (PBS, 6 × 3 min).
9. Stain (2% phosphotungstic acid adjusted to pH 6.8 with 1 M KOH).

Provided the antibodies are affinity purified or are monoclonal, the antigen–antibody interactions will be specific and indicative of the presence and localization of the antigen. This assumes that the antigen has not been leached or translocated from or within its biological matrix due to inappropriate fixation. The possibility of "unwanted" antibodies, from whole sera, reacting with cell constituents is discussed in the following section.

Double Labeling

Gold particles of varying sizes can be produced by different methodologies (Handley, 1989). When the gold particles vary in size, i.e., their coefficient of variation (CV) is greater than 15% (Slot and Geuze, 1985), the solutions are considered polydispersive in size; when the CV is less than 15% they are referred to as monodispersive. The production of monodispersive colloidal gold is dependent on the method for synthesis (Handley, 1989) and subsequent treatment (Bendayan, 1984). The ability to produce solutions of monodispersive gold particles is of paramount importance in multiple-labeling immunoelectron microscopy. The advantage of protein A is that it can be readily complexed to a gold probe of any size and possesses an affinity for most IgG of most mammals. Protein A–gold probes therefore provide a potential means whereby colocalization of antigens within a single sample can be achieved. In the literature there are numerous reports which claim success in double labeling of antigens within/on the one biological section. The techniques include colocal-

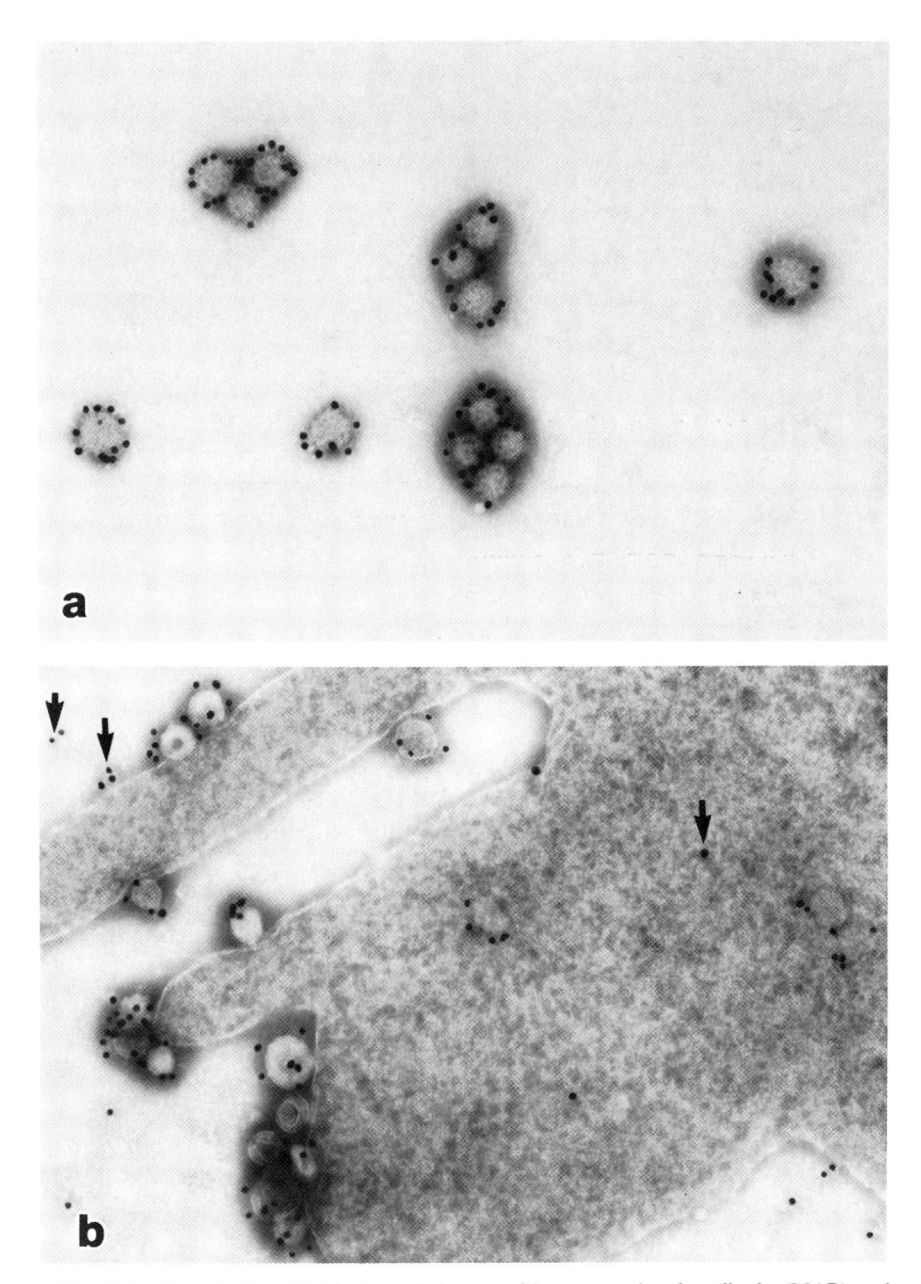

Fig. 2.1. Protein A–gold labeling of viruses with a monoclonal antibody (MAB) and whole serum. (a) Bluetongue virus (orbivirus), prepared via the grid-cell-culture technique (GCCT) (Hyatt *et al.*, 1987) and labeled with MAB (anti-VP7) and 12-nm protein A–gold. (b) Transmissible gastroenteritis virus (coronavirus) prepared via the GCCT and labeled with whole serum and 12-nm protein A–gold. The grid substrate in (b) possessed some adsorbed cellular material released from virus-infected cells. When heterogeneous antibodies are used (e.g., whole serum), nonspecific labeling follows (arrows); ×64,555.

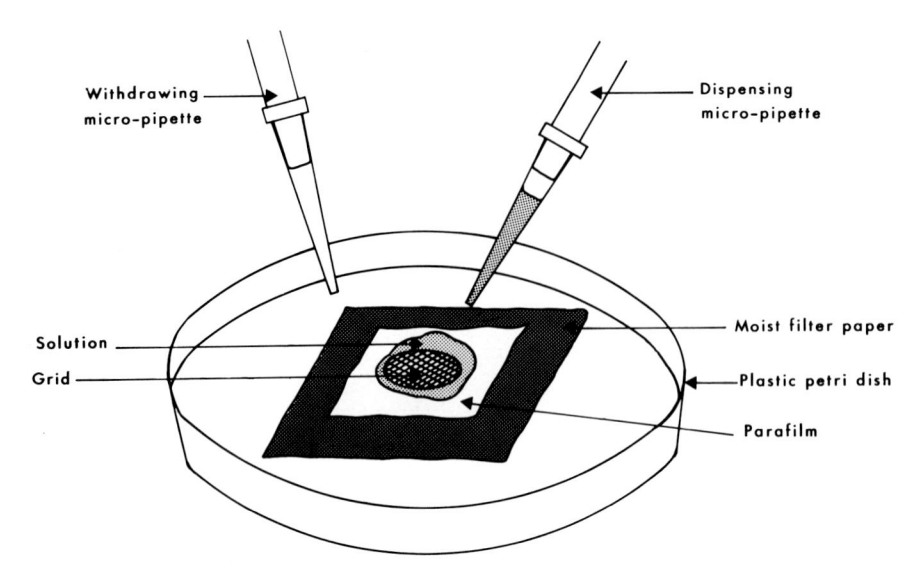

Fig. 2.2. Diagram of the method by which viruses (adsorbed to a grid substrate) were washed and incubated. All solutions were exchanged via micropipettes. For the incubation steps, the petri dish is placed in a hot room or incubator at 37°C (310 K); the water-saturated filter paper ensured a moist environment.

ization of antigens on one face of a section with probes of different sizes (Geuze et al., 1981; Roth, 1982; Hisano et al., 1984) or the labeling of both faces of a section (Bendayan, 1982).

In the first technique addition of each antibody is followed by protein A–gold probes of specific sizes in a defined sequence. The first protein A–gold probe to be applied is the smaller of the two (3–5 nm), followed by the larger probe (12–15 nm). When this sequence is followed, little cross-contamination occurs (Roth, 1982). If free protein A is added during the last minutes of step 7 (see above) little cross-contamination is reported to occur irrespective of which probe is applied first (Slot and Geuze, 1984). If, however, the order of probes is reversed without the addition of free protein A, cross-contamination can occur (Roth, 1982). The protocol for this form of double labeling is a simple extension of that outlined for single labeling.

    1–7. Same as on p. 22.
       8. Free protein A ($\cong$ 0.05 mg/ml) in the last few minutes of step 7.
       9. Wash (PBS, 6 × 3 min).
     10. Repeat steps 5 to 7.

While this procedure has enjoyed some success, there have been numerous reports of problems associated with it (Bendayan, 1982; Bendayan and Stephens, 1984; Hyatt et al., 1988). An alternative procedure

2. Nonspecific Binding and Cross-Contamination

is described by Bendayan (1982). In this procedure both sides of the section are used; that is, there are two independent labeling steps. This second technique avoids any cross-contamination between the different immunolabeling steps.

Not all double labeling is performed on biological sections. When only one aspect of a sample is available for labeling (for example, viruses adsorbed to a grid substrate), then successful double labeling, as illustrated in Fig. 2.3, may not be possible with either of the above methods. Hyatt *et al.* (1988), for example, found that protein A–gold could not be used for the multiple epitope mapping of Akabane virus as it resulted in colabeling of the primary antibody. Successful double labeling could be achieved only with specific IgG–gold probes (Fig. 2.3). For additional information on multiple labeling, the reader is referred to Doerr-Schott (1989).

SOURCES OF ERROR IN PROTEIN A–GOLD LABELING

Some potential problems associated with protein A–gold labeling are also associated with other immunocytochemical techniques, namely non-

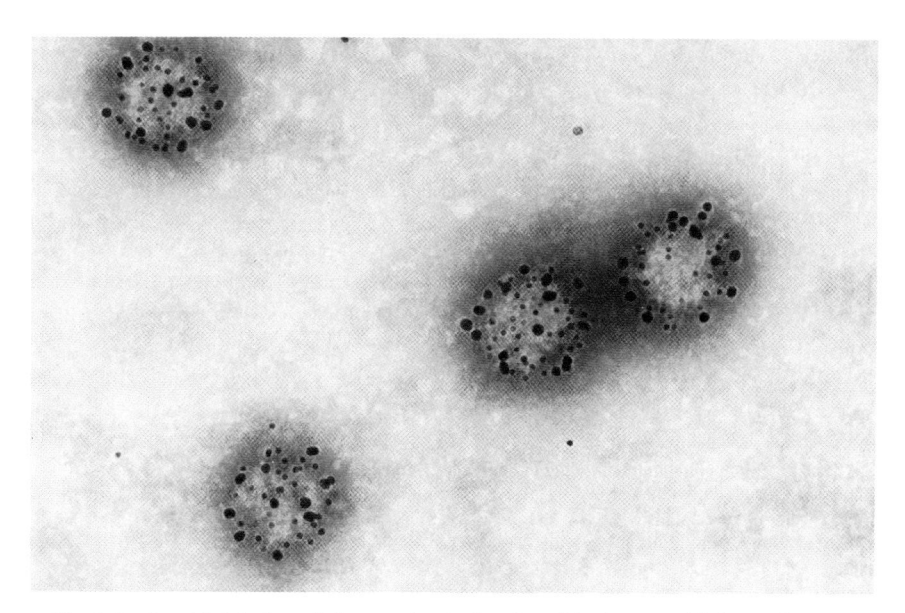

Fig. 2.3. Double labeling of viruses adsorbed to the grid substrate. Akabane virus (Hyatt *et al.*, 1988) labeled with monoclonal antibodies F5/D5 and A12/G3 complexed directly to 6- and 12-nm gold probes, respectively; ×133,562.

specific staining (for example, nonspecific antibodies and electrostatic attachment; Taylor, 1978; Behnke et al., 1986; Gosselin et al., 1986; Birrell et al., 1987). Double immunolabeling with the protein A method can induce further sources of error, namely cross-contamination of antibodies with protein A–gold. The problems of cross-contamination and nonspecific labeling are discussed below (also see Birrell and Griffith, 1989; Park et al., 1989).

Cross-Contamination

As stated previously, it is advisable to use monospecific antibodies raised in the appropriate mammalian species for the primary localization of antigens. From a practical viewpoint, affinity-purified and monoclonal antibodies generally yield specific labeling with a clean background (Fig. 2.1a). The labeling in Fig. 2.1b, on the other hand, was produced with a whole porcine serum as the primary antibody. Although 1% BSA was used to minimize nonspecific labeling (as per Fig. 2.1a), the background in Fig. 2.1b was considerably greater. An explanation for this is that the use of whole serum (containing numerous natural antibodies, targeted antibodies, and antibodies to any carrier molecules or extraneous material carried over or used in the immunization) with protein A–gold produces many immunocomplexes. The overall effect can result in the adsorption of unwanted antibodies and subsequent labeling of various cellular constituents in addition to the targeted antigen. If normal serum is used to block nonspecific binding of primary antibodies to the reactive sites of tissues, an effect similar to that described above may occur. In many instances the problem of such nonspecific binding may be reduced by decreasing the effective antibody concentration. If the studies involve infectious agents, hormones, or the like (i.e., cause-and-effect experiments), then nonspecific staining can be further reduced by preadsorption of the serum with the normal (nonexperimental) biological tissue.

Biological samples bathed in serum may also provide another source of error (for example, tissue culture cells and cells of the blood). It is not unreasonable to expect that during specimen preparation some antibodies may fortuitously bind to the sample surface. Such antibodies may provide additional binding sites for protein A. These samples should be thoroughly washed (e.g., with PBS) before labeling is attempted.

In double-labeling experiments cross-contamination is now recognized as a problem which arises from the affinity which protein A possesses for different regions of IgG antibodies, namely the Fc and Fab regions (Roth, 1982; Endresen, 1979; Zikan, 1980). Protein A can bind two IgG molecules; thus, after the primary antibody–protein A complex has been

2. Nonspecific Binding and Cross-Contamination

formed, free IgG binding sites can still be available on the bound protein A. Addition of the second antibody can result in its binding to the free sites on protein A molecules and target specified antigens. A second protein A–gold probe can potentially bind not only to any protein A–gold free IgG antibodies from the primary incubation (the Fc region of the secondary IgG complexes) but also to the Fab regions of IgG antibodies.

Reports which claim success with the "one-face" protein A–gold method use either a defined sequence of protein A–gold probes or free protein A (described above). If free protein A is not incorporated, then the success of the method depends on the smaller of the protein A–gold probes being used first. The small probes ($\cong$ 3 nm) have approximately one associated protein A molecule, whereas larger probes ($\cong$ 15 nm) can adsorb ~60 such molecules (Roth, 1982). If the larger of the gold probes is used for the visualization of the first antigen, many potential Fc binding sites (unoccupied protein A) are available for the second immunoglobulin and significant cross-contamination may therefore result. If excess free protein A is added, the primary antibody–protein A interactions may reach saturation and so minimize interactions with secondary protein A–gold probes. The binding of secondary antibodies to the existing protein A Fc binding sites should not be recognized by the secondary protein A–gold probe as it is assumed that IgG can bind only one protein A molecule. Cross-contamination can still occur, however, via interactions with the Fab regions of IgG. If cross-contamination is a continued problem, alternative techniques to the protein A method should be investigated. Such techniques generally involve direct labeling and species-specific antibody–gold indirect labeling techniques.

Nonspecific Binding

The use of small (3–6-nm) gold probes has the advantage of improving immunocytochemical resolution. Anyone who has worked with small gold probes has observed them (in some specimens) to be extremely "sticky" (Fig. 2.4a,b). Behnke *et al.* (1986), Birrell *et al.* (1987), and Birrell and Griffith (1989) have reported the problem to be largely associated with exposed areas (areas not covered with protein) on the gold particles themselves and secondarily with the method by which the colloidal gold is produced. Since small protein A–gold complexes are used widely, the problems associated with small probes merit discussion.

Flocculation of colloidal gold can be prevented by partial saturation of the particles (Goodman *et al.*, 1980; Horisberger and Vauthey, 1984). The stabilization of colloidal gold therefore does not ensure its saturation. It is because of this that excess unrelated protein (for example, BSA) is

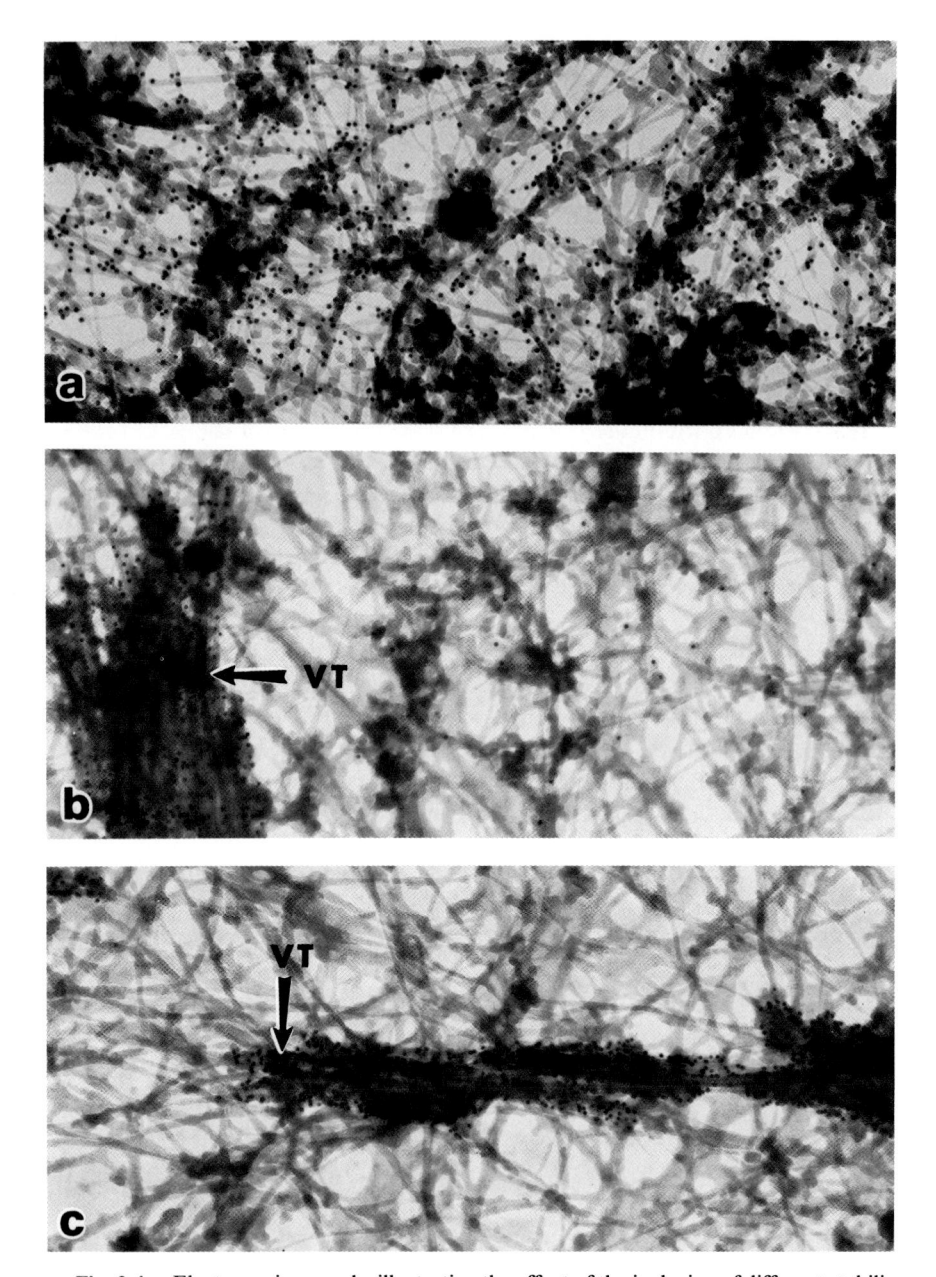

Fig. 2.4. Electron micrographs illustrating the effect of the inclusion of different stabilizers in the labeling protocol. All samples were critical-point dried from CO_2 after dehydration in alcohol. Samples were examined with a TEM at 50 kV. The labeling protocol was as described in the text. (a) Uninfected cytoskeleton of a tissue culture cell (Super-Vero-Porcine) incubated with 6-nm protein A–gold (1 : 10). No stabilizer was included within the

added to these protein–gold complexes; the subsequent complex is then assumed to be saturated. However, it would appear that some immunogold labeling patterns are independent of protein–protein interactions and result from interactions between specimen constituents and the gold particles (Behnke et al., 1986). The occurrence of such nonspecific binding is obvious in whole cytoskeletons (Fig. 2.4a) and thick polyethylene glycol-extracted sections (A. D. Hyatt, unpublished observations). Strategies in colloidal gold labeling of cytoskeletal elements are discussed at length by Birrell and Griffith (1989). Nonspecific binding is attributed to electrostatic interactions between exposed areas on the gold particles (areas of negative charge) and cationic structures within the specimen (Behnke et al., 1986).

The method used to prepare colloidal gold has also been considered a source for nonspecific labeling (Birrell et al., 1987). Birrell et al. found that 5-nm gold particles produced by the trisodium citrate–tannic acid method exhibited a higher degree of nonspecificity than particles of similar size produced by other methods. This nonspecificity was attributed to the chemical agents used in colloidal gold production (Birrell et al., 1987); for example, colloidal gold produced by the white phosphorus method was found to contain phosphate groups. Tannic acid may therefore also be a source of contamination for colloidal gold produced by the trisodium citrate–tannic acid procedure (Birrell et al., 1987; Birrell and Griffith, 1989). Such contaminants on gold particles may, as with tannic acid, form a series of complex interactions with biological specimens, particularly those fixed with an aldehyde and/or OsO_4 (Simionescu and Simionescu, 1976). Observations in this laboratory and by Birrell et al. (1987) and Birrell and Griffith (1989) that nonspecific labeling is more pronounced with smaller probes may be explained by the greater accessibility of these probes to tissue constituents within open and/or permeabilized specimens.

Nonspecific binding due to protein–protein and electrostatic interactions can be minimized by including in the washing and incubation steps a noncompeting protein which has a high affinity for gold particles. These proteins (e.g., bovine skin gelatin, fish gelatin, and BSA) will bind to pro-

---

washing and incubation steps; ×103,362. (b) Cytoskeleton of a bluetongue virus-infected cell (Super-Vero-Porcine) incubated with MAB (anti-VP3, 1 : 20) and 6-nm protein A–gold (1 : 20). All incubations done in the presence of 1% BSA. VT, Virus tubules; ×90,137. (c) Cytoskeleton of a bluetongue virus-infected cell incubated with MAB (anti-VP3, 1 : 20) and 6-nm protein A–gold (1 : 20) in the presence of 1% fish gelatin. VT, Virus tubules; ×90,137. The increase in background labeling in (a) is due to the absence of stabilizer and a concentrated solution of protein A–gold.

tein-reactive constituents within the specimen and will minimize electrostatic interactions by covering the exposed areas on the gold particles. The choice of an appropriate stabilizer will significantly reduce the level of nonspecific labeling, as is illustrated in Fig. 2.4c (compare with Fig. 2.4b). An alternative or concomitant procedure would involve the incubation of protein A–gold with the biological sample.

CONCLUDING REMARKS

Immunolabeling with protein A–gold has several distinct advantages: (1) it is easy and quick to prepare complexes of various sizes; (2) the complexes interact with antibodies from many mammalian species; and (3) the method is sensitive, clean, and can yield valuable information provided the parameters of immunolabeling are understood. In other words, the investigator must be familiar with the characteristics of the primary antibody, protein A–gold probes, and the specimen itself.

Generally, the use of the protein A–gold method should involve (1) the use of highly specific antibodies; (2) the use of such antibodies at their lowest effective concentration; (3) the absence of normal serum as a blocker; and (4), if using open matrix specimens, inclusion of an effective stabilizer in the incubation steps. These observations, if followed, will reduce the level of nonspecific labeling and cross-contamination. It should also be noted that at no stage during the labeling protocol should the sample be allowed to dry. In addition, the sample should be thoroughly rinsed between antibody and protein A–gold steps; this will remove unbound antibodies and protein A. Failure to comply with the above precautions will result in high backgrounds.

There are no detailed protocols for protein A–gold labeling which have universal applicability. For each different experiment, detailed protocols should be developed to produce optimum labeling characteristics. These protocols must be based on results of control immunocytochemical experiments. If protein A–gold is the label, and nonspecific binding and cross-contamination are persistent problems in spite of the addition of excess free protein A, high-affinity stabilizers, adsorption of contaminating antibodies, and colloidal gold, then alternative labeling techniques should be pursued and evaluated.

The author gratefully acknowledges Drs. B. T. Eaton and A. R. Gould for reading the manuscript and Mr. T. Wise for skillful technical assistance.

REFERENCES

Behnke, O., Ammitzboll, T., Jenssen, H., Klokker, M., Nilausen, K., Tranum-Jensen, J., and Olsson, L. (1986). Non-specific binding of protein-stabilized gold sols as a source of error in immunocytochemistry. *Eur. J. Cell Biol.* **41,** 326.

Bendayan, M. (1982). Double immunocytochemical labeling applying the protein A–gold technique. *J. Histochem. Cytochem.* **30,** 81.

Bendayan, M. (1984). Protein A–gold electron microscopic immunocytochemistry: Methods, applications, and limitations. *J. Electron Microsc. Tech.* **1,** 243.

Bendayan, M., and Stephens, H. (1984). Double labelling cytochemistry applying the protein A–gold technique. *Immunolabelling for Electron Microscopy* (J. M. Polak and I. M. Varndell, eds.), pp. 143–154. Elsevier, Amsterdam.

Biberfeld, P., Ghetie, V., and Sjöquist, J. (1975). Demonstration and assaying of IgG antibodies in tissues and on cells by labeled staphylococcal protein A. *J. Immunol. Methods* **6,** 249.

Birrell, G. B., Hedberg, K. K., and Griffith, O. H. (1987). Pitfalls of immunogold labeling: Analysis by light microscopy, transmission electron microscopy and photoelectron microscopy. *J. Histochem. Cytochem.* **35,** 843.

Birrell, G. B., and Griffith, O. H. (1989). Strategies on colloidal gold labeling of cell surfaces and cytoskeletal elements of cultured cells. In *Colloidal Gold: Principles, Methods, and Applications, Vol. 1* (M. A. Hayat, ed.), pp. 323–347. Academic Press, San Diego, California.

Doerr-Schott, J. (1989). Colloidal gold for multiple labeling method. In *Colloidal Gold: Principles, Methods, and Applications, Vol. 1* (M. A. Hayat, ed.), pp. 145–190. Academic Press, San Diego, California.

Endresen, C. (1979). The binding of protein A, of immunoglobulin G and of Fab and Fc fragments. *Acta Pathol. Microbiol. Scand., Sect. C* **87C,** 185.

Forsgren, A., and Sjöquist, J. (1966). "Protein A" from *S. aureus*. I. Pseudo-immune reaction with human gamma-globulin. *J. Immunol.* **97,** 822.

Frens, G. (1973). Controlled nucleation for the regulation of the particle size in monodisperse gold suspensions. *Nature (London), Phys. Sci.* **241,** 20.

Geuze, H. J., Slot, J. W., Peter, A., van der Ley, P. A., and Scheffer, R. C. (1981). Use of colloidal gold particles in double-labeling immunoelectron microscopy of ultrathin frozen tissue sections. *J. Cell Biol.* **89,** 653.

Goodman, S. L., Hodges, G. M., and Livingston, D. C. (1980). A review of the colloidal gold marker system. *Scanning Electron Microsc.* **2,** 133.

Gosselin, E. J., Cate, C. C., Pettengill, O. S., and Sorenson, G. D. (1986). Immunocytochemistry: Its evolution and criteria for its application in the study of Epon-embedded cells and tissue. *Am. J. Anat.* **175,** 135.

Goudswaard, J., van der Donk, J. A., Noordzij, A., van Dam, R. H., and Vaerman, J. P. (1978). Protein A reactivity of various mammalian immunoglobulins. *Scand. J. Immunol.* **8,** 21.

Handley, D. A. (1989). Methods for synthesis of colloidal gold. In *Colloidal Gold: Principles, Methods, and Applications, Vol. 1* (M. A. Hayat, ed.), pp. 13–32. Academic Press, San Diego, California.

Hisano, S., Adachi, T., and Daikoku, S. (1984). Immunolabeling of adenohypophysial cells with protein A–colloidal gold–antibody complex for electron microscopy. *J. Histochem. Cytochem.* **32,** 705.

Horisberger, M., and Vauthey, M. (1984). Labelling of colloidal gold with protein. A quantitative study using beta-lactoglobulin. *Histochemistry* **80,** 13.

Hyatt, A. D., Eaton, B. T., and Lunt, R. (1987). The grid-cell-culture technique: The direct examination of virus-infected cells and progeny viruses. *J. Microsc. (Oxford)* **145**, 97.

Hyatt, A. D., McPhee, D. A., and White, J. R. (1988). Antibody competition studies with gold-labelling immunoelectron microscopy. *J. Virol. Methods* **19**, 23.

Kronvall, G., Seal, U. S., Finstad, J., and Williams, R. C. (1970). Phylogenetic insight into evolution of mammalian Fc fragment of γG globulin using staphylococcal protein A. *J. Immunol.* **104**, 140.

Kronvall, G., Seal, U. S., Svensson, S., and Williams, R. C. (1974). Phylogenetic aspects of staphylococcal protein A-reactive serum globulins in birds and mammals. *Acta Pathol. Microbiol. Scand. Sect. B* **82B**, 12.

Langone, J. J. (1982). Protein A of *Staphylococcus aureus* and related immunoglobulin receptors produced by streptococci and pneumococci. *Adv. Immunol.* **32**, 157.

Lindmark, R., Thoren-Tolling, K., and Sjöquist, J. (1983). Binding of immunoglobulins to protein A and immunoglobulin levels in mammalian sera. *J. Immunol. Methods* **62**, 1.

Park, K., Park, H., and Albrecht, R. M. (1989). Factors affecting the staining with colloidal gold. In *Colloidal Gold: Principles, Methods, and Applications, Vol. 1* (M. A. Hayat, ed.), pp. 489–518. Academic Press, San Diego, California.

Roth, J. (1982). The preparation of protein A–gold complexes with 3nm and 15nm gold particles and their use in labelling multiple antigens on ultra-thin sections. *Histochem. J.* **14**, 791.

Roth, J. (1983). The colloidal gold marker system for light and electron microscopic cytochemistry. In *Techniques in Immunocytochemistry* (G. R. Bullock and P. Petrusz, eds.), Vol. 2, pp. 217–284. Academic Press, London and New York.

Simionescu, N., and Simionescu, M. (1976). Galloylglucoses of low molecular weight as mordant in electron microscopy. I. Procedure, and evidence for mordanting effect. *J. Cell Biol.* **70**, 608.

Slot, J. W., and Geuze, H. J. (1981). Sizing of protein A–colloidal gold probes for immunoelectron microscopy. *J. Cell Biol.* **90**, 533.

Slot, J. W., and Geuze, H. J. (1984). Gold markers for single and double immunolabelling of ultrathin cryosections. In *Immunolabelling for Electron Microscopy* (J. M. Polak and I. M. Varndell, eds.), pp. 129–142. Elsevier, Amsterdam.

Slot, J. W., and Geuze, H. J. (1985). A new method of preparing gold probes for multiple-labeling cytochemistry. *Eur. J. Cell Biol.* **38**, 87.

Stathis, E. C., and Fabrikanos, A. (1958). Preparation of colloidal gold. *Chem. Ind. (London)* **27**, 860.

Taylor, C. R. (1978). Immunoperoxidase techniques: Practical and theoretical aspects. *Arch. Pathol. Lab. Med.* **102**, 113.

Tschopp, J., Podack, E. R., and Müller-Eberhard, H. J. (1982). Ultrastructure of the membrane attack complex of complement: Detection of the tetramolecular C9-polymerizing complex C5b-8. *Proc. Natl. Acad. Sci. U.S.A.* **79**, 7474.

Zikan, J. (1980). Interactions of pig Fab gamma fragments with protein A from *Staphylococcus aureus*. *Folia Microbiol. (Prague)* **25**, 246.

Zsigmondy, R. (1901). *Z. Anal. Chem.* **40**, 697.

Zsigmondy, R. (1905). *Zur Erkenntnisse der Kolloide.* Fischer, Jena.

Zsigmondy, R., and Thiessen, P. A. (1925). *Das kolloidale gold.* Akad. Verlagsges., Leipzig.

3

Role of Tissue Processing in Colloidal Gold Methods

GEOFFREY R. NEWMAN
and
JAN A. HOBOT

Electron Microscopy Unit
University of Wales College of Medicine
Heath Park, Cardiff, Wales

INTRODUCTION
LR WHITE AND LOWICRYL K4M
LR WHITE OR LOWICRYL K4M?
MODERN ACRYLICS AND COLLOIDAL GOLD
REFERENCES

INTRODUCTION

A lot has been written about fixation in cytochemistry and immunocytochemistry, especially its requirement for the preservation of ultrastructure and tissue reactivity (for an excellent review of glutaraldehyde fixation, see Hayat, 1986). However, its relevance to the subsequent processes through which tissue may have to be taken prior to sectioning and colloidal gold labeling has often been ignored. Recently, tissue processing methods (some of them not optimal) have also been compared, but only after the tissue was heavily cross-linked by fixation in either glutaraldehyde alone (Bendayan *et al.*, 1987) or glutaraldehyde followed by postfixation in OsO_4 (Ring and Johanson, 1987).

Not surprisingly after such powerful fixation, little difference of real significance was seen in labeling between any of the different methods. In fact, high concentrations of glutaraldehyde and/or postosmication has been shown to damage antigenicity severely (Roth *et al.*, 1981; Hayat, 1986) but, regrettably, neither Bendayan *et al.* (1987) nor Ring and Johanson (1987) made any comparisons between fixation in high and low glutaraldehyde concentrations or varied the fixation times in their work, which makes their analyses of the role played by tissue processing and embedding uninterpretable. Postosmication also interferes with the photochemical visualization of colloidal gold (Danscher and Rytter Norgaard, 1983) for light microscopy.

It is, perhaps, more surprising that any labeling is seen at all after heavy fixation, and this testifies to the remarkable stability of tissue substances, some of which are apparently able to survive rigorous fixation and unoptimized processing regimes and remain recognizable to antibodies. Even so, the capacity to label antigen in high-concentration sites does not exclude the possibility that a serious loss of antigenicity has occurred and, therefore, its localization in low-concentration sites cannot be guaranteed. For example, using on-grid immunolabeling of the hormone somatostatin, no difference could be detected from normal in the number of gold particles over the endocrine storage granules of D cells whose hormone content had been profoundly reduced pharmacologically (Patel *et al.*, 1987). On the other hand, there may be a calculable minimal threshold for detectable antigen in resin sections (Kellenberger *et al.*, 1987). If antigen has been lost or denatured at sites where it is in low concentration, it may have fallen below the threshold level needed for its localization with colloidal gold.

Some antigens at the other extreme, for example, cell surface antigens associated with the identification of lymphocyte subsets, are apparently so fragile that, after even the mildest aldehyde fixation and embedding method, they cannot be localized at all by on-grid methods.

In between these extremes, there exist many antigens and lectin-reactive substances, in sites of low as well as high concentration, that for their localization require an appropriate fixation regime matched to an optimized tissue processing and embedding method. This provides the best chance for antigens and other reactive substances to be maximally retained in a form recognizable and available to antibodies or lectins, while they are set in adequate ultrastructural surroundings. This require-

3. Role of Tissue Processing in Colloidal Gold Methods

ment is partly imposed by the limited sensitivity of colloidal gold methods whose reactions are confined solely to the surface layer of sections (Newman and Hobot, 1987).

Theoretically, the perfect fixative would stabilize the structure of tissue, protecting it from the ravages of subsequent processing, while leaving it unaltered in its reactivity with lectins and antibodies. On the other hand, the protection of fixation would not be needed if the processing and embedding of tissue were completely nondeleterious, such that it could be sectioned and immunolabeled without restraint. Since, in practice, the achievement of this ideal is not yet possible, a compromise has to be struck between two opposing forces, either of which can lead to a loss of tissue reactivity and antigenicity.

Excessive exposure of the tissue to fixatives will often result in poor or nonexistent colloidal gold labeling, however effective the embedding technique. The reason is that the antigen (or lectin-reactive substance) has been radically changed, both chemically and physically, or that it has become inaccessible. However, if tissue is insufficiently stabilized (cross-linked) by fixation it will be eluted during dehydration, infiltration, and embedding, resulting in ultrastructural damage and loss or displacement of antigen and other substances. Even ultracryomicrotomy, which avoids the need for dehydration and embedding, requires tissue fixation to stabilize sections which have to be thawed before they can be immunolabeled. Antigen loss and ultrastructural damage will occur after thawing and during immunolabeling if the tissue is underfixed.

Specifically, then, the tested procedures involving high glutaraldehyde concentrations (>1%) for long durations (>2 hr) followed by postfixation in OsO_4 and embedding in a three-dimensionally cross-linked, hydrophobic, impermeable epoxide result in specimens that are stable in the electron beam and show satisfactory preservation of ultrastructure. However, this is achieved at the expense of antigen and lectin reactivity. Any effort to change this system in a minor way in order to obtain greater antigenicity is usually counterproductive. For example, when postosmication is omitted, even without reducing the degree of cross-linking with glutaraldehyde, the preservation of ultrastructure and tissue reactivity is seriously impaired. The reason for this is that dehydration, infiltration, and lengthy polymerization of epoxy resins at 60°C cause extraction of cellular materials from the unosmicated tissue.

Some antigens can survive osmication, but to reveal them when the tissue is embedded in epoxide, the sections often need subjective and capricious measures such as "etching" with hydrogen peroxide (Baskin et al., 1979) or "unmasking" with sodium metaperiodate (Bendayan and Zollinger, 1983). Even then, tables comparing the immunolabeling counts for sections of osmicated tissue in Epon treated with metaperiodate and those of the same but unosmicated tissue in Lowicryl K4M not treated with metaperiodate (Bendayan, 1984) clearly demonstrate the superiority of the Lowicryl method; the latter gives higher labeling intensities and lower background staining. In addition, glycoproteins and other lectin-reactive substances are destroyed by sodium metaperiodate.

At the other extreme, when postosmication is omitted, cryoultramicrotomy can provide very high tissue reactivity by taking advantage of the omission of dehydration, infiltration, and resin embedding, but only after sufficiently fixing the tissue to hold it together during thawing and immunolabeling. However, the ultrastructural preservation is often less than satisfactory. Cryoultramicrotomy is also expensive, inconvenient (difficult to store tissue blocks), inconsistent, and permits only very small pieces of tissue to be examined.

LR WHITE AND LOWICRYL K4M

Two high-sensitivity methods (LR White and Lowicryl K4M) for processing tissue into resin while retaining tissue structure and reactivity have consistently shown their value for use with colloidal gold techniques and are the subject of the following two chapters. Both methods use unosmicated tissue for optimal colloidal gold labeling and seek to reduce the extraction losses and protein conformational changes which occur during processing including embedding. This is accomplished by using mild, hydrophilic acrylics and modified tissue-processing techniques, which deviate markedly from methods dedicated purely to the preservation of traditional ultrastructure.

The first method increases antigenicity by taking advantage of the miscibility of the acrylic resin LR White with small amounts of water. The seriously extractive higher ethanols, used during dehydration, are avoided by infiltrating the tissue with LR White directly from, for example, 70% ethanol (Newman et al., 1982, 1983; Newman, 1987). Lectins can also be localized in nonosmicated LR White-embedded tissue (Ellinger and Pavelka, 1985; Jones and Stoddart, 1986). Nonetheless, osmicated tissue has sometimes been used to good effect in both plant (Craig and Miller, 1984) and animal (Graber and Kreutzberg, 1985; Ring and Johanson, 1987) immunocytochemical studies. Rapid infiltration and catalytic polymerization can be used to reduce the time that the tissue is in contact with the resin monomer, which is also extractive (Yoshimura et al., 1986; Newman and Hobot, 1987).

3. Role of Tissue Processing in Colloidal Gold Methods

The second method employs the reduction in the rate of chemical and physical reactions provided by low temperatures to protect unosmicated tissue from the deleterious effects of dehydration. Progressively lower temperatures are used as the concentration of the dehydrating solution is increased. The method also depends upon the low viscosity of the methacrylate-based Lowicryl resins at a low temperature ($-35°C$, 238 K) to infiltrate the tissue (Carlemalm *et al.*, 1982). Polymerization by "cold" UV light enables the temperature to be kept at $-35°C$ (238 K). The progressive lowering of temperature (PLT) method may also protect protein tertiary structure, which has been cited as being important in antibody recognition (Hayat, 1986). An increase in labeling intensities using PLT has been reported by Armbruster *et al.* (1983), as also for cryosubstitution methods followed by embedding in Lowicryl K4M (Carlemalm *et al.*, 1985).

The PLT method improves not only the signal-to-noise ratio of the labeling (i.e., lowers background) but also the structural preservation of the tissue (Roth *et al.*, 1981). For example, minor protein components of the mitochondrial membranes have been detected by modified PLT using protein A–gold techniques (Bendayan and Shore, 1982). The ability to use Lowicryl K4M with lectin–gold complexes has also been demonstrated (Roth, 1983), and PLT has provided new information on the role that might be played in glycosylation by the transtubular network of the Golgi complex (Roth *et al.*, 1985).

To generalize, "light" fixation may be critical to the maximum efficiency of colloidal gold labeling seen with either method. Complete aldehyde cross-linking removes some of the need for these newer embedding methods, because then the tissue has already been chemically and physically altered and the requirement to restrict extraction is also much reduced. The difficulty is to define "light" and "strong" in the context of fixation, so that, for example, what is minimally needed to preserve structural integrity for one embedding method may be compared objectively with what is required for another. It is further complicated by the variation in composition of fixatives (and their component parts) and differences in their storage and administration from one laboratory to another.

We have found that reproducibility of fixation can be guaranteed only when pure-grade reagents are used. Impure glutaraldehyde in particular can vary widely in the content of its impurities and cyclic components (Gillett and Gull, 1972; Hopwood, 1972). In addition, the fixative should be administered by vascular perfusion. Perfusion fixation is important for immunoelectron microscopy because, if properly performed, the fixative is delivered rapidly and intimately to all parts of the tissue almost at the same time (Hayat, 1981, 1989). There should be no fixative diffusion gradients such as occur during immersion fixation, and low fixative concentrations are easily administered without fear of poor or patchy penetration (Pignal *et al.*, 1982).

In recent experiments (our unpublished observations) we have obtained results that cannot be reproduced after epoxide embedding. We have shown that perfusion fixation of pancreas and pituitary of rat at 22°C with only 0.1% neutral-buffered, highly purified, monomeric glutaraldehyde can lead to interpretable ultrastructure and very high immunocolloidal gold labeling counts after as little as 15 min of fixation followed by PLT and Lowicryl embedding (Fig. 3.1 Top) or 60 min of fixation followed by partial dehydration and rapid LR White embedding (Fig. 3.1 Bottom). Following immunolabeling with antitrypsin antiserum and immunocolloidal gold, high counts are registered not only over the storage and Golgi sites but also over the endoplasmic reticulum, which is a low-concentration antigen site, probably associated with the manufacture of the enzyme.

After the pancreas is fixed in more concentrated solutions of glutaraldehyde, the endoplasmic reticulum shows very little immunoreactivity (Fig. 3.2a,b). The validity of the labeling of the endoplasmic reticulum is strongly supported by its also appearing in rapidly frozen, unfixed pancreas cryosubstituted into Lowicryl K4M (Fig. 3.3), a research method which has been shown to give maximum structural and antigenic preservation (Carlemalm *et al.*, 1985). All labeling of both high- and low-concentration antigenic sites is lost after absorption of the antiserum with purified trypsin.

Fig. 3.1. (Top) Unosmicated pancreas of rat fixed by vascular perfusion with 0.1% neutral-buffered glutaraldehyde for 15 min at 22°C and embedded in Lowicryl K4M by the progressive lowering of temperature (PLT) method. (Bottom) As above but fixed for 60 min and embedded in LR White by the cold catalytic method. The unsupported thin sections were immersed in rabbit antirat anionic trypsin antiserum (a gift from Dr. John Kay, Department of Biochemistry, University College Cardiff, Cardiff, UK) followed by goat antirabbit IgG coupled to 10-nm colloidal gold. Counterstained in uranyl acetate and lead citrate; ×25,000. In spite of minimal levels of glutaraldehyde cross-linking, recognizable organelles have been preserved by both methods. The zymogen granules are most susceptible to damage and shrinkage, probably due to extraction during processing, which is worse for LR White than for PLT/Lowicryl. As anticipated, a high intensity of labeling lies over the zymogen granules and the Golgi apparatus (G) of both, but more surprising is the heavy labeling of the rough endoplasmic reticulum (RER). That this is specific is strongly supported by absorption controls and by controls missing out the primary antibody, which were always almost free of labeling. The very low level of labeling seen over the nuclei and the mitochondria is further evidence. However, this labeling could still be the result of dislocation or relocation of antigen (trypsin enzyme) displaced during fixation and processing (see Fig. 3.3).

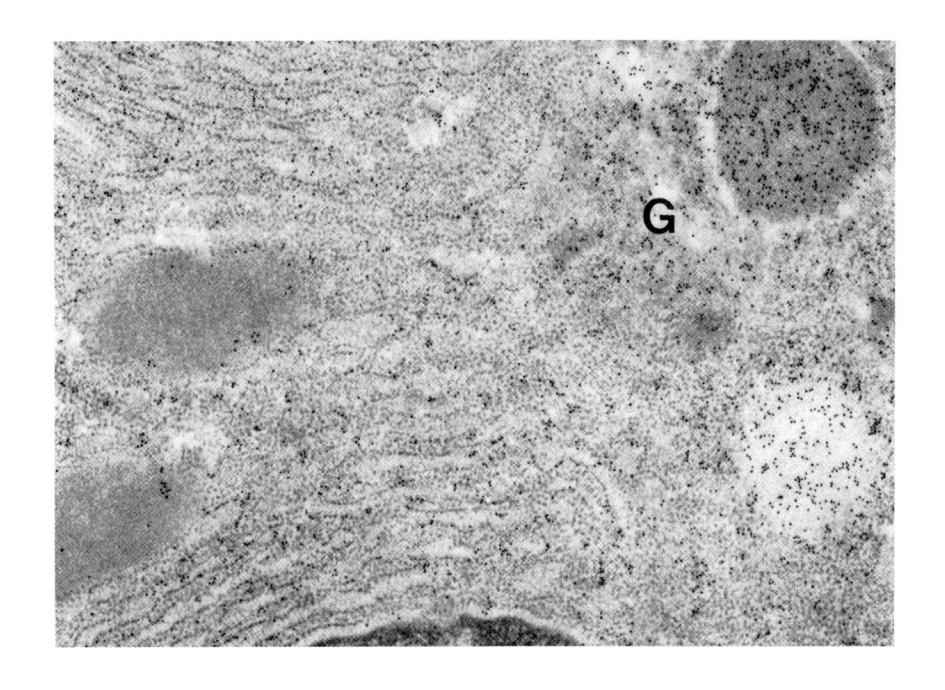

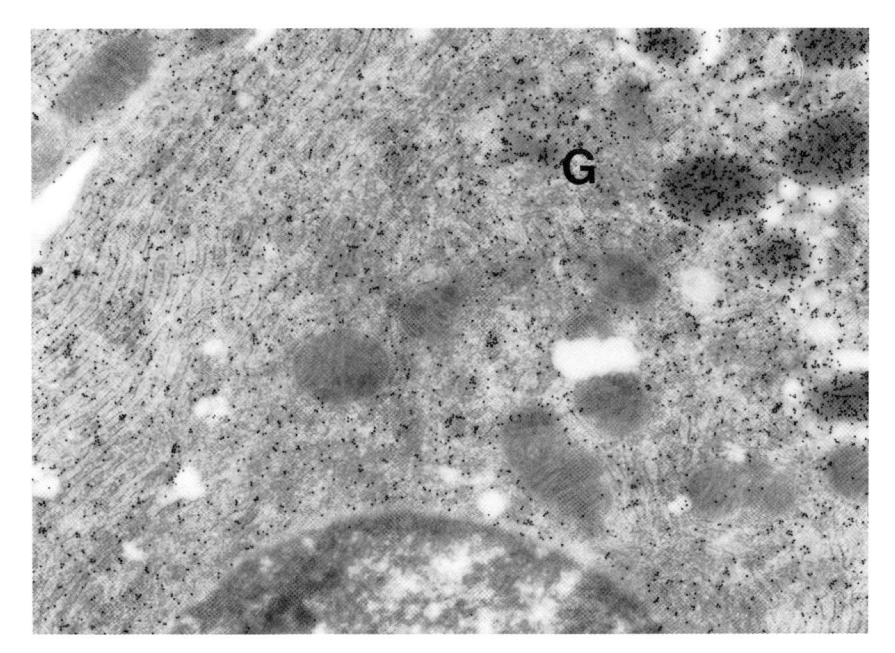

3. Role of Tissue Processing in Colloidal Gold Methods

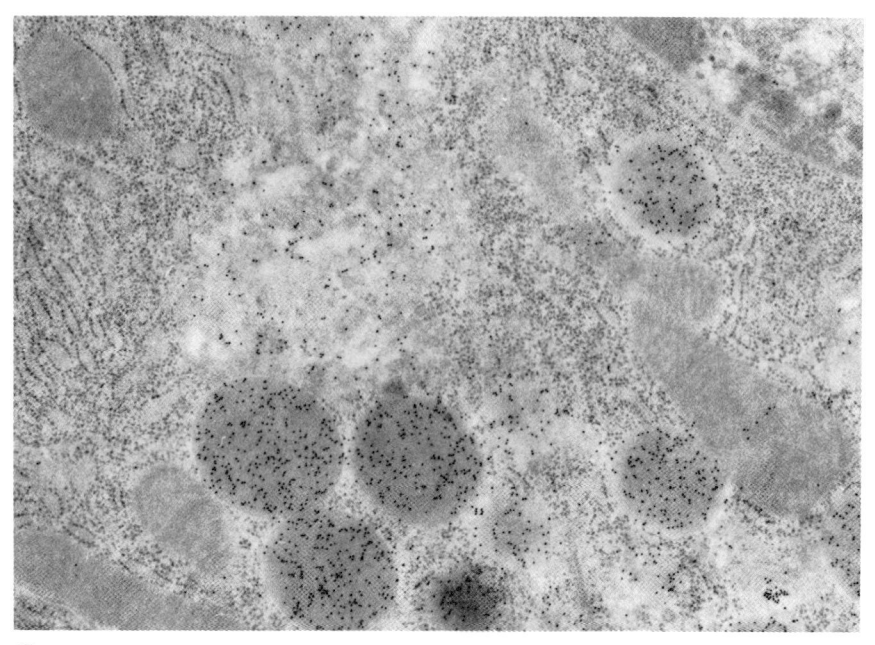

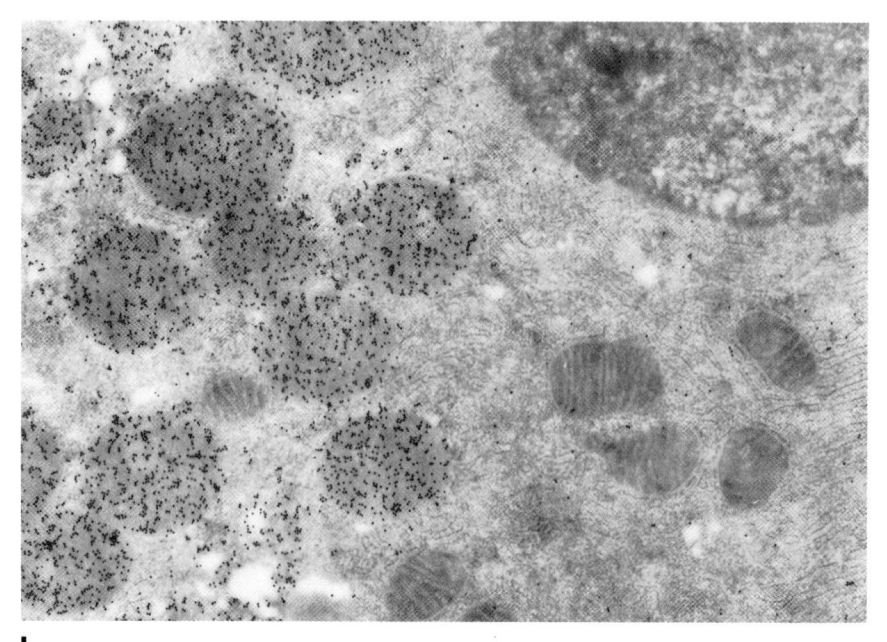

a

b

LR WHITE OR LOWICRYL K4M?

Both resins can be polymerized by heat, chemical catalysts, or UV light, but LR White becomes viscous and slow to infiltrate into tissue below −20°C (253 K). In order to polymerize K4M (or HM20) with heat or catalysts it is necessary to add to it an activator such as benzoyl peroxide, which is inconvenient because it is not supplied in the Lowicryl kits and must be handled with care. LR White is, therefore, most useful for embedding methods using heat or catalytic polymerization. Lowicryl, which is still quite mobile at −35°C (238 K), should be used for PLT methods and cryo(freeze)-substitution, using UV polymerization. LR White will cope better with many routine requirements such as large, immersion-fixed blocks and osmicated or pigmented tissues (although these may not give optimal labeling) and is therefore valuable in histopathology.

Much high-sensitivity research in cytochemistry and immunocytochemistry can be carried out after perfusion fixation using LR White. We have compared immunocolloidal gold labeling on LR White and Lowicryl K4M sections of pancreas fixed by vascular perfusion with 1% neutral-buffered, highly purified, monomeric glutaraldehyde at 22°C. The tissue was taken from the same rat and optimally embedded in each of the two resins. Identical immunomethods were used and all immunoreagents were kept at the same concentrations and temperature. Very little difference could be detected in either the ultrastructure (which was excellent in both cases) or the gold particle counts (Fig. 3.2a,b). It is, therefore, probably unimportant which of the two resins is chosen for embedding tissue which has been fixed by vascular perfusion for even a short time (15 min) in high (>1%) glutaraldehyde concentrations even if the glutaraldehyde is uncontaminated with impurities. However, this was not true after very minimal glutaraldehyde concentrations and fixation times (<0.1% for 15 min), where PLT and Lowicryl K4M gave superior ultra-

---

Fig. 3.2. Unosmicated pancreas of rat fixed by vascular perfusion in 1% neutral-buffered glutaraldehyde for 60 min at 22°C; × 25,000. (a) Embedded in Lowicryl K4M by the PLT method. (b) Embedded in LR White by the cold catalytic method. The unsupported sections were immunolabeled and counterstained exactly as in Fig. 3.1. The preservation of structure is improved in both (a) and (b) by the increased extent of glutaraldehyde cross-linking and there are now only minor differences between the sections. In (a) the intensity of labeling is still high but has dropped considerably in all compartments of the cell when compared with Fig. 3.1a. Most significantly, in both (a) and (b) the labeling on the endoplasmic reticulum is almost nonexistent, though background labeling levels are much the same as in Fig. 3.1. This result could be interpreted as being due to the increased stability of the tissue leading to reduced levels of displaced antigen in the RER. Figure 3.3 shows that this is probably not the case.

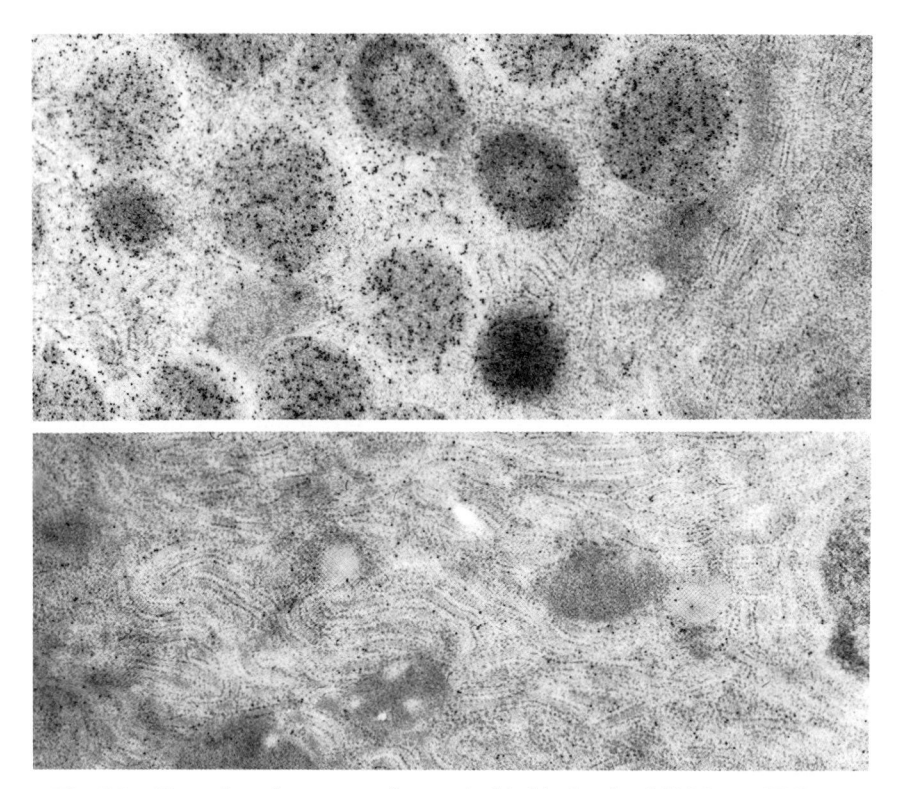

Fig. 3.3. Unosmicated pancreas of rat embedded in Lowicryl K4M at −35°C (238 K) after rapid freezing followed by cryosubstitution from 3% glutaraldehyde/acetone at −80°C (193 K). This technically complex method avoids the chemical changes that occur in tissue after conventional fixation and immobilizes and lowers the extraction of antigens. The unsupported thin sections were immunolabeled and counterstained exactly as in Figs. 3.1 and 3.2; ×25,000. As expected, there are very high immunolabeling intensities over the zymogen granules. More important, however, there is once again a considerable amount of colloidal gold closely following the contours of the rough endoplasmic reticulum, which lends strong support to the view that the RER labeling shown in Fig. 3.1 is genuine and not the product of a diffusion artifact. Once again, background levels of labeling over mitochondria and nuclei are low.

structural preservation and so would be the choice for difficult antigens which are very labile or sensitive to fixation.

To summarize, LR White is rapid and simple to use and is therefore a good workhorse for dealing with routine cytochemical and immunocytochemical problems, for example, in histopathology, especially where immersion fixation and concentrations of glutaraldehyde exceeding 0.5%

have been used. Loss of sensitivity associated with fixation and requiring a radical reduction in glutaraldehyde concentration to 0.2% or less, however, is better dealt with by PLT and Lowicryl K4M or cryosubstitution.

MODERN ACRYLICS AND COLLOIDAL GOLD

LR White and the Lowicryls were not designed for colloidal gold cytochemistry and immunocytochemistry. LR White was originally intended as a safer, less carcinogenic, alternative to epoxies and to bridge the gap between light microscopy and electron microscopy. Coincidentally, it has been found to be very suited to all immunolabels, and its excellent staining properties make it valuable for assessing, by light microscopy, the potential of embedded tissue for immunostaining or lectin reactivity. The Lowicryls were devised ultimately to examine the fine structure of minimally fixed tissue by scanning transmission electron microscopy, avoiding the use of osmium, uranium, lead, or any of the other heavy metals that are needed to increase the electron density of tissue when it is embedded in epoxide (Carlemalm and Kellenberger, 1982). Nonetheless, the methacrylate base of the Lowicryls makes them eminently suited to colloidal gold methods.

When taken together, these modern acrylic resins provide a virtually continuous sequence of embedding methods increasing in technical complexity and in the sensitivity allowed to immunocytochemical and cytochemical methods, for which they are almost ideal. Starting with simple heat polymerization of LR White, the sequence progresses through catalytic LR White polymerization methods, then PLT and the low-temperature Lowicryls, K4M and HM20, with UV light polymerization, and ends with complex pure research methods involving cryosubstitution and two relatively new, even lower-temperature acrylics, Lowicryls K11M and HM23. There is almost certainly a procedure here to cater for the most mundane or esoteric of requirements.

REFERENCES

Armbruster, B. L., Garavito, R. M., and Kellenberger, E. (1983). Dehydration and embedding temperatures affect the antigenic specificity of tubulin and immunolabeling by the protein A–colloidal gold technique. *J. Histochem. Cytochem.* **31**, 1380.

Baskin, D. G., Erlandsen, S. L., and Parsons, J. A. (1979). Influence of hydrogen peroxide or alcoholic sodium hydroxide on the immunocytochemical detection of growth hormone and prolactin after osmium fixation. *J. Histochem. Cytochem.* **27**, 1290.

Bendayan, M., and Shore, G. C. (1982). Immunocytochemical localisation of mitochondrial proteins in rat hepatocyte. *J. Histochem. Cytochem.* **30**, 139.

Bendayan, M., and Zollinger, M. (1983). Ultrastructural localisation of antigenic sites on osmium-fixed tissues applying the protein A–gold technique. *J. Histochem. Cytochem.* **31**, 101.

Bendayan, M. (1984). Protein A–gold electron microscopic immunocytochemistry: Methods, applications and limitations. *J. Electron Microsc. Tech.* **1**, 243.

Bendayan, M., Nanci, A., and Kan, F. W. K. (1987). Effect of tissue processing on colloidal gold cytochemistry. *J. Histochem. Cytochem.* **35**, 983.

Carlemalm, E., and Kellenberger, E. (1982). The reproducible observation of unstained embedded cellular materials in thin sections: Visualisation of an integral membrane protein by a new mode of imaging for STEM. *EMBO J.* **1**, 63.

Carlemalm, E., Garavito, R. M., and Villiger, W. (1982). Resin development for electron microscopy and an analysis of embedding at low temperature. *J. Microsc. (Oxford)* **126**, 123.

Carlemalm, E., Villiger, W., Hobot, J. A., Acetarin, J.-D., and Kellenberger, E. (1985). Low temperature embedding with Lowicryl resins: Two new formulations and some applications. *J. Microsc. (Oxford)* **140**, 55.

Craig, S., and Miller, C. (1984). LR White resin and improved on-grid immunogold detection of vicilin, a pea seed storage protein. *Cell Biol. Int. Rep.* **8**, 879.

Danscher, G., and Rytter Norgaard, J. O. (1983). Light microscopic visualisation of colloidal gold on resin-embedded tissue. *J. Histochem. Cytochem.* **31**, 1394.

Ellinger, A., and Pavelka, M. (1985). Post-embedding localisation of glycoconjugates by means of lectins on thin sections of tissues embedded in LR White. *Histochem. J.* **17**, 1321.

Gillett, R., and Gull, K. (1972). Glutaraldehyde—its purity and stability. *Histochemie* **30**, 162.

Graber, M. B., and Kreutzberg, G. W. (1985). Immunogold staining (IGS) for electron microscopical demonstration of glial fibrillary acidic (GFA) protein in LR White embedded tissue. *Histochemistry* **83**, 497.

Hayat, M. A. (1981). *Fixation for Electron Microscopy*. Academic Press, New York and London.

Hayat, M. A. (1986). Glutaraldehyde: Role in electron microscopy. *Micron Microsc. Acta* **17**, 115.

Hayat, M. A. (1989). *Principles and Techniques of Electron Microscopy: Biological Applications*, 3rd ed. Macmillan, London, and CRC Press, Boca Raton, Florida.

Hopwood, D. (1972). Theoretical and practical aspects of glutaraldehyde fixation. *Histochem. J.* **4**, 267.

Jones, C. J. P., and Stoddart, R. W. (1986). A post-embedding avidin–biotin peroxidase system to demonstrate the light and electron microscopic localisation of lectin binding sites in rat tubules. *Histochem. J.* **18**, 371.

Kellenberger, E., Durrenberger, M., Villiger, W., Carlemalm, E., and Wurtz, M. (1987). The efficiency of immunolabel on Lowicryl sections compared to theoretical predictions. *J. Histochem. Cytochem.* **35**, 959.

Newman, G. R., Jasani, B., and Williams, E. D. (1982). The preservation of ultrastructure and antigenicity. *J. Microsc. (Oxford)* **127**, Rp5.

Newman, G. R., Jasani, B., and Williams, E. D. (1983). A simple post-embedding system for the rapid demonstration of tissue antigens under the electron microscope. *Histochem. J.* **15**, 543.

Newman, G. R. (1987). Use and abuse of LR White. *Histochem. J.* **19**, 118.

Newman, G. R., and Hobot, J. A. (1987). Modern acrylics for post-embedding immunostaining techniques. *J. Histochem. Cytochem.* **35**, 971.

3. Role of Tissue Processing in Colloidal Gold Methods

Patel, Y. C., Ravazzola, M., Amherdt, M., and Orci, L. (1987). Somatostatin-14-like antigenic sites in fixed islet D-cells are unaltered by cysteamine: A quantitative electron microscopic immunocytochemical evaluation. *Endocrinology (Baltimore)* **120**, 1663.

Pignal, F., Maurice, M., and Feldman, G. (1982). Immunoperoxidase localisation of albumin and fibrinogen in rat liver fixed by perfusion or immersion: Effect of saponin on the intracellular penetration of labeled antibodies. *J. Histochem. Cytochem.* **30**, 1004.

Ring, P. K. M., and Johanson, V. (1987). Immunoelectron microscopic demonstration of thyroglobulin and thyroid hormones in rat thyroid gland. *J. Histochem. Cytochem.* **35**, 1095.

Roth, J., Bendayan, M., Carlemalm, E., Villiger, W., and Garavito, M. (1981). Enhancement of structural preservation and immunocytochemical staining in low temperature embedded pancreatic tissue. *J. Histochem. Cytochem.* **29**, 663.

Roth, J. (1983). Application of lectin gold complexes for electron microscopic localisation of glycoconjugates on thin sections. *J. Histochem. Cytochem.* **31**, 987.

Roth, J., Taatjees, D. J., Lucocq, J. N., Weinstein, J., and Paulson, J. C. (1985). Demonstration of an extensive trans-tubular network continuous with the Golgi apparatus stack that may function in glycosylation. *Cell (Cambridge, Mass.)* **43**, 287.

Yoshimura, N., Murachi, T., Heath, R., Kay, J., Jasani, B., and Newman, G. R. (1986). Immunogold electron-microscopic localisation of calpain 1 in skeletal muscle of rats. *Cell Tiss. Res.* **244**, 265.

4

LR White Embedding Medium for Colloidal Gold Methods

GEOFFREY R. NEWMAN

Electron Microscopy Unit
University of Wales College of Medicine
Heath Park, Cardiff, Wales

INTRODUCTION
PHYSICOCHEMICAL PROPERTIES AND ADVANTAGES OF LR
    WHITE
ROUTINE PROCESSING AND POLYMERIZATION BY HEAT
CATALYTIC POLYMERIZATION
    Rapid Infiltration and Catalytic Polymerization
        Tissue Blocks
        Cell Cultures
    Infiltration at 22°C and Catalytic Polymerization in the Cold
STORING LR WHITE BLOCKS
SECTIONING LR WHITE BLOCKS
    Semithin Sections
    Thin Sections
IMMUNOLABELING OF SECTIONS
    Semithin Sections

Thin Sections
    On-Grid Labeling
    Preembedding/On-Grid Labeling
  COUNTERSTAINING OF SECTIONS
    Semithin Sections
    Thin Sections
  CONCLUDING REMARKS
  REFERENCES

INTRODUCTION

In recent years the acrylic resin LR White (London Resin Company Ltd., Basingstoke, Hampshire, UK) has become established in cytochemistry and immunocytochemistry as an important embedding medium for use with colloidal gold and other markers. It has very wide application in both light and electron microscopy. Unfortunately, adaptations of the use of LR White to low-temperature embedding, comparing it with the Lowicryls, have tended to obscure its main strengths, only some of which it shares with them.

PHYSICOCHEMICAL PROPERTIES AND ADVANTAGES OF LR WHITE

LR White is a hydrophilic embedding medium, but it should not be inferred from the use of this term that the monomer is freely miscible with water, although it can polymerize in the presence of up to 12% by volume of water (B. Causton, personal communication, 1984). However, it does mean that, unlike the hydrophobic epoxides, sections of the polymer are freely permeable to aqueous solutions, even at neutral pH. In addition, they have less attraction for colloidal gold reagents, which are also hydrophilic, ensuring low nonspecific background labeling and obviating the need for subjective methods such as jet-washing. LR White is cross-linked which gives good sectioning properties and, along with the unsaturated organic component of the resin, confers a measure of electron-beam stability on sections (Causton, 1980). This, however, does not mean that there is no loss when the section is under electron bombardment.

As with many acrylic resins, the extent of cross-linking of LR White can be controlled throughout polymerization including the point where the resin is completely cross-linked. The hard (or original) resin, which is the one we have always recommended, would then, of course, become much too brittle. When in aqueous environments, sections of LR White

4. LR White Embedding Medium

swell enormously, although the extent to which they swell depends on the degree to which they are cross-linked. This swelling has implications for the handling of acrylic resin sections and might be a contributory factor in explaining their high immunosensitivity (Newman, 1987; Newman and Hobot, 1987).

LR White seems to be less extractive of and reactive with tissue components than are the more commonly used epoxides. Sections of LR White do not need "etching" (Baskin et al., 1979) or treatment with sodium metaperiodate (Bendayan and Zollinger, 1983), as epoxide sections often do, and so far we have also found enzyme digestion to be unnecessary. Finally, and again in common with other acrylic resins, LR White can be polymerized in a number of ways, using heat, chemical catalysts, or UV light, giving great versatility in its method of use. Only methods ensuring controlled polymerization by heat and catalysts are described here.

The fixation method used is the most important factor in deciding which of the three published approaches on the use of LR White it is best to follow (Newman et al., 1982, 1983a; Yoshimura et al., 1986; Newman and Hobot, 1987)

ROUTINE PROCESSING AND POLYMERIZATION BY HEAT

There is probably little point in using complex time-consuming methods to embed tissue which has been fully fixed (either by glutaraldehyde alone or mixtures of glutaraldehyde, formaldehyde, and/or picric acid), especially if it has been immersion fixed or postosmicated. Postosmication reduces the loss of lipids during dehydration and embedding, so preserving membranes (see Fig. 4.3b), and has been preferred by some authors when using LR White (Craig and Miller, 1984; Graber and Kreutzberg, 1985; Ring and Johanson, 1987) but there is a distinct risk that OsO_4 would denature or seriously damage what remains of the antigen (Roth et al., 1981). In addition, osmication interferes with the visualization of colloidal gold by photochemical methods (Danscher and Rytter Norgaard, 1983; Lackie et al., 1985). Therefore, osmium is best omitted. However, good structural preservation is still routinely obtainable, particularly if the serious tissue extraction damage caused, in the absence of osmication, by high concentrations of ethanol used during dehydration (Weibull et al., 1983) is avoided (Figs. 4.1 and 4.2).

The compatibility of LR White with small amounts of water means that tissue can be infiltrated from lower ethanol concentrations. It is important

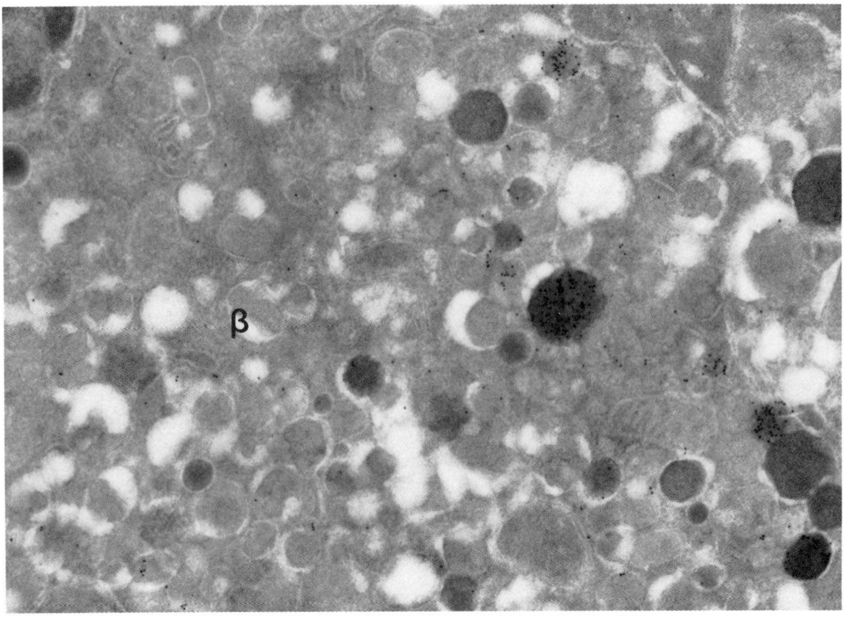

a

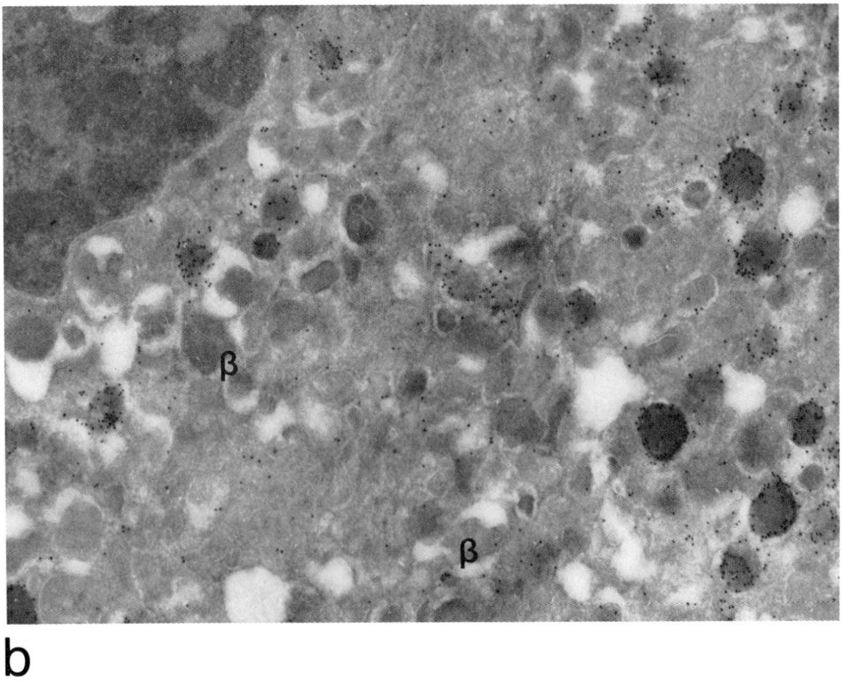

b

not to go above 70% ethanol prior to infiltration, and it is perfectly possible to infiltrate very small blocks of tissue from even lower ethanol dilutions. Partially dehydrated, nonosmicated tissue shows improved retention of cytosol elements and a considerable increase in antigenicity over that obtained in fully dehydrated tissue (Newman et al., 1982, 1983a). The tissue can be transferred directly into LR White resin from 70% alcohol, but an intermediate step of a 2 : 1 mixture of LR White : 70% ethanol (see caution below) is advisable in order to lessen osmotic shock and tissue shrinkage.

Infiltration of the resin into the tissue at room temperature and its slow polymerization in fully filled, tightly capped gelatin capsules at 50°C for 24 hr ensure controlled cross-linking and linearity of polymerization (B. Causton, personal communication, 1985). These properties are needed to produce the most sensitive response during immunolabeling (Newman et al., 1983a). A typical regime follows.

All procedures are carried out at room temperature unless otherwise specified. Following thorough glutaraldehyde fixation, small tissue blocks (2–3 mm^3) are rinsed free of fixative in 0.1 M Sorensen's phosphate buffer (pH 7.3), giving at least three changes of buffer over a minimum period of 4 hr. The tissue can then be placed into 50% ethanol for 15 min followed by two changes of 70% ethanol for 30 min each. (Tissues treated with fixatives containing picric acid should be placed directly into 70% ethanol without a buffer prewash, in order to wash out excess picric acid.)

A mixture of 1% highly purified monomeric glutaraldehyde and 0.2% picric acid in 0.1 M phosphate buffer (BGPA; Newman and Jasani, 1984a) is a very effective fixative for the immersion fixation of, for example, surgical biopsy specimens. It penetrates more rapidly than glutaraldehyde alone and appears to cross-link the tissue less strongly. The precipitative effect of picric acid, which probably improves the retention of proteins

---

Fig. 4.1. Surgically removed human neonatal pancreas fixed in neutral-buffered glutaraldehyde/picric acid (BGPA) and, without osmication, embedded by the routine method in LR White (see text). Thin sections containing endocrine and polycrine cells were immersed in (a) rabbit antiglucagon antiserum and (b) rabbit antiglicentin antiserum. Both primary antisera (kind gifts from Dr. A. J. Moody, NOVO Research Institute, Copenhagen, Denmark) were localized with goat antirabbit IgG coupled to 10-nm colloidal gold particles (Bioclinical Services Ltd., PO Box 129, Cardiff, UK). The sections were counterstained with uranyl acetate and lead citrate. The specificity and accuracy of the immunocolloidal gold method are illustrated by the way in which it has selected and labeled a few solitary endocrine storage granules from among the remainder, predominantly of the insulin-containing β type, which remain unlabeled. In (a) after antiglucagon, colloidal gold covers most of the exposed surface of the sectioned, immunolabeled granules, whereas in (b) after antiglicentin, the colloidal gold is often limited to the outer edge of the granule (see Ravazzola and Orci, 1980). β, β-granule; ×30,000.

β
β

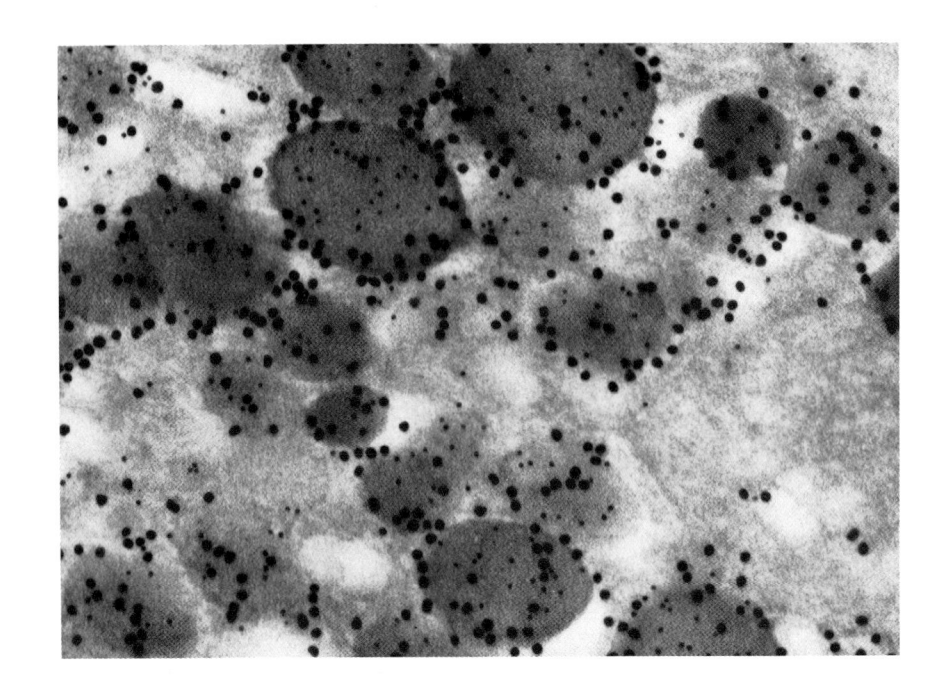

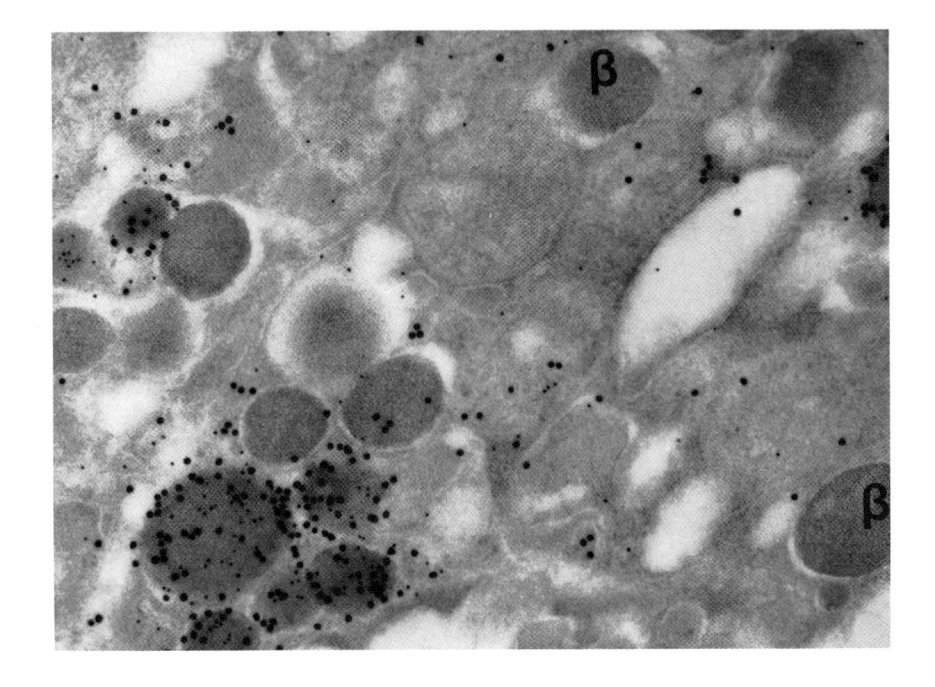

4. LR White Embedding Medium

during dehydration and embedding, may be responsible for the high levels of antigenicity and membrane clarity which are often obtained (Figs. 4.1 and 4.2). The stock solutions of phosphate buffer and saturated aqueous picric acid are easy to prepare and keep for months at room temperature. (If these solutions are refrigerated, the dihydrogen and disodium phosphate and picric acid can come out of solution.) A 50% stock solution of vacuum-distilled monomeric glutaraldehyde (Bio-Rad Polaron Division, Watford, UK) can be stored, preferably in small aliquots, for many months at $-20°C$ in a deep freeze without serious deterioration. Once diluted, however, cyclical forms of the aldehyde and its degradation products such as glutaric acid begin to accumulate, after which the composition and pH of the final fixative cannot be guaranteed (Hayat, 1981, 1986, 1989). The fixative mixture is prepared freshly for use as follows:

0.1 M Sorensen's phosphate buffer	83 ml
50% purified monomeric glutaraldehyde	2 ml
Saturated aqueous picric acid	15 ml

Small tissue blocks (2 mm^3) are fixed at 22°C for 2–4 hr. Tissue blocks are transferred directly into 70% ethanol to wash out both the unbound glutaraldehyde and the picric acid. Several changes of 70% ethanol are needed over a 4–6-hr period. The last change of 70% ethanol should be carefully and thoroughly decanted off the tissue so as to minimize carryover. The tissue blocks are placed in a 2:1 mixture of LR White:70% ethanol for 1 hr while being given gentle agitation on a rotary device (70% ethanol must be made from anhydrous 100% ethanol and only recently

---

Fig. 4.2. Neonatal pancreas as in Fig. 4.1. Double immunolabeling. Thin sections were immunolabeled using two primary antibodies, both of which were raised in rabbit. The sections were immersed first in unlabeled rabbit antiglicentin antiserum, which was localized with goat antirabbit IgG coupled to 20-nm colloidal gold (Bioclinical Services Ltd.). After blocking free valencies on the colloidal gold with normal rabbit serum, the sections were immersed in dinitrophenyl (DNP)-labeled rabbit antiglucagon antiserum (see Newman and Jasani, 1984b), which was localized with an anti-DNP antiserum coupled to 10-nm colloidal gold (Bioclinical Services Ltd.). Sections were counterstained with uranyl acetate and lead citrate. (Top) α-granules of an A-cell show the typical peripheral localization for glicentin with the 20-nm colloidal gold label and the more central localization for glucagon with the 5-nm colloidal gold label; ×75,000. (Bottom) Solitary granules in a polycrine cell are immunolabeled for both glicentin and glucagon in the same way as the granules of the A-cell, demonstrating that despite being set among β-granules they genuinely are α-granules (Newman et al., 1986). Sequential colloidal gold methods like this can run into problems of steric hindrance. Both labels are confined to the section surface so that the first may partially or wholly block the second (see Fig. 4.4 Bottom). Nonetheless, colocalization can be much more difficult to prove using serial sections; ×50,000. β, β-granule.

purchased LR White, i.e., less than 3 months old, should be used; otherwise the solution will not be miscible).

The blocks are transferred to pure LR White and given two changes of 1 hr each using the rotary device. They are left in a third change of LR White on the rotary device overnight and given a fourth change for 1 hr the following morning. Finally, individual tissue blocks can be dropped into gelatin capsules ("O"-gauge or smaller, which can also contain appropriate paper labels) that have been completely filled with fresh, hard-grade LR White. They are tightly capped and polymerized in an oven accurately set at 50°C for 24 hr. It is important to use gelatin capsules because they are impermeable to LR White (unlike some "BEEM"-type capsules, from which LR White will actually leak out) and completely exclude air. The presence of oxygen will totally inhibit polymerization. However, the tiny amount of air trapped inside the top of a gelatin capsule only leads to a little stickiness at the end distal to the tissue.

CATALYTIC POLYMERIZATION

The above very simple protocol will cope with the bulk of routine work but, where warranted, catalytic polymerization is an improvement. The latter has the attribute of being very rapid, although its main purpose is to reduce extraction. Tissue can be immersion fixed in formalin (or microwave formalin fixed for even faster results), dehydrated, infiltrated, and embedded so that blocks are ready for semithin and thin sectioning within

---

Fig. 4.3. Surgical biopsy of human kidney with dense-deposit disease fixed in neutral-buffered glutaraldehyde. (a) Embedded in LR White, without postosmication and after partial dehydration, by the rapid catalytic method (see text). (b) Embedded routinely in Epon 812 following postosmication and complete dehydration. Thin sections of the Epon-embedded kidney were "etched" in 10% hydrogen peroxide for 5 min (Baskin et al., 1979). These and untreated thin sections of LR White-embedded kidney were immunolabeled identically by immersing them in rabbit anti-human K-light-chain antiserum (Dako Ltd., High Wycombe, Buckinghamshire, UK) followed by goat anti-rabbit IgG coupled to 10-nm colloidal gold. Both were counterstained in uranyl acetate and lead citrate; ×25,000. (a) Shows good ultrastructure with intense gold label accurately localized to the dense deposits in the basement membrane and to vacuoles (arrows) in the cytoplasm of epithelial podocytes with fused foot processes. (b) The membranes are a little clearer but the gold label is distributed throughout the tissue and even over the resin in areas where there is no tissue. Only by accurate counting and comparisons of the numbers of gold particles over the dense deposits with those of other areas is it possible to prove that the dense deposits are immunoreactive. It is almost impossible to judge the immunoreactivity of unsuspected sites. No difference in the distribution or intensity of the label was seen when etching was omitted or when "unmasking" with sodium metaperiodate (Bendayan and Zollinger, 1983) was employed.

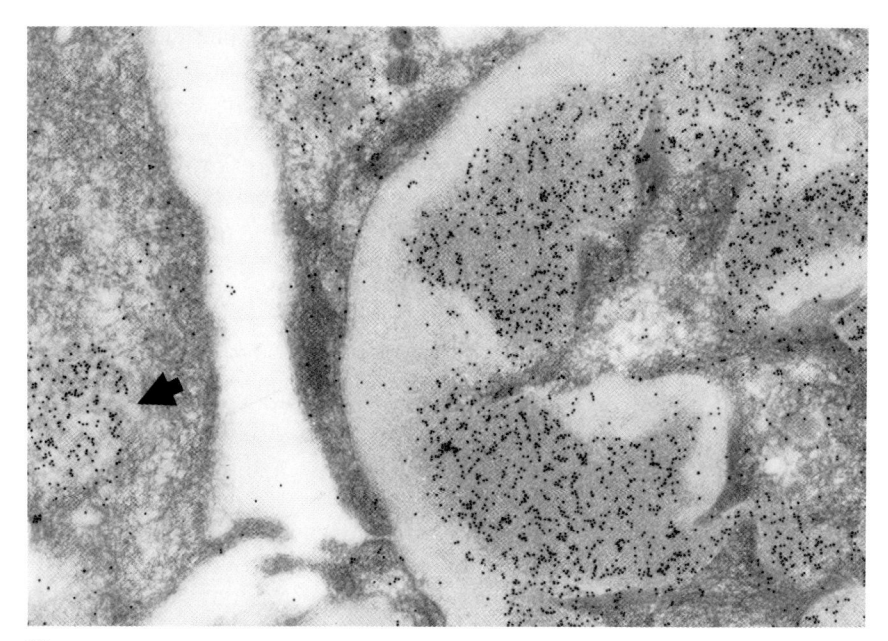

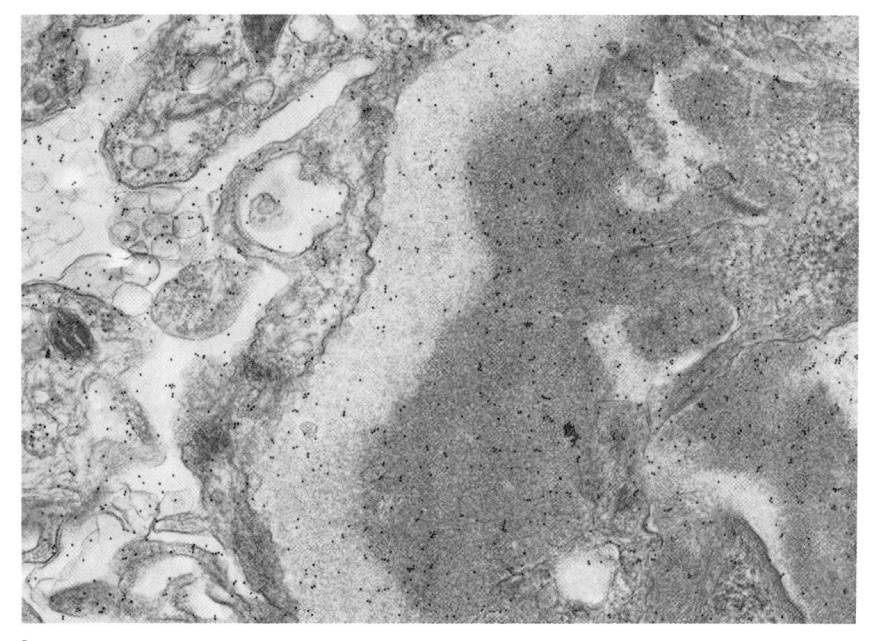

4. LR White Embedding Medium   55

4 hr, which can be important for urgent diagnosis. When glutaraldehyde is used, more time should be given for rinsing to make sure that the fixative is completely washed away. The rapid embedding method (see below) will improve the antigenic response only in those surgical biopsies which have not been subjected to autolysis, anoxia, or prolonged fixation (Figs. 4.3a and 4.4 Bottom). This method can also be used with perfusion-fixed tissue but the tissue must be well fixed and well washed (Fig. 4.4a).

Rapid Infiltration and Catalytic Polymerization

Tissue Blocks

The reason that the routine heat-polymerization embedding schedule above may not be suitable for mildly fixed tissues is the length of time that the tissue resides in monomeric LR White resin during infiltration and polymerization. Mild as acrylics are, prolonged exposure to them will still cause considerable extraction from the tissue, particularly where it is weakly cross-linked after formaldehyde or formaldehyde/glutaraldehyde fixation. By using warmth to reduce the time of infiltration followed by rapid catalytic polymerization, considerable gains in sensitivity can be achieved; e.g., the same labeling intensity has been achieved with dilutions of primary antibody 10 or sometimes 100 times greater than that used to immunolabel identical tissue embedded by the routine LR White method (Yoshimura *et al.*, 1986). As before, postosmication is avoided

---

Fig. 4.4. (Top) Pituitary of rat perfusion fixed in BGPA and embedded in LR White as in Fig. 4.3a. The thin section was immersed in rabbit antirat prolactin (a gift from the National Pituitary Hormone Distribution Programme, National Institute of Arthritis, Metabolic and Digestive Diseases (NIAMDD), Bethesda, MD, followed by goat antirabbit IgG coupled to 12-nm colloidal gold (Bioclinical Services Ltd.). Counterstained with uranyl acetate and lead citrate. PRL, Prolactin; ×27,000. (Bottom) Surgically removed human pituitary fixed in BGPA and embedded in LR White as above. Double immunolabeling. As in Fig. 4.2, the thin section was immunolabeled using two primary antibodies raised in rabbit. It was immersed first in rabbit antihuman prolactin (a gift from NIAMDD), which was localized with goat antirabbit IgG coupled to 12-nm gold. Then, after blocking spare valencies on the gold with normal rabbit serum, the section was immersed in DNP-labeled rabbit anti-Synacthen antiserum (an antibody against pure synthetic ACTH (1–24)—a kind gift from Professor Leslie Rees, St. Bartholomew's Hospital, West Smithfield, London, UK), which was localized by the DNP–hapten sandwich staining (DHSS) peroxidase/diaminobenzidine (DAB) method (Bioclinical Services Ltd.) (Newman and Jasani, 1984b). The DAB was made electron dense with gold chloride (Newman *et al.*, 1983b). No counterstaining. Steric hindrance is less of a problem with this method (cf. Fig. 4.2) because although the colloidal gold immunolabel is confined to the section surface, the DAB has been shown to be within the section (Newman and Hobot, 1987). ACTH, Adrenocorticotropic hormone; PRL, prolactin; ×13,500.

4. LR White Embedding Medium

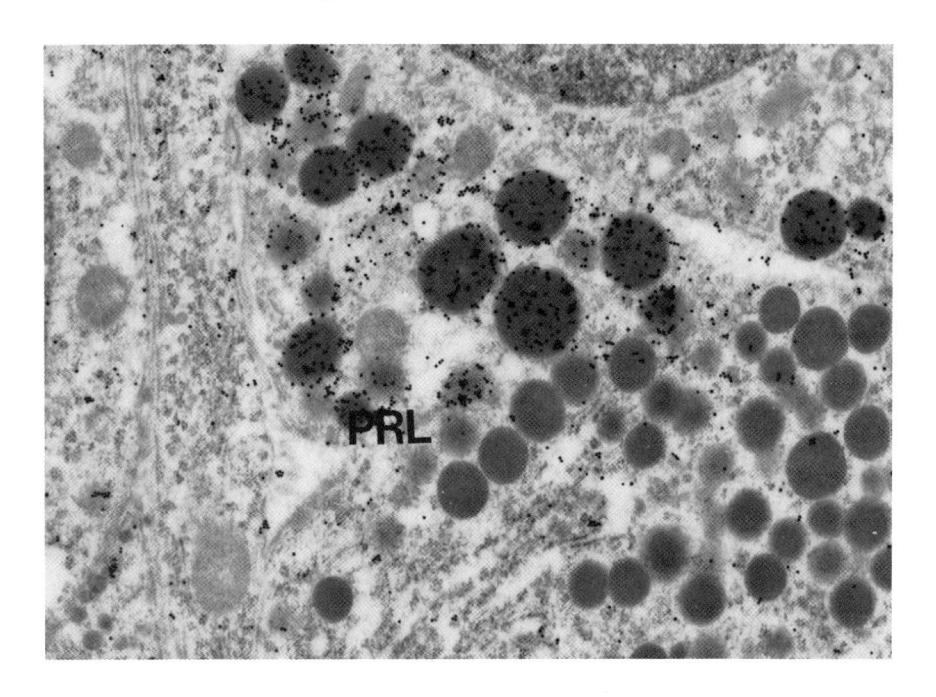

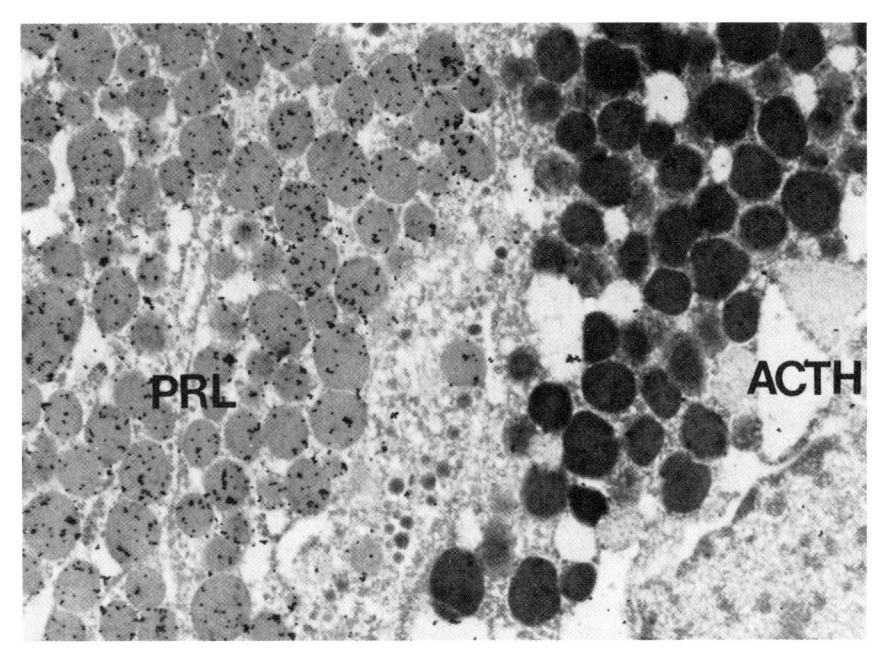

and the tissue is only partially dehydrated in 70% ethanol, observing the same precautions as for the previous embedding schedule.

The detailed method for heat-aided rapid infiltration by LR White and catalytic polymerization is as follows. All procedures take place at room temperature unless otherwise specified. Following rapid dehydration in 70% ethanol (one change in 50% ethanol for 10 min, three changes in 70% ethanol for 10 min each), small tissue blocks (<1 mm^3) are transferred to a 2:1 mixture of LR White and 70% ethanol and gently agitated on a rotary device for 20 min. The tissue blocks are given three changes of pure, hard-grade LR White at 50°C in an accurately set oven for 20 min each change.

The blocks are placed into gelatin capsules (O-gauge or smaller) containing appropriate paper labels and a mixture of recently purchased (<3 months old) LR White resin and the manufacturer's accelerator in the proportions of 1.5 µl accelerator to 1 ml of LR White hard-grade resin. Small aliquots of the mixture should be accurately prepared (too much accelerator leads to overrapid polymerization with the production of damagingly high temperatures and very brittle blocks), and when the accelerator is added it should be vigorously stirred (but not in a way that admits oxygen because that would inhibit polymerization). The polymerization mixture will begin to gel after ~7 min, so it should be dispensed rapidly into the awaiting gelatin capsules.

Polymerization is complete in ~45 min. A transient rise in temperature to 50–60°C usually occurs. The end of the capsule distal to the tissue should retain a small amount of unpolymerized LR White. If it does not, it may be that the monomer was already partly polymerized; i.e., it had been kept too long or unrefrigerated. If a large proportion or all of the LR White fails to polymerize and turns pink, it is usually due to either the introduction of oxygen or deterioration in the accelerator through having been kept too long or unrefrigerated. Because of its large volume, the catalyzed LR White which remains unused may get very hot as it polymerizes but this does not mean that the same high temperature is occurring in the much smaller volume of the gelatin capsules. Nevertheless, for this reason, the embedding of large blocks of tissue by this method is not recommended.

Thin sections should be tested for stability in the electron microscope beam. If they prove to be unstable, future sections from the same block can be mounted on plastic/carbon-coated grids, or additional blocks can be further polymerized by placing them in an oven at 50°C for 1–2 hr. This procedure can be repeated until satisfactory sections are obtained. If overpolymerized, however, the blocks will become very hard and brittle.

4. LR White Embedding Medium

Blocks which have been removed from their gelatin capsules or from which sections have already been cut and which have, therefore, been exposed to the air will have a limited repolymerization potential.

Cell Cultures

The catalytic polymerization method, as above, is also very useful for embedding fixed cell cultures (Fig. 4.5), particularly if the cells are grown on "quick-release" Thermanox plastic cell culture coverslips (Flow, Rickmansworth, UK) (Wynford-Thomas *et al.*, 1986). The cells are fixed with an aldehyde (OsO_4 is omitted) and, after brief rinsing, are immunolabeled before embedding (preembedding immunostaining) or after embedding (on-grid immunostaining) or both. The fixed cells on the coverslips are dehydrated in 50% ethanol for 5 min followed by 70% ethanol (two changes for 10 min each) by transferring the coverslips between appropriate solutions. An intermediate step of 2 parts LR White (recently purchased) to 1 part 70% ethanol (prepared from 100% ethanol) will help to reduce shrinkage artifact. Infiltration by LR White takes place at 22°C, where the very thin cell layers are completely impregnated after two changes lasting for 10 min each.

In situations where the antigen is sensitive to ethanol, acetone may be used but dehydration should, in this case, be complete, i.e., up to 100% acetone, and at least four changes of LR White given in order to wash away thoroughly any remaining traces of acetone. Acetone is a radical scavenger and will prevent polymerization if traces of it are left in the tissue.

The coverslips should be cut into six or eight pieces, making sure that they are kept cell surface uppermost. The polymerization mixture is then prepared as above, 1.5 µl of accelerator per 1 ml of LR White, and dispensed rapidly to flat-bottomed BEEM-type polypropylene capsules from which the lids have been removed. A piece of coverslip is then sited on the bottom of a foil flat-embedding dish (cells uppermost) and an LR White/accelerator-filled capsule is inverted over it. This is repeated until all the pieces have been embedded. It is important to remember that the resin/accelerator mixture begins to gel after ~7 min, so the number of pieces that can be embedded at any one time is limited and speed of operation is essential.

Residual LR White/accelerator mixture must be poured around the base of the capsules to take up any space caused by polymerization shrinkage. A slight rise in temperature will occur and polymerization is complete in ~45 min. The capsules and the coverslip pieces pull away easily from the base of the foil dishes and the plastic of the coverslip can

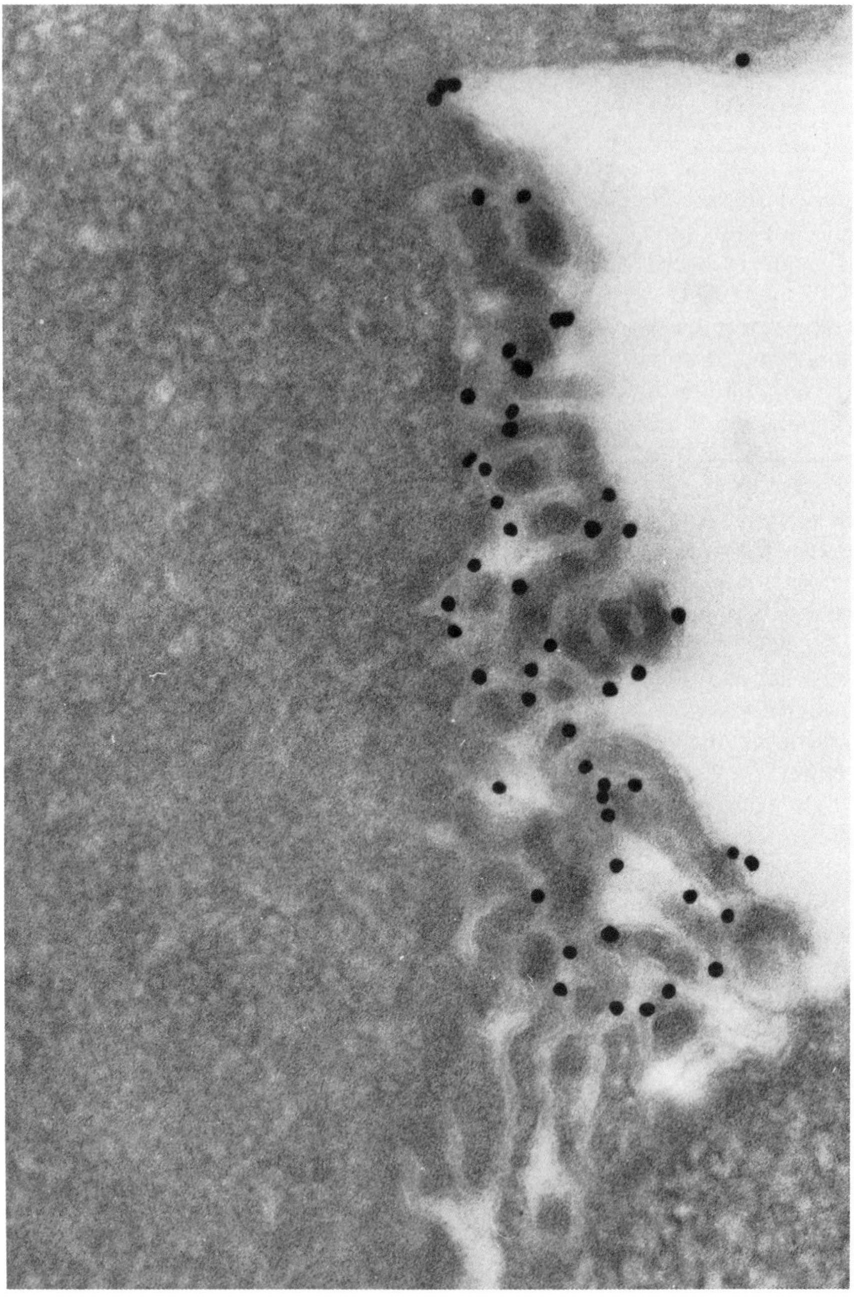

4. LR White Embedding Medium

be eased away, with the edge of a razor blade, from the polymerized LR White, leaving the cells behind, embedded in the resin. Further polymerization is possible in an accurately set 50°C oven, if the piece of coverslip has not been removed.

Infiltration at 22°C and Catalytic Polymerization in the Cold

It is most desirable that the tissue is minimally cross-linked by fixatives, commensurate with its having discernible ultrastructure (see Fig. 4.7), in order to preserve maximally its reactivity to antibodies and lectins. To achieve this, certain animal tissues can be perfusion fixed with very low concentrations of glutaraldehyde (<0.1%). The tissue fixed in this way and embedded by the rapid method described above can show a serious form of polymerization artifact (Fig. 4.6). The reason for this is that the tissue retains reactive groups (for example, amines) which, independent of any added accelerator, exert catalytic effects on the polymerization of the resin. During infiltration the use of heat causes the resin around the outside of the tissue blocks partly to polymerize. When the accelerator is now added it can only penetrate the outside of the blocks, which show an exaggerated response, polymerizing so rapidly that great contraction of the outside occurs while the inside becomes swollen.

Excellent preservation of the ultrastructure is, however, obtainable if a longer, slightly modified approach is employed (Fig. 4.7). New ultrastructural findings on, for example, the appearance of the transtubular Golgi apparatus network and the bacterial cell envelope, thought to be dependent on low-temperature methods (PLT/Lowicryl) (Hobot *et al.*, 1984; Roth *et al.*, 1985) and ultracryomicrotomy using unfixed frozen-hydrated sections (Dubochet *et al.*, 1983), have also been recorded using this modified approach (Newman and Hobot, 1987). The structure of the bacterial cell wall is particularly sensitive to processing and embedding and is altered if the cells are fully dehydrated (Newman and Hobot, 1987).

The modified method follows. All steps are carried out at room temperature unless otherwise specified. After minimal perfusion fixation, the un-

---

Fig. 4.5. Preembedding immunolabeling of the surface of cultured A431 cells (human carcinoma cell line) fixed in neutral-buffered glutaraldehyde (Wynford-Thomas et al., 1986). The cells were incubated with a monoclonal antibody, R1, against the exterior domain of the epidermal growth factor (EGF) receptor (kindly donated by Dr. M. D. Waterfield, Imperial Cancer Research Fund (ICRF), London, UK) followed by goat anti-mouse IgG coupled to 15-nm colloidal gold (Bioclinical Services Ltd.). They were embedded in LR White by the rapid catalytic method. Counterstained in uranyl acetate and lead citrate; ×100,000.

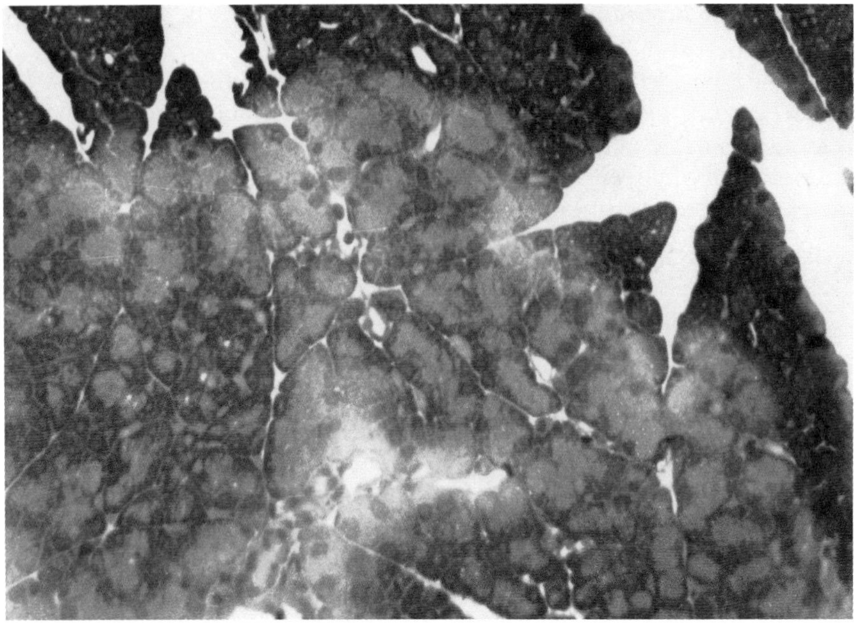

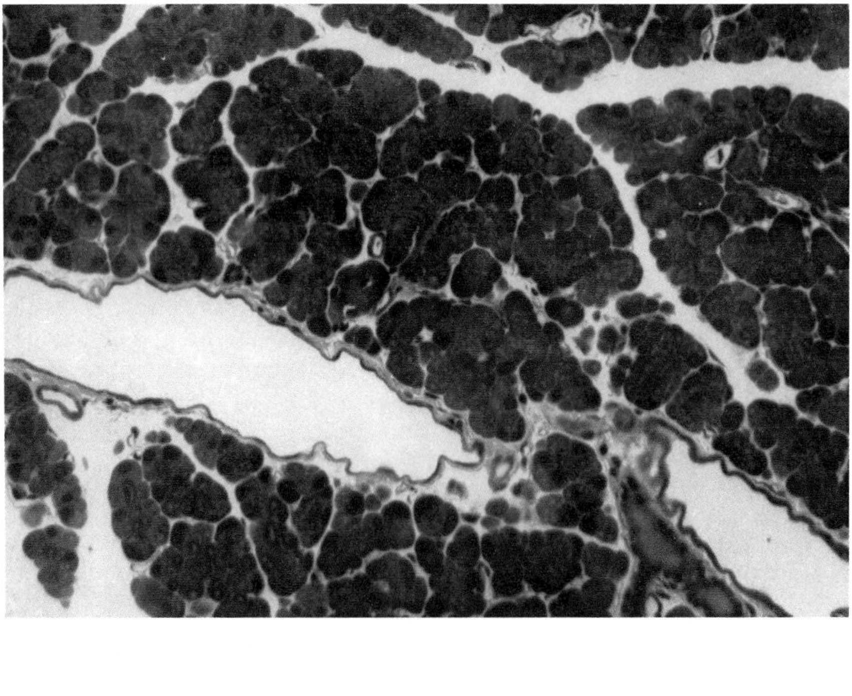

4. LR White Embedding Medium

osmicated tissue blocks are washed free of fixative, partially dehydrated through 50 and 70% ethanol, and gently agitated in a 2:1 mixture of LR White and 70% ethanol on a rotary device for 30 min. The tissue blocks should be kept as small as possible and no bigger than 1 mm^3 in size and smaller than this in at least one dimension.

To avoid tissue-induced polymerization, infiltration is best completed at 22°C in four changes (30 min each) of pure hard-grade LR White, on a rotary device. For difficult tissues (e.g., tissues with a high collagen, lipid, or chitin content) gentle heat at 35°C may be necessary during the resin changes. Polymerization is slowed, by conducting it at 0°C, to give the accelerator/LR White mixture time to diffuse into the center of the tissue block.

A mixture of 1.5 μl of accelerator per 1 ml of recently purchased LR White is used. Aliquots of 5 or 10 ml of LR White should be equilibrated at 0°C in a refrigerator while everything else is assembled and made ready. The correct amount of accelerator is quickly added to the resin, stirred, and the mixture rapidly transferred into O-gauge (or smaller) gelatin capsules containing appropriate labels. The tissue blocks must also be rapidly dropped into the capsules, which are quickly capped and returned to the refrigerator. By working swiftly with amounts of resin small enough for only 5 to 10 gelatin capsules at a time, temperature rises are kept minimal. An aluminum block, drilled to hold gelatin capsules tightly so as to act as a heat sink, is very useful though not essential. If it is loaded with capsules and precooled to 0°C it will very effectively keep temperatures down while the capsules are being filled and conduct away heat during polymerization. Polymerization will take up to 4 hr at 0°C, thus giving time for the catalyzed LR White mixture to diffuse into the center of the tissue block before gelling begins.

As the LR White is now exposed to the influence of air for a longer time before polymerization, more of it will remain unpolymerized at the top of the capsules than that found in the previous method. About 25–30% of the monomer at the end of the capsule distal to the tissue will turn pink and remain liquid. If it does not, the resin may have incorrectly polymerized. We have found it convenient to leave the blocks to polymerize over-

---

Fig. 4.6. Unosmicated pancreas of rat, perfusion fixed in 0.1% neutral-buffered glutaraldehyde for 1 hr. Semithin sections stained in toluidine blue; ×250. (Top) Embedded by the rapid catalytic method at room temperature (see text). Peripherally the tissue is shrunken, its increased density resulting in dark staining; centrally it is swollen and incompletely polymerized as shown by its patchy, light staining. (Bottom) Embedded by the cold catalytic method (see text). No polymerization artifact; the tissue is evenly stained throughout.

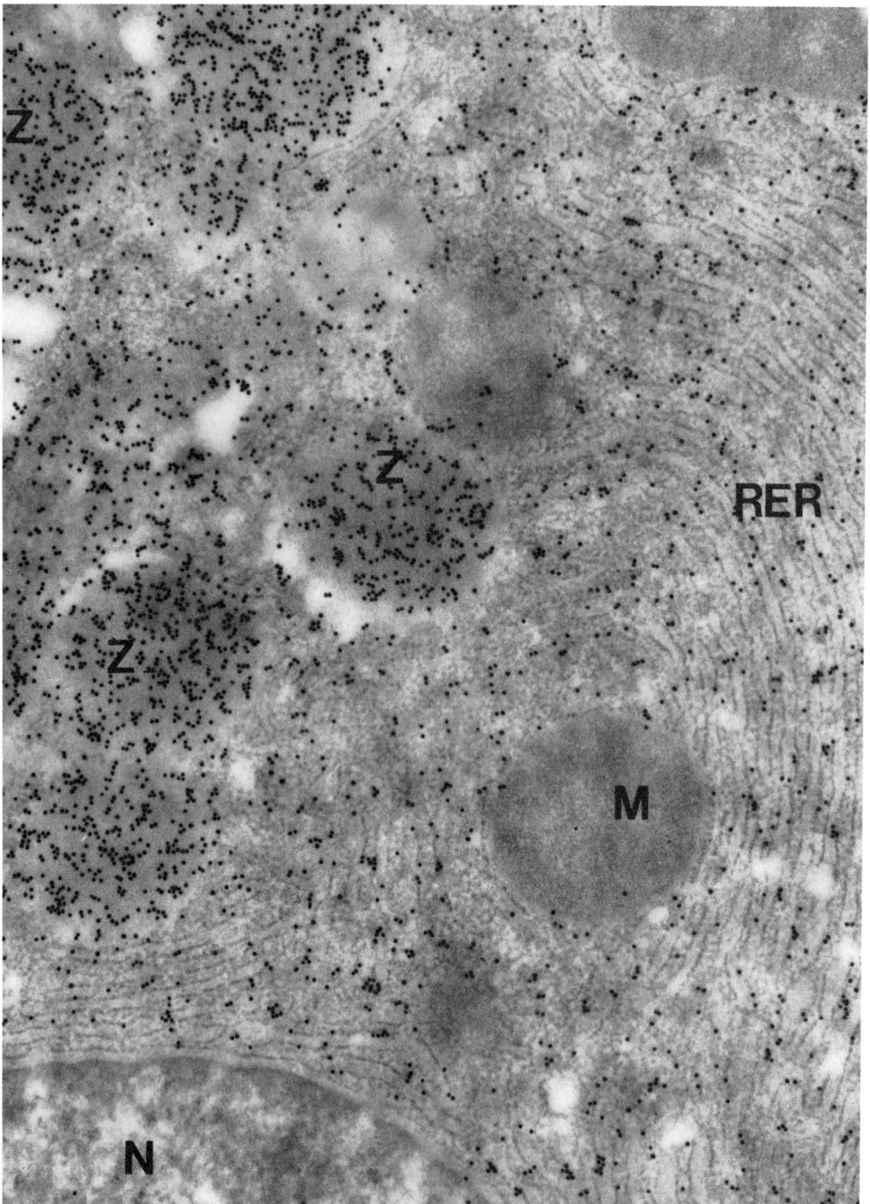
Z
Z
RER
Z
M
N

4. LR White Embedding Medium

night in a refrigerator set at 0°C and to remove them, unopened, to a −40°C deep freeze the following morning for storage.

The LR White resin of such freshly polymerized blocks is minimally cross-linked at this stage and it is unlikely that thin sections of them would be electron beam resistant without further polymerization. However, semithin sections taken from newly polymerized blocks are very sensitive to antibodies and immunolabeling, where the colloidal gold can be made visible photochemically in the light microscope (Danscher and Rytter Norgaard, 1983).

Thin sections can also be cut but they need to be supported on Formvar–carbon films, which limits immunolabeling or lectin–gold methods to one side of the grid. Maximum sensitivity by immersing the grids in the immunoreagents cannot be obtained and double-labeling methods employing gold particles of different sizes as labels for each of the two sides of the grid, the two-face technique (Bendayan, 1982), are impossible to carry out. As with the previous rapid method, the electron beam stability of sections can be increased by further polymerization, which is easily acheived by heating the unopened blocks. It is likely that even keeping the blocks at 22°C allows polymerization to continue, hence our preference for storing them in a deep freeze. Exposing the tissue-containing end of blocks to air restricts the extent to which they can be further polymerized.

STORING LR WHITE BLOCKS

In the majority of cases, little or no measurable change has been observed in LR White-embedded tissue on normal storage at room temperature, particularly in unopened blocks. Sections obtained from soft animal tissues embedded in LR White in 1981 still produce satisfactory immuno-

---

Fig. 4.7. An acinar cell from the exocrine pancreas of the rat fixed and embedded as in Fig. 4.6b. The thin section was immersed in rabbit antirat anionic trypsin (a gift from Dr. John Kay, Department of Biochemistry, University College Cardiff, Cardiff, UK) followed by goat antirabbit IgG coupled to 10-nm colloidal gold. Counterstained in uranyl acetate and lead citrate. Despite very minimal cross-linking in fixative and no osmication, many ultrastructural features are still visible. Worst affected are the zymogen storage granules. Gold labeling is intense, even over the rough endoplasmic reticulum (RER), which is a site of low antigen concentration. The RER labeling is almost lost when higher concentrations of glutaraldehyde are used (see Chapter 3). M, Mitochondria; N, nucleus; RER, rough endoplasmic reticulum; Z, zymogen granule; ×50,000.

labeling for hormones and enzymes. However, there is evidence to suggest that some deterioration in the antigenic response (although not the ultrastructure) of LR White-embedded tissue stored at room temperature can occur, at least with regard to immunoglobulins. We have noticed a decline in the immunolabeling of intracytoplasmic immunoglobulins (light and heavy chains) over a period of 6–12 months in opened or unopened blocks of human tonsil tissue. Although the exact reasons for this decline are not known, it could be due to a chemical change or simply the uncontrolled continuance of polymerization. The latter theory is supported by the fact that storage in a deep freeze at $-20°C$ appears to arrest this decline. However, as yet insufficient time has elapsed for us to be categorical.

In an effort to characterize any measurable change, we have stored variously fixed and embedded tissues aerobically and anaerobically, refrigerated and at room temperature, but, because this is a very long-term and difficult experiment during the course of which other factors may also change, no firm data are available yet.

When an important experiment is planned, it is probably prudent to leave at least some blocks unopened and unsectioned after each LR White embedding of tissues. Some blocks should be reserved anyway in case further polymerization is required. If these are stored in a deep freeze, continuity of tissue reactivity with antibodies and lectins is assured and any decline in immunolabeling can be ascribed to some other factor.

SECTIONING LR WHITE BLOCKS

Semithin Sections

Semithin sections (1–2 μm) are easily cut dry on a glass or diamond knife. The block face may be kept large (3 mm or more if Ralph knives are used) and should be smoothed to remove all pits and blemishes by the initial cutting of very thin sections (100 nm). When taking a semithin section, one or two very thin sections left on the knife edge will force it to lift and will reduce its rolling up and creasing. It can then be picked up with fine forceps and floated on a droplet of distilled water on a glass slide. No organic solvents should be used to unwrinkle the sections, either dissolved in the floating solution or as vapors wafted over them, because such reagents will very quickly dissolve the resin. Gentle heat (<60°C) alone is recommended, preferably in an oven. Thick sections (1–2 μm) sometimes adhere poorly to glass slides and come away from the slide during immunolabeling or counterstaining. Glass slides which

4. LR White Embedding Medium

have been dipped in chrome–alum gelatin (Pappas, 1971) will provide much greater adhesion (Newman and Jasani, 1984a) and, because of the hydrophobic nature of this coating, droplets of water stand up with a very high meniscus which is ideal for stretching sections, especially when warmed to get rid of creases.

Thick sections (1–2 μm) lack resolution and are not very economic of valuable tissue. Thinner sections adhere to glass slides more firmly, although it is still a good idea to use chrome-gel-coated slides except when protease digestion is planned. We have found that 350-nm sections are quite sufficiently thick to give good immunolabeling and counterstaining (Newman *et al.*, 1986; Wynford-Thomas *et al.*, 1986). Sections should be cut onto distilled water held by a "boat" attached to the glass or diamond knife. They are picked up on the blade of a cleaned pair of forceps, floated off onto a droplet of distilled water, and dried using gentle heat, as explained for the thick sections above.

Thin Sections

Thin sections are cut in the same way as the 350-nm semithin sections described above. Large block faces are perfectly feasible for thin sectioning (<3 mm) but care should be exercised over the level of the water held by the boat. The meniscus must be kept below the knife edge, so that the water just approaches up to it. If the water level in the boat is too high, water will flood onto the hydrophilic face of the LR White block. For this reason glass knives, which are easier to wet than diamond knives, are preferred. The knife angle is the same as for Epon 812 (Hayat, 1989). However, it is better to cut at slightly higher speeds than would be appropriate for Epon blocks.

LR White has a slightly lower refractive index than Epon so that sections of the former are a little thinner than those of a similar interference color cut from Epon. Pale gold sections are normally thin enough for good resolution but robust enough to resist the electron beam. When unsupported, they should be mounted on the polished (shiny) side of the grids, to which they adhere much more firmly than the dull side, unlike epoxide sections. If the sections are to be used for immunolabeling, it is conventional to employ nickel or gold grids, which do not interact with the immunoreagents. Just before the sections are picked up, the grids should be dip-rinsed in ethanol or acetone, blotted on filter paper, and dip-rinsed in distilled water. Delicate sections (very thin or underpolymerized) can be mounted on plastic/carbon-coated grids for extra support. Glow discharging will reduce the hydrophobia of carbon-coated grids, making it much easier to pick up the thin sections. Sections on coated grids are restricted in the number of techniques that can be applied to them (see page 65). The beam stability of thin sections can be improved by applying a fine carbon coating after they have been immunolabeled and counterstained.

IMMUNOLABELING OF SECTIONS

Semithin Sections

LR White, unlike other resins, is entirely compatible with peroxidase/diaminobenzine (DAB) methods, making it particularly useful for assessing the immunoreactivity or lectin reactivity of large sections of tissue with the light microscope before fining down to a smaller area for the electron microscope (Newman et al., 1986). This used to be a facility restricted to preembedding methods, but with the advent of good-quality antibodies, high postembedding sensitivity, and very efficient, light-insensitive photochemical methods for amplifying DAB (DAB Enhancement Kit, Amersham International, Amersham, UK) or for visualizing colloidal gold ("IntenSE," Janssen, Beerse, Belgium), postembedding immunostaining of semithin sections of LR White-embedded tissue is often significantly more sensitive than that of either paraffin wax or frozen (cryostat) sections. Sections of unosmicated LR White-embedded tissues should not be etched in sodium ethoxide, as semithin epoxide sections are, and we have not found it necessary to digest them with trypsin or any other protease in order to improve immunolabeling.

Thin Sections

On-Grid Labeling

As with semithin sections, thin sections of unosmicated LR White-embedded tissues should not be digested, etched, or treated with sodium metaperiodate. The high level of tissue reactivity in LR White means that, with economic dilutions and quite short incubation times (15–60 min), monoclonal and polyclonal primary antisera and lectins have given good results with all colloidal gold and peroxidase/DAB labeling methods (Newman et al., 1983a; Newman and Jasani, 1984a,b; Newman et al., 1986). Diaminobenzidine, made electron dense with gold chloride (Newman et al., 1983b), is found inside the sections, in contrast to the colloidal gold, which is always superficial (Newman and Hobot, 1987). This means that double immunolabeling, especially where both antigens occupy the same site, employing a sequential method in which immunocolloidal gold is used first followed by immunoperoxidase (Fig. 4.4 Bottom) (Newman

4. LR White Embedding Medium

et al., 1986), does not suffer seriously from the problems of steric hindrance inherent in using two sizes of colloidal gold (Fig. 4.2).

Preembedding/On-Grid Labeling

LR White embedding also provides great versatility for preembedding immunolocalization to be followed by sensitive on-grid localization of other antigens or lectins by using either colloidal gold particles or different sizes or an entirely different label such as peroxidase or ferritin. Provided no OsO_4 fixation has been used, photochemical visualization of colloidal gold or DAB is perfectly feasible in the preembedding mode.

COUNTERSTAINING OF SECTIONS

Semithin Sections

Sections of LR White-embedded tissue on glass slides can be stained by all the common dyes, such as toluidine blue and methyl green, without interference from background staining of the resin. Even hematoxylin and eosin will stain formaldehyde-fixed tissue well, although glutaraldehyde-fixed tissue sometimes stains rather weakly. Tissue embedded in LR White (see pages 49 and 54) can be very receptive of counterstains. So, in order to avoid impossibly short staining times, dilutions of the standard concentrations may be needed. Alcoholic solutions of stains and any other organic solvent solutions must be avoided (or used with extreme caution) because the LR White and the tissue embedded in it may be partially or wholly dissolved (Newman and Jasani, 1984a). For example, hematoxylin should be differentiated in acid water not acid alcohol, and eosin solutions should be aqueous not alcoholic. In addition, after staining and rinsing in distilled water, the sections are air dried, not dehydrated in an alcoholic gradient, before mounting and coverslipping.

Mountants containing xylene will cause severe wrinkling of LR White sections. Gurr's Neutral Mounting Medium (British Drug Houses, Poole, Dorset, UK) not only clears the section for viewing without wrinkling but also very effectively preserves toludine blue metachromasia. Some workers have suggested mounting LR White sections in LR White monomer or epoxide. In our experience, although this is a useful method, sections tend to bleach rapidly on storage.

Thin Sections

Sections of LR White-embedded tissue mounted on grids are counterstained, immediately after immunolabeling, with aqueous uranyl acetate and lead citrate, in the same way as epoxide sections of osmicated tissue (see Hayat, 1989). Care should be exercised not to overstain or the colloidal gold label may be obscured. Uranyl acetate on its own will provide some contrast and we have found that immersion in 2% aqueous uranyl acetate for 5 min is sufficient at 22°C for most tissues. The rinsing in distilled water is critical. Too long a rinse will destain the sections; too short a rinse will leave the uranyl acetate within the section to crystallize out in the form of a thin needlelike contamination, when the section is air dried. In general, three accurately timed 30-sec changes of distilled water are enough, but some trial and error may be needed.

Lead citrate (Reynolds, 1963) will provide much higher levels of contrast, particularly of membranes, when used after uranyl acetate, although some mental adjustment may have to be made by the observer to get used to the different appearance of membranes in unosmicated tissue. Only two 30-sec rinses need be given following uranyl acetate, after which a 30-sec immersion of the section in lead citrate will usually be enough. Some extrareceptive tissues will require that the lead citrate solution is diluted (1 : 10). Grids can now be more thoroughly rinsed than after uranyl acetate alone (three changes, 1 min each, of distilled water), and after air drying (see below) they are ready for viewing.

CONCLUDING REMARKS

Fixation is optional, but antigenicity is "inversely proportional to the glutaraldehyde concentration and duration of fixation (Hayat, 1986). Good preservation of ultrastructure can be obtained in the absence of postfixation in OsO_4, often accompanied by significantly increased immunolabeling and avoiding the requirement of metaperiodate treatment. LR White facilitates less deleterious dehydration by being able to infiltrate tissue from alcohol solutions of 70% or less. The lower the alcohol dilution prior to infiltration, the greater the retention of antigens and other cellular substances, but the smaller will have to be the tissue blocks. Partial dehydration is difficult to employ and largely ineffectual after postosmication.

When LR White is polymerized with heat or catalysts, polymerization is faster and it is easier to control the extent of cross-linking than when UV light is used. Embedding in hydrophilic, beam-stable acrylics like LR White makes other subjective measures, such as "etching" and "jet-washing" of sections, unnecessary. After wetting or soaking at any stage, LR White sections on grids should be treated carefully. The resin will have swollen and distended and in this condition the sections are easily

4. LR White Embedding Medium

damaged. For example, jet-washing or blot-drying them between stages during immunolabeling can cause them to become seriously folded or torn. They become even more swollen and easily creased in the presence of detergents such as Tween or Triton, whose addition to immunocolloidal gold reagents has been advocated for the lowering of nonspecific background labeling.

When immunolabeling and counterstaining are finished, the grids should be immersed in droplets of double-distilled water, carefully transferring them between droplets, to rinse them clean. After any procedure which requires that the grids are dried, they should be clamped in the jaws of fine forceps and air-dried gently at 22°C, without blotting. If correctly polymerized, the LR White of newly cut thin sections will not be completely cross-linked. When viewing such sections in the electron microscope for the first time, they should be equilibrated at high kV (80–120) and low magnification (<300×) with low illumination and the electron beam well spread. The section should be allowed to clear before going to higher magnifications and illuminations. The electron beam will complete the polymerization and cross-linking processes so that sections that have been directly exposed to it cannot be immunolabeled again or easily recounterstained.

I am greatly indebted to my staff, in particular Miss Alison Bowdler and Mrs. Sim Singhrao, who have tested and now use all the above techniques routinely and from whom I have "borrowed" some pictures.

REFERENCES

Baskin, D. G., Erlandsen, S. L., and Parsons, J. A. (1979). Influence of hydrogen peroxide or alcoholic sodium hydroxide on the immunocytochemical detection of growth hormone and prolactin after osmium fixation. *J. Histochem., Cytochem.* **27**, 1290.

Bendayan, M. (1982). Double immunocytochemical labeling applying the protein A–gold technique. *J. Histochem. Cytochem.* **30**, 81.

Bendayan, M., and Zollinger, M. (1983). Ultrastructural localisation of antigenic sites on osmium-fixed tissue applying the protein A–gold technique. *J. Histochem. Cytochem.* **31**, 101.

Causton, B. E. (1980). The molecular structure of resins and its effect on the epoxy embedding resins. *Proc. R. Microsc. Soc.* **15**, 185.

Craig, S., and Miller, C. (1984). LR White resin and improved on-grid immunogold detection of vicilin, a pea seed storage protein. *Cell Biol. Int. Rep.* **8**, 879.

Danscher, G., and Rytter Norgaard, J. O. (1983). Light microscopic visualisation of colloidal gold on resin-embedded tissue. *J. Histochem. Cytochem.* **31**, 1394.

Dubochet, J., McDowall, A. W., Menge, B., Schmid, E. N., and Lickfeld, K. G. (1983). Electron microscopy of frozen-hydrated bacteria. *J. Bacteriol.* **155**, 381.

Graber, M., and Kreutzberg, G. W. (1985). Immunogold staining (IGS) for electron microscopical demonstration of glial fibrillary acidic (GFA) protein in LR White embedded tissue. *Histochemistry* **83**, 497.

Hayat, M. A. (1981). *Fixation for Electron Microscopy.* Academic Press, New York and London.

Hayat, M. A. (1986). Glutaraldehyde: Role in electron microscopy. *Micron Microsc. Acta* **17**, 115.

Hayat, M. A. (1989). *Principles and Techniques of Electron Microscopy,* 3rd ed. Macmillan, London, and CRC Press, Boca Raton, Florida.

Hobot, J. A., Calemalm, E., Villiger, W., and Kellenberger, E. (1984). Periplasmic gel: New concept resulting from the reinvestigation of bacterial cell envelope ultrastructure by new methods. *J. Bacteriol.* **160**, 143.

Lackie, P. M., Hennessy, R. J., Hacker, G. W., and Polak, J. M. (1985). Investigation of immunogold–silver staining by electron microscopy. *Histochemistry* **83**, 545.

Newman, G. R., Jasani, B., and Williams E. D. (1982). The preservation of ultrastructure and antigenicity. *J. Microsc. (Oxford)* **127**, Rp5.

Newman, G. R., Jasani, B., and Williams E. D. (1983a). A simple post-embedding system for the rapid demonstration of tissue antigens under the electron microscope. *Histochem. J.* **15**, 543.

Newman, G. R., Jasani, B., and Williams, E. D. (1983b). Metal compound intensification of the electron density of diaminobenzidine. *J. Histochem. Cytochem.* **31**, 1430.

Newman, G. R., and Jasani, B. (1984a). Immunoelectron microscopy: Immunogold and immunoperoxidase compared using a new post-embedding system. *Med. Lab. Sci.* **41**, 238.

Newman, G. R., and Jasani, B. (1984b). Post-embedding immunoenzyme techniques. In *Immunolabelling for Electron Microscopy* (J. S. Polak and I. M. Varndell, eds.), pp. 53–70. Elsevier, Amsterdam.

Newman, G. R., Jasani, B., and Williams, E. D. (1986). Multiple hormone storage by "polycrine" cells in the pancreas (from a case of nesidioblastosis). *Histochem. J.* **18**, 67.

Newman, G. R. (1987). Use and abuse of LR White. *Histochem. J.* **19**, 118

Newman, G. R., and Hobot, J. A. (1987). Modern acrylics for post-embedding immunostaining techniques. *J. Histochem. Cytochem.* **35**, 971.

Pappas, P. W. (1971). The use of a chrome alum-gelatin (subbing) solution as a general adhesive for paraffin sections. *Stain Technol.* **46**, 121.

Ravazzola, M., and Orci, L. (1980). Glucagon and glicentin immunoreactivity are topologically segregated in the a-granule of the human pancreatic A-cell. *Nature (London)* **284**, 66.

Reynolds, E. S. (1963). The use of lead citrate at high pH as an electron opaque stain in the electron microscope. *J. Cell Biol.* **17**, 208.

Ring, P. K. M., and Johanson, V. (1987). Immunoelectron microscopic demonstration of thyroglobulin and thyroid hormones in rat thyroid gland. *J. Histochem. Cytochem.* **35**, 1095.

Roth, J., Bendayan, M., Carlemalm, E., Villiger, W., and Garavito, M. (1981). Enhancement of structural preservation and immunocytochemical staining in low temperature embedded pancreatic tissue. *J. Histochem. Cytochem.* **29**, 663.

Roth, J., Taatjees, D. J., Lucocq, J. N., Weinstein, J., and Paulson, J. C. (1985). Demonstration of an extensive transtubular network continuous with the Golgi apparatus stack that may function in glycosylation. *Cell (Cambridge, Mass.)* **43**, 287.

Weibull, C., Christiansson, A., and Carlemalm, E. (1983). Extraction of membrane lipids during fixation, dehydration and embedding of *Alcholeplasma laidlawii*-cells for electron microscopy. *J. Microsc. (Oxford)* **129**, 201.

4. LR White Embedding Medium

Wynford-Thomas, D., Jasani, B., and Newman, G. R. (1986). Immunohistochemical localisation of cell surface receptors using a novel method permitting simple, rapid and reliable using a novel method permitting simple, rapid and reliable LM/EM correlation. *Histochem. J.* **18**, 387.

Yoshimura, N., Murachi, T., Heath, R., Kay, J., Jasani, B., and Newman, G. R. (1986). Immunogold electron-microscopic localisation of calpain 1 in skeletal muscle of rats. *Cell Tiss. Res.* **244**, 265.

5

Lowicryls and Low-Temperature Embedding for Colloidal Gold Methods

JAN A. HOBOT

Electron Microscopy Unit
University of Wales College of Medicine
Heath Park, Cardiff, Wales

INTRODUCTION
ADVANTAGES OF LOWICRYLS
FIXATION AND CHOICE OF EMBEDDING
    TEMPERATURE
THE PROGRESSIVE LOWERING OF TEMPERATURE
    TECHNIQUE
  Methods of Achieving Low Temperatures
  Choice of Dehydration Agent
  Lowicryls K4M and HM20
  Composition and Handling of Lowicryls
PROTOCOLS
  Progressive Lowering of Temperature Technique
  Embedding at 0°C (273 K) or 22°C (295 K)
  Embedding at −20°C (253 K) or −10°C (263 K)
  Rapid Embedding

POLYMERIZATION
  Ultraviolet Light
  Chemical Catalysts
  Heat
IMMUNOLABELING
  Lowicryl K4M or HM20?
CRYOSUBSTITUTION
  Methods for Achieving Substitution Temperatures
  Methods of Freezing and Tissue Preparation
  Preparation of Substitution Media
  Infiltration and Embedding
  Lowicryls K11M and HM23
  Composition and Handling
  Substitution Procedures
  Lowicryl K11M: Embedding and Polymerization
  Lowicryl HM23: Embedding and Polymerization
  Polymerization Methods
  Immunolabeling
  Future Developments
FREEZE-DRYING AND LOW-TEMPERATURE
    EMBEDDING
GENERAL COMMENTS
  Storage and Thin Sectioning of Lowicryl Blocks
  Methods for Immunolabeling of Thin Sections
  Counterstaining of Thin Sections
REFERENCES

INTRODUCTION

Since the time of their introduction (Carlemalm *et al.*, 1980; Kellenberger *et al.*, 1980), the use of Lowicryl resins (Chemische Werke Lowi, Waldkraiburg, West Germany) has spread widely. Many publications have appeared that deal mainly with Lowicryl K4M coupled to on-section colloidal gold techniques. However, many of these methodologies differ from the original low-temperature protocols advocated for the Lowicryls, usually with no reasons given for the changes introduced. Even so, they demonstrate the tremendous versatility for electron microscopy. I will deal with only some of the modifications of the basic Lowicryl protocols, discussing mainly the original ideas and procedures concerning Lowicryl K4M and HM20. I will also present information on cryosubstitution, freeze-drying, and the use of the new Lowicryls, K11M and HM23. The adaptability of Lowicryls to suit various preparative methods for immunolabeling and structural preservation will also be discussed.

5. Lowicryls and Low-Temperature Embedding

The advantages of low-temperature embedding in reducing the adverse effects of organic solvents and resins on biological tissue have been known for some time. Carlemalm *et al.* (1982) pointed out earlier work by other groups which had shown that enzymes present in organic solvent concentrations as high as 70–90% maintained their native structure and activity at very low temperatures, $-50°C$ (223 K) to $-100°C$ (173 K). This led to the development of the Lowicryl resins, which are methacrylate mixtures having properties particularly suited to low temperatures: miscibility with different organic solvents, low viscosity, and ability to be polymerized by UV light. Lowicryl K4M is polar, while Lowicryl HM20 is apolar (Carlemalm *et al.*, 1982). Experiments by Carlemalm *et al.* (1982) with crystals of aspartate aminotransferase and catalase present in various organic solvents or embedded in the Lowicryl resins (K4M or HM20) demonstrated by x-ray crystallography that at low temperatures of $-35°C$ (238 K) to $-50°C$ (223 K), a higher preservation of molecular order was maintained when compared with results obtained at higher temperatures, $0°C$ (273 K) to $20°C$ (293 K). In addition, low-temperature embedding in hydroxypropylmethacrylate/butylmethacrylate mixtures as well as in Lowicryl HM20 or K4M showed an improved ultrastructural preservation of thylakoid membranes from pea or spinach chloroplasts when compared with conventional embedding (Weibull *et al.*, 1980).

Kellenberger *et al.* (1980) and Carlemalm *et al.* (1982) processed glutaraldehyde-fixed tissues through increasing concentrations of organic solvents as the temperature was progressively lowered and then embedded them in the Lowicryls at temperatures of $-35°C$ (238 K) to $-50°C$ (223 K). This procedure is termed the progressive lowering of temperature (PLT) technique. The PLT method has given good ultrastructural preservation of different tissues (Armbruster *et al.*, 1982; Carlemalm *et al.*, 1982; Roth *et al.*, 1981) and has provided new structural data in two different systems. The first system is the bacterial cell envelope (Hobot *et al.*, 1984). The PLT embedding was compared with conventional embedding techniques or with embedding in the Lowicryls at $0°C$ (273 K) or $22°C$ (295 K). The cells prepared with PLT were the only ones that showed a uniform layer of material between the outer and inner membranes of the bacterial cell envelope. This evidence led Hobot *et al.* (1984) to suggest a new model for gram-negative cell envelope organization based on the concept of a periplasmic gel. The efficacy of the PLT was substantiated by a similar appearance of the bacterial cell envelope in frozen-hydrated sections (Dubochet *et al.*, 1983). The second system was the Golgi apparatus in the rat liver embedded in Lowicryl K4M using the PLT method (Roth *et al.*, 1985); the results showed the presence of an extensive transtubular network which was continuous with the Golgi stacks.

The PLT method employs low temperature to reduce protein denaturation and to maintain a degree of hydration, which may be important in preserving protein structural conformations. Furthermore, the Lowicryls can be used with very low glutaraldehyde concentrations (0.1%) for very short times (15 min). Extraction of soluble tissue components is much reduced at lower temperatures. This allows for the detection by colloidal gold methods of antigens at low-concentration secondary sites (i.e., sites where much less antigen is present than in the main, primary sites). The sensitivity of antigenic sites to high concentrations of fixative and to room temperature dehydration by organic solvents and/or organic resins can therefore be overcome by low-temperature methods. Antigenic sites very sensitive to chemical fixatives can be processed by cryosubstitution in Lowicryls at even lower temperatures.

ADVANTAGES OF LOWICRYLS

The use of Lowicryl resins is not restricted to low-temperature embedding. It is part of their versatility that they, unlike other resins, can be used for infiltration of tissue over a wide range of temperatures, 20°C (293 K) to −70°C (203 K), and polymerized within this range by either UV light or chemical means. Heat polymerization at 65°C (338 K) is also possible. Based on this flexibility, Kellenberger *et al.* (1980) were able to set out a study program to better understand the effects of different steps of specimen preparation on the final appearance of the ultrastructure. For the first time it is possible to study the effects of the following variables.

1. Concentration of fixative and the duration of fixation.
2. Importance of the remaining water in the resin (vis-à-vis hydration shells).
3. Temperature during embedding.
4. Temperature and mode of polymerization.
5. Polarity of the resin.

One variable at a time can be changed while the others are kept constant, thus enabling its effects on the ultrastructure and antigenic sensitivity to be gauged.

When embedding specimens in Lowicryls, they are stabilized enough by cross-linking them with an aldehyde alone. This is advantageous for on-section immunolabeling techniques with colloidal gold, as OsO_4 can mask or destroy antigenic sites. This approach is also advantageous for observing unstained thin sections in the scanning transmission electron microscope (STEM). The PLT technique gives improved structural pres-

5. Lowicryls and Low-Temperature Embedding

ervation and was used by Carlemalm and Kellenberger (1982) to study unstained biological structures. No heavy metals such as osmium, uranyl, and lead, which are necessary to give contrast in the conventional transmission electron microscope (CTEM), were used. The contrast is given solely by the tissue itself, being the ratio of elastically to inelastically scattered electrons, known as *Z* contrast (Carlemalm and Kellenberger, 1982). The design of the Lowicryl resins is such that they give a low cross-scattering ratio because of their low density relative to the higher density (higher atomic number or *Z*) of tissue (for review, see Carlemalm *et al.*, 1985a). Using *Z* contrast, the transmembrane proteins of the septate junction from the testis of *Drosophila melanogaster* have been made visible (Carlemalm and Kellenberger, 1982; Garavito *et al.*, 1982), as has the uniform, matter-containing layer of gram-negative bacterial cell envelopes (Hobot *et al.*, 1984). The *Z* contrast can be applied advantageously to the identification of very small (1 to 3 nm) gold particles with the STEM (not seen easily in the CTEM) which have been linked to Fab fragments for very specific antigenic site localization or linked to undecagold complexes of size 0.8 nm (Hainfeld, 1987; Chapter 21, this volume).

The original aim, therefore, of introducing the Lowicryls together with low-temperature embedding procedures was to achieve the best possible preservation of ultrastructure and to observe it unstained and directly in the STEM. However, it was soon found that Lowicryl K4M had excellent properties for on-section colloidal gold labeling techniques (Roth *et al.*, 1981), combining the advantages of both good structural preservation and satisfactory labeling. Presently, these methods are extended to cryosubstitution and the introduction of two new Lowicryl resins, K11M and HM23.

FIXATION AND CHOICE OF EMBEDDING TEMPERATURE

Low concentrations of fixative (<0.2% glutaraldehyde) for short times (15 min) should be followed by PLT embedding protocols. This method reduces extraction of soluble tissue components and therefore preserves both the ultrastructure and the reactivity of low-concentration secondary antigen sites.

Higher glutaraldehyde concentrations or longer fixation times often improve the structural preservation of tissue components, but the antigenicity of secondary sites may be lost. Room temperature protocols can be used in a similar way to those set out for LR White (Chapter 4, this volume), reserving the partial dehydration method for just Lowicryl K4M but not for Lowicryl HM20. However, with UV light polymerization, small blocks of tissue no larger than 0.5 mm^3 are recommended (Chiovetti, 1982), although larger (1–2 mm^3) well-polymerized blocks have been produced (J. A. Hobot, unpublished observations).

The use of UV light polymerization systems means that aldehyde fixation alone can be used to cross-link and stabilize the tissue. Further protection from OsO$_4$ postfixation is not necessary as damaging heat polymerization (above 60°C, 333 K) is not used. When postfixation with OsO$_4$ is chosen, then UV light polymerization cannot be used, as osmium will interfere with UV penetration into the tissue, resulting in a poorly polymerized resin block. The same holds true for certain pigmented specimens. For these and osmicated tissues, it is necessary to use either chemical polymerization, which can cover a range of −35°C (238 K) to 20°C (293 K), or heat polymerization.

Fixation with OsO$_4$ often severely affects antigenic sites and can reduce labeling intensity (Roth *et al.*, 1981; Bendayan, 1984). It can also lead to fragmentation of polypeptides (Emerman and Behrman, 1982; Baschong *et al.*, 1984). Removal of the osmium with sodium metaperiodate "unmasks" the antigenic sites (Bendayan and Zollinger, 1983) but restores only some of the antigenicity. The degree of restoration is dependent on the protein (Bendayan, 1984; Bendayan *et al.*, 1987) and whether it is present in high enough amounts. However, previous experiments (Bendayan, 1984; Bendayan *et al.*, 1987) were carried out using Epon-embedded tissues, which were fixed in 1% glutaraldehyde and postfixed in OsO$_4$. As a result, the low-concentration secondary antigen sites were almost certainly lost along with any sensitive primary antigens and glycoproteins. Such losses distort the distribution pattern for any of the proteins under study. For high immunolabeling sensitivity and for good structural preservation, acrylic resin embedding methods using only aldehyde fixation should be used. Where aldehyde fixation itself affects the antigen, cryomethods such as cryosubstitution or freeze-drying should be employed. Here, no chemical fixation is required.

THE PROGRESSIVE LOWERING OF TEMPERATURE TECHNIQUE

Methods of Achieving Low Temperatures

Several methods are available for reaching the low temperatures required for progressive low-temperature embedding. The most sophisticated method is to use a low-temperature embedding apparatus (LTE 020,

5. Lowicryls and Low-Temperature Embedding

Balzers Union, Liechtenstein). Four temperature chambers, each capable of holding four specimen vials plus two for precooling, can be preset to the required temperatures ranging from 0°C (273 K) to −50°C (223 K). Stirrer controls allow for gentle mixing of specimens throughout the infiltration and embedding steps.

The second method involves the use of refrigerators (chest-freezer type if possible) for the different temperatures required for PLT, namely 0°C (273 K), −20°C (253 K), and −35°C (238 K). Agitation (gentle swirling action) for a few seconds can be carried out every 15 min, although for best exchange of solutions at a low temperature, continuous agitation (similar to that of the Balzers unit) gives better PLT embedding. For example, a small table for holding vials can be placed at an angle in the freezer and is then connected by a Teflon rod plus springs, via a hole in the side of the freezer, to a variable-speed motor outside.

The third method uses specific mixtures which give the required low temperatures. For 0°C (273 K), use ice; for −20°C (253 K), use an ice plus sodium chloride mixture in a ratio of 3 : 1 (w/w); for −35°C (238 K), use crushed ice plus crushed $CaCl_2 \cdot 6H_2O$, well mixed in a ratio of 1 : 1.44 (1581 g : 2259 g). Mixtures of o-xylene plus m-xylene in crushed dry ice have also been suggested for −35°C (238 K) to −50°C (223 K); however, as xylene vapors are toxic, the last method is not strongly recommended. The specimen vials can be precooled and transferred from one mixture to the next during the infiltration steps in two aluminum blocks in which holes have been drilled.

Choice of Dehydration Agent

The common dehydrating agents (ethanol, methanol, and acetone) are miscible with the Lowicryls. However, ethylene glycol and dimethylformamide (DMF), although miscible with the polar Lowicryl K4M, do not mix with the apolar Lowicryl HM20. All these organic solvents can be used for room temperature or 0°C (273 K) protocols and, except for acetone and ethylene glycol, for PLT embeddings down to −35°C (238 K) and lower temperatures. Acetone is unsuitable for the PLT technique, as 70% acetone has a freezing point of about −27°C (246 K), and PLT requires the 70% organic solvent step to be at −35°C (238 K). Ethylene glycol can be used, but only at concentrations down to 90%, where the freezing point is −50°C (223 K); the freezing point of 100% ethylene glycol is higher, −12°C (261 K). The concentration of organic solvent used in the PLT technique has been carefully chosen such that as the temperature is lowered, the freezing point of the organic solvent concentration just used is below that of the next temperature step (see graph 1 in the Lowicryl booklet; Chiovetti, 1982).

Ethanol is the most commonly used organic solvent with the Lowicryls at all temperatures, although excellent results in terms of structural preservation and labeling intensity have also been achieved using methanol, ethylene glycol, and DMF (Roth *et al.*, 1981). Bayer *et al.* (1985) observed that after dehydration in DMF, polysaccharides are not precipitated out, and they obtained good preservation and labeling of bacterial extracellular capsules using PLT and Lowicryl K4M.

Lowicryls K4M and HM20

Each Lowicryl K4M and HM20 kit contains a cross-linker, a monomer, and an initiator for UV polymerization. An instruction manual is also included. It is the monomer which gives each particular Lowicryl resin its polar or apolar properties. The cross-linker and initiator are the same for both. The initiator (labeled C) is benzoin methyl ether and is for UV light polymerization at low temperatures, $-35°C$ (238 K) to $-50°C$ (223 K). The initiators to use for polymerization at other temperatures will be given in sections dealing with those particular procedures. For the chemical compositions of the monomers and cross-linker, see Carlemalm *et al.* (1982).

Composition and Handling of Lowicryls

The relevant amounts of cross-linker and then monomer are weighed out in a vial and mixed gently so as not to introduce air into the resin mixture, as oxygen will inhibit polymerization of any acrylic resin. The best way to mix is to bubble a small stream of dry nitrogen through the resin via a Pasteur pipette. Besides mixing the resin, it will prevent oxygen dissolving in the mixture. Other methods are to stir gently with a glass rod or, with the lid on, slowly rock the vial to avoid air bubbles. Once cross-linker and monomer are mixed, the initiator is added and mixed as above. This resin mixture is now used for the embedding protocols either as 100% resin or mixed with an organic solvent to give 1 : 1 or 2 : 1 mixtures. When mixing, carry out the procedure in a well-ventilated fume cupboard, always using gloves for all steps involving resin handling, as methacrylates can cause sensitive individuals to develop eczema.

The ratio of cross-linker to monomer depends on the hardness of the resin block required for the specimen being embedded. Increasing the amount of cross-linker produces harder blocks. For Lowicryl K4M, it is recommended to use 1.35 g cross-linker plus 8.65 g monomer with 50 mg initiator C; the cross-linker concentration can be raised from 4 to 18% by weight. For Lowicryl HM20, use 1.5 g cross-linker plus 8.5 g monomer with 50 mg initiator C; the cross-linker concentration can be varied from 5 to 17% by weight. In practice, it may probably be advantageous to use amounts which give slightly harder blocks than listed here, especially for Lowicryl K4M (even up to a cross-linker concentration of 20% by weight).

PROTOCOLS

Following fixation, low-temperature embedding is carried out by infiltration with an organic solvent, normally ethanol but methanol or DMF can be used, followed by organic solvent–resin mixtures and pure resin (DMF is immiscible with Lowicryl HM20; use it only with K4M). The temperature is progressively lowered to $-35°C$ (238 K) but can, in the case of Lowicryl HM20, be lowered at the second 100% organic solvent exchange to $-50°C$ (223 K) for all subsequent infiltration and UV light polymerizaton procedures.

Progressive Lowering of Temperature Technique

1. 30% Organic solvent, 0°C (273 K), 30 min
2. 50% Organic solvent, $-20°C$ (253 K), 1 hr
3. 70% Organic solvent, $-35°C$ (238 K), 1hr
4. 100% Organic solvent, $-35°C$ (238 K), 1 hr
5. 100% Organic solvent, $-35°C$ (238 K), 1 hr

Infiltration is continued at $-35°C$ (238 K) for all steps with increasing concentrations of Lowicryl K4M or HM20 resin in organic solvents as follows:

6. Resin + 100% organic solvent (1 : 1), 1 hr
7. Resin + 100% organic solvent (2 : 1), 1 hr
8. Pure resin, 1 hr
9. Pure resin, overnight
10. Next morning, fresh pure resin, 1 hr

After infiltration, the resin is polymerized in gelatin capsules by indirect UV light for 24 hr at $-35°C$ (238 K), followed by further hardening by direct UV light at room temperature for 3 days to improve sectioning quality. However, blocks are already sectionable after 24 hr at $-35°C$ (238 K). (For conditions of UV light polymerization, see page 86.)

As acrylic resins can tolerate small amounts of water, the last solvent concentration used need not be 100% but can be reduced to 90%. Then all the steps are as above with the intervening Lowicryl resin–organic solvent mixtures being resin + 90% organic solvent in the proportions of 1 : 1 and then 2 : 1.

The real flexibility of the Lowicryl resins is that they can be used at any embedding temperature, allowing for analyses of parameters involved in specimen preparation.

Embedding at 0°C (273 K) or 22°C (295 K)

1. 50% Organic solvent, 10 min
2. 70% Organic solvent, 10 min
3. 90% Organic solvent, 10 min
4. 100% Organic solvent, three changes of 30 min each
5. Lowicryl resin plus 100% organic solvent (1 : 1), 1 hr
6. Lowicryl resin plus organic solvent (2 : 1), 1 hr
7. Pure resin, 1 hr
8. Pure resin, overnight
9. Next morning, fresh pure resin, 1 hr

After infiltration, the resin is polymerized in gelatin capsules by indirect UV light for 24 hr followed by direct UV light for 48–72 hr. The following three points are important: (1) The initiator used is not initiator C (benzoin methyl ether), which is supplied with the Lowicryl kits, but benzoin ethyl ether, which is used in the same proportion. The amounts of cross-linker and monomer are also the same as given before (see section on Composition and Handling). (2) The times of infiltration are shortened when compared to PLT, as at higher temperatures exchange of solutions occurs faster, especially if the specimen vials are on a bench rotamixer operating at low speeds. (3) It is possible to transfer specimens from 90% solvent (three changes of 30 min each) directly into the 1 : 1 and 2 : 1 mixtures of Lowicryl and 90% solvent, and finally into pure resin.

Embedding at −20°C (253 K) or −10°C (263 K)

The protocols for embedding at −20°C (253 K) or −10°C (263 K) are very different from the PLT procedure. Dehydration in 100% organic solvent is not completed at −35°C (238 K), so the tissue is not protected by this low temperature from the adverse effects of the solvent and resin. Carlemalm *et al.* (1982) cite the observations of earlier workers that in general the amplitude of molecular thermal vibrations is reduced at low temperatures but particularly so below −30°C (243 K). The point is of great importance for embedding into the Lowicryls following cryosubstitution (see section on Cryosubstitution). Thus, although there may be some gain by using $-20°C$ (253 K) or $-10°C$ (263 K) embedding protocols over room temperature methods, they do not necessarily confer the same improvements that can be achieved by PLT.

The effectiveness of using lower temperatures in order to obtain substantial improvements in immunolabeling efficiencies over room temperature procedures has still to be evaluated. Interestingly, however, Acetarin et al. (1986) commented that no substantial improvement in ultrastructural preservation would probably be reached by extending PLT embeddings at $-35°C$ (238 K) down to a temperature of $-60°C$ (213 K). Furthermore, Armbruster et al. (1983) have shown some advantages of PLT embedding into Lowicryl K4M for colloidal gold immunolabeling of antitubulin in isolated microtubules when compared with results from embedding in Lowicryl K4M at 0°C (273 K) or 20°C (293 K). It is therefore important for the beginner to use the PLT at $-35°C$ (238 K) with its accredited advantages.

Rapid Embedding

Rapid methods for embedding into Lowicryl K4M have been published (Altman et al., 1984; Simon et al., 1987). The aim was to introduce a fast technique of embedding into Lowicryl K4M such that results would be available quickly for routine diagnostic immunoelectron microscopy. Altman et al. (1984) achieved rapid embedding into Lowicryl K4M at 20°C (293 K) in 4 hr. High glutaraldehyde (3%) and formaldehyde (3%) concentrations were used. Direct UV light polymerization was required at 0°C (273 K) for 1 hr or less. Simon et al. (1987) used a rapid method for Lowicryl K4M at $-10°C$ (263 K). With UV light polymerization, their rapid method took 48 hr. It should be noted that PLT also takes 48 hr, because after the initial indirect UV light polymerization at $-35°C$ (238 K) for 24 hr, the resin blocks are ready to be sectioned. In addition, in the case of Simon et al. (1987), there were very low labeling intensities of antiamylase on zymogen granules of rat pancreas when compared with earlier results of others (Bendayan, 1984; Roth et al., 1981). The method of Altman et al. (1984) produced good labeling.

When preservation of ultrastructure, good labeling intensities, and fast processing are needed (e.g., in a clinical situation), comparisons should first be made with embedding into LR White. LR White embedding can take as little as 2.5 hr at room temperature and should be considered before deciding on a routine, fast embedding procedure. The LR White methods use chemical polymerization instead of UV light setups. Good labeling and ultrastructure are produced, especially when chemical polymerization is carried out at 0°C (273 K), slightly lengthening the processing time to 4–5.5 hr (Newman and Hobot, 1987).

POLYMERIZATION

The Lowicryls can be polymerized in three different ways: by UV light, by chemical catalysts, or by heat.

Ultraviolet Light

Under the conditions specified for polymerization of the Lowicryls at low temperature (Carlemalm et al., 1982; Chiovetti, 1982), there is a uniform and controlled polymerization of the resins. There is no significant shrinkage of the specimen, and when small volumes are polymerized (1–5 ml), "no significant temperature increase occurred during polymerization" (Carlemalm et al., 1982). The temperature rise at −35°C (238 K) has been measured to be no more than 2°C (275 K) (Weibull, 1986).

Two initiators can be used for UV light polymerization: initiator C (benzoin methyl ether), which is supplied with each Lowicryl kit and can be used in the temperature range of −35°C (238 K) to −10°C (263 K), and benzoin ethyl ether, which is used in the temperature range −10°C (263 K) to 30°C (303 K). The amounts added to the resin mixtures are the same for both and the same for whichever temperature is chosen for polymerization, i.e., 50 mg per 10 g of resin mixture. It should be noted that oxygen will inhibit any form of polymerization of the Lowicryl resins, so great care must be taken in mixing the various components.

Samples are best polymerized in small gelatin capsules (size 0, volume 0.68 ml) to minimize against heat rise or shrinkage artifacts. A wire holder/stand to take the capsules is recommended, as this allows UV light to reach all sides of the capsules (see Fig. 1 in Chiovetti, 1982). The actual polymerization is carried out in a polymerization chamber (see Fig. 2 in Chiovetti, 1982). It consists of a box that fits into a deep freeze (chest type is better, because, when opened, upright freezers warm very quickly). The box is lined on all surfaces with UV reflective material, usually aluminum foil. A hole in the top allows for the positioning of the UV source (two 15-W fluorescent tubes, e.g., Philips TLD 15-W/05 or similar type). It is positioned ~30–40 cm above the specimen capsules. If a weaker source of UV light is used, the distance between the lamp and the specimen must be reduced (e.g., for two 6-W UV lamps a distance of ~22–24 cm has been found to give good polymerization; J. A. Hobot, unpublished observation).

In all cases the wavelength of the light must be 360 nm, and in all cases the UV light is indirect. Diffuse UV light is formed by having a right-angle deflector lined with aluminum foil positioned just below the UV source. By means of a trial run, the distance above the specimen capsules can be adjusted. If the polymerization is too rapid, it will cause shrinkage and deformation along the sides of the capsules, or small bubbles will be seen at the bottom of the capsules. In this situation, the distance between lamps and capsules is increased. UV light polymerization is not used for osmium-fixed specimens or for certain pigmented specimens as these can absorb UV light and so give incomplete polymerization. In these situations either chemical or heat polymerization procedures should be employed.

To avoid temperature rises inside the deep freeze caused by heat emission from the UV lamps (especially two 15-W lamps), it may be necessary to have the source placed external to the deep freeze. A hole cut in the deep freeze top for part insertion of the light source is recommended. Weaker UV sources, e.g., two small 6-W UV lamps (of the hand-held variety), may be used inside a deep freeze with no temperature rise occurring within the body of the freezer or in the polymerization chamber, but this also may depend on the particular freezer model used. Conversely, a cold room which has a temperature range of $-30°C$ (243 K) to $-40°C$ (233 K) could be used. Alternatively, there are commercially available UV polymerization chambers on the market.

Chemical Catalysts

Although used infrequently, details of chemical and heat polymerization of the Lowicryls are given by Acetarin and Carlemalm (1982). As UV light is not used in this form of resin hardening, the chemical initiation of free-radical polymerization is produced by an organic peroxide, the reactivity of which is dependent on temperature. Therefore, at low or room temperature it is necessary to add an activator which catalyzes the release of radicals by the peroxide. The amounts of each will vary depending on the temperature chosen for polymerization (Acetarin and Carlemalm, 1982).

For chemical polymerization, the initiator is dibenzoyl peroxide (DBP) and the activator is N,N-dimethylparatoluidine (DMpT). The amounts of each will vary for the temperature of polymerization, the type of resin, and the gel time selected (i.e., working time available before the whole resin mixture begins to gel and becomes too viscous and unworkable). Acetarin and Carlemalm (1982) have chosen amounts of DBP and DMpT which give relatively long gel times (up to 30 hr), allowing for overnight infiltration with these resin mixtures. For Lowicryl K4M at $-35°C$ (238 K) use 0.4% (by weight) DBP plus 0.25% DMpT; at room temperature use 0.05% DBP plus 0.03% DMpT. For Lowicryl HM20 at −35°C (238 K) use 1% DBP plus 0.6% DMpT; at room temperature use 0.08% DBP plus 0.5% DMpT.

When preparing the resin, mix cross-linker and monomer, and then divide the resin mixture into two; to one portion add DBP and to the other DMpT. Equilibrate to the temperature chosen for infiltration and polymerization and, just before use, mix the two portions together. The chemical polymerization is initiated as soon as the portions are mixed. As oxygen can inhibit polymerization, it is best to mix all resin components/mixtures by bubbling dry nitrogen gently through them. Polymerization is carried out in small gelatin capsules placed in appropriate holes in an aluminum block, which acts as a heat sink to minimize heat rise during polymerization. Many factors can affect chemical polymerization, including the nature of the embedded material, presence of osmium, and heat (Acetarin and Carlemalm, 1982; Acetarin et al., 1986). Because of this uncertainty, it is best to use the method of UV light polymerization whenever possible with the Lowicryls.

Heat

Heat polymerization requires the addition of the initiator DBP. The heat itself will produce enough free radicals for polymerization to start. For heat polymerization in the range of 50°C (323 K) to 60°C (333 K), add 0.2% (by weight) of DBP for Lowicryl K4M and 0.5% for Lowicryl HM20. Mix all resin components by bubbling a stream of dry nitrogen through the components, and polymerize in small gelatin capsules (three-quarters full as the resin expands at the start of polymerization) placed in an aluminum block acting as a heat sink (Acetarin and Carlemalm, 1982).

IMMUNOLABELING

When rat pancreas is fixed by vascular perfusion for 15 min with 0.1% glutaraldehyde and embedded with Lowicryl K4M by the PLT method, a surprisingly good level of ultrastructural preservation is seen (Fig. 5.1a–d). There is intense specific labeling for trypsin over the zymogen granules, trypsin storage sites (Fig. 5.1a–c), much of the rough endoplasmic reticulum (Fig. 5.1a,d), lumen contents (Fig. 5.1c), and Golgi apparatus (Fig. 5.1a,b). On the other hand, when the same tissue is perfusion fixed for 15 min (Fig. 5.2 Top) or for 1 hr (Fig. 5.2 Bottom) with 1% glutaraldehyde, it shows very little labeling over the rough endoplasmic reticulum. There is still intense labeling over the zymogen granules and the Golgi apparatus.

5. Lowicryls and Low-Temperature Embedding

Ultrastructural preservation is good in all preparations (Figs. 5.1 and 5.2) It seems, therefore, that high concentrations of glutaraldehyde (1%) affect antigens at low-concentration secondary sites, such as rough endoplasmic reticulum, preventing them from being detected. This is probably due to changes in the antigen caused by strong cross-linking by glutaraldehyde. The same would hold true for many proteins present in the tissue in low concentrations. It is in these areas that the PLT technique has the most to offer colloidal gold methodologies. Where low-concentration antigenic sites are not involved, or where the antigen is not affected by fixation or sensitive to electron microscopic preparative techniques, other embedding methods at higher temperatures can be equally effective.

a

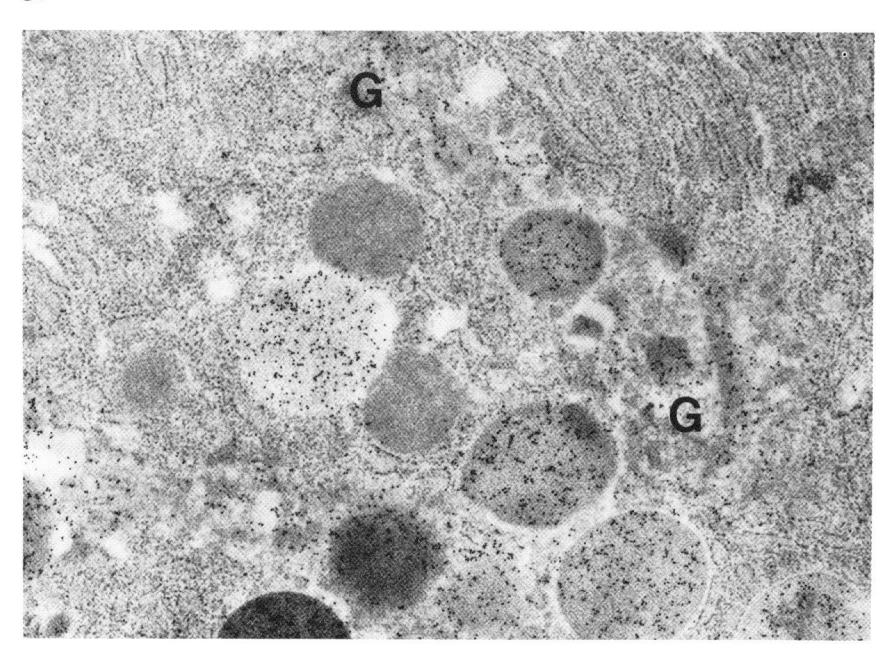

Fig. 5.1. Thin sections of rat pancreas fixed with 0.1% glutaraldehyde in buffer (pH 7.0) by vascular perfusion for 15 min. The specimens were embedded by PLT in Lowicryl K4M. Grids were immersed in rabbit antirat anionic trypsin (a gift from Dr. J. Kay), followed by goat antirabbit IgGs tagged to 10-nm colloidal gold (GAR-10) (Bioclinical Services, Cardiff, UK). There is intense labeling of the zymogen granules in (a)–(c), while in (d) prolific label is seen along the length of the rough endoplasmic reticulum (RER). Lumen (L) contents have been retained and appear labeled (c), as does the Golgi apparatus (G) in (a) and (b). Background is low over the mitochondria (M). Desmosomes (D) are frequently seen. Sections stained with uranyl and lead acetate; ×24,500. (Figure continued on next page.)

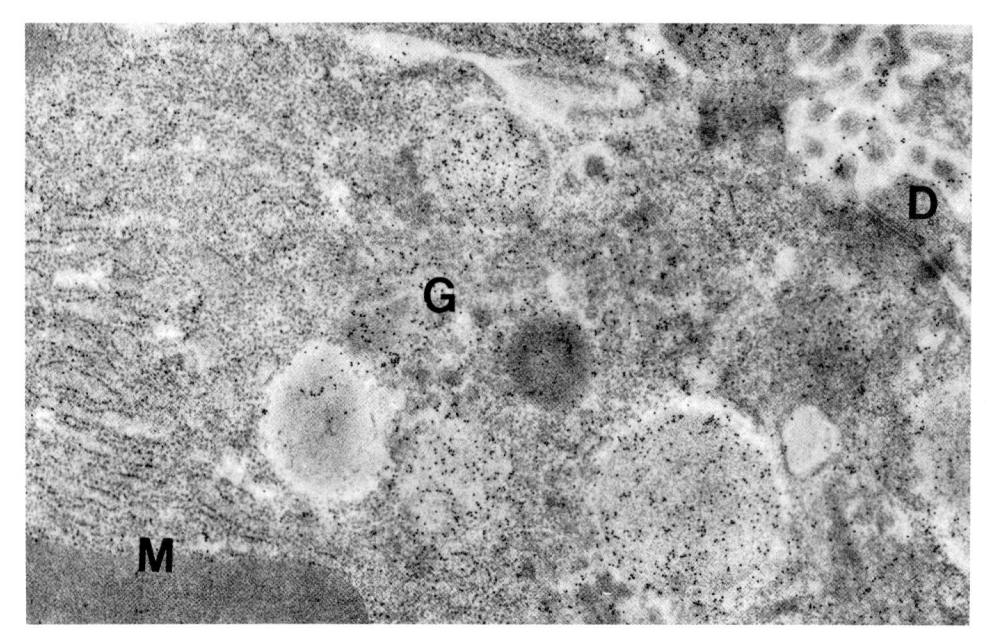

Fig. 5.1. (*Continued*)

5. Lowicryls and Low-Temperature Embedding

d

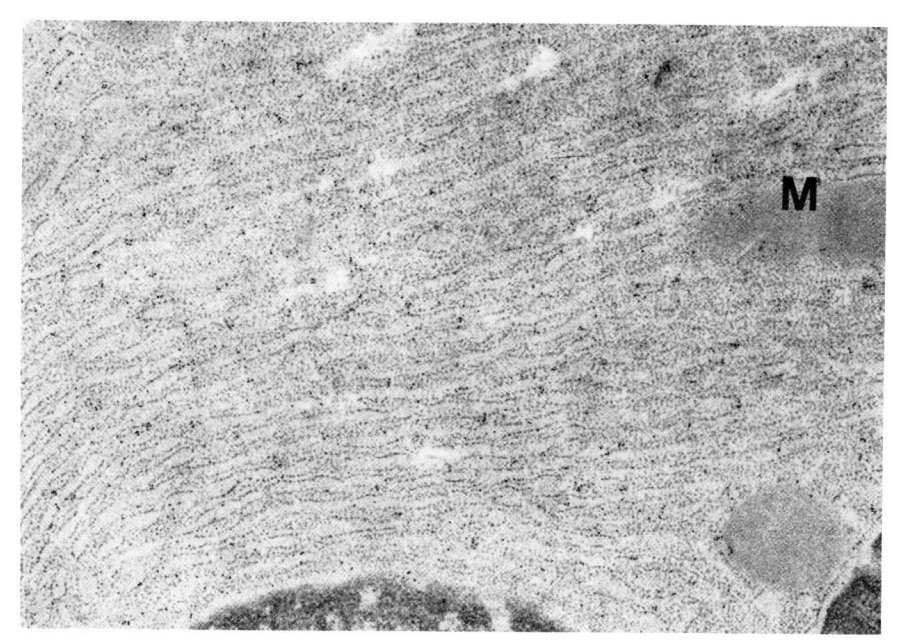

Fig. 5.1. (*Continued*)

Lowicryl K4M or HM20?

The overwhelming majority of scientists use only Lowicryl K4M for on-section colloidal gold labeling. There is little or no mention of any results with Lowicryl HM20 (Lin and Langenberg, 1983, 1984). It seems that Lowicryl K4M has been chosen because of its polar properties, and Lowicryl HM20, being apolar, has been grouped in with the apolar epoxide and polyester resins. When Lowicryl K4M sections are compared with Epon sections, in both cases the tissue having been fixed with an aldehyde only, the specificity and intensity of labeling over specific sites are generally the same in both for certain antigens (Roth *et al.*, 1981; Bendayan, 1984). However, background labeling over nonspecific sites is much lower on the Lowicryl K4M sections. This improved specific labeling and the superior ultrastructural preservation obtained in Lowicryl K4M are probably the reasons for this resin being adopted as the more satisfactory for colloidal gold methods.

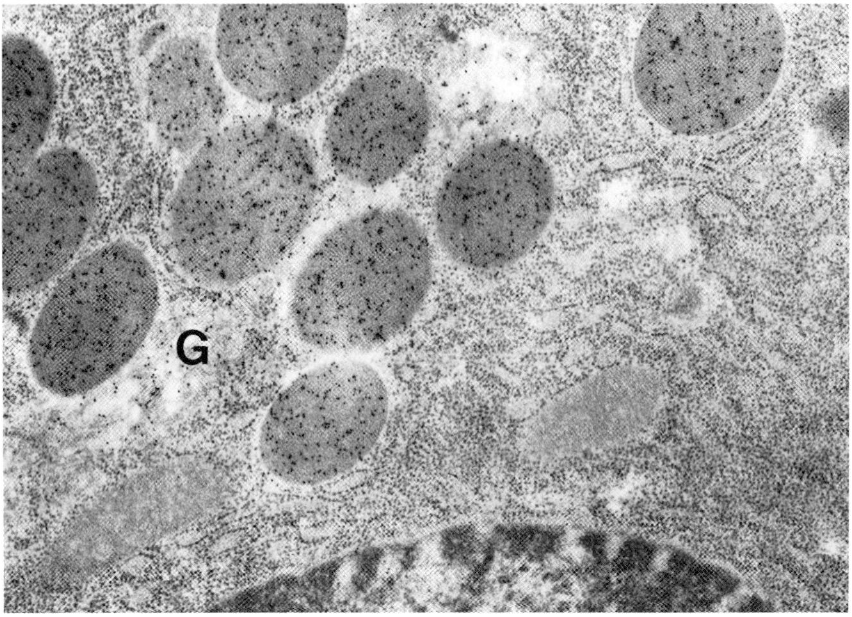
G

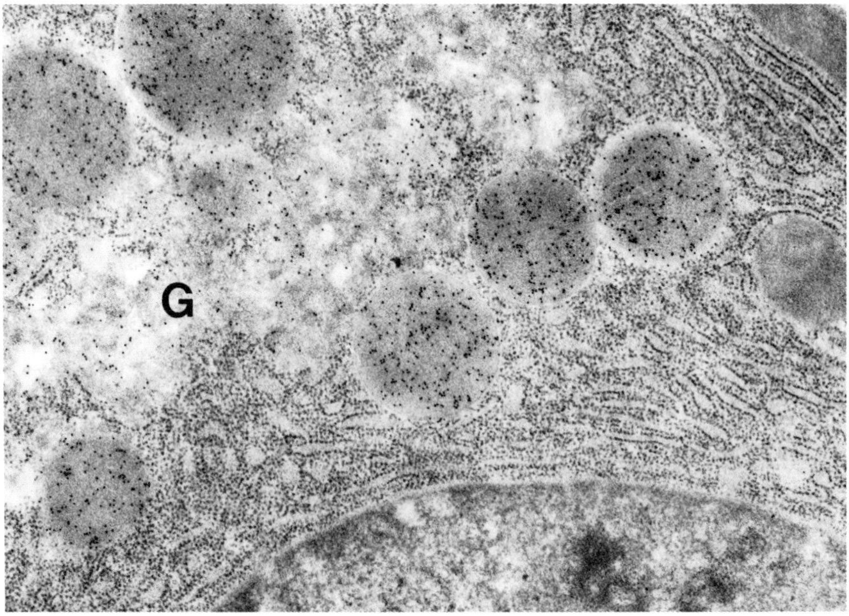
G

5. Lowicryls and Low-Temperature Embedding

Why low background staining is obtained with Lowicryl K4M is unclear. It has been suggested (Dürrenberger, 1989) that there may be components in some sera which interact with apolar resin surfaces and subsequently attract protein A–gold particles, thus increasing nonspecific labeling. Also, a specimen-related surface relief on sections of Lowicryls K4M and HM20 may be a factor in explaining their high immunolabeling efficiency with colloidal gold methods (Kellenberger et al., 1987), but this does not explain why similar intensities of specific gold label (Roth, et al., 1981; Bendayan, 1984) are obtained on sections of Epon which show no appreciable surface relief (Bendayan and Zollinger, 1983). Some of these points are discussed by Kellenberger et al. (1987). Therefore, with Lowicryl K4M one achieves excellent structural preservation in the absence of harmful fixation with OsO_4. Moreover, low labeling backgrounds and the ability to vary aldehyde fixation regimes to include room and low-temperature embedding protocols, or the avoidance of chemical fixation altogether with cryotechniques, make Lowicryl K4M highly suitable for gold labeling methods.

In fact, Lowicryl HM20 also qualifies on all the points raised in the preceding paragraph. In addition, it can be used at lower temperatures ($-50°C$, 223 K) than Lowicryl K4M ($-35°C$, 238 K), has a much lower viscosity and so allows for a more efficient infiltration/resin exchange with the specimen, and further is easier to section. Methods prepared by Dürrenberger (1989) overcome the higher background staining which may sometimes be a problem with apolar resins. Protein A–gold is precoupled with IgGs from the specific antiserum which is to be used and this complex is employed directly on the sections of tissue embedded in Lowicryl HM20 (Kellenberger et al., 1987; Dürrenberger et al., 1988; Dürrenberger, 1989). However, if this procedure is not carried out under carefully controlled conditions, it results in aggregation of gold–IgG complexes, leading to clumps of gold on the section. The specificity of localization, an important factor in the efficacy of colloidal gold methods, can then be obscured.

To avoid the possible loss of specificity and the many steps in preparing the protein A–gold/IgG complex (incubation, centrifugation, and sonication), it is simpler to use an indirect immunoprocedure. For example, primary antibody followed by colloidal gold complexed to specific antibodies raised against the species of the primary antibody can be used. Protein

---

Fig. 5.2. Thin sections of rat pancreas fixed with 1% glutaraldehyde in buffer (pH 7.0) by vascular perfusion for (Top) 15 min and (Bottom) 60 min and embedded by PLT in Lowicryl K4M. Immunolabeling as in Fig. 5.1. There is intense labeling of the zymogen granules and the Golgi apparatus (G) with antirat anionic trypsin. However, there is no labeling of the RER to be seen in either micrograph. Sections stained with uranyl and lead acetate; ×24,500.

A–gold alone can be used, provided the primary antiserum used allows for low backgrounds. This can be achieved by preparing sera in phosphate-buffered saline (PBS) (pH 7.4) containing 0.6% bovine serum albumin (BSA) as a blocking agent.

Some results with Lowicryl HM20 are presented below. Rat pancreas was fixed by vascular perfusion with 1% glutaraldehyde for 15 min, followed by immersion fixation for 45 min. It was then embedded in either Lowicryl K4M (Fig. 5.3) or Lowicryl HM20 (Fig. 5.4) by the PLT method and yielded good ultrastructural preservation in both resins. Identical methods were used for immunolabeling of sections of this tissue embedded in each of the two resins. The intensity of labeling over the zymogen granules and Golgi apparatus way very similar in both, with no aggregation or clumps of gold and with low backgrounds over nonspecific sites

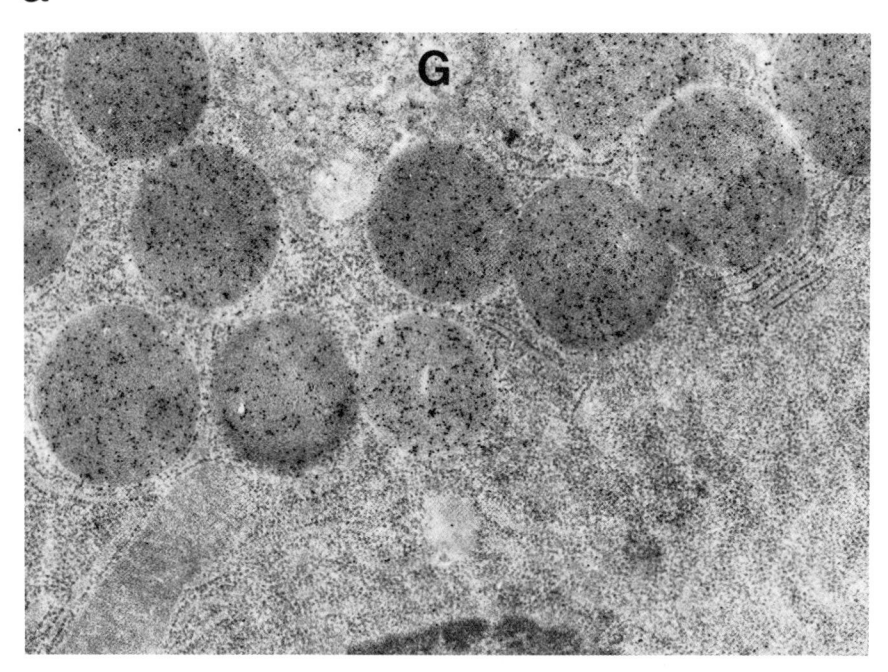

Fig. 5.3. Thin sections of rat pancreas fixed with 1% glutaraldehyde in buffer (pH 7.0) by vascular perfusion for 1 hr and embedded by PLT in Lowicryl K4M. Immunolabeling was by immersion in rabbit antirat anionic trypsin (in PBS containing 0.6% BSA) followed by (a) GAR-10 or (b) protein A coupled to 10-nm colloidal gold (PAG-10). There is approximately the same level of labeling over the zymogen granules in both preparations. The Golgi apparatus (G) is also labeled. The background is negligible. Sections stained with uranyl and lead acetate; ×24,500.

5. Lowicryls and Low-Temperature Embedding

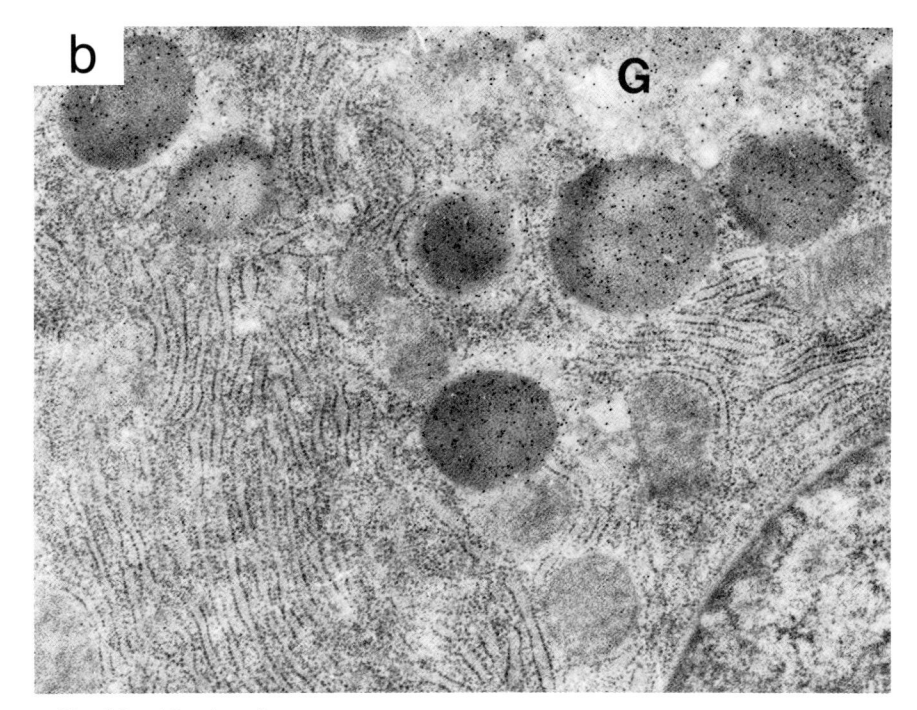

Fig. 5.3. (*Continued*)

(Figs. 5.3 and 5.4). Such a result readily demonstrates the possibility of using Lowicryl HM20 for colloidal gold labeling methods.

CRYOSUBSTITUTION

Cryosubstitution (or freeze substitution) is a method which does not require any chemical fixation of tissue and so avoids its damaging effects on sensitive and labile antigenic sites. It uses instead a physical fixation, in which the rapid freezing of the specimen immobilizes all the cellular components. After freezing, the ice in the specimen is replaced by an organic solvent which is miscible with the embedding resin. The two most commonly used organic solvents are acetone and methanol.

Removal of the ice by the organic solvent, i.e., substitution, is carried out at very low temperatures, −80°C (193 K) to −90°C (183 K). At these temperatures, the activation energy for a rearrangement of water and solute molecules is minimal, so that the specimen is in a "stable" condition (see Plattner and Bachmann, 1982, and Hayat, 1989, for excellent reviews

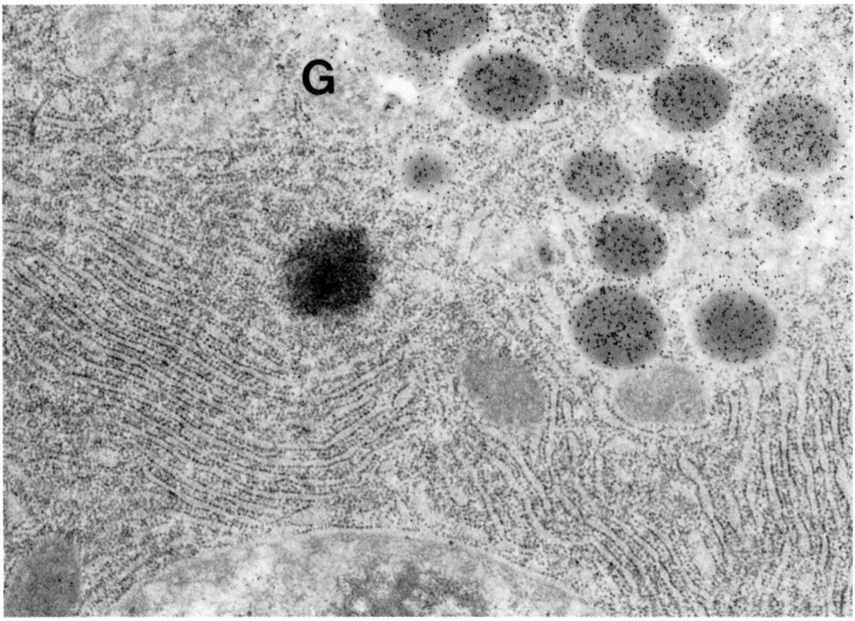
G

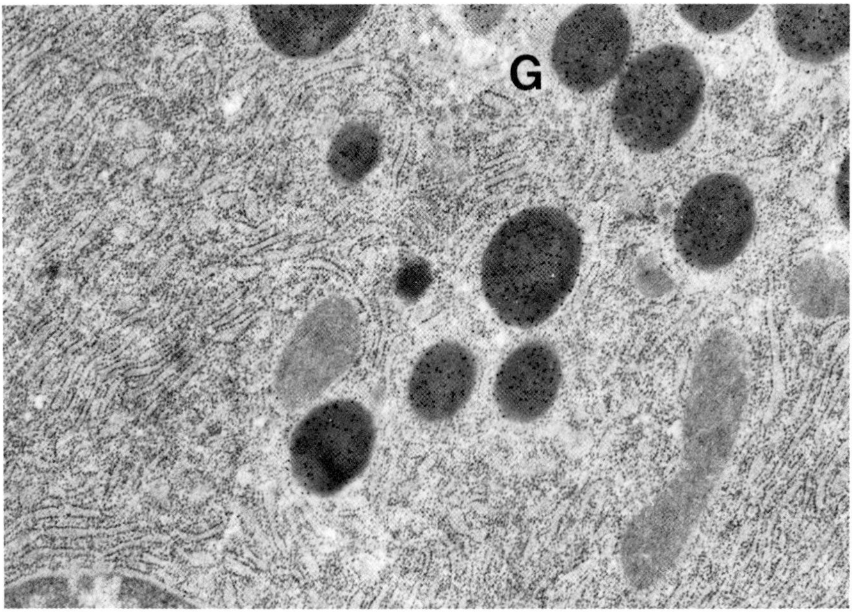
G

of cryomethods). As the temperature rises above −80°C (193 K), the activation energy begins to increase, and collapse phenomena (e.g., aggregation or precipitation of cellular components) can occur in the temperature range of −58°C (215 K) to −10°C (263 K) (MacKenzie, 1972). However, although biophysical data indicate that the tissue may now be entering a dangerous temperature zone, the temperatures are still so low that no adverse effects on ultrastructure are seen and any solvent-induced rearrangements are kept to a minimum. Ice recrystallization also poses no great problems (Steinbrecht, 1982).

Practical results with cryosubstitution point to new aspects of ultrastructure (Hoch and Howard, 1980; Ebersold et al., 1981; Howard, 1981; Dahmen and Hobot, 1986) and to the preservation of very sensitive structures such as bacterial DNA. Prokaryotic DNA, which, unlike eukaryotic DNA, has little protein associated with it, cannot be chemically fixed by aldehydes or OsO_4 alone and so is prone to collapse (aggregation or precipitation) in any organic solvent. This collapse will occur whether room or low-temperature embedding protocols are used but does not occur in cryosubstitution procedures (Hobot et al., 1985).

The appearance of the very sensitive prokaryotic DNA is the same whether substituted with OsO_4–acetone mixtures into Epon (Hobot et al., 1985) or with aldehyde–acetone or just acetone alone into the Lowicryls (Hobot et al., 1987). The same holds true for proteoglycans within cartilage matrix (Hunziker and Herrmann, 1987). Fixatives (aldehydes, OsO_4, and uranyl acetate) therefore are included in the organic solvent solution used for substitution in order to help reduce the adverse effects that could occur on prolonged exposure of biological tissue to organic solvents. This is especially necessary above −30°C (243 K), when it is usually essential to have OsO_4 present as one of the fixatives. Fixatives are not really necessary, however, if the temperature is kept below −30°C (243 K) and the specimen finally embedded in Lowicryl K4M (Hobot et al., 1987), Lowicryl HM20 (Humbel et al., 1983), or Lowicryls K11M and HM23 (Carlemalm et al., 1985b).

The level of ultrastructural preservation of cryosubstituted samples is most probably very similar whether they are embedded at a high temperature of 60°C (333 K) or embedded in the Lowicryls at a low temperature, −35°C (238 K) (Hobot et al., 1985, 1987; Hippe and Hermanns, 1986).

Fig. 5.4. Thin sections of rat pancreas fixed with 1% glutaraldehyde in buffer (pH 7.0) by vascular perfusion for 1 hr and embedded by PLT in Lowicryl HM20. Immunolabeling was as in Fig. 5.3 followed by (Top) GAR-10 or (Bottom) PAG-10. There is approximately the same level of labeling over the zymogen granules in both preparations and the intensity is similar to that of Lowicryl K4M-embedded samples (Fig. 5.3). The Golgi apparatus (G) is also labeled. The background is negligible. Sections stained with uranyl and lead acetate; ×24,500.

The important difference is that with high temperatures OsO_4 must generally be used, whereas with low temperatures OsO_4 is not necessary and even aldehydes can be avoided. It is therefore of great value in immunolabeling when choosing cryosubstitution to avoid procedures involving OsO_4 by using the Lowicryls at low temperatures (i.e., below $-30°C$, 243 K).

The first results showing excellent ultrastructural preservation involving cryosubstitution and low-temperature embedding into the Lowicryls were presented by Muller and his group in 1980–1981. They later showed that immunolabeling was also possible with this technique (Humbel et al., 1983). Improved labeling intensity is achieved by using cryosubstitution methods when compared with low-temperature or conventional embedding techniques (Carlemalm et al., 1985b; Hunziker and Herrmann, 1987). Also, accurate immunolocalization of structures not preserved well with other methods can be obtained (Björnsti et al., 1986; Hobot et al., 1987).

Methods for Achieving Substitution Temperatures

A temperature of $-35°C$ (238 K) is achieved, as discussed earlier, by using a chest-type deep freeze. Temperatures of $-80°C$ (193 K) can be reached by using dry ice (solid CO_2, $-78°C$, 195 K) and adding acetone to lower the temperature to $-80°C$ (193 K), although it is possible to substitute quite well at $-78°C$ (195 K). The dry ice (crushed or pellets) is kept in large Dewars with sealable tops, and the tubes containing the specimens are placed directly in the dry ice or in aluminum blocks which have suitable holes drilled in them. Conical centrifuge tubes (polypropylene) of 50-ml volume have been found to be very convenient (J. Bibby Ltd., Stone, Staffs, UK; Cat. No. 25330B with caps); 10–15 ml of substitution medium is added to these tubes (containing molecular sieve).

All solutions and aluminum blocks are cooled prior to use. Specimens are transferred by precooled tweezers from liquid nitrogen storage or from the freezing apparatus into the cold substitution medium. The Dewars can be placed in the freezer if room allows. In order to raise the temperature to $-35°C$ (238 K), the tubes are taken from the dry ice and placed in the freezer or, if a slower rise in temperature is required, left in the aluminum blocks at $-35°C$ (238 K). Following equilibration to $-35°C$ (238 K), the specimens are kept at this temperature for 6 hr, after which they are transferred to appropriate embedding vials for further processing. Alternatively, a cryosubstitution unit can be used (Reichert-Jung, Austria; Balzers Union, Liechtenstein). It has automatic temperature controls and allows temperatures of $-90°C$ (183 K) to be reached easily; it can also control the duration of the temperature rise.

Methods of Freezing and Tissue Preparation

There are four main methods of freezing: (1) plunging into liquid propane or ethane, (2) jet-freezing with liquid propane, (3) slam-freezing on a metal block (usually copper) cooled by either liquid nitrogen or liquid helium, and (4) high-pressure freezing. Each has produced good results, and systems (1) to (3) can be set up in a laboratory if a good and competent workshop is available. Alternatively, commercial freezing systems are also manufactured. Hayat (1989) and Menco (1986) have reviewed in detail various methods of freezing.

The freezing must achieve very fast cooling rates so as to minimize ice-crystal formation. Slam-freezing on a cold metal block provides the fastest cooling rates (Escaig, 1982; Bald, 1985). The largest depth of well-frozen material when measured from the freezing edge is 10–12 μm for metal block, 5–6 μm with liquid coolant, or up to 0.5 mm with high-pressure freezing. However, this technique has a slower cooling rate and may not be useful for observing rapid biochemical or physiological processes (Menco, 1986). For immunotechniques, it is important to have as much well-frozen tissue as possible. Suspensions or cell monolayers are better frozen using method (1) or (2) above. In skillful hands, however, each technique can yield good results, but the slam method gives overall the greatest depth of well-frozen tissue and is suitable for coping with most biological tissues.

However, having a good apparatus for freezing is just the beginning of cryowork. The specimen under study must be prepared properly and carefully. The aim is to preserve structure as near to the situation in life as possible. Protocols must in each case be carefully thought out such as to minimize any trauma to the tissue. Pretreatment of tissue with cryoprotectants should be avoided, as it can cause structural alterations and is not necessary because of the rapid freezing methods employed. Very wet tissue or specimens will freeze badly and should be avoided. Excess fluid or water can be removed by touching the specimen with the edge of a small piece of filter paper, but care must be taken not to allow any air drying to occur or drying artifacts will form.

It is advisable to use very small pieces of tissue (1 mm^3 or less), but if larger pieces are needed, it is important to ensure that the tissue has all excess fluid removed and is thinly sliced for freezing. Suspensions of cell cultures, cell isolates, or bacteria can be pelleted or filtered, smeared onto very thin filter paper (or cigarette paper, diameters up to 5 mm), and frozen. Frozen specimens can be stored under liquid nitrogen until required for processing. Care should be taken not to allow water vapor to condense and collect on the frozen samples during handling and transfer operations.

Cryosubstitution gives excellent structural preservation and very good immunolocalization, especially for sensitive and labile antigenic sites. However, much patience and dedication may be needed, as it will take time to learn to use this technique properly and reproducibly. Also, each specimen freezes differently, and probably several freezing attempts will be necessary before a satisfactory method of preparation and freezing is found for a particular sample. Hence, cryosubstitution as an embedding method for colloidal gold labeling, although a most sophisticated method, is also technically the most difficult and should be used when all the other techniques have failed.

Preparation of Substitution Media

Following freezing, the ice is removed by substitution with an organic solvent, generally 100% methanol (freezing point, −94°C, 179 K) or 100% acetone (freezing point, −95°C, 178 K). Glutaraldehyde at a final concentration of 3% (v/v) may be added to the organic solvent; it can be omitted but has not yet been seen to affect immunolabeling adversely (Carlemalm et al., 1985; Hobot et al., 1987; Hunziker and Herrmann, 1987) and may give some stability to ultrastructure in the temperature range −50°C (223 K) to −30°C (243 K) (Humbel and Muller, 1984). To 9.4 ml of organic solvent, 0.6 ml of 50% glutaraldehyde is added. The solution is cooled to −35°C (238 K) and in the case of acetone added to molecular sieve (0.4-nm pore size, perlform or small pellets). This is necessary because without molecular sieve acetone cannot effectively dissolve ice in the presence of even small amounts of water (1%). Methanol, however, can substitute ice even in the presence of 10% water (Humbel et al., 1983). Methanol can be used with or without molecular sieve (but, if used, only with 0.3-nm pore size and prepared as above). With molecular sieve present, acetone is a very effective substitution medium and gives good results (Hobot et al., 1985; Dahmen and Hobot, 1986).

Infiltration and Embedding

Substitution of specimens in the substitution medium of choice is carried out at −80°C (193 K) to −90°C (183 K) for 64–88 hr. The temperature is then raised (duration of rise is 1 hr) to −35°C (238 K) and kept at this temperature for 6 hr. All subsequent processing is carried out at this temperature. After 6 hr, the specimens are treated as follows (use the same organic solvent as was present in the substitution medium):

1. 100% Solvent, 1 hr
2. Lowicryl (K4M or HM20) plus 100% solvent (1 : 1), 1 hr

3. Lowicryl plus 100% solvent (2 : 1), 1 hr
4. Pure Lowicryl, 1 hr
5. Pure Lowicryl, overnight
6. Fresh pure Lowicryl, 2 hr

Samples are polymerized in gelatin capsules by indirect UV light for 24 hr at −35°C (238 K), followed by direct UV light for 72 hr at room temperature. The resins are prepared as described earlier.

Lowicryls K11M and HM23

The full potential of cryosubstitution can be realized only with embedding media that can be used at temperatures as low as −80°C (193 K). These are now available in kit form as the polar Lowicryl K11M and apolar Lowicryl HM23 (Chemische Werke Lowi, Waldkraiburg, West Germany). Their chemical composition has been published by Acetarin *et al.* (1986) and shows them to be modifications of their two predecessors, such that they can be used at very low temperatures. Lowicryl HM23 can be polymerized by UV light at temperatures down to −70°C (203 K), Lowicryl K11M down to −60°C (213 K). Both have low viscosities at these temperatures, similar in value to those of Lowicryls HM20 and K4M at −35°C (238 K) (Acetarin and Carlemalm, 1985). Substitution following rapid freezing can be in organic solvent alone; no fixative is needed.

Composition and Handling

Both kits are supplied with a cross-linker, monomer, and photoinitiator for UV light polymerization. The cross-linker and monomer are first weighed and mixed together, after which the initiator is added. As polymerization takes longer at these low temperatures, it is most important to avoid inhibition by oxygen. Therefore, all resin components must be mixed by bubbling with a gentle stream of dry nitrogen.

For Lowicryl K11M, use 1 g of cross-linker plus 19 g of monomer; to this add 0.1 g of photoinitiator C (benzoin methyl ether). For Lowicryl HM23, use 1.1 g of cross-linker plus 18.9 g of monomer; to this add 0.1 g of photoinitiator C (benzoin methyl ether) for UV polymerization down to −50°C (223 K). For UV polymerization in the range −50°C (223 K) to −70°C (203 K), add 0.1 g of the photoinitiator J (Igracure 651). The hardness of the two resins can be modified, but only by making small changes in the amounts of cross-linker and monomer used (Acetarin *et al.*, 1986). These authors recommended that the resins be dried over molecular sieve (4 Å Linde type) prior to use, and care should be taken at the very low temperatures used to avoid condensation of moisture. The dust from the molecular sieve must be allowed to settle before use.

Substitution Procedures

Lowicryls K11M and HM23 are still relatively new in the field of electron microscopy, and their use is far from routine; therefore, little has been published about them (Carlemalm *et al.*, 1985b; Acetarin *et al.*, 1986). Basically, there is no "standard recipe" for their application. Given below are protocols based on early experiences of cryosubstitution with Lowicryl HM23 (J. A. Hobot, unpublished results, 1983). These protocols can be changed as the technique of handling these resins becomes more refined. Methods of freezing and tissue preparation are as discussed in the previous section on cryosubstitution into Lowicryl K4M and HM20. The substitution medium can be either 100% acetone or 100% methanol; no fixative is needed. Substitution is carried out for 64–88 hr at −80°C (193 K).

Lowicryl K11M: Embedding and Polymerization

1. Raise the temperature to −60°C (213 K) and add Lowicryl K11M plus organic solvent (1 : 1) for 8 hr at −60°C (213 K).
2. Lowicryl K11M plus organic solvent (2 : 1) for 8 hr at −60°C (213 K).
3. Pure Lowicryl K11M overnight at −60°C (213 K).
4. Fresh pure Lowicryl K11M for 8 hr at −60°C (213 K).
5. Polymerization by indirect UV light (360-nm long wavelength) for 3 days at −60°C (213 K), followed by 2–3 days at room temperature with direct UV light.

It may be possible to use Lowicryl K11M and organic solvent mixtures at −80°C (193 K) before raising the temperature to −60°C (213 K), especially with a 1 : 1 mixture. A close watch should be kept on the mixture's viscosity.

Lowicryl HM23: Embedding and Polymerization

1. Lowicryl HM23 plus organic solvent (1 : 1) for 8 hr at −80°C (193 K).
2. Lowicryl HM23 plus organic solvent (2 : 1) for 8 hr at −80°C (193 K).
3. Pure Lowicryl HM23 overnight at −80°C (193 K).
4. Fresh pure Lowicryl HM23 for 8 hr at −80°C (193 K).

5. Lowicryls and Low-Temperature Embedding

5. Polymerization by indirect UV light (360 nm, long wavelength) for 6 days at −70°C (203 K), followed by 2–3 days at room temperature with direct UV light to improve the sectioning quality of the blocks.

Shorter infiltration times for bacteria (2 hr) have recently been found successful (Dürrenberger *et al.*, 1988). If infiltration is incomplete with either resin, the times may be prolonged with further changes of pure resin, although this may cause adverse effects on ultrastructure and/or antigenicity.

Polymerization Methods

Polymerization by indirect UV light at −70°C (203 K) or −60°C (213 K) is carried out with the samples in small gelatin capsules placed in a polymerization chamber using exactly the same setup as described earlier for Lowicryls K4M and HM20. If access to either a freezer or cold room in the temperature range of −70°C (203 K) to −60°C (213 K) is not possible, the CS-Auto Substitution Apparatus (manufactured by Reichert-Jung, Austria) allows for both substitution and infiltration at these low temperatures (and avoids the problem of moisture condensation) and allows subsequent UV light polymerization at −70°C (203 K) to −60°C (213 K). It is hoped that technical problems do not restrict the use of these very interesting Lowicryls.

Immunolabeling

Earlier in this chapter, it was emphasized that with low concentrations of glutaraldehyde and short times of fixation, combined with optimal embedding procedures (e.g., PLT), low concentrations of antigen at secondary sites of localization could be detected as well as the primary sites, where the antigen was present in higher concentrations (Fig. 5.1). Increasing the concentration of glutaraldehyde reduced or removed labeling from these secondary sites (Fig. 5.2). However, this leaves open the possibility that the tissue remains underfixed and that antigen moves from its primary sites, after processing, to other sites (Hayat, 1986). With cryosubstitution, no chemical fixation is necessary. In addition, processing in organic solvents at a very low temperature, −80°C (193 K), stops the movement of soluble components/ions and the aggregation of tissue components.

Sections of rat pancreas cytosubstituted with 3% glutaraldehyde in acetone as the substitution medium and embedded in Lowicryl K4M at −35°C (238 K) show intense labeling with antitrypsin antibodies (Fig. 5.5). Immunolabeling was carried out on-section in exactly the same way

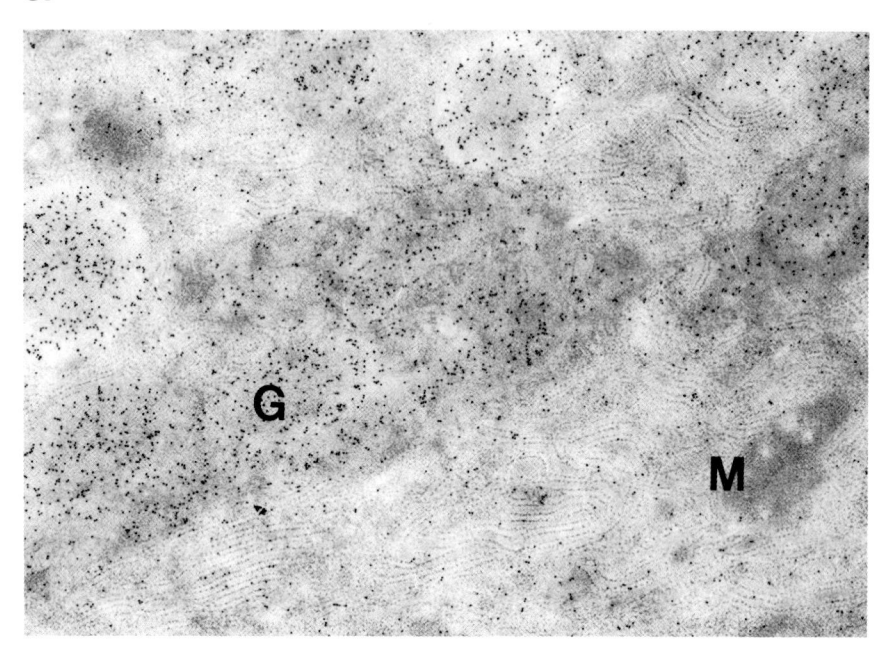

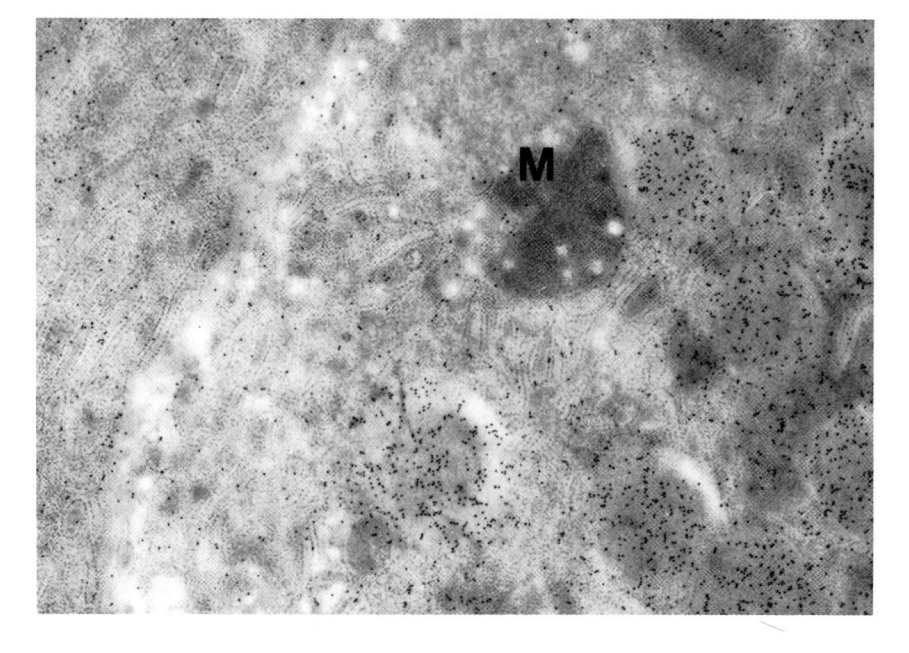

a

G

M

b

M

as for Figs. 5.1–5.4. There is a high level of labeling over the zymogen granules, the Golgi apparatus, and the rough endoplasmic reticulum, a low-concentration secondary site, along which the gold particles are specifically lined up. This result strongly supports the concept that both the high concentration of fixative and prolonging the fixation time reduce the ability to detect low-concentration secondary sites by immunolabeling.

Another advantage of cryosubstitution is in the preservation of structures which chemical fixation and subsequent processing cause to be reorganized or lost, for example, bacterial DNA. Immunolabeling studies can now be carried out on the location and distribution of DNA within the bacterial cell and on the role of potential protein partners (Björnsti et al., 1986; Hobot et al., 1987; Dürrenberger et al., 1988). In Fig. 5.6a, the ribosomes of the cytoplasm can be clearly defined from an area free of ribosomes, called the ribosome-free space. This is equivalent to the bacterial nucleoid which contains DNA (Hobot et al., 1985). Indirect immunolabeling with anti-single-stranded DNA (ssDNA) antibodies and protein A–gold particles (15 nm) shows this species of DNA to be present on the periphery of the bacterial nucleoid. The bulk of nonmetabolizing DNA is in the rest of the bacterial nucleoid, as shown by labeling with antibodies directed against either double-stranded DNA (dsDNA) (Fig. 5.6b) or both dsDNA and ssDNA (Fig. 5.6c). As a comparison, Fig. 5.6d shows labeling directed against ribosomal RNA using anti-RNA antibodies (Hobot et al., 1987).

Future Developments

An exciting prospect has been opened up for examining resin-embedded thin sections by the introduction of Lowicryls K11M and HM23. After freezing, under the best possible conditions, in the absence of any pretreatments by chemical fixatives or cryoprotectants, biological tissue can be processed at very low temperatures ($-80°C$, 193 K). Now the adverse effects of organic solvents and, hopefully, organic resins should be substantially reduced. Processing treatments (fixation and dehydra-

---

Fig. 5.5. Thin sections of rat pancreas prepared by cryosubstitution and embedded in Lowicryl K4M. The pancreas was removed quickly from a freshly killed male Wistar rat (last sample 10 min after death) and frozen by slam-freezing on a copper block cooled by liquid nitrogen (Reichert-Jung, Austria; KF80 freezer plus MM80 slam attachment). Immunolabeling was with rabbit antirat anionic trypsin followed by GAR-10 as per the method for Figs. 5.1–5.4. In both (a) and (b) there is intense labeling of zymogen granules. The Golgi apparatus (G) is also labeled (a), with accurate localization along the RER. There is no label over the mitochondria (M), denoting low background levels. Sections stained with uranyl and lead acetate; ×24,500.

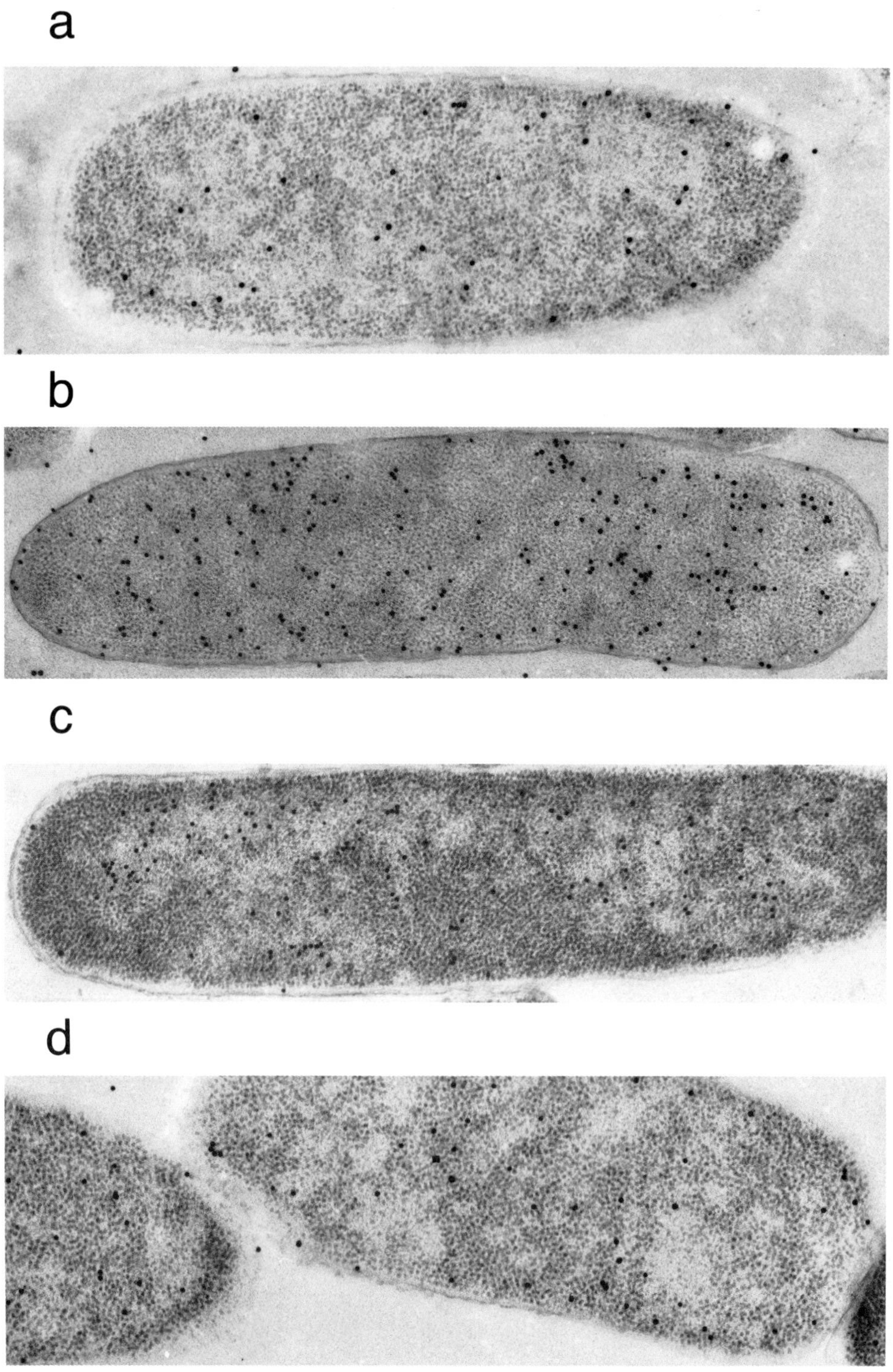
a
b
c
d

tion), known to affect very sensitive and labile antigenic sites, are minimized or excluded.

First results of cryosubstitution with Lowicryls K11M and HM23 have shown great promise (Carlemalm et al., 1985b; Dürrenberger et al., 1988). Bacteria, frozen by the copper block method and substituted into Lowicryl HM23 by the protocol detailed in the foregoing section, show good cytoplasmic infiltration and the grainy fibrillar aspect of their DNA plasm, which is seen only in cryosubstituted cells (Fig. 5.7) (Hobot et al., 1985, 1987). Immunolabeling is now also possible with Lowicryl HM23 (Dürrenberger, 1989; Dürrenberger et al., 1988). This resin has advantages over Lowicryl K11M similar to those of Lowicryl HM20 over Lowicryl K4M. Also, lipid loss is greatly reduced with acetone (5% loss at these low cryosubstitution temperatures) (Weibull et al., 1984), when compared with PLT or room temperature embedding techniques (Weibull et al., 1983; Weibull and Christiansson, 1986). This opens up new possibilities for the study of membrane structure together with localization studies of sensitive membrane receptor sites and protein-mediated membrane functions.

FREEZE-DRYING AND LOW-TEMPERATURE EMBEDDING

Freeze-drying has been coupled to low-temperature embedding in Lowicryl K11M (Edelman, 1986) and Lowicryl HM20 (Wroblewski and Wroblewski, 1986) for microanalysis. Like cryosubstitution, freeze-drying does not require chemical fixation. It has the advantage of avoiding organic solvents. The next step, embedding with an organic resin, is the same. The disadvantage of freeze-drying is that collapse phenomena (ag-

---

Fig. 5.6. Thin sections of *Escherichia coli* B prepared by cryosubstitution and embedded in Lowicryl K4M. All sections stained with uranyl and lead acetate; ×32,000. (a) *E. coli* B labeled with mouse monoclonal anti-single stranded DNA (ssDNA) followed by protein A coupled to 15-nm colloidal gold (PAG-15). The gold label is on the periphery of the ribosome-free area, or nucleoid, which has a cleft, lobular appearance and a grainy and fine fibrillar appearance to the DNA-plasm. (b) *E. coli* B labeled with a polyclonal antibody from MRL/Mp mouse serum directed against dsDNA, followed by PAG-15. The gold label is located over the ribosome-free area, which appears grayish in this preparation because of the phenomenon of serum staining (see Hobot et al., 1987). (c) *E. coli* B labeled with mouse IgM monoclonal antibodies directed against both ssDNA and dsDNA, bridged by goat anti-mouse IgM, IgG antibody to PAG-15. The gold label is over the ribosome-free area but tends to be toward the periphery of this area. (d) *E. coli* B labeled with mouse monoclonal antibodies directed against rRNA, followed by PAG-15. The gold label is only over the cytoplasm or ribosomal areas. From Hobot et al., 1987; courtesy of the American Society for Microbiology.

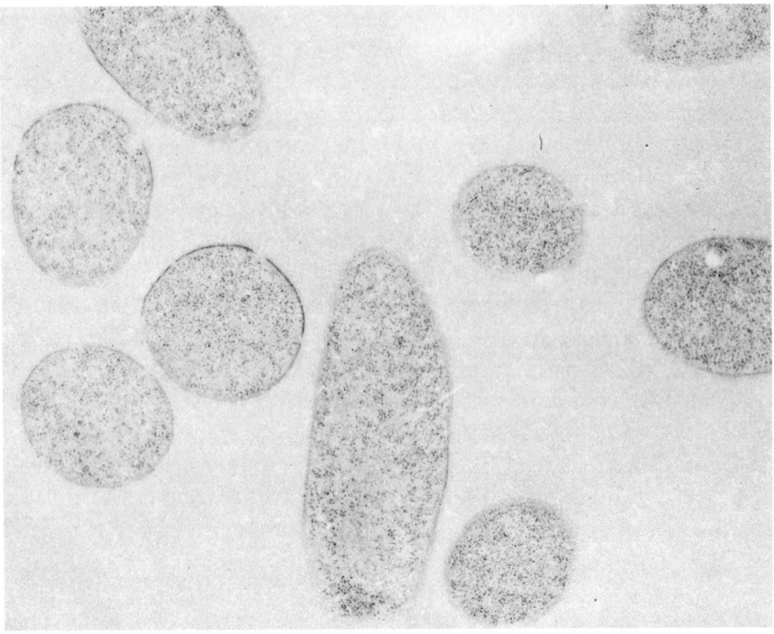

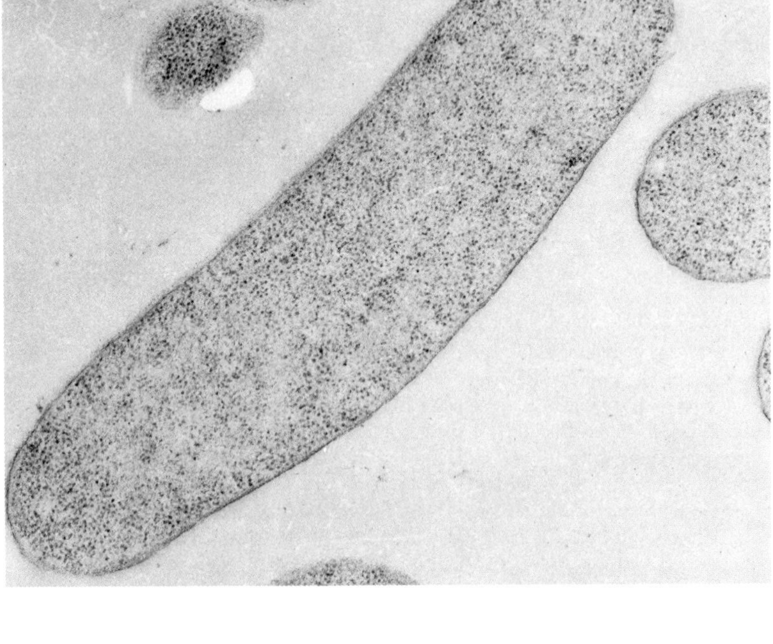

5. Lowicryls and Low-Temperature Embedding

gregation and precipitation artifacts) as discussed for cryosubstitution can occur, with the problem that soluble cytoplasmic components can collapse onto cellular structures when dry. Reintroducing a "wet" environment by infiltration with pure resin may possibly cause further changes in the location of cellular components. Some of these problems have been discussed by Chiovetti *et al.* (1987), who describe in detail the technique of freeze-drying and low-temperature embedding. Results of immunolabeling with colloidal gold have been presented (Jorgensen and McGuffee, 1987). These studies have been restricted to muscle cells (Chiovetti *et al.*, 1987; Jorgensen and McGuffee, 1987), but in order to judge the importance of the technique other tissues (including the bacteria and their sensitive DNA plasm) need to be prepared by freeze-drying and their labeling intensities, for primary and secondary sites, compared with those obtained by other methods.

Some of these problems have already been noticed by Chiovetti *et al.* (1987), and their interest in the future in coupling this technique to the Lowicryl resins K11M and HM23 has been perhaps to reduce collapse or drying artifacts by using lower temperatures. Chiovetti *et al.* (1987) chose $-20°C$ (253 K) for Lowicryl K4M embedding, as it seemed that at lower temperatures, $-35°C$ (238 K), Lowicryl K4M was of too high a viscosity to give reasonable infiltration of freeze-dried tissue. There is some barrier to complete penetration or movement of this resin caused by the way the tissue has been dried, which raises questions about the use of epoxides with this method because of their higher viscosities at room temperature compared with that of Lowicryl K4M at $-35°C$ (238 K).

GENERAL COMMENTS

Storage and Thin Sectioning of Lowicryl Blocks

There are no problems of long-term storage of Lowicryl blocks. However, it is advisable to store open and trimmed capsules of Lowicryl K4M in a bench desiccator, as this polar resin readily takes up water (as is obvious if proper care is not taken during sectioning).

Blocks which are being prepared for sectioning can have their excess plastic cut away with a razor blade, but only peripheral areas, clear of tissue, so as to lessen vibrational damage to tissue–resin interfaces. Final

Fig. 5.7. *Escherichia coli* B prepared by cryosubstitution and embedded in Lowicryl HM23. The cells are well embedded, and the DNA-plasm within the ribosome-free areas is grainy and fine fibrillar in appearance. Sections stained with uranyl and lead acetate; (Top) ×21,500; (Bottom) ×32,000.

trimming should always be with sharp glass knives for both the block face and the sides of the pyramid. Smaller block faces give better sections with less compression or chatter artifacts than larger faces. Lowicryl HM20 is easily sectioned, and variable cutting speeds of choice can be used. The water in the trough should have an optimal meniscus. Lowicryl K4M, the more polar of the two resins, is also the more difficult to section. It is very important to have the water level slightly below the knife edge. Indeed, the larger the block face, the lower this level should be. If this is not correctly adhered to, the block face of the Lowicryl K4M specimen will immediately flood with water as soon as it touches the knife edge. Cutting speeds for Lowicryl K4M are usually in the range of 2–5 mm/sec; the speed can be increased for larger block faces. Both glass and diamond knives can be used for cutting thin sections from Lowicryl resins. Lowicryls K11M and HM23 should be approached in the same way.

Sections must be picked up on the shiny surface of copper, nickel, or gold grids and air dried. Any grid with a carbon-coated collodion or Formvar plastic film can be used. Nickel or gold grids (high-transmission hexagonal is preferred over square mesh) are used for immunolabeling, as copper tends to oxidize with long exposure to aqueous solutions. This adverse quality of copper grids can be overcome by providing a protective coat by dipping the grids in 0.05% collodion (in amylacetate) and quickly drying the grids on Whatman No. 50 hardened filter paper (Whatman Laboratories Ltd., Maidstone, Kent, UK). The grids can then be used naked or with a 2% collodion film placed on them (Hobot *et al.*, 1987).

The stability of sections on naked grids in the electron microscope can be improved, if required, by coating the grids plus sections with a thin coat of carbon in a vacuum-coating unit. This carbon coating should be applied only after all immunolabeling and contrast staining have been done. Areas of interest should first be viewed in the electron microscope at low magnifications to stabilize the sections in the beam and then switched gradually to the desired magnification.

Methods for Immunolabeling of Thin Sections

Lowicryl sections can be labeled by totally immersing the grid in the solution or by flotation of the grid on top of the solution (30–50-μl drops). In the former method, both sides are labeled and washing is also by immersion, shaking the excess off gently between each exchange but not so as to cause the grid to dry out. Naked grids without any plastic support films are used. The latter method allows for only one surface of the section to be labeled; plastic support films can be used, and grids are gener-

5. Lowicryls and Low-Temperature Embedding

ally washed by gentle jet washes. However, the immersion method yields clean grids with negligible background labeling and higher specific labeling intensities than the flotation method. No superimpositional effects of gold deposited on both sides of the section can be seen (Figs. 5.1–5.5). The technique is also more reproducible in that it is free of the variations in "strength" of jet washes. Variable jet washes could dislodge IgGs and IgG–gold complexes, so lowering the intensity of specific labeling. An incubation step involving 1% glutaraldehyde could fix the IgG–gold complexes to the antigenic site (Kellenberger et al., 1987), but this becomes an additional handling step.

Counterstaining of Thin Sections

Relatively short times can be used to counterstain Lowicryl sections. All staining protocols (at room temperature) use the flotation method, followed by a gentle jet wash with distilled water (a few drops, 0.5 ml, gently passed over both sides of the grid using a 500-µl micropipette). Grids are air dried.

When using uranyl acetate on unlabeled sections, treatment with 2% uranyl acetate in distilled water for 1 min (maximum, 10 min) will suffice. Following immunolabeling, treatment with 4% uranyl acetate for 20 min is recommended. Here, if the immersion method of immunolabeling has been used, the grids are immersed in the droplets. In the case of lead acetate (Millonig, 1961), treatment for 30–60 sec is sufficient, whether sections have been immunolabeled or not. The stock solution is prepared by dissolving 20 g of NaOH and 1 g of potassium–sodium tartrate in 50 ml of distilled water. One milliliter of this solution is added to 5 ml of 20% (w/v) lead acetate, $Pb(CH_3COO)_2 \cdot 3H_2O$, in distilled water, and the mixture is stirred for 5 min. It can be diluted 5–10 times with distilled water (staining times given here are for undiluted preparations). The mixture is filtered through a Millipore filter (pore size, 0.2 µm) and stored in tubes under paraffin. Lead acetate gives a finer-grain staining than lead citrate.

The writing of this chapter would not have been possible without working in Edward Kellenberger's group in Basel and without having had such able colleagues to learn from about the Lowicryls. I am greatly indebted to Edward Kellenberger, Eric Carlemalm, Werner Villiger, Jean-Dominique Acetarin, and Robert Chiovetti (the editor of the Lowicryl instruction manual). I would also like to thank Geoffrey R. Newman for critical reading of the chapter and many helpful discussions.

REFERENCES

Acetarin, J.-D., and Carlemalm, E. (1982). *Chemical Polymerisation Methods for Methacrylate*, pp. 140–141. Appendix in Carlemalm *et al.* (1982).

Acetarin, J.-D., and Carlemalm, E. (1985). Lowicryl HM23 and K11M: Two new embedding resins for very low temperature embedding. In *Lowicryl Letters, No. 33* (E. Carlemalm, W. Villiger, and E. Kellenberger, eds.), pp. 2–4. Chemische Werke Lowi GmbH, P.O. Box 1660, D-8264 Waldkraiburg, Federal Republic of Germany.

Acetarin, J.-D., Carlemalm, E., and Villiger, W. (1986). Developments of new Lowicryl resins for embedding biological specimens at even lower temperatures. *J. Microsc. Oxford* **143**, 81.

Altman, L. G., Schneider, B. G., and Papermaster, D. S. (1984). Rapid embedding of tissues in Lowicryl K4M for immunoelectron microscopy. *J. Histochem. Cytochem.* **32**, 1217.

Armbruster, B. L., Carlemalm, E., Chiovetti, R., Garavito, R. M., Hobot, J. A., Kellenberger, E., and Villiger, W. (1982). Specimen preparation for electron microscopy using low temperature embedding resins. *J. Microsc. (Oxford)* **126**, 77.

Armbruster, B. L., Garavito, R. M., and Kellenberger, E. (1983). Dehydration and embedding temperature affect the antigenic specificity of tubulin and immunolabelling by the protein A–colloidal gold technique. *J. Histochem. Cytochem.* **31**, 1380.

Bald, W. B. (1985). The relative merits of various cooling methods. *J. Microsc. (Oxford)* **140**, 17.

Baschong, W., Baschong-Prescianotto, C., Wurtz, M., Carlemalm, E., Kellenberger, C., and Kellenberger, E. (1984). Preservation of protein structures for electron microscopy by fixation with aldehydes and/or OsO_4. *Eur. J. Cell Biol.* **35**, 21.

Bayer, M. E., Carlemalm, E., and Kellenberger, E. (1985). Capsule of *Escherichia coli* K29: Ultrastructural preservation and immunoelectron microscopy. *J. Bacteriol.* **162**, 985.

Bendayan, M., and Zollinger, M. (1983). Ultrastructural localisation of antigenic sites on osmium-fixed tissues applying the protein A–gold technique. *J. Histochem. Cytochem.* **31**, 101.

Bendayan, M. (1984). Protein A–gold electron microscopic immunocytochemistry: Methods, applications and limitations. *J. Electron Microsc. Tech.* **1**, 243.

Bendayan, M., Nanci, A., and Kan, F. W. K. (1987). Effect of tissue processing on colloidal gold cytochemistry. *J. Histochem. Cytochem.* **35**, 983.

Björnsti, M. A., Hobot, J. A., Kelus, A. S., Villiger, W., and Kellenberger, E. (1986). New electron microscopic data on the structure of the nucleoid and their functional consequences. In *Bacterial Chromatin* (C. O. Gualerzi and C. L. Pon, eds.), pp. 64–81. Springer-Verlag, Berlin.

Carlemalm, E., Garavito, R. M., and Villiger, W. (1980). Advances in low temperature for electron microscopy. *Electron Microsc., Proc. Eur. Congr., 7th, 1980* Vol. 2, p. 656.

Carlemalm, E., and Kellenberger, E. (1982). The reproducible observation of unstained embedded cellular material in thin sections: Visualisation of an integral membrane protein by a new mode of imaging for STEM. *EMBO J.* **1**, 63.

Carlemalm, E., Garavito, R. M., and Villiger, W. (1982). Resin development for electron microscopy and an analysis of embedding at low temperature. *J. Microsc. (Oxford)* **126**, 123.

Carlemalm, E., Colliex, C., and Kellenberger, E. (1985a). Contrast formation in electron microscopy of biological material. *Adv. Electron. Electron Phys.* **63**, 269.

Carlemalm, E., Villiger, W., Hobot, J. A., Acetarin, J.-D., and Kellenberger, E. (1985b). Low temperature embedding with Lowicryl resins: Two new formulations and some applications. *J. Microsc. (Oxford)* **140**, 55.

5. Lowicryls and Low-Temperature Embedding

Chiovetti, R. (1982). *Instructions for Use, Lowicryl K4M and Lowicryl HM20*. Chemische Werke Lowi GmbH, PO Box 1660, D-8264 Waldkraiburg, Federal Republic of Germany.

Chiovetti, R., McGuffee, L. J., Little, S. A., Wheeler-Clark, E., and Brass-Dale, J. (1987). Combined quick freezing, freeze-drying, and embedding tissue at low temperature and in low viscosity resin. *J. Electron Microsc. Tech.* **5**, 1.

Dahmen, H., and Hobot, J. A. (1986). Ultrastructural analysis of *Erysiphe graminis* Haustoria and subcuticular stroma of *Venturia inaequalis* using cryosubstitution. *Protoplasma* **131**, 92.

Dubochet, J., McDowall. A. W., Menge, B., Schmid, B. N., and Lickfeld, K. G. (1983). Electron microscopy of frozen-hydrated bacteria. *J. Bacteriol.* **155**, 381.

Dürrenberger, M. (1989). Removal of background label in immunocytochemistry with the apolar Lowicryls by using washed protein A–gold precoupled antibodies in a one-step procedure. *J. Electron. Microsc. Tech.* **11**, 109.

Dürrenberger, M., Björnsti, M. A., Uetz, T., Hobot, J. A., and Kellenberger, E. (1988). The intracellular location of the histone-like protein HU in *Escherichia coli*. *J. Bacteriol.* **170**, 4757.

Ebersold, H. R., Cordier, J. L., and Luthy, P. (1981). Bacterial mesosomes: Method dependent artifacts. *Arch. Microbiol.* **130**, 19.

Edelman, L. (1986). Freeze-dried embedded specimens for biological microanalysis. In *Scanning Electron Microscopy* (O. Johari, ed.), Pt. 4, p. 1337. Scanning Electron Microsopy Inc., AMF O'Hare, Chicago, Illinois.

Emerman, M., and Behrman, E. J. (1982). Cleavage and crosslinking of proteins with osmium(VIII) reagents. *J. Histochem. Cytochen.* **30**, 395.

Escaig, J. (1982). New instruments which facilitate rapid freezing at 83 K and 6 K. *J. Microsc. (Oxford)* **126**, 221.

Garavito, R. M., Carlemalm, E., Colliex C., and Villiger, W. (1982). Septate junction ultrastructure as visualised in unstained and stained preparations. *J. Ultrastruct. Res.* **80**, 344.

Hainfeld, J. F. (1987). A small gold-conjugated antibody label: Improved resolution for electron microscopy. *Science* **236**, 450.

Hayat, M. A. (1986). Glutaraldehyde: Role in electron microscopy. *Micron Microsc. Acta* **17**, 115.

Hayat, M. A. (1989). *Principles and Techniques of Electron Microscopy*, 3rd ed. Macmillan, London, and CRC Press, Boca Raton, Florida.

Hippe, S., and Hermanns, M. (1986). Improved structural preservation in freeze-substituted sporidia of *Ustilago avenae*—a comparison with low temperature embedding. *Protoplasma* **135**, 19.

Hobot, J. A., Carlemalm, E., Villiger, W., and Kellenberger, E. (1984). Periplasmic gel: New concept resulting from the reinvestigation of bacterial cell envelope ultrastructure by new methods. *J. Bacteriol.* **160**, 143.

Hobot, J. A., Villiger, W., Escaig, J., Maeder, M., Ryter A., and Kellenberger, E. (1985). Shape and fine structure of nucleoids observed on sections of ultrarapidly frozen and cryosubstituted bacteria. *J. Bacteriol.* **162**, 960.

Hobot, J. A., Björnsti, M. A., and Kellenberger, E. (1987). Use of on-section immunolabelling and cryosubstitution for studies of bacterial DNA distribution. *J. Bacteriol.* **169**, 2055.

Hoch, H. C., and Howard, R. J. (1980). Ultrastructure of freeze-substituted hyphae of the basidiomycete *Laetisaria arvalis*. *Protoplasma* **103**, 281.

Howard, R. J. (1981). Ultrastructural analysis of hyphal tip growth in fungi: Spitzenkorper, cytoskeleton and endomembranes after freeze-substitution. *J. Cell Sci.* **48**, 89.

Humbel, B., Marti, T., and Muller, M. (1983). Improved structural preservation by combining freeze-substitution and low temperature embedding. *Beitr. Elektronenmikrosk. Direktabb. Oberfl.* **16**, 585.

Humbel, B., and Muller, M. (1984). Freeze substitution and low temperature embedding. *Electron Microsc., Proc. Eur. Congr., 8th, 1984* Vol. 3, p. 1789.

Hunziker, E. B., and Herrmann, W. (1987). In situ localization of cartilage extracellular matrix components by immunoelectron microscopy after cytochemical tissue processing. *J. Histochem. Cytochem.* **35**, 647.

Jorgensen, A. O., and McGuffee, L. J. (1987). Immunoelectron microscopic localisation of sarcoplasmic reticulum proteins in cryofixed, freeze-dried, and low temperature embedded tissue. *J. Histochem. Cytochem.* **35**, 723.

Kellenberger, E., Carlemalm, E., Villiger, W., Roth, J., and Garavito, R. M. (1980). *Low Denaturation Embedding for Electron Microscopy of Thin Sections.* Chemische Werke Lowi GmbH, P.O. Box 1660, D-8264 Waldkraiburg, Federal Republic of Germany.

Kellenberger, E., Durrenberger, M., Villiger, W., Carlemalm, E., and Wurtz, M. (1987). The efficiency of immunolabel on Lowicryl sections compared to theoretical predictions. *J. Histochem. Cytochem.* **35**, 959.

Lin, N., and Langenberg, W. G. (1983). Immunohistochemical localisation of barley stripe mosaic virions in infected wheat cells. *J. Ultrastruct. Res.* **84**, 16.

Lin, N., and Langenberg, W. G. (1984). Chronology of appearance of barley stripe mosaic virus protein in infected wheat cells. *J. Ultrastruct. Res.* **89**, 309.

MacKenzie, A. P. (1972). Freezing, freeze-drying and freeze-substitution. In *"Scanning Electron Microscopy* (O. Jahari, ed.), Vol. 2, pp. 273–280. Scanning Electron Microscopy Inc., AMF O'Hare, Chicago, Illinois.

Menco, B. P. M. (1986). A survey of ultra-rapid cryofixation methods with particular emphasis on applications to freeze-fracturing, freeze-etching and freeze-substitution. *J. Electron Microsc. Tech.* **4**, 177.

Millonig, G. (1961). A modified procedure for lead staining of thin sections. *J. Biophys. Biochem. Cytol.* **11**, 736.

Newman, G. R., and Hobot, J. A. (1987). Modern acrylics for post-embedding immunostaining techniques. *J. Histochem. Cytochem.* **35**, 971.

Plattner, H., and Bachmann, L. (1982). Cryofixation: A tool in biological ultrastructural research. *Int. Rev. Cytol.* **79**, 237.

Roth, J., Bendayan, M., Carlemalm, E., Villiger, W., and Garavito, R. M. (1981). Enhancement of structural preservation and immunocytochemical staining in low temperature embedded pancreatic tissue. *J. Histochem. Cytochem.* **29**, 663.

Roth, J., Taatjees, D. J., Lucocq., J. N., Weinstein, J., and Paulson, J. C. (1985). Demonstration of an extensive transtubular network continuous with the Golgi apparatus stack that may function in glycosylation. *Cell (Cambridge, Mass.)* **43**, 287.

Simon, G. T., Thomas, J. A., Chorneyko, K. A., and Carlemalm, E. (1987). Rapid embedding in Lowicryl K4M for immunoelectron microscopic studies. *J. Electron Microsc. Tech.* **6**, 317.

Steinbrecht, R. A. (1982). Experiments on freezing damage with freeze-substitution using moth antennae as test objects. *J. Microsc. (Oxford)* **125**, 187.

Weibull, C., Carlemalm, E., Villiger, W., Kellenberger, E., Fakan, J., Gautier, A., and Larsson, C. (1980). Low-temperature embedding procedures applied to chloroplasts. *J. Ultrastruct. Res.* **73**, 233.

Weibull, C., Chrisriansson, A., and Carlemalm, E. (1983). Extraction of membrane lipids during fixation, dehydration and embedding of *Acholeplasma laidlawii* cells for electron microscopy. *J. Microsc. (Oxford)* **129**, 201.

5. Lowicryls and Low-Temperature Embedding

Weibull, C., Villiger, W., and Calemalm, E. (1984). Extraction of lipids during freeze-substitution of *Acholeplasma laidlawii* cells for electron microscopy. *J. Microsc. (Oxford)* **134**, 213.

Weibull, C. (1986). Temperature rise in Lowicryl resins during polymerization by ultraviolet light. *J. Ultrastruct. Mol. Struct. Res.* **97**, 207.

Weibull, C., and Christiansson, A. (1986). Extraction of proteins and membrane lipids during low temperature embedding of biological material for electron microscopy. *J. Microsc. (Oxford)* **142**, 79.

Wroblewski, J., and Wroblewski, R. (1986). Why low temperature embedding for x-ray microanalytical investigations? A comparison of recently used preparation methods. *J. Microsc. (Oxford)* **142**, 351.

6

The Enzyme–Gold Cytochemical Approach: A Review

Moise Bendayan

Department of Anatomy
Université de Montréal
Montréal, Québec, Canada

INTRODUCTION
PREPARATION OF THE ENZYME–GOLD COMPLEXES
ASSAYS FOR ASSESSING THE BIOLOGICAL ACTIVITY OF ENZYME–GOLD COMPLEXES
TISSUE PROCESSING
   Cytochemical Labeling
      Procedure
   Cytochemical Controls
QUANTITATIVE EVALUATIONS
APPLICATIONS
   Cytochemistry of Nucleic Acids
      RNAse–Gold Complex
      DNAse–Gold Complex
   Cytochemistry of Extracellular Matrix Components
      Elastase–Gold Complex
      Collagenase–Gold Complex
   Cytochemistry of Polysaccharides and Glycoconjugates
      Amylase–Gold Complex

Glucosidase–Gold and Galactosidase–Gold Complexes
  Mannosidase–Gold Complex
  Neuraminidase–Gold and Hyaluronidase–Gold Complex
 Cytochemistry of Fungal and Plant Cell Wall Components
  Xylanase–Gold Complex
  Cellulase–Gold Complex
  Chitinase–Gold Complex
  Exoglucanase–Gold Complex
  Pectinase–Gold Complexes
 Cytochemistry of Phospholipids
  Phospholipase A2–Gold Complex
 CONCLUSIONS
 REFERENCES

INTRODUCTION

The first objective of microscopy having been achieved, cell biologists sought more detailed information from morphological sciences. Indeed, the fine ultrastructural organization of the various cellular compartments has been well described; their specific role, however, still constitutes the final goal in cell biology. Elucidation of the chemical composition of cellular structures as well as assignment of particular sites to specific components should allow a better understanding of cell function. Through the years, histochemistry has made great progress in providing information concerning the nature of the tissue and cellular components. "Microchemistry" originated in 1826 when François-Vincent Raspail, a French botanist, practiced chemical analysis of tissues in combination with microscopic examination (Raspail, 1830). From the classical general histochemical techniques, the approach has progressed to highly specific tools with the development of new techniques. In 1941, Coons (Coons *et al.*, 1941) introduced the immunological approach in the field of histochemistry, thus initiating a new era in microchemistry. The techniques were first applied in light microscopy and then brought to electron microscopy with the development of electron-dense markers.

One of the most recent advances has been the introduction of colloidal gold particles as a marker for electron microscopy (Faulk and Taylor, 1971). This marker was found to be adequate for a large variety of techniques, since it is detectable with light and electron microscopy, by transmission and scanning modes. Various probes displaying highly specific binding properties could be used in combination with this marker. The use of colloidal gold with immunoglobulins and related immunoreagents such as protein A or protein G led to the immunogold techniques (Faulk and Taylor, 1971; Romano *et al.*, 1974; Geoghegan and Ackerman, 1977;

6. The Enzyme–Gold Cytochemical Approach

Roth *et al.*, 1978; De Mey, 1983; Bendayan, 1984a, 1987; 1989), the lectin–gold approach for the detection of glycoconjugates (Roth, 1983; Benhamou, 1989), and the avidin–biotin–gold system (Tolson *et al.*, 1981). The common basis of these techniques is the affinity properties existing between each probe and its binding site in the tissue. Along with this progress, the affinity exhibited by enzyme molecules toward their substrates was tested as a possible basis for using enzymes as histochemical probes. It was applied for the development of a novel approach: the enzyme–gold approach (Bendayan, 1981). Indeed, we took advantage of the biological properties of the enzymes for the *in situ* localization of substrate molecules. The use of the colloidal gold marker in this system allowed for application of the technique at the electron microscopic level with very high resolution (Bendayan, 1981, 1984b). It was later also used with light microscopy (Danscher and Norgaard, 1983; Coulombe and Bendayan, 1989). Enzymes tagged with electron-dense colloidal gold particles constitute a specific probe for the cytochemical detection of corresponding substrate molecules. First introduced for the localization of nucleic acids, the approach has rapidly expanded for the detection of a large variety of substrate molecules. The present review updates the basis of the technique and its applications. It demonstrates the potential of the approach and the good specificity and excellent resolution of the labeling that are obtained for a large variety of substrate molecules.

PREPARATION OF THE ENZYME–GOLD COMPLEXES

Introduced in immunoelectron microscopy by Faulk and Taylor in 1971, colloidal gold had since proved to be the electron-dense marker of choice for postembedding cytochemistry. Because of its particulate nature and its small size (~3 nm), it allows accurate high-resolution identification of labeled structures, quantitative evaluation of intensities of labeling, and double-labeling experiments. Extensive characterization of this marker and detailed methods for its preparation are given in several chapters of this series of books (Bendayan, 1989; Handley, 1989).

For its application in the enzyme–gold affinity technique, enzyme molecules are adsorbed at the surface of the gold particles through noncovalent electrostatic interactions. To obtain a stable protein–gold complex, various factors should be taken into consideration, in particular the pK_I and the molecular weight of the protein. Thus, for each enzyme–gold complex to be prepared, specific conditions for optimal binding must be determined. The optimal pH value for binding each enzyme to the colloidal gold should also be determined, as described previously (Horisberger et al., 1975; Bendayan, 1989), taking into account the pK_I of each enzyme. The minimal amount of the enzyme can also be determined as described previously (Geoghegan and Ackerman, 1977; Bendayan, 1989). In practice, for the preparation of each enzyme–gold complex, 10 to 20% excess of the minimal amount of the particular enzyme should be added to the colloidal gold suspension. This should be brought to the pH value optimal for efficient binding of the enzyme (this pH value has no relation to the optimal pH value required for enzymatic activity). The colloidal gold is rapidly added with continuous stirring to the well-dissolved protein. After mixing, two drops of a 1% solution of polyethylene glycol (PEG) (MW 20,000) may be added for further stabilization of the complex. This complex is centrifuged using a fixed-angle rotor (Beckman Ti-50) to recover the enzyme–gold complex. The speed and time of centrifugation vary according to the size of the colloidal gold particles as described previously (Bendayan, 1989).

At the end of the centrifugation, three phases are obtained: a clear supernatant containing the free enzyme; a dark sediment at the bottom of the tube, which corresponds to the enzyme–gold complex; and a black spot, which corresponds to the metallic gold not complexed to the protein. The supernatant must be aspirated as completely as possible and discarded, since it contains the free enzyme, which, if remains with the complex, will compete with it for the binding sites and lower the efficiency of the probe. The sediment, corresponding to the enzyme–gold complex, should be recovered and resuspended in an appropriate buffer containing 0.02% PEG. The nature of this buffer may vary from one enzyme to another. At this stage of the procedure, each enzyme–gold complex should be considered independently. Indeed, each enzyme–gold complex should be resuspended in a buffer at a pH value compatible with its enzymatic activity. Since labeling will take place through the enzyme–substrate interactions, conditions should be made appropriate for optimal enzymatic activity. The stock solution of the enzyme–gold complex is kept at 4°C until its use for labeling. The complex is diluted with the same buffer and should be centrifuged at a low speed (~2500 rpm) for 5 min prior to performing the labeling, in order to remove small aggregates which may form with time.

As for the preparation of any protein–gold complex, the glassware should be scrupulously cleaned and if possible (but not absolutely required) siliconized. The enzyme–gold complexes should not be frozen. Although the enzyme–gold complexes are very stable, the enzymatic activity of some of them has been found to be labile, decreasing rapidly even when they are kept at 4°C. Thus, most of the complexes should be used in the first week after their preparation.

ASSAYS FOR ASSESSING THE BIOLOGICAL ACTIVITY OF ENZYME–GOLD COMPLEXES

It is of prime importance to demonstrate that the enzymes, once tagged with colloidal gold particles, retain their biological activity. Because of the presence of colloidal particles, it is difficult in many cases to apply current enzymatic assays. Depending on the type of enzyme under study, we have applied four different approaches in order to assess the biological activity of the enzyme–gold complexes.

1. Use of radiolabeled substrates. This was done for the nuclease–gold complexes (RNase–gold and DNase–gold). The complexes were exposed to radiolabeled substrate molecules in *"in vitro"* assays for the determination of the rate of hydrolysis (M. Bendayan and V. Gisiger, unpublished results).

2. Colorimetric determinations. These were performed to assess the enzymatic activity of the mannosidase–gold complex (I. Londoño and M. Bendayan, unpublished results), using a spectrophotometric assay with p-nitrophenyl-α-D-mannoside as substrate according to the protocol of Faber and Glew (1983).

3. Biological assays. Biological assays can be done *in vitro* by incubating cell suspensions or cells in culture with the enzyme–gold complex. The cells will react to the presence of the enzyme in a detectable way. Such a biological assay was performed to assess the enzymatic activity of the phospholipase–gold complex, which in the presence of erythrocytes induces hemolysis. The degree of hemolysis was found to be similar to that elicited by the native enzyme, demonstrating the retention of enzymatic activity after complexing the enzyme to gold particles (Coulombe *et al.*, 1988). This test can also be performed *in vivo* by injecting the complex, which will induce detectable changes in blood composition (Ackerman and Wocken, 1981).

4. Use of immobilized substrate preparations. This can be performed on electron microscopic preparations of purified substrates. Fixed and embedded substrate preparations can be exposed to the enzyme–gold complex and examined for labeling activity. This allows for the testing of various substrate molecules as well as for the evaluation of optimal conditions of labeling (Coulombe *et al.*, 1988). The approach can also be

performed on nitrocellulose paper after immobilizing the substrate molecules and exposing the strip of paper to the enzyme–gold complex (I. Londoño, P. Coulombe, and M. Bendayan, unpublished results).

TISSUE PROCESSING

In the enzyme–gold postembedding method the enzyme–gold complexes are applied to thin tissue sections. For optimal results, the labeling should occur over well-identifiable structures. However, like any other cytochemical technique, this approach is confronted by the fact that good preservation of cellular structures is not always compatible with retention of the biological properties of the macromolecules. In the present case, the tissue components should retain their three-dimensional configuration in order to be recognized by the enzymes. Several studies have demonstrated that fixation and embedding alter the biological configuration of macromolecules to different extents, depending on the nature of the molecules. Thus, for each substrate to be revealed, protocols for optimal labeling should be worked out (Bendayan *et al.*, 1987). These are defined as conditions that allow significant specific labeling over well-preserved structures. Fixation can be carried out with glutaraldehyde, formaldehyde, or a mixture of these two aldehydes at different concentrations (Hayat, 1986, 1989). Since retention of the biological configuration also depends on the concentration of the fixative solutions, these should be kept as low as possible. In general, postfixation with OsO_4 is avoided. In some cases, however, it is possible to postfix the tissue with OsO_4; in others, it is required to prevent extraction of the substrate molecules. In these cases, treatment of the tissue sections with an oxidizing agent may be needed before labeling (Bendayan and Zollinger, 1983; Bendayan, 1984b, 1989).

The embedding protocol is as crucial as the fixation and can totally inhibit labeling or modify its pattern. Several protocols have been tested to define the best embedding condition for cytochemistry. As is true for fixation, each class of substrate molecules may require a different embedding for optimal labeling. Among the most common media used, we can mention the epoxies, methacrylates (Lowicryl and glycol methacrylate), and acrylics (LR resins). Sections obtained through cryoultramicrotomy can also yield excellent results.

Cytochemical Labeling

After fixation and embedding under optimal conditions, thin sections are mounted on Parlodion (or any other supporting film)–carbon-coated

6. The Enzyme–Gold Cytochemical Approach

nickel grids and processed for cytochemical labeling. The enzyme–gold approach is a direct one-step technique. Its principle is described in Fig. 6.1. Upon incubation of the tissue section with the enzyme–gold complex, the enzyme molecules adsorbed to the gold particles interact with the surface of the section, binding their specific substrate molecules. The gold particles enable the detection of the complex and reveal the site of the substrate. The gold particles being very small and the technique performed in a single step, the resolution obtained is excellent.

The exact mechanism by which labeling of substrate molecules occurs is still poorly understood and subject to investigation. The normal binding of an enzyme to its substrate is a reversible reaction. Upon hydrolysis, the enzyme moves out from its binding site in order to proceed with the hydrolytic activity. Many data support the fact that the interaction between the enzyme–gold complex and the tissue section components takes place between the enzyme and its substrate through the enzymatic activity of the complex. Indeed, it has been shown that the enzyme molecules retain their biological activity and remain enzymatically active after adsorption to the gold particles. Furthermore, the facts that specific enzyme inhibitors do interfere with the labeling and that the conditions for specific labeling follow those required for optimal enzymatic activity support the concept of enzymatic interactions between the enzyme–gold and the substrate molecules present in the tissue sections. In addition, although the enzyme–gold complexes are extremely stable, their enzymatic activity is labile and they lose quite rapidly the property of providing specific labeling. Thus, these data support the concept of interactions occurring between the enzyme and its substrate. The reason(s) why labeling does occur may lie in additional interactions that could occur between the colloidal gold particles and the resin at the site of enzyme–substrate binding. Also, the substrate is fixed and immobilized in the resin and may retain the enzyme at its site of action.

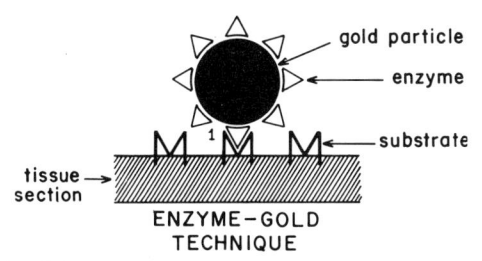

Fig. 6.1. Principle of the enzyme–gold approach. The technique is carried out in one step. Upon incubation of the tissue section with the enzyme–gold complex, the enzyme molecules surrounding the gold particle interact with their substrate molecules present at the surface of the tissue section. The gold particle enables the direct localization of these substrate molecules. (From Bendayan, 1984b.)

Procedure

The labeling is carried out according to the following protocol. Thin tissue sections mounted on nickel grids are incubated by floating them, tissue sections down, on a drop of 0.01 M phosphate-buffered saline (PBS) at a pH value optimal for the enzyme under study. If the phosphate buffer is not adequate for the enzyme, another appropriate buffer should be used. The grids are then transferred to a drop of the enzyme–gold complex brought to the appropriate dilution. Incubation is usually carried out at room temperature for 30 min. The length and temperature of incubation can vary from one enzyme–gold complex to another and also according to the preparation of the tissue. Thin cryosections, for example, should be incubated for a very short time (5 min) at room temperature. Also, if incubations are carried out at 37°C, they should be for a very short time. After this incubation, the grids should be jet washed with the same buffer, rinsed with distilled water, and dried. Staining with uranyl acetate and lead citrate can be performed.

When the tissue has been postfixed with OsO_4, treatment with a strong oxidizing agent could be performed. A saturated solution of sodium metaperiodate has been found to give optimal results (Bendayan and Zollinger, 1983; Bendayan, 1984a,b, 1989). For labeling of glycogen deposits in liver tissue and ribonucleic acids in various cell types, this treatment was required in order to obtain specific and intense labeling (Bendayan, 1984b; Bendayan and Puvion, 1984). For other systems, such as the localization of phospholipids (Coulombe *et al.*, 1988) and plant cell wall components (Benhamou and Ouellette, 1986, 1987; Bendayan and Benhamou, 1987), this treatment was not required. When performed, this treatment consists of a 30- to 60-min incubation of thin sections in an aqueous saturated solution of sodium metaperiodate at room temperature, followed by a thorough wash with distilled water, transfer to the buffer solution, and then to the enzyme–gold complex. This procedure improves the quality of the ultrastructural preservation, resulting in better labeling resolution. Furthermore, it allows labeling of tissues that have been processed for routine electron microscopy; this has great advantages for clinical specimens and diagnostic purposes.

Cytochemical Controls

The labeling obtained through application of the enzyme–gold complexes should be assessed for specificity. This is done by several control experiments, the most commonly used being those listed below. For many reasons, however, not all controls can be performed for each en-

6. The Enzyme–Gold Cytochemical Approach

zyme–gold complex, but the most important ones should always be carried out. Application of any of the control experiments should significantly reduce the labeling.

1. Incubation of thin tissue sections with an enzyme–gold complex to which the corresponding specific substrate (~1 mg/ml) has been added. This will involve the enzyme–gold complex in a competition experiment.

2. Incubation of thin tissue sections with the enzyme–gold complex under nonoptimal conditions for the enzymatic activity (at 4°C and inappropriate buffer).

3. Incubation of thin tissue sections with an enzyme–gold complex, the enzymatic activity of which has been abolished (e.g., by heat treatment).

4. Incubation of thin tissue sections with the enzyme–gold complex in the presence of specific enzyme inhibitors.

5. Incubation of thin tissue sections with an enzyme–gold complex in the presence of a specific antibody directed against the enzyme.

6. Incubation of thin tissue sections with a protein–gold complex devoid of enzymatic activity. Complexes such as albumin–gold or protein A–gold can be used. This will assess the nonspecific binding of protein–gold complexes to the particular tissue section. In general, all the resins tested yielded very low background staining. It does, however, vary between resins; Lowicryl yields one of the lowest amount of background staining.

7. Incubation of thin tissue sections with the unlabeled enzyme for a time long enough to extract the substrate. This is followed by incubation of the sections with the enzyme–gold complex. Since the substrate molecules have been removed from the section, the enzyme–gold complex should not yield any specific labeling, or if it does, it should be of low intensity. In some cases, the extraction of the substrate can be performed prior to embedding of the tissue (Bendayan and Puvion, 1983). After fixation, the tissue blocks are incubated with the native enzyme and then processed for embedding, sectioning, and labeling. We must take into consideration that after fixation, total extraction of the substrate from the tissue can be difficult and thus this control experiment may not be successful in all cases.

8. Incubation of thin sections from tissues which are known not to contain the specific substrate molecules.

The overall results obtained by using the controls mentioned above should demonstrate the specificity of the labeling.

QUANTITATIVE EVALUATIONS

One of the advantages of the colloidal gold marker is its particulate nature, which allows for easy quantitative evaluation. This is done in order to determine the intensities of the labeling over cellular or extracellular compartments or along membranes. Comparative tests should be performed with great care after verifying that all the conditions of labeling are identical: use of the same enzyme–gold complex simultaneously on all sections, uniformity of the size of the gold marker, and identical processing of the tissues. Problems such as variations in the thickness of the sections, diffusion of the marker, and accessibility of the binding sites can be ruled out when postembedding colloidal gold labeling is performed (Bendayan, 1984a,b). Indeed, as demonstrated by the study of the profile view of labeled sections, which has been described in detail (Bendayan, 1984a,b; 1989), only binding sites exposed by the cutting procedure can be labeled. Considering the nature of the results obtained from quantitative evaluations, these represent intensities of labeling which are an indication of the amounts of substrate molecules; only relative comparisons between intensities of labeling can be considered.

Quantitation can be performed in relation to labeled areas of cellular compartments or in relation to length of labeled membranes. In the former case, the surface of each compartment (S_a) is evaluated by a morphometric approach, and then the number of gold particles (N_i) present over the same area is counted. The density of labeling (N_s) corresponds to the ratio between the number of particles and the surface: $N_s = N_i/S_a$. For the density of labeling of membranes (N_l), the length of the specific membrane (L_a) is first evaluated and the number of particles (N_i) present over the membrane is then counted. The density corresponds to the ratio: $N_l = N_i/L_a$. A sufficient number of areas or a sufficient length of membrane should be evaluated to obtain representative values.

APPLICATIONS

Since its introduction in 1981 by Bendayan for the localization of nucleic acids with the RNase–gold and DNase–gold complexes, the enzyme–gold approach has been expanded significantly to various fields of research and has been applied for the localization of a large variety of substrate molecules. These comprise the nucleic acids, the components of the extracellular matrix, plant cell walls, complex carbohydrates and polysaccharides, and phospholipids in cellular structures and membranes. Although recently introduced, the approach has already reached the level of a currently used high-resolution specific cytochemical tech-

6. The Enzyme–Gold Cytochemical Approach

nique that is applied in eukaryotic as well as prokaryotic cells in control, experimental, and pathological conditions.

Cytochemistry of Nucleic Acids

Two enzymes involved in the hydrolysis of nucleic acids, the pancreatic RNase and DNase, were used for the localization of the corresponding substrate molecules in thin tissue sections. RNase–gold and DNase–gold complexes were prepared and applied to thin tissue sections. These complexes were the first to be applied and were used to establish this novel technical approach. Several reasons led us to choose these complexes for the demonstration of the technique. Following the pioneer work done by the group of Bernhard and that of Gauthier as reviewed by Moyne (1980), the ultrastructural localization of nucleic acids was elucidated, which allowed for an easy evaluation of the efficiency of our labeling. It was also well established that after fixation and processing of the tissue for electron microscopy, these components are not extracted. Furthermore, both enzymes, RNase and DNase, are well characterized with respect to their biophysical characteristics and enzymatic activities. This knowledge is helpful in establishing optimal conditions for the preparation of the complexes and their application. (See Bendayan, 1981, 1982, 1984b.)

RNase-Gold Complex

RNase (EC 3.1.27.5) purified from porcine pancreas was tagged with colloidal gold particles and applied directly to tissue sections. The optimal pH for complexing RNase to colloidal gold particles was found to be 9.0–9.2. After centrifugation, the complex was recovered and resuspended in PBS containing 0.02% PEG (PBS + PEG) (pH 7.2–7.4). Incubations were systematically performed at room temperature for 30 min at the same pH values. Optimal results were obtained on thin sections of the tissue fixed with 1% glutaraldehyde, postfixed or not with 1% OsO_4, and embedded in an epoxy resin. Sections were treated with a saturated solution of sodium metaperiodate, followed by a thorough wash with distilled water and a short preincubation with PBS prior to the 30-min exposure to the RNase–gold complex. (See Bendayan, 1981, 1982, 1984b; Bendayan and Puvion, 1983, 1984.)

The labeling was obtained mainly over the ribosomal particles of the rough endoplasmic reticulum, the polysomes, and the nuclei. Mitochondria and chloroplasts were only weakly labeled. Other organelles such as lysosomes, peroxysomes, and secretory granules were devoid of labeling. In the case of the nucleus, the labeling was systematically more intense

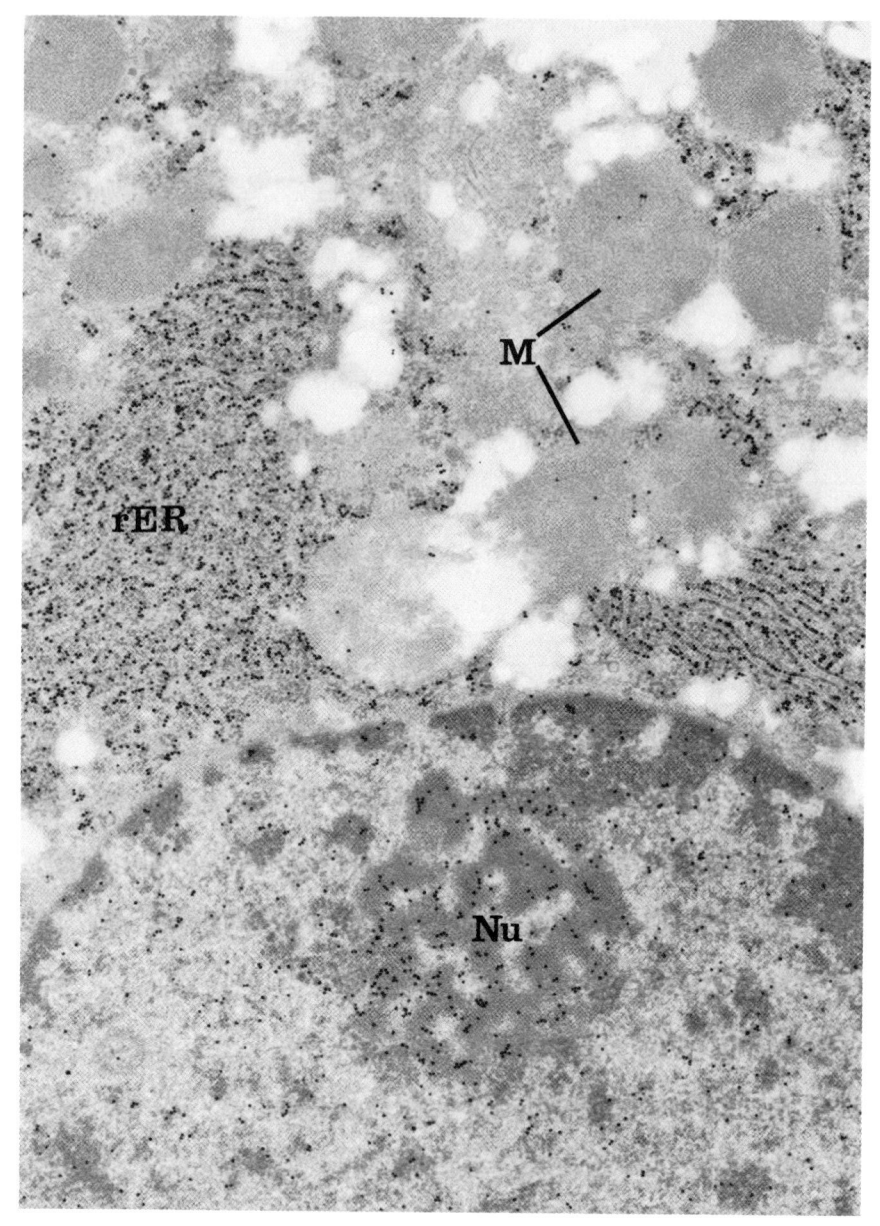

Fig. 6.2. RNase–gold labeling. Rat hepatocyte fixed with 1% glutaraldehyde and embedded in Lowicryl. The labeling is particularly intense over the rough endoplasmic reticulum (rER) and the nucleolus (Nu). Mitochondria (M) appear devoid of labeling. The clear areas present in the cytoplasm are regions where glycogen deposits have been leached from the tissue because of the omission of osmium tetroxide in the fixation protocol; ×25,000.

6. The Enzyme–Gold Cytochemical Approach

over the granular regions of the nucleolus, the fibrillar region being preferentially labeled at the edge of the cords. With respect to the nucleoplasm, labeling was concentrated at the border between the condensed and the dispersed chromatin. Perichromatin granules were also found to be preferentially labeled, the gold particles being located mostly over the clear halo which surrounds these structures. The labeling of the ribosomal particles as well as the detailed labeling obtained over the nucleus gave us a good indication of the resolution achieved by the technique. In addition, excellent preservation of the ultrastructure demonstrates the quality of the results that can be achieved with this approach. (See Figs. 6.2 and 6.3.)

DNase–Gold Complex

DNase I (EC 3.1.21.1) purified from calf thymus was tagged with colloidal gold particles to form a DNase–gold complex. The optimal pH value for complexing this enzyme with gold particles was found to be 6.0. After centrifugation, the complex was resuspended in PBS + PEG at pH 6.0. The incubation of the tissue sections was also performed at this pH value. The tissue sections were preincubated with PBS at pH 6.0 before exposing them to the DNase–gold complex. Optimal conditions of fixation and embedding were found to be very specific; they correspond to fixation with 1% glutaraldehyde and embedding in methacrylate resins (either glycol methacrylate or Lowicryl K4M). The intensity of labeling was further enhanced with a short incubation (5 to 10 min) at 37°C in a 0.1% DNase solution before exposing the sections to the DNase–gold complex. When these optimal conditions were used, the nuclei were intensely labeled, particularly over the peripheral and perinucleolar chromatin. Some internal nucleolar clumps linked to the perinucleolar chromatin by thin labeled fibers were also intensely labeled. The dispersed chromatin and the fibrillar cords of the nucleoli were labeled to a lesser extent. The labeling present over the cytoplasm was low. (See Bendayan, 1981, 1982, 1984b; Bendayan and Puvion, 1984.)

The specificity of the labeling obtained with both the RNase–gold and DNase–gold complexes was assessed following the different protocols described above. All the protocols were used to assess the labeling of the RNase–gold complex and showed major and significant reductions of the labeling (Fig. 6.4). The substrate was extracted by exposing the tissue to the RNase solution at both the pre- and postembedding steps. Both experiments resulted in the disappearance of ribosomal particles and absence of labeling (Fig. 6.4). For the DNase–gold complex, all the control experiments performed also confirmed the specificity of the labeling. However, total abolition of the labeling by long incubation of the tissue

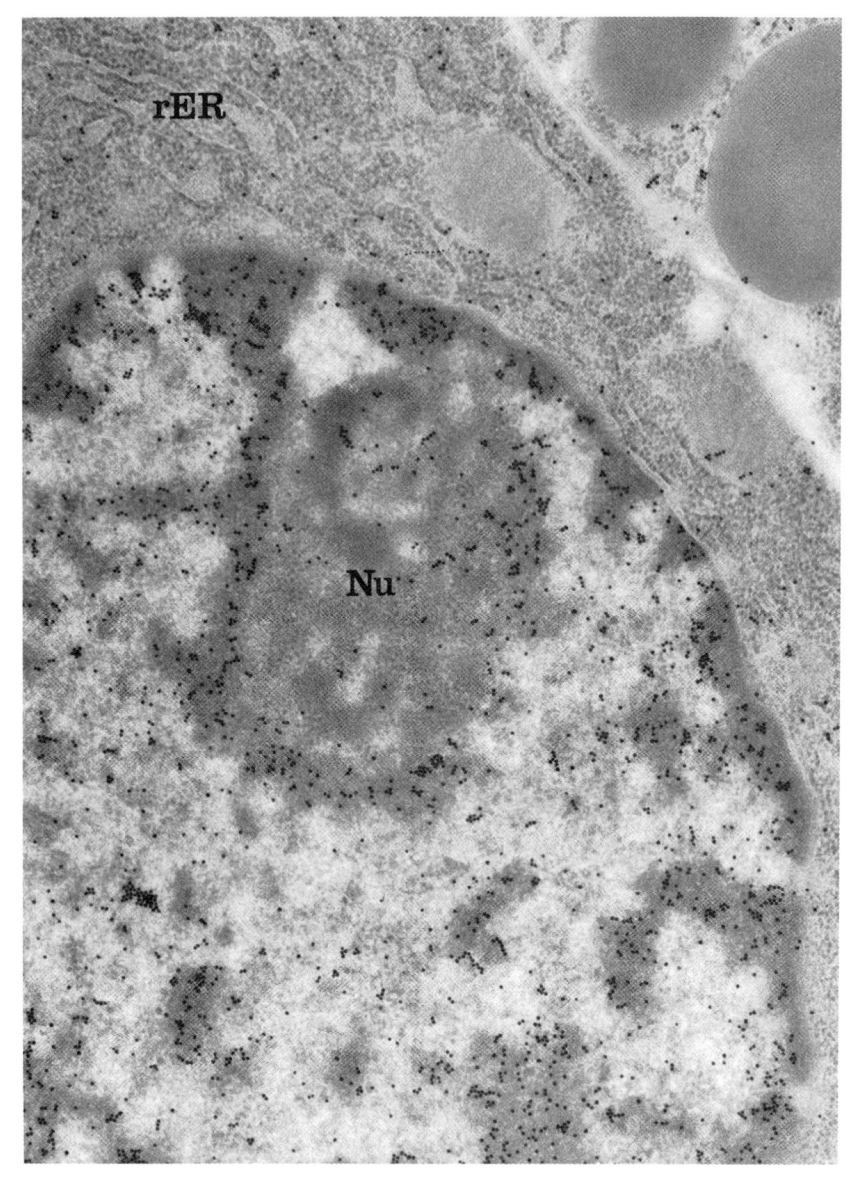

Fig. 6.3. DNase–gold labeling. Rat pancreatic tissue fixed with 1% glutaraldehyde and embedded in glycol methacrylate. The labeling is particularly intense over the dense chromatin at the periphery of the nucleus and around the nucleolus (Nu). Few gold particles are seen over the rough endoplasmic reticulum (rER); ×30,000.

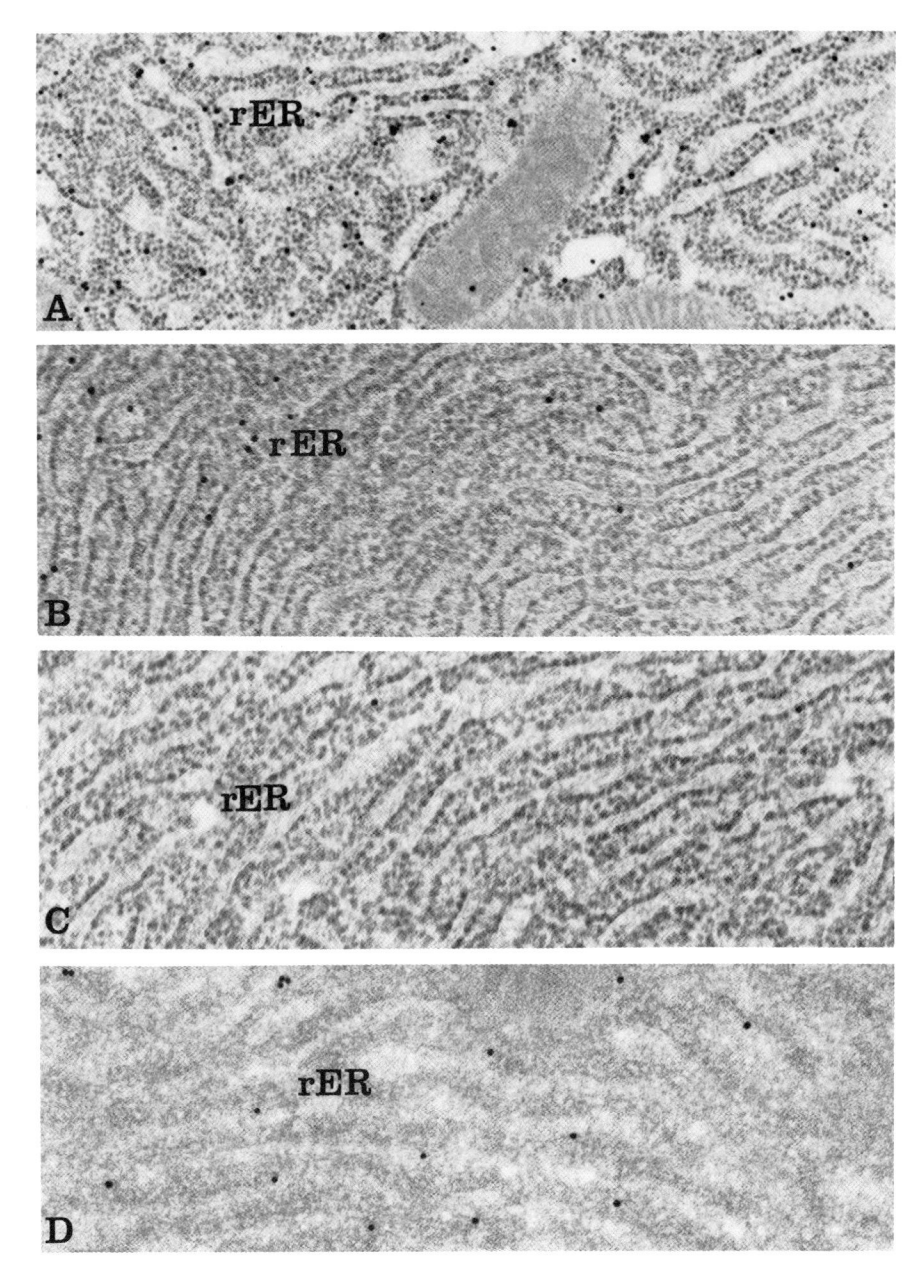

Fig. 6.4. Control experiments performed with the RNase–gold complex. The addition of RNA to the RNase–gold complex (A) or of a specific inhibitor (heparin) (C) resulted in a significant reduction of the labeling present over the rough endoplasmic reticulum (rER). A similar result was obtained when the incubation was performed at low temperature (B). Treatment of the tissue section with RNase also prevented labeling (D); ribosomes in this case were extracted from the tissue. A, ×35,000; B, ×35,000; C, ×50,000; D, ×50,000.

TABLE 6.1
Density of Labeling Obtained with the RNase–Gold and DNase–Gold Complexes over Various Cellular Compartments of the Acinar Cells

	Gold particles per μm^2 ± SEM	
	RNase–Gold	DNase–Gold
Rough endoplasmic reticulum	232.9 ± 22.2	5.2 ± 0.5
Golgi apparatus	1.6 ± 0.4	4.1 ± 0.8
Secretory granules	1.6 ± 0.3	2.6 ± 0.3
Mitochondria	2.6 ± 0.3	30.3 ± 3.3
Nucleus		
Dense chromatin	22.3 ± 3.5	311.4 ± 15.9
Dispersed chromatin	39.8 ± 4.1	70.5 ± 0.7
Perinucleolar chromatin	28.1 ± 1.9	378.0 ± 23.5
Nucleolus	95.7 ± 9.1	39.3 ± 2.9

in a solution of DNase was not achieved, most probably due to the inability of the DNase to extract the total amount of DNA after tissue fixation.

The intensities of labeling obtained on pancreatic cells with both the RNase–gold and DNase–gold complexes were evaluated as described previously (Bendayan, 1984b). The results, reported in Tables 6.1 and 6.2, demonstrate the distribution of the labeling for each cellular compartment and confirm the subjective observations. The quantitative evaluation has also made it possible to establish the specificity of the labeling (Table 6.2).

With the reliability of the approach established, both probes are being

TABLE 6.2
Density of Labeling Obtained with the RNase–Gold and DNase–Gold Complexes over Various Cellular Compartments of the Acinar Cells Under Control Conditions

	Gold particles per μm^2 ± SEM		
	RNase–Gold + RNA	DNase–Gold + DNA	Albumin–Gold
Rough endoplasmic reticulum	8.0 ± 1.7	1.6 ± 0.3	1.3 ± 0.1
Golgi apparatus	2.2 ± 0.6	2.3 ± 0.4	1.6 ± 0.4
Secretory granules	1.4 ± 0.4	1.9 ± 0.2	1.3 ± 0.2
Mitochondria	1.2 ± 0.3	2.1 ± 0.4	1.1 ± 0.6
Nucleus			
Dense chromatin	1.8 ± 0.3	1.3 ± 0.1	2.7 ± 0.8
Dispersed chromatin	7.4 ± 0.9	4.5 ± 0.6	1.6 ± 0.2
Nucleolus	4.4 ± 0.9	2.8 ± 0.4	1.4 ± 0.4

used to label corresponding nucleic acids in animal and plant tissues under normal and pathological conditions as well as in prokaryotic cells and viral particles. The technique was also combined with autoradiography in order to discriminate newly synthesized molecules. The latter study has demonstrated the complementarity of the enzyme–gold and autoradiographic approaches.

Cytochemistry of Extracellular Matrix Components

Elastase–Gold Complex

Elastase (pancreatopeptidase E) (EC 3.4.21.36) purified from porcine pancreas tagged with colloidal particles allows labeling of elastin on tissue sections. The optimal pH for complexing this enzyme with colloidal gold particles was found to be 8.5. After centrifugation, the complex was resuspended in PBS + PEG at pH 8.5. All incubations were performed at this pH value, which corresponds to the optimal pH for enzymatic activity. Sections of tissues comprising muscular or elastic arteries fixed with 1% glutaraldehyde and embedded in epoxy resin, incubated with the elastase–gold complex, demonstrated labeling by gold particles over the elastic laminae, either in the tunica intima or throughout the tunica media. No labeling was detected anywhere else on the tissue sections. (See Bendayan, 1984b; Fig. 6.5.)

Collagenase–Gold Complex

Following the same protocol, collagenase (clostridiopeptidase A) (EC 3.4.24.3) was complexed with colloidal gold particles and brought to pH 7.3–7.5. After centrifugation, the complex was resuspended in PBS + PEG at pH 7.3. All incubations were performed at this pH value. For optimal labeling, the tissues were fixed with 1% glutaraldehyde and embedded in methacrylate resins. The labeling was obtained over the bundles of collagen fibers. In some instances, the labeling appeared to follow the characteristic striations of the fibrils. A certain level of staining was also systematically observed over the cytoplasm and nucleoplasm. (See Bendayan, 1984b; Fig. 6.6.)

Cytochemistry of Polysaccharides and Glycoconjugates

Glycoconjugates are widely distributed in cellular as well as extracellular compartments and play important roles in several cellular events as well as in intercellular interactions. The enzyme–gold approach has been applied successfully for the detection of such components. Techniques available up to now for revealing glycoconjugates took advantage of

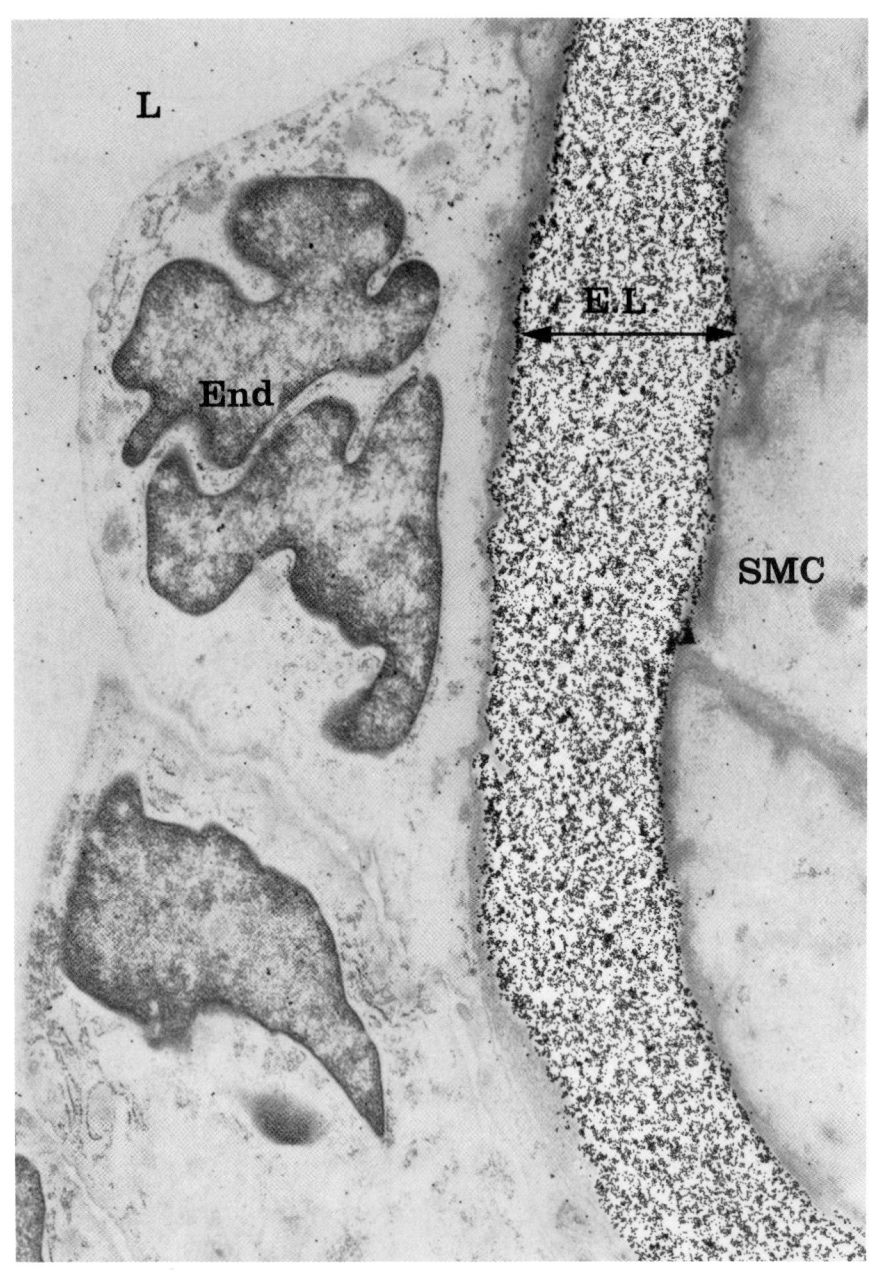

Fig. 6.5. Elastase–gold labeling. Rat kidney, muscular artery fixed with 1% glutaraldehyde and embedded in Epon. The labeling is present over the internal elastic lamina (E.L.). The smooth muscle cells (SMC), endothelial cells (End), and lumen (L) of the vessel are devoid of labeling; ×15,000.

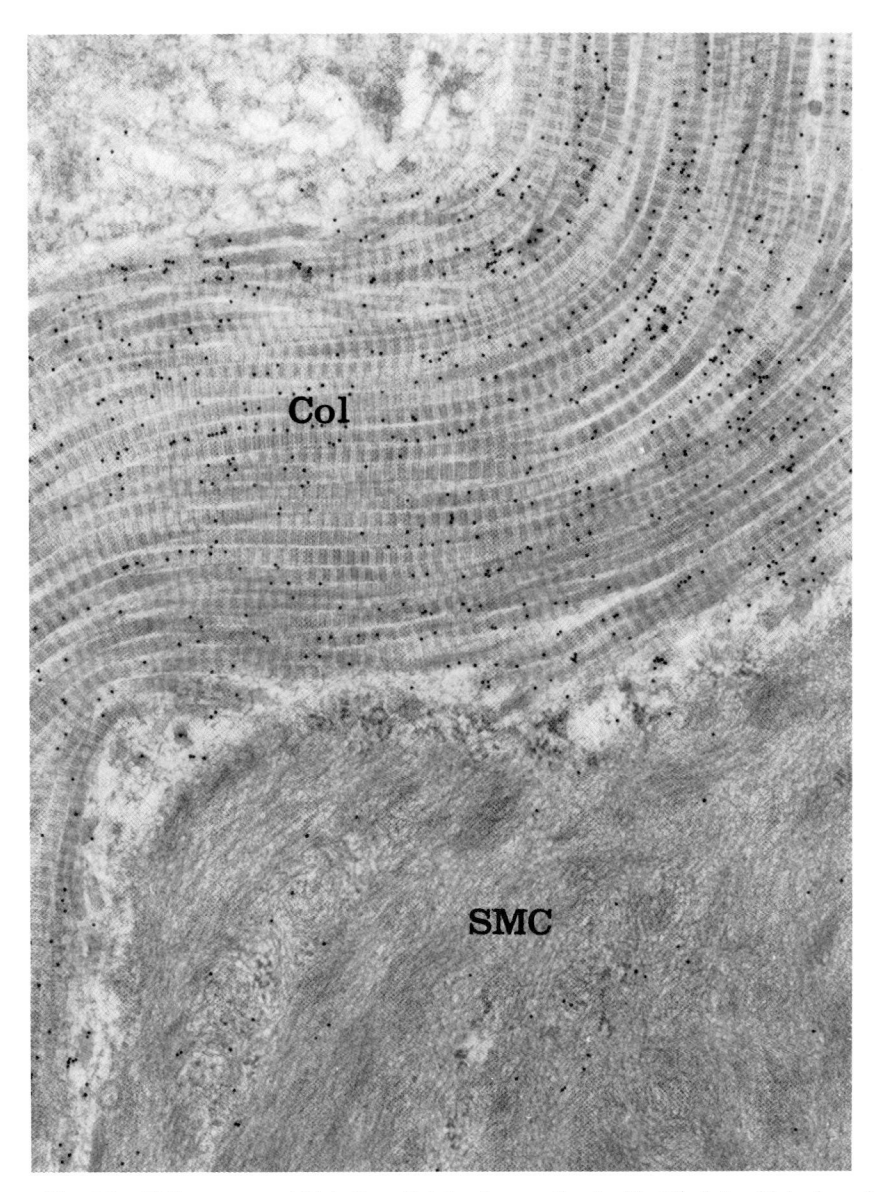

Fig. 6.6. Collagenase–gold labeling. Rat duodenum, fixed with 1% glutaraldehyde and embedded in glycol methacrylate. The labeling is present over the collagen fibrils (Col). In some instances the gold particles seem to follow the pattern of striation of the fibrils. Few particles are seen over the smooth muscle cells (SMC); ×28,000.

chemical reactivities of carboxyl, ester sulfate, or other groups (reviewed by Pearse, 1985; Thomopoulus *et al.,* 1987). Also, indirect methods making use of loss of reactivity after enzyme digestion were applied for such purposes (Pearse, 1985). More recently, because of their high specific sugar-binding properties, lectins have been used in cytochemistry for the detection of glycoconjugates (Roth, 1983; Benhamou, 1989). The enzyme–gold approach brings new possibilities, and it was found to complement the results obtained with lectin cytochemistry. Several enzymes directed against either polysaccharides or glycoconjugates tagged with colloidal gold particles were used for the ultrastructural detection of the corresponding substrate components.

Amylase–Gold Complex

The endoglycosidase α-amylase, 1,4-α-D-glucan glucanohydrolase (EC 3.2.1.1), purified from porcine pancreas was complexed with colloidal gold particles at pH 7.3. The amylase–gold complex was resuspended in PBS + PEG at pH 7.3. Incubation of the tissue sections was performed at the same pH value. The α-amylase hydrolyzes internal α1,4-glucan bonds (Glcα1 → 4Glcα1 → 4Glc) in oligosaccharides which contain at least three glucose units, such as starch, amylose, amylopectin, and glycogen. The amylase–gold complex was applied for the detection of glycogen deposits in rat liver (Bendayan, 1984b) as well as in fungal cells (Benhamou and Ouellette, 1987). Glycogen deposits in hepatocytes are known to be retained in the tissue only if double fixation is performed. The liver tissue must be postfixed with OsO_4 for optimal labeling. Embedding in epoxy resin was found adequate. The tissue sections were first incubated in a saturated solution of sodium metaperiodate for 30 to 60 min, rinsed with distilled water, transferred to PBS, and then transferred to the amylase–gold complex. This procedure leads to an intense and specific labeling of the glycogen deposits (Bendayan, 1984b). For fungal cells, the pretreatment with periodate was not required (Benhamou and Ouellette, 1987). (See Fig. 6.7.)

Glucosidase–Gold and Galactosidase–Gold Complexes

The β-glucosidase, β-D-glucoside glucohydrolase (EC 3.2.1.21), from sweet almond and the β-galactosidase, β-D-galactoside galactohydrolase (EC 3.2.1.23), from bovine testes were complexed to colloidal gold particles at pH 9.3. The glucosidase–gold complex was resuspended in PBS + PEG at pH 7.4 and the incubation of the tissue sections was performed at the same pH value. The β-glucosidase is specific for the β-D configuration of terminal β-D-glucoside, β-D-galactoside, and β-D-fucoside resi-

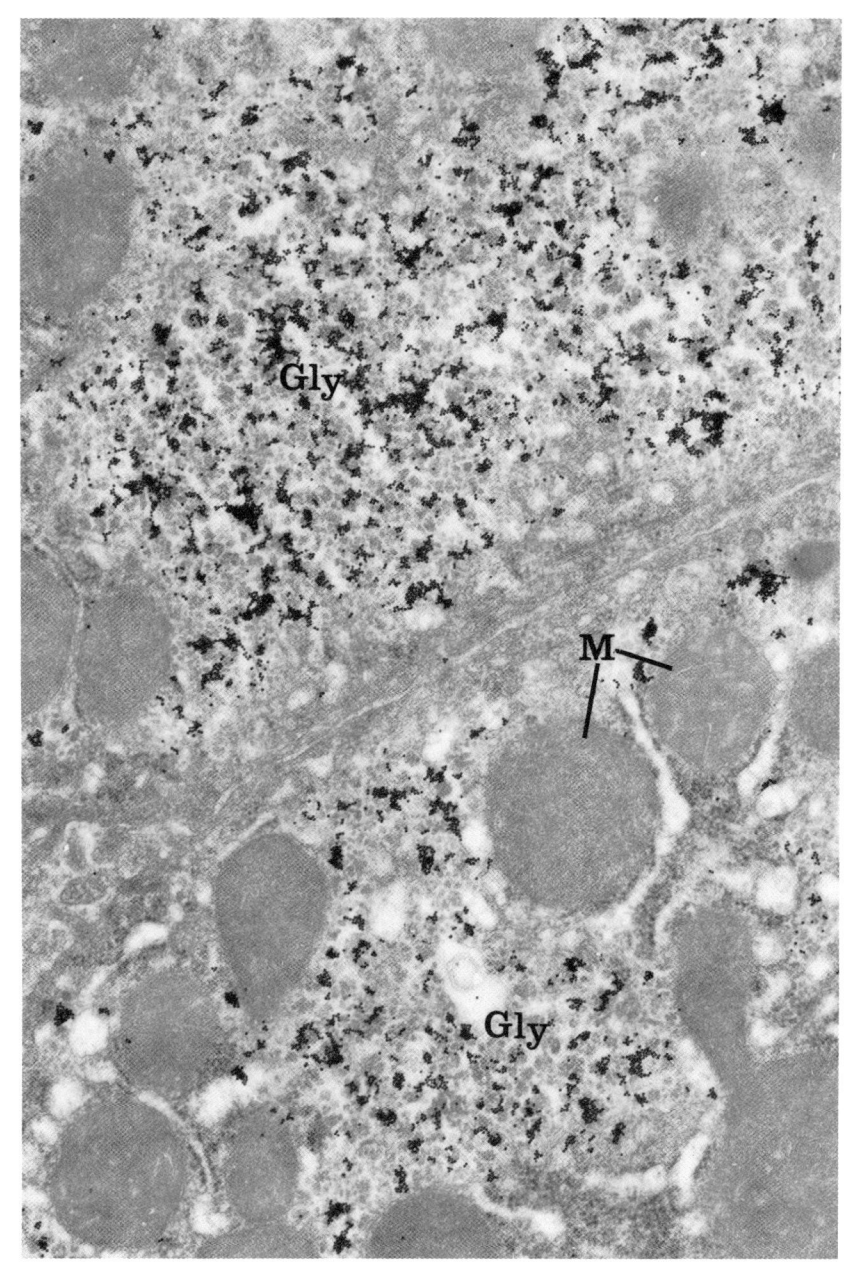

Fig. 6.7. Amylase–gold labeling. Rat liver tissue fixed with 1% glutaraldehyde, postfixed with 1% OsO_4, and embedded in Epon. The tissue sections were treated with a saturated solution of sodium metaperiodate for 60 min. The labeling is intense over the glycogen deposits (Gly). Few particles are seen over mitochondria (M) and other cellular structures; ×23,000.

dues. Once it was applied on thin sections, from either animal or plant tissues, several structures showed specific labeling. Conditions for tissue preparation were different for animal and plant specimens. For animal tissues, omission of osmium in the fixation protocol and embedding in Lowicryl were required, whereas for plant tissues, osmication and embedding in epoxy resin were adequate. Such discrepancies between animal and plant or fungal tissues, already detected for labeling with lectins (Benhamou, 1989), must be due to the amounts of sugar residues existing in the cell wall of plant parenchyma. (See Vian, 1986; Bendayan and Benhamou, 1987.)

Mannosidase–Gold Complex

The α-mannosidase, α-D-mannoside mannohydrolase (EC 3.2.1.24), purified from jack bean was complexed to colloidal gold particles at pH 6.0. The complex was resuspended in PBS + PEG at pH 5.0 and incubation of the tissue sections was performed at the same pH value. The enzyme is specific for the nitrophenyl-D-mannoside and α-(1–2), (1–3), (1–6) oligomannoside residues. When applied to animal tissue sections, it gave labeling over various cellular and extracellular compartments, in particular the rough endoplasmic reticulum and the dense chromatin in the nucleus. In the extracellular space, the basal laminae were specifically labeled. For optimal labeling, the tissue must be processed at low temperatures with embedding in Lowicryl or glycol methacrylate resins. In a particular study (Londoño and Bendayan, 1987), the labeling with mannosidase–gold was compared to that obtained with the use of concanavalin A (Con A) agglutinin specific for mannoside residues. The results showed good correlations. However, in the rough endoplasmic reticulum the labeling obtained with the mannosidase–gold was restricted to the cytosolic side of the membrane, whereas that obtained with Con A was found at both the cytosolic and luminal sides of the reticulum. (See Fig. 6.8.)

Neuraminidase–Gold and Hyaluronidase–Gold Complexes

The neuraminidase (sialidase; EC 3.2.1.18) purified from *Clostridium perfringens* and the hyaluronidase (EC 3.2.1.35) from bovine testes were complexed to colloidal gold particles at pH 6.0 and 7.5, respectively. After centrifugation, both complexes were resuspended in 0.05 M acetate buffer (pH 5.0), containing 0.02% PEG. These conditions are optimal for the biological activity of both enzymes. The neuraminidase hydrolyzes 2–3, 2–6, 2–8 glycosidic linkages joining final nonreducing N- or O-acylneuraminyl residues to galactose and N-acetylhexosamine in oligosaccharides, glycoproteins, and glycolipids. Hyaluronidase, on the other hand,

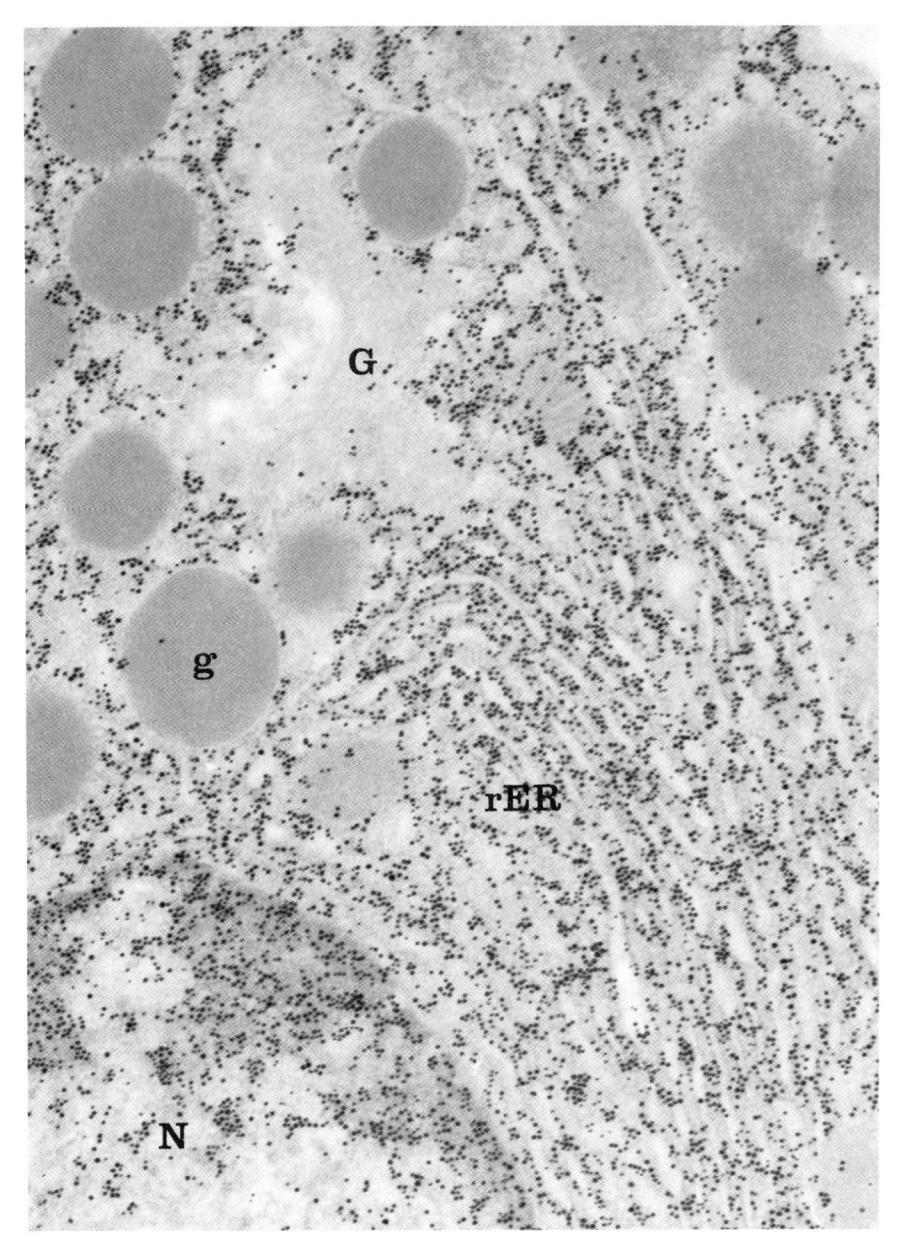

Fig. 6.8. Mannosidase–gold labeling. Rat pancreatic tissue fixed with 1% glutaraldehyde and embedded in Lowicryl. The labeling is particularly intense over the rough endoplasmic reticulum (rER) and the dense chromatin in the nucleus (N). The Golgi apparatus (G) and the secretory granules (g) are devoid of labeling; ×30,000.

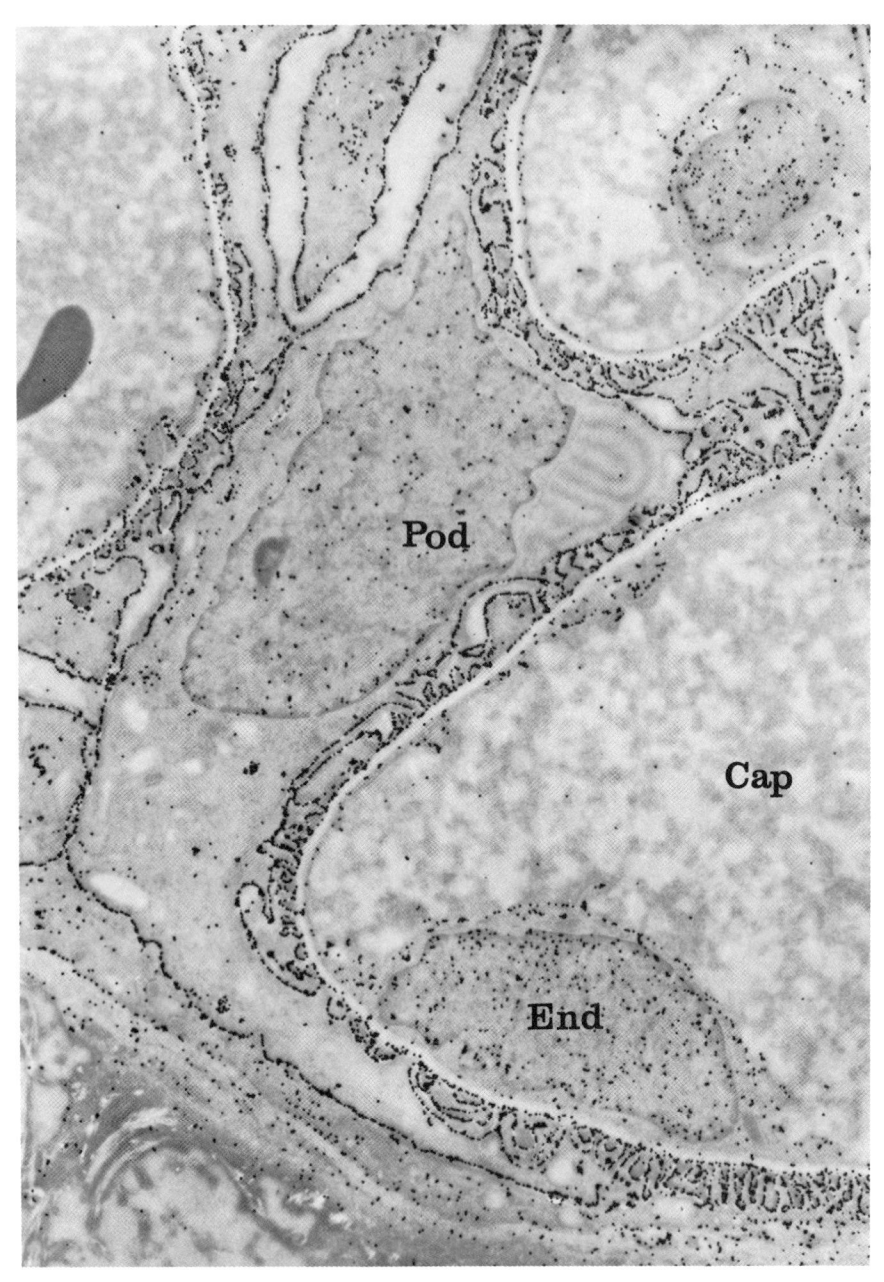

Fig. 6.9. Hyaluronidase–gold labeling. Rat kidney, fixed with 1% glutaraldehyde and embedded in Lowicryl. The labeling by 40-nm gold particles appears to follow the plasma membrane of the epithelial cells or podocytes (Pod) and to a lesser extent also that of the endothelial cells (End). Labeling is also present over the nuclei. Capillary lumen (Cap) and the urinary space are devoid of labeling; ×12,000.

hydrolyzes 1,4-*O*-glycosidic linkages between *N*-acetyl-β-D-glucosamine, *N*-acetyl-β-D-galactosamine, or galactosamine sulfate and glucuronic acid in hyaluronate, chondroitin, chondroitin 4- and 6-sulfates, and dermatan. When applied to animal tissue sections, these two complexes yield specific labeling, which, however, shows variations according to the cells and tissues examined. This was expected, particularly in cells secreting glycoproteins of different types (digestive enzymes from pancreas versus mucin from duodenal goblet cells) or tissues at different stages of differentiation. Again, a comparative evaluation of the labeling obtained with the neuraminidase–gold and with the *Limax flavus* agglutinin–gold complex (specific for neuraminic acid) showed similar patterns of labeling with variations at the level of the nuclei. For hyaluronidase–gold, the pattern of labeling also differed according to the cell type under study. As with previous enzymes, labeling with neuraminidase–gold and hyaluronidase–gold was obtained only on Lowicryl-embedded tissues. The use of an epoxy resin prevented any specific labeling. (See Londoño and Bendayan, 1988; Fig. 6.9.)

Cytochemistry of Fungal and Plant Cell Wall Components

Xylanase–Gold Complex

Endo-1,4-β-xylanase (EC 3.2.1.8) specific for 1,4-β-D-xylopyranose linkages was complexed with colloidal gold particles at pH 7.2, and the complex was resuspended in 0.05 *M* acetate buffer (pH 5.0). (See Vian *et al.*, 1983; Ruel and Joseleau, 1984.)

Cellulase–Gold Complex

Cellulase, a mixture of cellobiohydrolases I and II (EC 3.2.1.91) and endo-1,4-β-D-glucanase (EC 3.2.1.4) isolated from *T. reesei*, was complexed with colloidal gold particles at pH 4 and the complex was resuspended in 0.05 *M* citrate buffer (pH 4.5). (See Berg *et al.*, 1988.)

Chitinase–Gold Complex

Chitinase (EC 3.2.1.14) purified from *Streptomyces griseus* specific for the α-1,4-acetamido-2-deoxy-D-glucoside links in chitin and chitodextrin was complexed to colloidal gold at pH 7.0 and the complex was resuspended in 0.05 *M* PBS (pH 6.0–6.3). (See Chamberland *et al.*, 1985; Benhamou and Ouellette, 1987.)

Exoglucanase–Gold Complex

β-1,4-exoglucanase purified from *Trichoderma harzianum* Rufai was complexed with colloidal gold at pH 9.0 and the complex was resuspended in PBS + PEG at pH 6.0. (See Benhamou *et al.*, 1987.)

Pectinase–Gold Complexes

Several enzymes in the group of pectinases were complexed to colloidal gold. The pectinesterase (EC 3.1.1.11) hydrolyzing the methyl ester groups in polygalacturonic acid was complexed at pH 7.5 and resuspended in PBS + PEG at pH 7.4. The polygalacturonase (EC 3.2.1.15) hydrolyzing nonesterified portions of polygalacturonide chain 1,4-α-D-glycosidic links, purified from *Aspergillus niger,* was complexed at pH 7.2. The complex was then resuspended in PBS + PEG at pH 7.2. The pectinlyase purified from *Aspergillus japonicus,* which hydrolyzes the 1,4-α-D-glycosidic bonds of esterified polygalacturonide chains, was complexed at pH 8.2, and the complex was resuspended in PBS + PEG at pH 8.0. (See Benhamou and Ouellette, 1986.)

Application of all these complexes to sections of plant tissue yielded labeling over extracellular regions, particularly over the primary or secondary cell walls, depending on the specimen studied. Little labeling was found over intracellular regions.

Cytochemistry of Phospholipids

Phospholipase A_2–Gold Complex

The differential distribution of phospholipids among and within various cell membranes is an important feature in cell biology, since they appear to play major roles in various cellular processes. So far, their study has been mainly restricted to biochemical and biophysical techniques; cytochemical approaches have not succeeded or are being confronted by serious limitations for high-resolution studies. This is mainly due to difficulties in retaining these components during tissue preparation. The enzyme–gold approach was thus tested by applying a phospholipase–gold complex for the specific labeling of its corresponding substrate molecules. The phospholipase A_2 (phosphatide 2-acylhydrolase, EC 3.1.1.4) purified from bee venom was complexed with colloidal gold particles at pH 8.0. After centrifugation, the complex was resuspended in PBS + PEG containing 0.05% polyvinylpyrrolidone (MW 10,000) at pH 8.0. This corresponds to the optimal pH value for this enzyme's activity. The enzyme hydrolyzes the *sn*-2 fatty acid from *sn*-3 glycerophospholipids, converting them to their lyso derivatives. The complex was used for either preembedding or postembedding labeling of various classes of phospholipids at the light and electron microscopic levels and backscattered scanning modes. It was also used for freeze-fracture label replicas and thin-sectioned fractured-labeled specimens (Coulombe *et al.,* 1988). (See Fig. 6.10.)

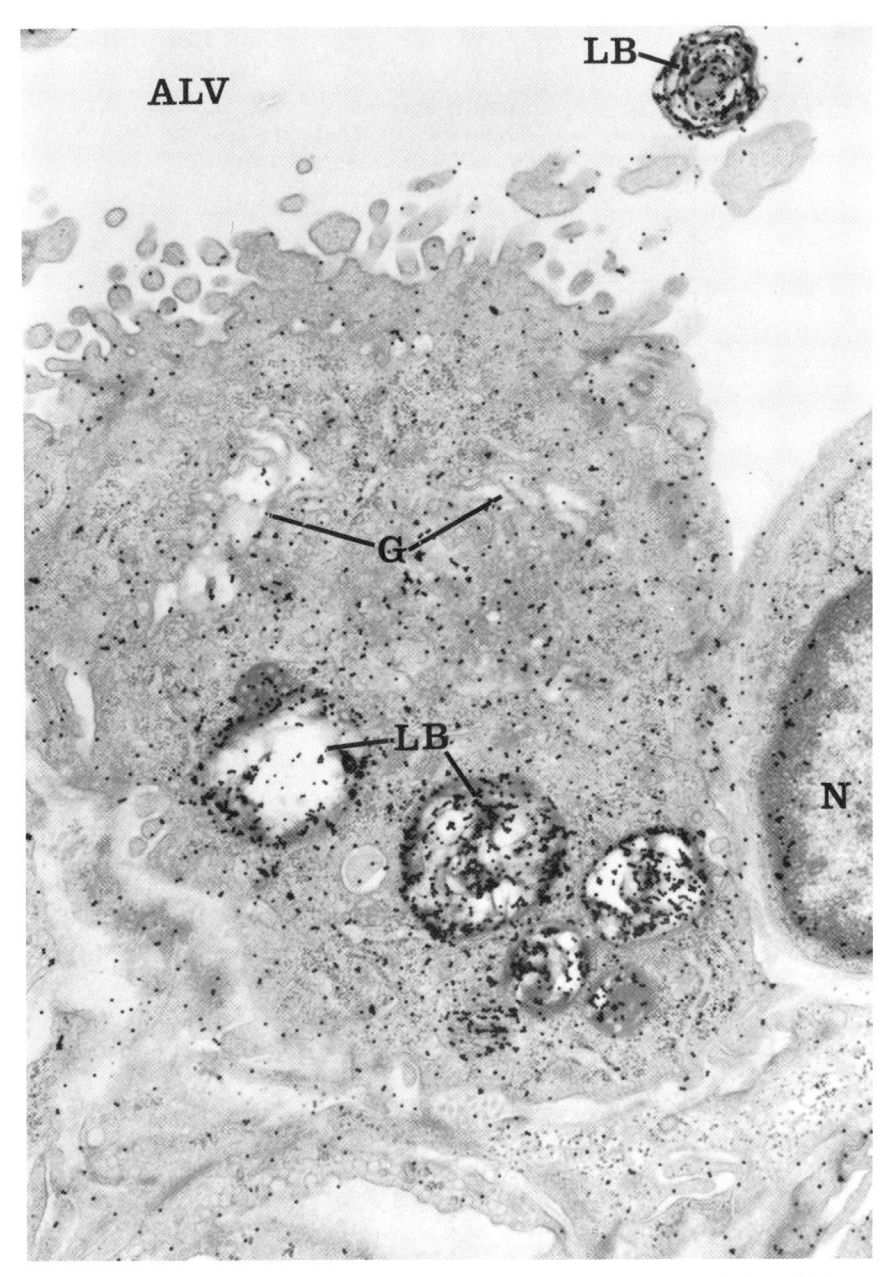

Fig. 6.10. Phospholipase A$_2$–gold labeling. Rat lung tissue fixed with 1% glutaraldehyde, postfixed with 1% OsO$_4$, and embedded in Epon. The labeling is particularly intense over the lamellar bodies (LB) of the type II pneumocyte. All the membranes show labeling by gold particles, including those of the Golgi (G), while the alveolar space (ALV) and the extracellular connective tissue are almost devoid of labeling; ×30,000.

TABLE 6.3
Density of Phospholipase A$_2$–Gold Labeling over the Various Membrane Compartments of Thin-Sectioned Lung Type II Pneumocytes[a]

Membrane Compartments	Labeling Density[b]
Nucleus	
Inner membrane	4.95 ± 0.21
Outer membrane	3.19 ± 0.20
Plasma membrane	
Basolateral	2.58 ± 0.14
Apical	1.65 ± 0.14
Mitochondria	
Outer membrane	3.27 ± 0.10
Cristae	2.42 ± 0.08
Rough endoplasmic reticulum	3.60 ± 0.13
Golgi apparatus	2.80 ± 0.15
Lamellar bodies	12.22 ± 0.92
Collagen fibers[c]	0.42 ± 0.05

[a] From Coulombe et al. (1988).
[b] Values are expressed as number of gold particles per micrometer length of membrane; mean values ± SEM.
[c] The collagen fibers provide an estimation of the background.

Conditions for obtaining optimal results were systematically investigated, testing various fixatives and embedding conditions. High levels of phospholipid labeling were obtained when the tissue was fixed with glutaraldehyde and postfixed with OsO$_4$ prior to embedding in epoxy resins. The results showed specific, high-resolution labeling of all cellular membranes but to different degrees, depending on the cellular or subcellular compartment studied. This was confirmed by quantitative evaluations of density of labeling (Table 6.3) (Coulombe et al., 1988). Furthermore, quantitative evaluations revealed changes in phospholipid distribution among cellular membranes in experimental conditions (Coulombe and Bendayan, 1989). These changes confirm biochemical data showing general increases of total phospholipids and demonstrate the reliability of the approach in addition to its specificity and resolution.

CONCLUSIONS

This review of the enzyme–gold approach indicates its applications to a large variety of substrate molecules at the light and electron microscopic levels. The approach is appropriate for the study of thin sections and fractured-labeled replicas as well as for scanning electron microscopy by the backscattered mode. These applications demonstrate the versatility of the approach. The colloidal gold marker provides resolution which corresponds to the highest achieved by electron microscopy today. The approach is based on the affinity existing between an enzyme and its corresponding substrate, the interaction of which confers the level of specificity achieved by the approach. The theoretical principle of the technique corresponds to the binding of the enzyme to particular intramolecular linkages in its corresponding substrate molecule. The main concern is retention of the substrate molecules in the tissues through the various preparative operations. This is the reason for the determination of optimal conditions of labeling for each system studied. The approach corresponds to a one-step cytochemical technique with direct accessibility to the substrate molecules. Due to its theoretical principle, the technique has been shown to be complementary to other methods. Indeed, comparative evaluations of the enzyme–gold and lectin–gold approaches showed very good correspondence. The variations observed are probably due to differences in the principles governing each approach (Londoño and Bendayan, 1987). The enzymes recognize linkages between residues, while the lectins are directed toward the residues themselves when present in a particular position in the molecule. Similarly, it was also demonstrated that the enzyme–gold approach is complementary to autoradiography (Bendayan and Puvion, 1983).

The author wishes to express appreciation to his two close collaborators who have contributed to the work on the enzyme–gold technique, Doctors Irene Londoño and Pierre Coulombe. He is also grateful to Diane Gingras, Cécile Venne, and Jean Léveillé for their excellent technical assistance. The Medical Research Council is credited for continuous support. The author was previously a Scholar and is now a Scientist of the Medical Research Council of Canada.

REFERENCES

Ackerman, G. A., and Wocken, K. W. (1981). Histochemical evidence for the differential surface labeling, uptake and intracellular transport of a colloidal gold-labeled insulin complex by normal human blood cells. *J. Histochem. Cytochem.* **29,** 1137.
Bendayan, M. (1981). Ultrastructural localization of nucleic acids by the use of enzyme–gold complexes. *J. Histochem. Cytochem.* **29,** 531.
Bendayan, M. (1982). Ultrastructural localization of nucleic acids by the use of enzyme–gold complexes: Influence of fixation and embedding. *Biol. Cell.* **43,** 153.
Bendayan, M., and Puvion, E. (1983). Ultrastructural detection of RNA: Complementarity of high-resolution autoradiography and of RNase–gold method. *J. Ultrastruct. Res.* **83,** 274.

Bendayan, M., and Zollinger, M. (1983). Ultrastructural localization of antigenic sites on osmium-fixed tissues applying the protein A–gold technique. *J. Histochem. Cytochem.* **31**, 101.

Bendayan, M. (1984a). Protein A–gold electron microscopic immunocytochemistry: Methods, applications and limitations. *J. Electron Microsc. Tech.* **1**, 243.

Bendayan, M. (1984b). Enzyme–gold electron microscopic cytochemistry: A new affinity approach for the ultrastructural localization of macromolecules. *J. Electron Microsc. Tech.* **1**, 349.

Bendayan, M., and Puvion, E. (1984). Ultrastructural localization of nucleic acids through several cytochemical techniques on osmium-fixed tissues: Comparative evaluation of the different labeling. *J. Histochem. Cytochem.* **32**, 1185.

Bendayan, M. (1987). Introduction of protein G–gold complex for high resolution immunocytochemistry. *J. Electron Microsc. Tech.* **6**, 7.

Bendayan, M., and Benhamou, N. (1987). Ultrastructural localization of glucoside residues on tissue sections by applying the enzyme–gold approach. *J. Histochem. Cytochem.* **35**, 1149.

Bendayan, M., Nanci, A., and Kan, F. W. K. (1987). Effect of tissue processing on colloidal gold cytochemistry. *J. Histochem. Cytochem.* **35**, 983.

Bendayan, M. (1989). Protein A–gold and protein G–gold postembedding immunoelectron microscopy. In *Colloidal Gold: Principles, Methods, and Applications, Vol. 1* (M. A. Hayat, ed.), pp. 33–94. Academic Press, San Diego, California.

Benhamou, N., and Ouellette, G. B. (1986). Use of pectinases complexed to colloidal gold for the ultrastructural localization of polygalacturonic acids in the cell walls of the fungus *Ascocalyx abietina*. *Histochem. J.* **18**, 95.

Benhamou, N., and Ouellette, G. B. (1987). Ultrastructural study and cytochemical investigation, by means of enzyme–gold complexes, of the fungus *Ascocalyx abietina*. *Can. J. Bot.* **65**, 168.

Benhamou, N., Chamberland, H., Ouellette, G. B., and Pauzé, F. J. (1987). Ultrastructural localization of β-(1-4)-D-glucans in two pathogenic fungi and in their host tissues by means of an exoglucanase–gold complex. *Can. J. Microbiol.* **33**(5), 405.

Benhamou, N. (1989). Preparation and application of lectin–gold complexes. In *Colloidal Gold: Principles, Methods, and Applications, Vol. 1* (M. A. Hayat ed.), pp. 95–143. Academic Press, San Diego, California.

Berg, R. H., Erdos, G. W., Gritzall, M., and Brown, R. D., Jr. (1988). Enzyme–gold affinity labelling of cellulose. *J. Electron Microsc. Tech.* **8**, 371.

Chamberland, H., Charest, P. M., Ouellette, G. B., and Pauzé, F. J. (1985). Chitinase–gold complex used to localize chitin ultrastructurally in tomato root cells infected by *Fusarium oxysporum* f.sp. *radicis-lycopersici*, compared with a chitin specific gold-conjugated lectin. *Histochem. J.* **17**, 313.

Coons, A. H., Creech, H. J., and Jones, R. N. (1941). Immunological properties of an antibody containing a fluorescent group. *Proc. Soc. Exp. Biol. Med.* **47**, 200.

Coulombe, P. A., Kan, F. W. K., and Bendayan, M. (1988). Introduction of a high-resolution cytochemical method for studying the distribution of phospholipids in biological tissues. *Eur. J. Cell Biol.* **46**, 564.

Coulombe, P. A., and Bendayan, M. (1989). Cytochemical demonstration of an increased phospholipid content in cell membranes in chlorphentermine-induced phospholipidosis. *J. Histochem. Cytochem.* **37**, 139.

Danscher, G., and Norgaard, J. (1983). Light microscopic visualization of colloidal gold on resin-embedded tissue. *J. Histochem. Cytochem.* **31**, 1934.

De Mey, J. (1983). Colloidal gold probes in immunocytochemistry. In *Immunocytochemistry: Practical Applications in Pathology and Biology* (J. M. Polak and S. Van Noorden, eds.), pp. 821–112. Wright, Bristol.

Faber, C. N., and Glew, R. H. (1983). α-D-Mannosidase. In *Methoden der Enzymatischen Analyse* (H. U. Bergmeyer, ed.), Vol. 4, pp. 230–240. Verlag Chemie, Weinhein, West Germany.
Faulk, W. P., and Taylor, G. M. (1971). An immunocolloid method for the electron microscope. *Immunochemistry* **8**, 1081.
Geoghegan, W. D., and Ackerman, G. A. (1977). Adsorption of horse-radish peroxidase, ovomucoid and anti-immunoglobulin to colloidal gold for the indirect detection of concanavalin A, wheat germ agglutinin, and goat antihuman immunoglobulin G on cell surfaces at the electron microscopic level: A new method, theory and application. *J. Histochem. Cytochem.* **25**, 1187.
Handley, D. A. (1989). Methods for synthesis of colloidal gold. In *Colloidal Gold: Principles, Methods, and Applications, Vol. 1* (M. A. Hayat, ed.), pp. 13–32. Academic Press, San Diego, California.
Hayat, M. A. (1986). *Basic Techniques for Transmission Electron Microscopy*. Academic Press, Orlando, Florida.
Hayat, M. A. (1989). *Principles and Techniques of Electron Microscopy: Biological Applications*, 3rd ed. Macmillan, London, and CRC Press, Boca Raton, Florida.
Horisberger, M., Rosset, J., and Bauer, H. (1975). Colloidal gold granules as markers for cell surface receptors in the scanning electron microscope. *Experientia* **31**, 1147.
Londoño, I., and Bendayan, M. (1987). Ultrastructural localization of mannoside residues on tissue sections: Comparative evaluation of the enzyme–gold and the lectin–gold approaches. *Eur. J. Cell Biol.* **45**, 88.
Londoño, I., and Bendayan, M. (1988). High-resolution cytochemistry of neuraminic and hexuronic acid-containing macromolecules applying the enzyme–gold approach. *J. Histochem. Cytochem.* **36**(8), 1005.
Moyne, G. (1980). Methods in ultrastructural cytochemistry of the cell nucleus. *Prog. Histochem. Cytochem.* **13**, 1.
Pearse, A. G. E. (1985). *Histochemistry: Theoretical and Applied*, Vol. 2. Churchill-Livingstone, Edinburgh and London.
Raspail, F. V. (1830). *Essai de chimie microcopique appliquée à la physiologie*. Paris.
Romano, E. L., Stolinski, C., and Hughes-Jones, N. C. (1974). An antiglobulin reagent labelled with colloidal gold for use in electron microscopy. *Immunochemistry* **17**, 521.
Roth, J., Bendayan, M., and Orci, L. (1978). Ultrastructural localization of intracellular antigens by the use of protein A–gold complex. *J. Histochem. Cytochem.* **26**, 1974.
Roth, J. (1983). Application of lectin–gold complexes for electron microscopic localization of glycoconjugates on thin sections. *J. Histochem. Cytochem.* **31**, 987.
Ruel, K., and Joseleau, J. P. (1984). Use of enzyme–gold complexes for the ultrastructural localization of hemicelluloses in the plant cell wall. *Histochemistry* **81**, 573.
Thomopoulos, G. N., Schlultze, B. A., and Spicer, S. S. (1987). Post-embedment staining of complex carbohydrates: Influence of fixation and embedding procedures. *J. Electron Microsc. Tech.* **5**, 17.
Tolson, N. D., Boothroyd, B., and Hopkins, C. R. (1981). Cell surface labelling with colloidal gold particulates: The use of avidin and staphylococcal protein A-coated gold in conjunction with biotin and Fc-bearing ligands. *J. Microsc. (Oxford)* **123**, 215.
Vian, B., Brillonet, J. M., and Satiat-Jeunemaître, B. (1983). Ultrastructural visualization of xylans in cell walls of hardwood by means by xylanase–gold complex. *Biol. Cell.* **49**, 179.
Vian, B. (1986). Ultrastructural localization of carbohydrates. Recent developments in cytochemistry and affinity methods. *NATO ASI Ser., Ser. H* **1**, 49–57.

7

Preparation and Application of Lipoprotein–Gold Complex

DEAN A. HANDLEY
and
CYNTHIA M. ARBEENY

Monoclonal Antibody Department
Sandoz Research Institute
East Hanover, New Jersey
Department of Medicine
Albert Einstein College of Medicine
Bronx, New York

INTRODUCTION
PREPARATION OF LIPOPROTEINS
  Labeling Procedure
VISUALIZATION OF THE LIPOPROTEIN–GOLD
  COMPLEX
APPLICATION OF THE HDL–GOLD COMPLEX
APPLICATION OF THE LDL–GOLD COMPLEX
APPLICATIONS OF VERY LOW DENSITY
  LIPOPROTEIN–GOLD COMPLEX
SUMMARY
REFERENCES

INTRODUCTION

Colloidal gold is extensively used as an ultrastructural electron-dense probe to investigate a wide range of cell structure–function relationships. A variety of different methods of colloidal gold formation are available that afford the investigator the use of particles of different sizes. In addition, the methodology of colloidal gold labeling is well established and applicable to a variety of ligands, such as proteins, lectins, enzymes, antibodies, and hormones (Handley and Chien, 1983, 1987). The labeling process occurs by electrostatic adsorption of the ligand, is complete within seconds, and apparently does not involve chemical changes to the ligand (Horisberger, 1979; Warchol *et al.*, 1982). However, in the case of lipoproteins (both high-density and low-density lipoproteins), the labeling procedure is not as straightforward or as routine as it is for other ligands. The difficulties stem from the fact that lipoproteins are actually biological colloids and the labeling represents colloid–colloid interactions. Second, lipoproteins do not form a monolayer around the colloidal gold particle and, therefore, labeling saturations cannot be determined by electrolyte-induced flocculation tests. Third, lipoprotein–gold complexes are multivalent conjugates, necessitating special considerations in interpreting and evaluating receptor-mediated interactions. Addressed will be the critical aspects of the preparation of lipoproteins necessary to labeling, conditions required for optimal labeling, and methods to verify the ultrastructural integrity of the lipoprotein–gold complex.

PREPARATION OF LIPOPROTEINS

Following isolation by conventional ultracentrifugation techniques, lipoproteins must be extensively dialyzed in 0.05% EDTA (w/v) at pH 7.4. Only freshly isolated lipoproteins should be used, since storage results in lipid peroxidation causing clumping of the particles. Preparations of high-density lipoproteins (HDL) or low-density lipoproteins (LDL) should be visualized by negative staining to ensure that they are monodispersed and free of fused aggregates (Figs. 7.1a and 7.2a). Warming to 37°C for 1 hr with brief tub sonication will help achieve disaggregation, as will dilution with water. Once this has been achieved, the HDL or LDL preparations should be adjusted to pH 5.2–5.5 with dilute acetic acid. This pH adjustment will be stable for only 4–8 hr, necessitating that labeling with colloidal gold be completed in that time frame. At pH 4.5–5.2, LDL and HDL are at their pk_1 and will exhibit a tendency to form a fine suspension, which is unsuitable for labeling. At the ultrastructural level, the LDL sus-

Fig. 7.1. Preparative steps in the labeling of human LDL to colloidal gold. (a) Purified and extensively diluted (~5 μg protein/ml) human LDL at pH 5.5 in 0.05% EDTA in water viewed by negative staining showing that the LDL are monodispersed, spherical, 22.0-nm particles. (b) Labeling reaction with equal volumes of 20-nm colloidal gold (prepared by the method of Frens, 1973) and LDL prepared by standard isolation techniques (~1 mg protein/ml), showing excess of lipoprotein. (c) Labeling reaction of equal volumes of LDL and colloidal gold carried out at an acidic pH (4.5), which produced fused chains of LDL (arrowheads) and resulted in excessive clumping and aggregation of the LDL with the gold. (d) Gold–LDL complex showing nearly total consumption of the LDL as part of the labeling reaction. Marker bar, 0.1 μm.

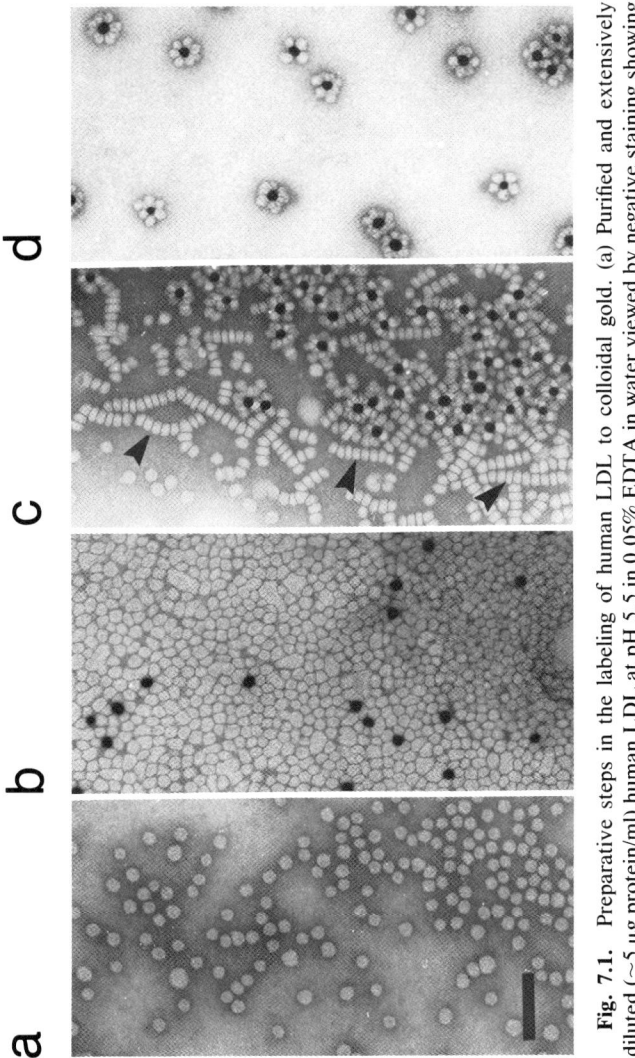

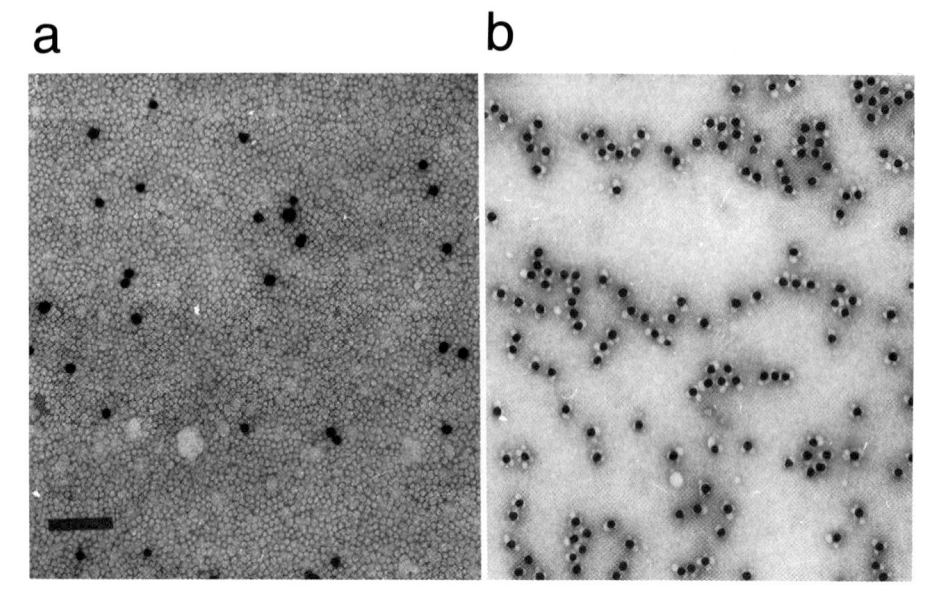

Fig. 7.2. Preparative steps in the labeling of human HDL to colloidal gold. (a) Purified human HDL at pH 5.5 in 0.05% EDTA reacted with an equal volume of 20-nm colloidal gold (prepared according to Frens, 1973), showing the vast excess of HDL. The HDL is seen as monodispersed, 8–10-nm spherical particles. (b) Final reaction mixture of diluted HDL showing nearly complete consumption of the lipoprotein as labeled to the colloidal gold. About three to five HDL bind to each centrally positioned gold particle. Marker bar, 0.1 μm.

pension is seen as fused lipoproteins exhibiting chain formation (Fig. 7.1c). However, this pH-induced suspension of the lipoprotein is fully reversible by elevating the pH to the 5.2–5.5 range.

Labeling Procedure

Prior to labeling, the pH of the colloidal gold (initially at 6.2–6.5 after citrate synthesis; Frens, 1973) must also be adjusted to pH 5.2–5.5 with acetic acid. Similar to the HDL or LDL adjustment, this pH will persist for only 6–8 hr and then a more basic pH (6.0–6.5) will return.

Two methods can be used for labeling: dilution of the lipoprotein to achieve direct labeling saturation of the gold or labeling with excess with lipoproteins followed by centrifugation to remove excess unlabeled ligands (Handley *et al.*, 1981a,b). In the first method, individual reactions are carried out by labeling a fixed volume of colloidal gold with HDL or LDL and then visualizing the reaction mixture by negative staining. With

7. Preparation and Application of Lipoprotein–Gold Complex

commonly used preparations of HDL or LDL, there is usually a vast excess of lipoproteins compared to the number of gold particles when equal volumes of each are reacted (Figs. 7.1b and 7.2a). The excess lipoproteins can be sequentially reduced by dilution of the HDL or LDL preparation with water adjusted to pH 5.2 with acetic acid, and a fixed volume of this diluted preparation is then added to the colloidal gold. Dilution of the lipoprotein should continue until the resulting reaction exhibits nearly complete consumption of the lipoproteins in the labeling process (as assessed by negative staining) (Figs. 7.1d and 7.2b). Negative staining can also disclose inadequacies in pH adjustment, as lipoprotein preparations that are too acidic (and therefore positively changed) will form massive aggregates with the gold (Fig. 7.1c), while basic preparations essentially remain unlabeled. Should either of these conditions exist, the pH of the lipoprotein stock should be readjusted accordingly, but only using small increments (0.1 pH unit). The pH of the colloid solution can be measured using pH paper or with a pH meter having a gel-filled electrode.

Once the optimal dilution of the lipoprotein preparation is achieved so that labeling with a fixed volume of colloidal gold consumes the available LDL or HDL, an equal volume of 1% bovine serum albumin (w/v) should be added. The albumin is necessary to quench (by adsorption) exposed surfaces of the colloidal gold that are not interacting with the lipoprotein. This consideration reflects the fact that lipoproteins are biological colloids and when labeled to colloidal gold do not form a complete monolayer but instead cluster around the gold particle (Handley, 1987). Consequently, the LDL–gold or HDL–gold complex is vulnerable to electrolytes and albumin must be added to "fill in" the unlabeled areas on the gold colloid. Carbohydrate stabilizers (polyethylene glycol, PEG, MW 20,000) have a tendency to induce clumping of the lipoprotein–gold complex and should not be used.

In the second method, the colloidal gold is added to an excess of lipoprotein (with the same pH requirements as above), labeling is allowed to proceed for 10 min with gentle mixing, and the lipoprotein–gold complexes are separated from unlabeled lipoproteins by centrifugation (9,000–15,000 g, 10–20 min) on to a 35% sucrose cushion (designed to prevent centrifugal collapse of the complex) (Handley et al., 1981b). Following centrifugation, the supernatant is removed, and the lipoprotein–gold complex is quenched with albumin and dialyzed against a physiological buffer to remove the sucrose. The amount of labeled lipoprotein can be estimated by determining the number of colloidal gold particles (Ackerman et al., 1983) and multiplying this number by the valency of the number of lipoproteins bound to each colloid (assessed by negative staining).

VISUALIZATION OF THE LIPOPROTEIN–GOLD COMPLEX

We routinely use negative staining of lipoprotein–gold complexes on Butvar-coated grids (Handley and Olsen, 1979). We find films cast by this resin to be of high strength, electron transparent, and easily wettable. As mentioned above, pH inadequacies that would interfere with the labeling reaction can be readily discerned by the use of negative staining. Negative staining is sufficiently sensitive to reveal adequate lipoprotein labeling with a slight excess of lipoprotein (Mommaas-Kienhuis et al., 1985a,b) or grossly inadequate labeling reactions, unfit for cell receptor studies (Chen and Abel, 1985).

In addition to visualization by negative staining, the lipoprotein–gold complex can be observed by thin-section electron microscopy. This can be achieved by the use of lipophilic staining to enhance the ultrastructural density of the lipoprotein (Handley and Chien, 1981; Handley et al., 1981b, 1983a,b) with either oxidized ruthenium red (Handley and Chien, 1981) or tannin (Handley et al., 1983b). These enhancement stains afford direct visualization of the lipoproteins (as complexed to the gold) during receptor binding and endocytosis and following lysosomal delivery (Handley et al., 1981a,b, 1983a,b).

The advantages of such staining procedures are confirmation of the structural integrity of the lipoprotein–gold complex during various phases of cellular processing and direct visualization of lipoproteins that are not gold labeled (Handley et al., 1983b; Arbeeny et al., 1987).

APPLICATION OF THE HDL–GOLD COMPLEX

The HDL–gold complex (Handley, 1987) has been used to study receptor binding and internalization in a variety of cell types, such as porcine granulosa cells (Chen and Abel, 1985), mouse peritoneal macrophages (Robenek and Schmitz, 1985, 1988; Schmitz et al., 1985b), human monocytes (Schmitz et al., 1985a), and fibroblasts (Robenek and Severs, 1984). These studies have demonstrated that the HDL receptor is located in a coated region of the cell membrane, distinct from the LDL receptor, and that internalization involves coated vesicle delivery (Fig. 7.3a) to lysosomal-like structures (Fig. 7.3b). Following internalization, HDL are resecreted from normal human monocytes and mouse peritoneal macrophages (Schmitz et al., 1985a). The cellular interaction with HDL may be modified in certain disease states (Schmitz et al., 1985a) or following exposure to other lipoproteins such as LDL (Schmitz et al., 1985b).

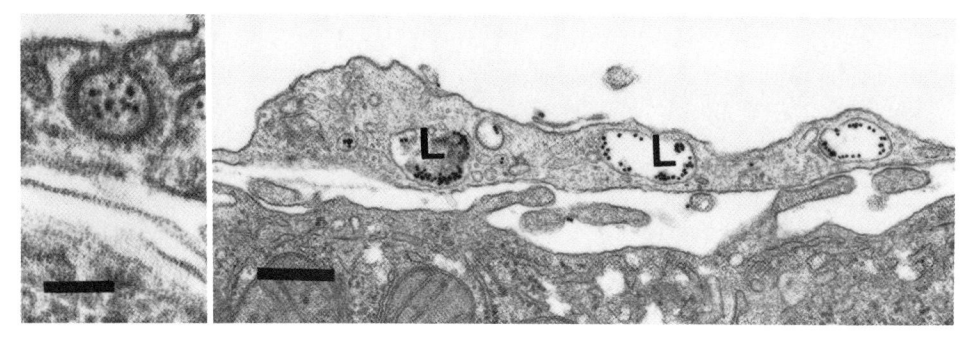

Fig. 7.3. Uptake of HDL–gold complex by sinusoidal endothelial cells in perfused rat livers. Uptake is via coated vesicles (a) that deliver the HDL–gold complex to structures resembling lysosomes (L) (b). Marker bars are 0.1 μm (a) and 0.2 μm (b).

APPLICATION OF THE LDL–GOLD COMPLEX

More papers have been published using the LDL–gold complex as a probe to study receptor-mediated binding than any other single mammalian protein. The LDL–gold complex was developed as an ultrastructural method (Handley et al., 1981b) to permit visualization of the LDL receptor-mediated pathway in fibroblasts (Fig. 7.4a–d) that has been described biochemically (Anderson et al., 1976; Goldstein et al., 1979). Colloidal gold was used instead of ferritin because of its greater electron density and the ease with which complexing of colloidal gold to LDL could be achieved. The LDL–gold complex was also applied to study the LDL pathway in sinusoidal endothelial cells (Handley and Chien, 1983; Handley et al., 1983a), the effect of chloroquine on cellular processing of LDL (Handley et al., 1985), and the ability of ethinyl estradiol to increase the number of LDL receptors in the livers of treated rats (Handley et al., 1981a) (Fig. 7.5a–d). The LDL–gold complex has been used to study receptor interactions in cultured endothelial cells (Grünwald et al., 1985; Havekes et al., 1985; Mommaas-Kienhuis et al., 1985a), capillary endothelium in perfused lung (Nistor and Simionescu, 1986), isolated liver cells (Mommaas-Kienhuis et al., 1985b), squamous carcinoma cells (Vermeer et al., 1985), granulosa cells (Paavola et al., 1985), fibroblasts (Robenek and Hesz, 1983; Robenek et al., 1982, 1983, 1984; Hesz et al., 1985), and macrophages (Mommaas-Kienhuis et al., 1985c). Other studies have evaluated chemical modification such as acetylation (Robenek et al., 1984, 1985; Havekes et al., 1985; Hesz et al., 1985; Mommaas-Kienhuis

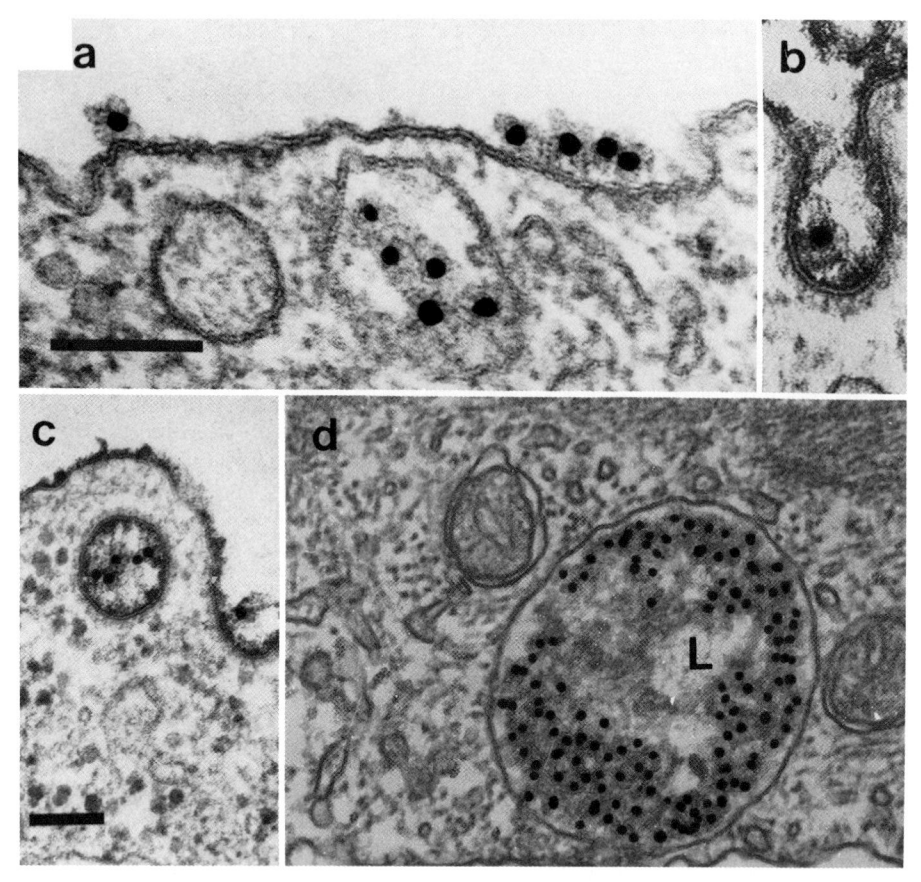

Fig. 7.4. The pathway of LDL–gold complex uptake by cultured human fibroblasts grown in lipoprotein-deficient medium (Handley et al., 1981b). Cell cultures were stained en bloc with tannic acid to permit direct visualization of the LDL (Handley et al., 1983b). Shown are (a) initial surface binding to the fibroblast after incubation at 4°C for 4 hr, (b) coated pit formation after warming to 37°C for 5 min, (c) coated vesicle containing the LDL–gold complex, and (d) lysosomal (L) accumulation after incubation for 1 hr at 37°C. Marker bar, 0.2 μm.

et al., 1985b,c; Robenek and Schmitz, 1985) and cationization (Robenek and Severs, 1984). Disruption of cellular internalization by the lysosomotrophic amine chloroquine prevents dissociation of LDL from the receptor and leads to accumulation of the ligand–receptor complex (Handley and Chien, 1983; Handley et al., 1985).

APPLICATIONS OF VERY LOW DENSITY LIPOPROTEIN–GOLD COMPLEX

Labeling of very low density lipoprotein (VLDL) remnants (50 nm in diameter) or very low density lipoproteins (62 nm in diameter) presents a special problem in that conjugation techniques used for HDL or LDL

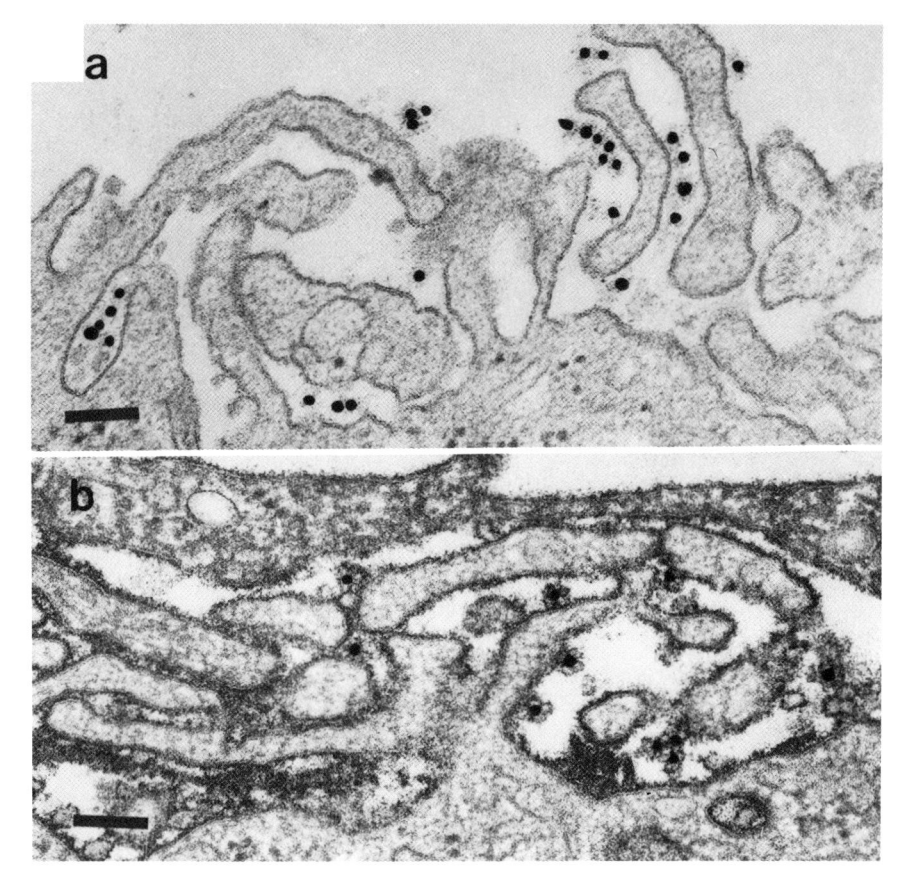

Fig. 7.5. The pathway of LDL–gold complex uptake by the perfused liver from ethinyl estradiol-treated rats (Handley et al., 1981a). The LDL–gold complex was perfused in a recirculating system for 5–60 min and the livers processed by conventional electron microscopic methods (a) or with en bloc staining using oxidized ruthenium red (Handley and Chien, 1981) (b). The LDL–gold complex is seen as randomly distributed along the hepatocyte microvillus (a and b), but only with the ruthenium red staining can the LDL be directly visualized (b). After 30 min of perfusion, uptake is to multivesicular bodies (MVB) (c), which at 60 min resemble mature lysosomes (L) (d). Marker bar, 0.2 μm. (Figure continued next page.)

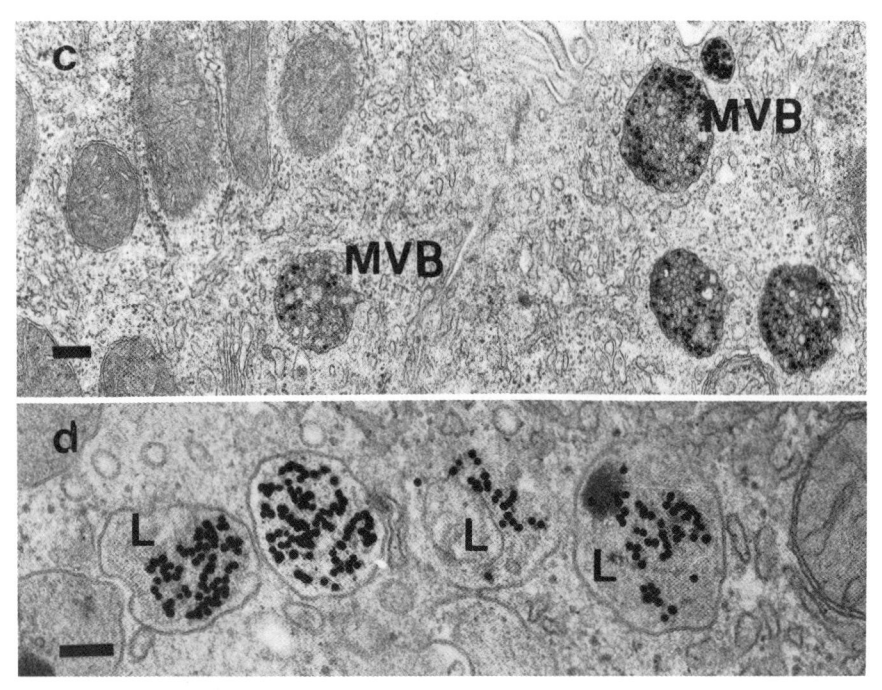

Fig. 5.7. (*Continued*)

produce exceptionally large complexes. To overcome this, we developed a monomeric or dimeric surface tagging technique (Fig. 7.6a,b). In this method, one to three small gold particles (5–10 nm in diameter) are adsorbed to each lipoprotein, followed by albumin to quench the labeling reaction (Handley and Chien, 1983). This surface-labeling technique has recently been applied to study β-VLDL internalization in macrophages (Robenek *et al.*, 1987) and LDL uptake with pulmonary capillary endothelium (Nistor and Simionescu, 1986). Although surface tagging will modify morphology, such conjugates appear to exhibit the same functional processing as unlabeled β-VLDL (Robenek *et al.*, 1987).

SUMMARY

The gold labeling methodology has been successfully extended to the major classes of lipoproteins (HDL, LDL, VLDL, and remnants). Albeit certain special considerations in the labeling procedure, conjugation (clustering of lipoproteins around the gold), or surface tagging (adsorption

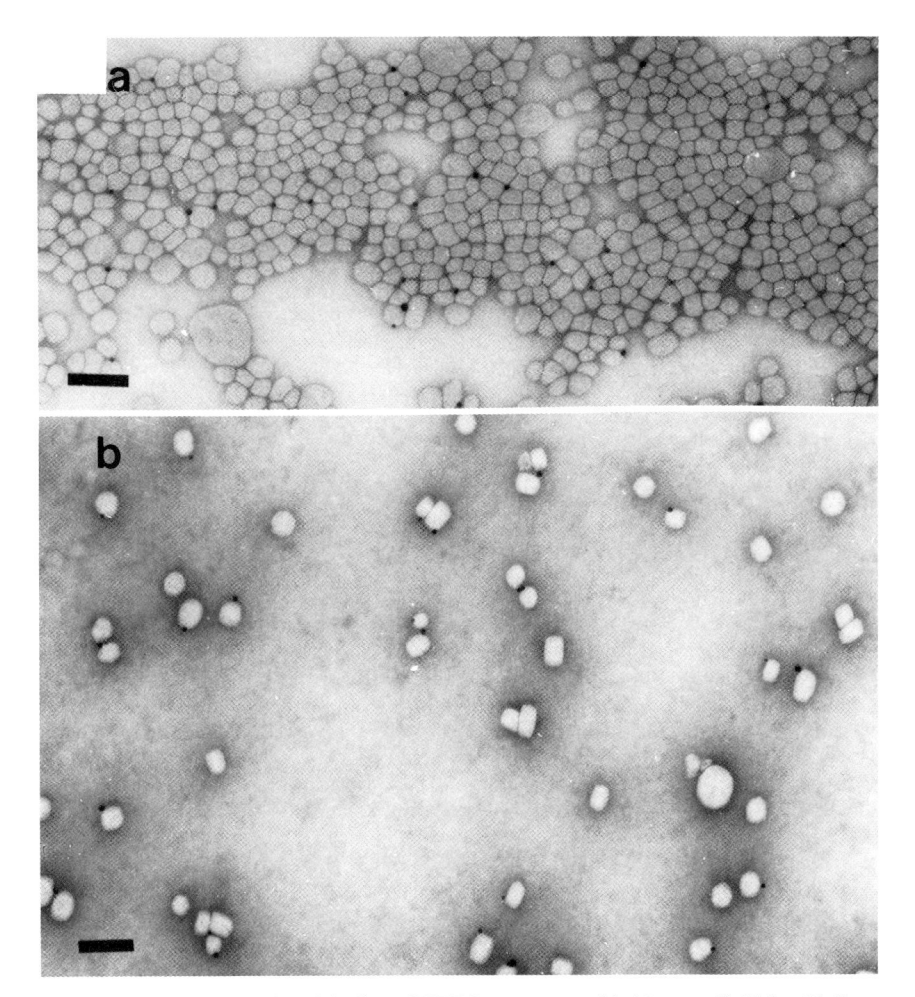

Fig. 7.6. Dimeric surface labeling of VLDL remnants with 10-nm colloidal gold. Reaction of equal volumes of VLDL remnants and colloidal gold shows a substantial excess of the VLDL remnants (a). By successive dilution of the remnant preparation with water, a labeling reaction mixture can be achieved where 75% of the remnants are labeled with one gold particle (b). The reaction is then quenched with 1% bovine serum albumin (BSA). Marker bars, 0.1 μm.

of several gold particles to the lipoprotein surface) are simple to use and readily verifiable by electron microscopy. The resulting lipoprotein–gold complexes retain the qualitative properties of the native lipoproteins and serve as a valuable ultrastructural probe for examining receptor-mediated processing of metabolically important ligands.

I am indebted to Carol Patterson for preparation of this chapter.

REFERENCES

Ackerman, G. A., Yang, J., and Wolken, K. W. (1983). Differential surface labeling and internalization of glucagon by peripheral leukocytes. *J. Histochem. Cytochem.* **31**, 433.

Anderson, R. G. W., Goldstein, J. L., and Brown, M. S. (1976). Localization of low density lipoprotein receptors on plasma membranes of normal fibroblasts and their absence in cells from a familial hypercholesterolemia homozygote. *Proc. Natl. Acad. Sci. U.S.A.* **73**, 2434.

Arbeeny, C. M., Rifici, V. A., Handley, D. A., and Eder, H. A. (1987). Determinants of uptake of very low density lipoprotein remnants by the perfused rat liver. *Metab. Clin Exp.* **36**, 1106.

Chen, T. T., and Abel, J. H. (1985). Receptor-mediated uptake of lipoproteins by cultured porcine granulosa cells. *Eur. J. Cell Biol.* **39**, 410.

Frens, G. (1973). Controlled nucleation for the regulation of the particle size in monodisperse gold solutions. *Nature (London), Phys. Sci.* **241**, 20.

Goldstein, J. L., Anderson, R. G. W., and Brown, M. S. (1979). Coated-pits, coated vesicles, and receptor-mediated endocytosis. *Nature (London)* **279**, 679.

Grünwald, J., Hesz, A., Robenek, H., Brücker, J., and Buddecke, E. (1985). Proliferation, morphology, and low-density lipoprotein metabolism of arterial endothelial cells cultured from normal and diabetic minipigs. *Exp. Mol. Pathol.* **42**, 60.

Handley, D. A., and Olsen, B. K. (1979). Butvar B-98 as a thin support film. *Ultramicroscopy* **4**, 479.

Handley, D. A., and Chien, S. (1981). Oxidation of ruthenium red for use as an intercellular tracer. *Histochemistry* **71**, 249.

Handley, D. A., Arbeeny, C. M., Eder, H. A., and Chien, S. (1981a). Hepatic binding and internalization of low density lipoprotein–gold conjugates in rats treated with 17 α-ethinyl estradiol. *J. Cell Biol.* **90**, 778.

Handley, D. A., Arbeeny, C. M., Witte, L. D., and Chien, S. (1981b). Colloidal gold–low density lipoprotein conjugates as membrane receptor probes. *Proc. Natl. Acad. Sci. U.S.A.* **78**, 368.

Handley, D. A., and Chien, S. (1983). Colloidal gold: A pluripotent receptor probe. *Proc. Soc. Exp. Biol. Med.* **174**, 1.

Handley, D. A., Arbeeny, C. M., and Chien, S. (1983a). Sinusoidal endothelial endocytosis of low density lipoprotein–gold conjugates in perfused livers of ethinyl-estradiol treated rats. *Eur. J. Cell Biol.* **30**, 266.

Handley, D. A., Arbeeny, C. M., Witte, L. D., Goodman, D. S., and Chien, S. (1983b). Ultrastructural visualization of low-density lipoproteins during receptor binding and cellular endocytosis. *J. Ultrastruct. Res.* **83**, 43.

Handley, D. A., Arbeeny, C. M., and Witte, L. D. (1985). Intralysosomal accumulation of colloidal gold–low density lipoprotein conjugates in chloroquine-treated fibroblasts. *Proc.—43rd Annu. Meet., Electron Microsc. Soc. Am.* pp. 546–547.

Handley, D. A. (1987). Receptor-mediated binding, endocytosis and cellular processing of macromolecules conjugated with colloidal gold. *Scanning Microsc.* **1**, 339.

7. Preparation and Application of Lipoprotein–Gold Complex 161

Handley, D. A., and Chien, S. (1987). Colloidal gold labeling studies related to vascular and endothelial function, hemostasis and receptor-mediated processing of plasma macromolecules. *Eur. J. Cell Biol.* **43**, 163.

Havekes, L., Mommaas-Kienhuis, A. M., Schouten, D., De Wit, E., Scheffer, M., and Van Hinsbergh, V. W. M. (1985). High-affinity uptake and degradation of acetylated low density lipoprotein by confluent human vascular endothelial cells. *Atherosclerosis* **56**, 81.

Hesz, A., Robenek, H., Ingolic, E., Roscher, A., Krempler, F., Sandhofer, F., and Kostner, G. M. (1985). Demonstration of receptor binding of two apo-B containing lipoproteins by differential labelling with colloidal gold. *Eur. J. Cell Biol.* **37**, 229.

Horisberger, M. (1979). Evaluation of colloidal gold as a cytochemical marker for transmission and scanning electron microscopy. *Biol. Cell.* **36**, 253.

Mommaas-Kienhuis, A. M., Krijbolder, L. H., Van Hinsbergh, V. W. M., Daems, W. T., and Vermeer, B. J. (1985a). Visualization of binding and receptor-mediated uptake of low density lipoproteins by human endothelial cells. *Eur. J. Cell Biol.* **36**, 201.

Mommaas-Kienhuis, A. M., Nagelkerke, J. F., Vermeer, B. J., Daems, W. T., and van Berkel, T. J. C. (1985b). Visualization of the interaction of native and modified low density lipoproteins with isolated rat liver cells. *Eur. J. Cell Biol.* **38**, 42.

Mommaas-Kienhuis, A. M., Van der Schroeff, J. G., Wijsman, M. C., Daems, W. T., and Vermeer, B. J. (1985c). Conjugates of colloidal gold with native and acetylated low density lipoproteins for ultrastructural investigations on receptor-mediated endocytosis by cultured human monocyte-derived macrophages. *Histochemistry* **83**, 29.

Nistor, A., and Simionescu, N. (1986). Uptake of low density lipoproteins by hamster lung. *Am. Rev. Respir. Dis.* **134**, 1266.

Paavola, L. G., Strauss, J. F., III, Boyd, C. O., and Nestler, J. E. (1985). Uptake of gold- and [^{3}H]cholesteryl linoleate-labeled human low density lipoprotein by cultured rat granulosa cells: Cellular mechanisms involved in lipoprotein metabolism and their importance to steroidogenesis. *J. Cell Biol.* **100**, 1235.

Robenek, H., Rassat, J., Hesz, A., and Grünwald, J. (1982). A correlative study on the topographical distribution of the receptors for low density lipoprotein (LDL) conjugated to colloidal gold in cultured human skin fibroblasts employing thin section, freeze-fracture, deep-etching, and surface replication techniques. *Eur. J. Cell Biol.* **27**, 242.

Robenek, H., and Hesz, A. (1983). Dynamics of low-density lipoprotein receptors in the plasma membrane of cultured human skin fibrolasts as visualized by colloidal gold in conjunction with surface replicas. *Eur. J. Cell Biol.* **31**, 275.

Robenek, H., Hesz, A., and Rassat, J. (1983). Variability of the topography of low-density lipoprotein (LDL) receptors in the plasma membrane of cultured human skin fibroblasts as revealed by gold–LDL conjugates in conjunction with the surface replication technique. *J. Ultrastruct. Res.* **82**, 143.

Robenek, H., and Severs, N. J. (1984). Double labeling of lipoprotein receptors in fibroblast cell surface replicas. *J. Ultrastruct. Res.* **87**, 149.

Robenek, H., Schmitz, G., and Assmann, G. (1984). Topography and dynamics of receptors for acetylated and malondialdehyde-modified low density lipoprotein in the plasma membrane of mouse peritoneal macrophages as visualized by colloidal gold in conjunction with surface replicas. *J. Histochem. Cytochem.* **32**, 1017.

Robenek, H., and Schmitz, G. (1985). Receptor domains in the plasma membrane of cultural mouse peritoneal macrophages. *Eur. J. Cell Biol.* **39**, 77.

Robenek, H., Schmitz, G., and Greven, H. (1987). Cell surface distribution and intracellular fate of human β-very low density lipoprotein in cultured peritoneal macrophages: A cytochemical and immunocytochemical study. *Eur. J. Cell Biol.* **43**, 110.

Robenek, H., and Schmitz, G. (1988). Ca^{++} antagonists and ACAT inhibitors promote cholesterol efflux from macrophages by different mechanisms. *Arteriosclerosis* **8,** 57.

Schmitz, G., Assmann, G., Robenek, H., and Brennhausen, B. (1985a). Tangier disease: A disorder of intracellular membrane traffic. *Proc. Natl. Acad. Sci. U.S.A.* **82,** 6305.

Schmitz, G., Robenek, H., Lohmann, U., and Assmann, G. (1985b). Interaction of high density lipoproteins with cholesteryl ester-laden macrophages: Biochemical and morphological characterization of cell surface receptor binding, endocytosis and resecretion of high density lipoproteins by macrophages. *EMBO J.* **4,** 613.

Vermeer, B. J., Wijsman, M. C., Mommaas-Kienhuis, A. M., Ponec, M., and Havekes, L. (1985). Modulation of low density lipoprotein receptor activity in squamous carcinoma cells by variation in cell density. *Eur. J. Cell Biol.* **38,** 353.

Warchol, J. B., Brelinska R., and Herbert, D. C. (1982). Analysis of colloidal gold methods for labeling proteins. *Histochemistry* **76,** 567.

8

Preparation and Application of Albumin–Gold Complex

SERGIO VILLASCHI

Department of Surgery
Second University of Rome "Tor Vergata"
Rome, Italy

INTRODUCTION
PREPARATION OF ALBUMIN–GOLD COMPLEX
  Preparation and Characterization of Gold Markers
  Determination of Stabilizing Albumin Concentration and Preparation of Gold Probes
  Storage of Albumin–Gold Complex
  Working Protocol
APPLICATION OF ALBUMIN–GOLD COMPLEX
REFERENCES

INTRODUCTION

Albumin is the most abundant of the serum proteins and has multiple metabolic functions. It is the plasma protein that exerts the greatest oncotic pressure and thus plays an important role in the regulation of the permeability of the capillary blood vessels (Drinker, 1927; Danielli, 1940). Studies suggest that this function is mediated by interaction of albumin with the endothelial cell coat, but the physicochemical bases of this interaction are still poorly understood (Mason *et al.*, 1977; Curry and Michel, 1980; Schneeberger and Hamelin, 1984).

Albumin also acts as a carrier for many endogenous as well as exogenous molecules, including bilirubin, fatty acids, hormones, toxins, carcinogens, and various drugs (Muller and Wallert, 1979). In this role albumin may mediate the interaction between cells and albumin-bound substances (Weisiger *et al.*, 1981, 1982; Ockner *et al.*, 1983; Hütter *et al.*, 1984). Thus, interactions between albumin and various cells appear to play an important role in the physiopathology of the entire organism.

Unfortunately, albumin is electron lucent and cannot be detected with the electron microscope. Indirect visualization by immunoperoxidase techniques is mainly qualitative and lacks the desired precision owing to artifactual diffusion of the reaction product and the amplifying effect of the cytochemical reaction (Courtoy *et al.*, 1983).

Colloidal gold is a marker which meets many requirements necessary for the precise ultrastructural localization, distribution, and quantitation of macromolecules in living cells and in fixed cells and tissues. It is a noncytotoxic marker which adsorbs many proteins without affecting their bioactivity (Horisberger *et al.*, 1975; Geoghegan and Ackerman, 1977; Horisberger and Rosset, 1977; Goodman *et al.*, 1981). Furthermore, gold markers may be prepared in a wide range of sizes, including particles of relatively small size (3–5 nm). This property, coupled with the electron density of gold markers, allows sharp and easy localization of the protein–gold complex.

PREPARATION OF ALBUMIN–GOLD COMPLEX

Albumin can be bound to colloidal gold particles (gold markers) of different sizes to give albumin–gold complexes (gold probes). While larger particles, ~50 nm, are suitable for scanning electron microscopy, small particles are more conveniently used in the transmission electron microscope, where finer resolution can be achieved.

Preparation and Characterization of Gold Markers

The techniques used in the various preparations of gold markers are fully described in elsewhere in this book. Here we will briefly summarize the method of Mühlpfordt (1982), which in our hands has given consistently good results with gold particles of average diameter ~6 ± 1.5 nm and use of the technique of Frens (1973) to obtain 15-nm particles.

8. Preparation and Application of Albumin–Gold Complex

In order to achieve good results, the following precautions should be taken: (1) All solutions must be prepared in double glass-distilled water (DDW). (2) The glassware must be absolutely clean; this can be achieved by washing in soap and hot water and then, after careful rinsing in tap water, further thorough rinsing in DDW.

A stock solution of 1% (w/v) chloroauric acid (hydrogen tetrachloroaurate trihydrate, $HAuCl_4 \cdot 3H_2O$; Sigma, St. Louis, MO) is prepared in DDW. This solution is very stable and can be kept for months at 4°C (227 K) in a clean, dark, well-closed vial. A 100-ml solution of 0.01% (w/v) chloroauric acid is prepared by dilution of the stock solution with DDW. This diluted solution is brought to a boil under reflux in a 500-ml Erlenmeyer flask on a hot plate equipped with a magnetic stirrer. The flask should be siliconized to prevent the precipitation of metallic gold on its walls or on the surface of the solution. When the solution boils, a reducing solution is rapidly added with vigorous stirring. To obtain 6-nm particles the reducing solution is prepared by mixing, just before use, 2 ml of 1% (w/v) Na-citrate dihydrate in DDW with 0.45 ml of 1% freshly prepared tannic acid in DDW. In a few seconds the solution turns a wine red color and after boiling under reflux for 5–10 min the reaction is complete.

To prepare 15-nm particles add to the boiling chloroauric solution 2 ml of 1% Na-citrate dihydrate solution and boil gently under reflux for 15 min. After boiling, the gold sol is cooled on ice, dialyzed overnight at 4°C (277 K) against DDW, and filtered through a 0.22-μm Millipore filter (Millipore, Bedford, MA). The dialysis water must be precooled before starting dialysis at 1 : 60 (v/v) with one change after ~1 hr. This step eliminates any residual salts in the gold solution. The colloidal gold, which can be prepared in larger amounts, may be stored at 4°C (277 K) in sterile polypropylene bottles for at least 6 months if filtered through a 0.22-μm Millipore filter and maintained under sterile conditions.

The size and shape of the gold particles are characterized at the electron microscope. Formvar-coated grids are laid on a small drop of the colloidal gold for 15 min. After drying with filter paper, the grids are examined under the electron microscope. The gold particles, which must be monodispersed, are photographed and the average diameter is calculated.

Determination of Stabilizing Albumin Concentration and Preparation of Gold Probes

Colloidal gold is a sol formed by negatively charged particles which is unstable in salt solutions (Weiser, 1949; Geoghegan and Ackerman, 1977). Flocculation induced by salts results in a change in color of the gold suspension from red to violet-blue. This color change is very evident and can be detected visually. Flocculation can be also estimated spectrophotometrically by determining the optical density at A_{max} 520 nm (Horisberger and Rosset, 1977) or at 580 nm (Geoghegan and Ackerman, 1977) (Fig. 1).

Gold markers have the characteristic of binding proteins noncovalently without changing their bioactivity. This adsorption, which leads to the formation of a "protein–gold complex," stabilizes the gold sol, allowing the use of gold probes at physiologic salt concentrations. Adsorption of albumin on gold particles is influenced by the pH and ionic concentration of the solution as well as the amount of added protein. Maximal binding and stability of the protein–gold complex is obtained at pH values close to the pI of the protein or slightly higher (Geoghegan and Ackerman, 1977). Since the pI of albumin ranges between 4.7 and 5.3 (Peters, 1975), the optimal pH of the solution must be between 5.3 and 6.0.

The gold solution is brought to the optimal pH by addition of 1 part 70 mM phosphate buffer (pH 6.2) to 9 parts dialyzed gold solution. The resulting 7 mM working solution brings the pH to its optimal value (~5.4 as determined in our laboratory with pH paper) without impairing the binding of albumin to the gold particles (Goodman *et al.*, 1981). It is essential to buffer the gold just before use or some flocculation will occur.

To find the minimal amount of albumin necessary to stabilize a given batch of gold sol, a salt-free solution of albumin must be used. To ensure this, a 0.1% albumin solution is prepared in DDW with gentle mixing to prevent the formation of foam. Ten milliliters of 0.1% albumin are sufficient for 100 ml of gold. The albumin solution is then filtered through a 0.22-μm Millipore filter and poured into a washed, boiled, and cooled dialysis tube, which must be tightly closed without air spaces to prevent dilution of the albumin solution, and dialyzed for 2–3 hr at 4°C (277 K) against four changes of precooled DDW (1 : 400 v/v). After dialysis, the solution is kept in a sterile siliconized Vacutainer (Becton Dickinson, Rutherford, NJ) or equivalent sterile glass vial.

The optimal (minimal) amount of albumin required to stabilize the gold colloid is determined in test tubes by adding 1 ml of buffered gold solution to 100 μl of serial dilutions of albumin. After 5 min, 100 μl of 10% NaCl solution is added to the test tubes to flocculate the unstable particles. After 1 hr the tubes are observed against white filter paper to check the differences in color: gold solutions stabilized by a sufficient amount of albumin maintain a red color, while the flocculated ones turn initially violet-blue and finally pale blue. Flocculation also changes the absorbance curve, inducing a decrease of the peak at A_{max} 510–550 and an increase of absorbance at 580 nm (Fig. 8.1). The first tube in which the color does not change from red to violet-blue or the A_{max} does not decrease contains

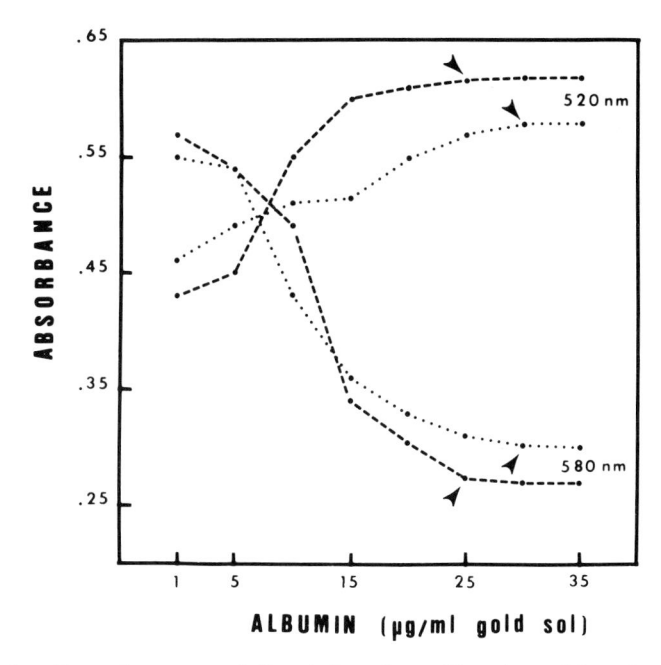

8. Preparation and Application of Albumin–Gold Complex

Fig. 8.1. Absorption curves of albumin bound to gold markers of 6 nm (· · ·) and 15 nm (---) at A_{max} 520 and 580 nm. An increase in absorbance at A_{max} indicates an increase in the stabilization of colloidal gold by albumin which corresponds to a decrease of absorbance at 580. The arrows indicate the minimal amount of albumin necessary to stabilize 1 ml of gold solution.

the minimal amount of albumin necessary to stabilize 1 ml of that gold solution. The minimal amount of stabilizing albumin must be found for each new batch of gold since it may vary in relation to the concentration of the gold as the average diameter and shape of the particles.

Once the minimum amount of albumin is calculated, a larger volume of gold marker can be stabilized. The amount of albumin required will be the minimum already determined plus 20% excess to ensure that all possible protein is adsorbed on the gold particles. The albumin solution is filtered through a 0.22-μm Millipore filter and placed in a beaker on a magnetic stirrer and the buffered gold sol is added while gently stirring. After 5 min, a 2.5% solution of Carbowax 20-M (FLUKA, Ronkonkoma, NY) or polyethylene glycol (MW 20,000; Serva, Garden City Park, NY) is added through a 0.22-μm filter to reach a final concentration of 0.4 mg/ml. The addition of Carbowax further stabilizes the gold probes (Geoghegan and Ackerman, 1977; Horisberger and Rosset, 1977).

Unbound albumin, which can compete with the gold probe, is removed by centrifugation at 65,000 g for 6-nm particles and 20,000 g for 15-nm particles for 45 min at 4°C (277 K). The supernatant containing the excess unbound albumin is discarded but extreme care should be taken not to disturb the loose red sediment, which contains the stabilized albumin–gold complex. The sediment is resuspended in phosphate-buffered saline (PBS) containing 4 mg/ml Carbowax 20-M and recentrifuged under the same conditions. The black spot which may be visible on one side near the bottom of the tube is metallic gold that is not stabilized and must be discarded. After the second centrifugation the sediment, which represents the concentrated, washed, albumin–gold complex (gold probe), may be used after dilution with PBS or any other physiological saline solution containing 0.4 mg/ml Carbowax 20-M.

It is recommended that the concentration of different batches of gold probe be standardized by measuring the A_{max} of a diluted aliquot of the sediment (10 –20 µl is usually sufficient) and then diluting it according to the experimental design. An optical density between 2.5 and 10 at A_{max} 510–550 is suitable for binding the albumin–gold complex to living cells, but other dilutions can be used for different applications.

The presence of albumin molecules on the gold probe can be confirmed in the electron microscope. A Formvar-coated grid is layered over a drop of diluted (1 : 20 is usually suitable) albumin–gold complex for 30 min. After drawing off the excess probe with filter paper, the grid is negatively stained for 30 min to 2 hr on a drop of 1% phosphotungstic acid brought to pH 7 with KOH. The negatively stained particles will show a white halo around them, which represents the albumin molecules. Bioactivity of albumin–gold probes may be tested by precipitation with antibody raised against the albumin bound to the gold.

Storage of Albumin–Gold Complex

There is evidence that proteins bound to colloidal gold retain, with few exceptions, their bioactivity and stability for a long time (Horisberger *et al.*, 1975; Horisberger and Vauthey, 1984). However, since small chemical modifications of the albumin molecule bound to gold particles may change the interaction between the probes and endothelial cells (see the section on Application of Albumin–Gold Complex) and denaturation of albumin, with consequent chemical modifications of the molecule, cannot be excluded even at 4°C (277 K), we recommend using the probes within 2 or 3 days after preparation. Furthermore, there is the possibility of some protein dissociation from the gold probes (Goodman *et al.*, 1981; Warchol *et al.*, 1982). For this reason the probes should be stored after the first centrifugation and centrifuged the second time immediately before use in order to eliminate the dissociated albumin. In this case the sediment is resuspended in PBS containing 0.4 mg/ml of Carbowax 20-M to ~10% of the initial volume of gold solution. This volume is not critical since the albumin–gold must be recentrifuged and its final concentration determined immediately before use. For storage the probes must be sterilized by filtration through a 0.22-μm Millipore filter and kept at 4°C (277 K) in sterile siliconized vials. We found Vacutainers convenient for this purpose.

Working Protocol

For all steps use only DDW and absolutely clean glassware.

1. Prepare 100 ml of 0.01% (w/v) chloroauric acid ($HAuCl_4 \cdot 3H_2O$; Sigma, St Louis, MO) by diluting a stock solution of 1% (w/v) chloroauric acid in DDW. The stock solution can be stored for months in dark, well-closed vials at 4°C (277 K).

2. Bring the 0.01% solution to a boil in a siliconized 500-ml Erlenmeyer flask.

3. Once the gold solution boils, stir vigorously and add rapidly a freshly prepared reducing solution. For 6-nm particles add a solution obtained by mixing 2 ml of 1% (w/v) Na-citrate dihydrate and 0.45 ml of freshly prepared 1% (w/v) tannic acid and boil under reflux for 5–10 min. For 15-nm particles add 2 ml of 1% Na-citrate dihydrate and boil under reflux for 15 min.

4. Cool the gold sol on ice and dialyze in a washed, boiled, and cooled dialysis tube overnight at 4°C (277 K) against precooled DDW. Store at 4°C (277 K) under sterile conditions after filtration through a 0.22-μm filter in a polypropylene bottle or siliconized glass vial.

5. Prepare a 0.1% (w/v) albumin solution in DDW and dialyze at 4°C (277 K) for 2 hr against four changes of DDW (1 : 400 v/v).

6. Determine the minimum amount of albumin necessary to stabilize the gold sol by titration: 100-μl volumes of DDW containing increasing amounts of albumin are added to 20 clean test tubes. We use a linear dilution of the 0.1% albumin solution with increases of a factor of 5. The first tube will contain 1 μl of the albumin solution + 99 μl of DDW, the second 5 μl of albumin + 95 of DDW, the third 10 μl + 90 μl, the fourth 15 μl + 85 μl, and so on up to the last, which will contain 95 μl of albumin and 5 μl of DDW.

Add 1 ml of 7 mM phosphate-buffered colloidal gold. Buffer directly before use by adding 1 part 70 mM phosphate buffer (pH 6.2) to 9 parts gold.

After 2 min add 100 µl of 10% NaCl solution, stir, and after 1 hr check the stability of the probe by observing the tubes against white filter paper: the minimum amount of stabilizing albumin will be the first tube whose color does not turn from red to violet-blue. This may be confirmed spectrophotometrically by reading at A_{max} 510–550.

7. Now a larger amount of albumin–gold probe may be prepared. Buffer the gold solution just before using (see step 6) and add, while gently stirring, to 0.22-µm filtered 0.1% albumin in 20% excess of its minimal stabilizing amount. After 5 min, further stabilize the probe by adding a 0.22-µm filtered solution of 2.5% Carbowax 20-M (polyethylene glycol, MW 20,000) to reach a final concentration of 0.4 mg/ml (0.016 ml for every milliliter of albumin–gold solution). Stir for an additional 5 min.

8. Centrifuge 45 min at 4°C (277 K) at 65,000 g for 6-nm and at 20,000 g for 15-nm gold particles; discard the supernatant carefully so as not to remove the loose red sediment. Resuspend in PBS containing 0.4 mg/ml Carbowax 20-M and centrifuge again in new tubes. Do not transfer the metallic gold spot which may have precipitated on one side of the bottom of the tube.

9. Check the concentration of the gold by diluting a small aliquot (10–20 µl) and reading the A_{max} between 510 and 550 nm.

10. Dilute the albumin–gold complex in PBS–Carbowax 20-M to the concentration, expressed as optical density at A_{max}, required for the experimental design; an optical density between 2.5 and 10 is usually adequate for binding purposes. Filter through a 0.22-µm Millipore filter (type Millex-GV) immediately before using.

11. Albumin–gold probes should be used immediately after the second centrifugation. If necessary, they may be stored for a few days after the first centrifugation under sterile conditions, provided the second centrifugation is done directly before use.

APPLICATION OF ALBUMIN–GOLD COMPLEX

As already pointed out, colloidal gold particles have a high electron density. This characteristic, together with their small size, permits very fine localization of the albumin–gold complex at the ultrastructural level. Although it cannot be excluded that free albumin may be handled by cells differently than gold-bound albumin, as some evidence indicates in sinusoidal cells of bone marrow (Geoffroy and Becker, 1986), it has been shown that differences in the albumin molecule itself adsorbed to the gold particles may change the interactions between the probes and endothelial cells. In fact, glycosylation of albumin greatly enhances the binding, up-

8. Preparation and Application of Albumin–Gold Complex 171

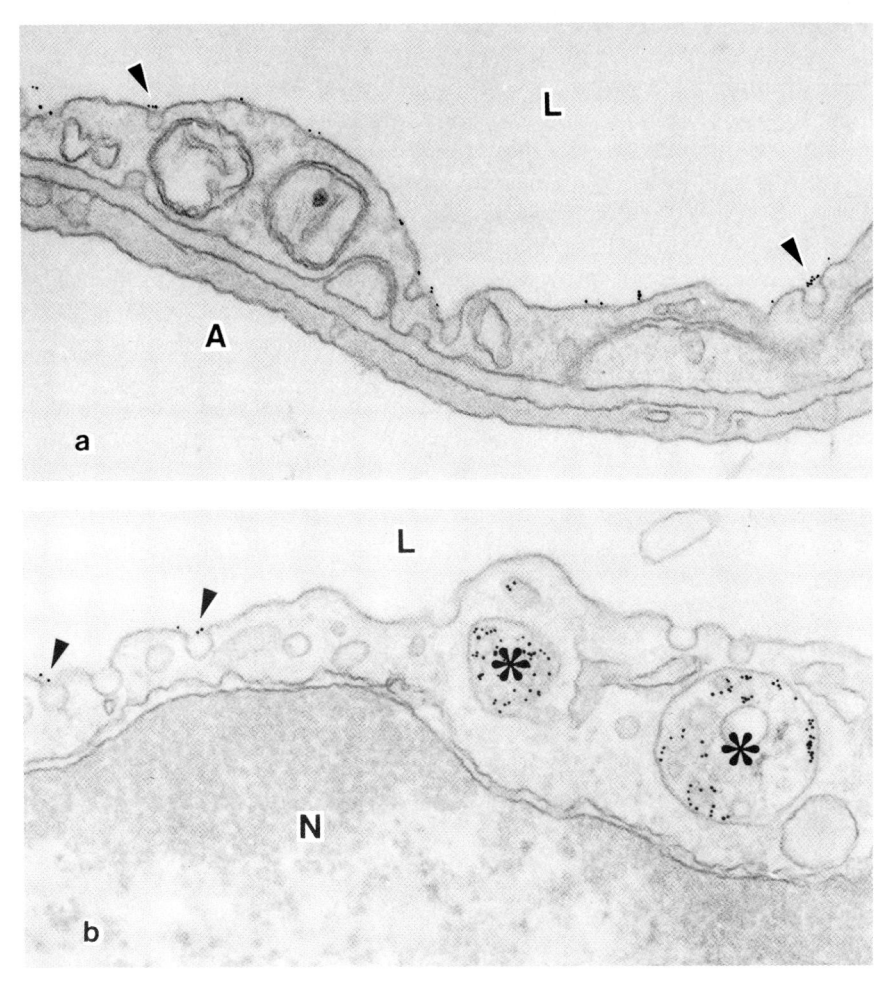

Fig. 8.2. (a) Isolated rat lung perfused at 4°C (277 K) for 15 min with physiologic solution containing 4% dextran 70 followed by 2 min of bovine serum albumin–gold complexes (optical density at $A_{max} = 8.0$). Excess gold probes were washed out for 5 min before fixation and processing for electron microscopy. Note the gold probes scattered on the plasmalemma and on luminal vesicles (arrowheads). No internalization occurred as a consequence of the low perfusion temperature. (b) Isolated rat lung perfused as in (a) but at 37°C (310 K) for 1 hr before fixation. Some particles are still visible on the luminal vesicle (arrowheads) but many particles are internalized in multivesicular bodies (asterisks). L, Lumen; A, air space; N, nucleus. Lead citrate stain; both parts ×54,000.

take, and transport of albumin–gold complex in capillary endothelial cells of isolated perfused rat lungs (Villaschi et al., 1986) (Figs. 8.2 and 8.3). Thus albumin–gold probes are a powerful tool for studying the interactions between different albumins and various cell types.

Interactions between albumin and endothelial cells have been studied in various capillary beds by means of gold probes in isolated or *in situ* perfused organs (Geoffroy and Becker, 1984, 1986; Ghitescu and Fixman, 1984; Ghitescu et al., 1986; Villaschi et al., 1986; De Bruyn and Cho, 1987). Pulse chase types of protocols or continuous exposure to the albumin–gold probes can be used. In the sinusoidal endothelia albumin binds mainly to bristle-coated pits and is internalized into the lysosomal apparatus of the cells. In continuous endothelium of capillary beds albumin is preferentially located on luminal vesicles and intercellular clefts and is partly internalized into lysosomes (Fig. 8.2) and partly transported into the interstitial space.

The author wishes to thank Dr. G. G. Pietra for his support in the preparation of this report. The original work on which this paper is based was supported by grant HL-32482 from the National Institute of Health to Dr. G. G. Pietra.

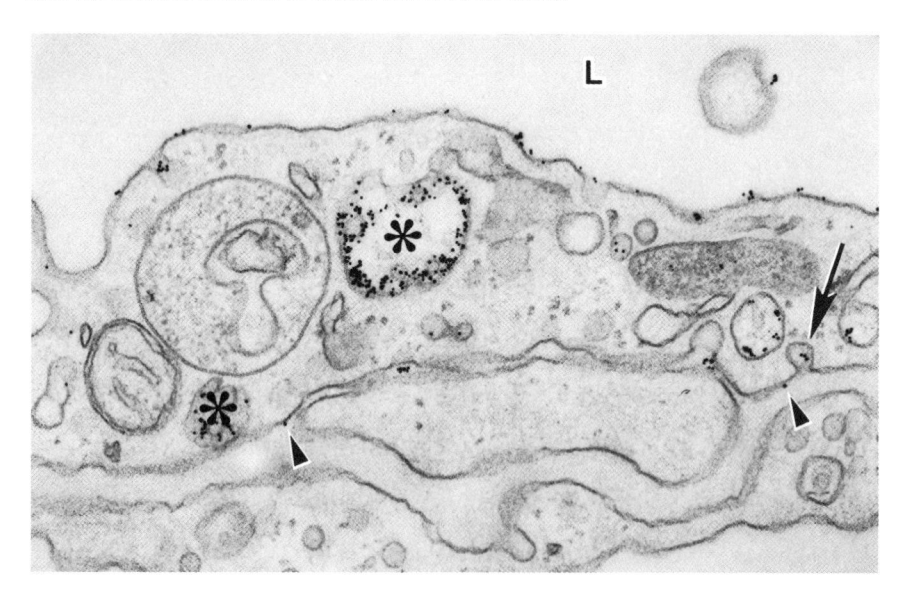

Fig. 8.3. Isolated rat lung perfused as in Fig. 2b but glycosylated albumin bound to colloidal gold instead of native albumin. Glycosylation of albumin enhanced binding of the probes to the cell surface and subsequent internalization in multivesicular bodies (asterisks). Note the presence of particles in an albuminal vesicle (arrow) as well as in the interstitial space (arrowheads). L, lumen. Lead citrate staining; ×58,500.

REFERENCES

Courtoy, P. J., Picton, D. H., and Farquhar, M. G. (1983). Resolution and limitations of the immunoperoxidase procedure in the localization of extracellular matrix antigens. *J. Histochem. Cytochem.* **31**, 945.

Curry, F. E., and Michel, C. C. (1980). A fiber matrix model of capillary permeability. *Microvasc. Res.* **20**, 96.

Danielli, J. F. (1940). Capillary permeability and edema in the perfused frog. *J. Physiol. (London)* **98**, 109.

De Bruin, P. P. H., and Cho, Y. (1987). Kinetic analysis of new formation, internalization and turnover of bristle coated pits of myeloid sinuses. *Lab. Invest.* **56**, 616.

Drinker, C. K. (1927). The permeability and diameters of the capillaries in the web of the brown frog *(Rana temporaria)* when perfused with solutions containing pituitary extract and horse serum. *J. Physiol. (London)* **63**, 249.

Frens, G. (1973). Controlled nucleation for the regulation of the particle size in monodispersed gold solutions. *Nature (London), Phys. Sci.* **241**, 20.

Geoffroy, J. S., and Becker, R. P. (1984). Endocytosis by endothelial phagocytes: Uptake of bovine serum albumin–gold conjugates in bone marrow. *J. Ultrastruct. Res.* **89**, 223.

Geoffroy, J. S., and Becker, R. P. (1986). Uptake of formaldehyde-denatured bovine serum albumin (Fm-BSA) and BSA bound to colloidal gold (Au-BSA) involves the same scavenger receptor on marrow sinusoidal cells. *Anat. Rec.* **214**, 41A.

Geoghegan, W. D., and Ackerman, A. (1977). Adsorption of horseradish peroxidase, ovomucoid and antiimmunoglobulin to colloidal gold for the indirect detection of concanavalin A, wheat germ agglutinin and goat antihuman immunoglobulin G on cell surfaces at the electron microscopic level: A new method, theory and application. *J. Histochem. Cytochem.* **25**, 1187.

Ghitescu, L., and Fixman, A. (1984). Surface charge distribution on the endothelial cell of liver sinusoids. *J. Cell Biol.* **99**, 639.

Ghitescu, L., Fixman, A., Simionescu, M., and Simionescu, N. (1986). Specific binding sites for albumin restricted to plasmalemmal vesicles of continuous capillary endothelium: Receptor-mediated transcytosis. *J. Cell Biol.* **102**, 1304.

Goodman, S. L., Hodges, G. M., Tejdosiewicz, L. K., and Livingston, D. C. (1981). Colloidal gold markers and probes for routine application in microscopy. *J. Microsc. (Oxford)* **123**, 201.

Horisberger, M., Rosset, J., and Bauer, H. (1975). Colloidal gold granules as markers for cell surface receptors in the scanning electron microscope. *Experientia* **31**, 1147.

Horisberger, M., and Rosset, J. (1977). Colloidal gold, a useful marker for transmission and scanning electron microscopy. *J. Histochem. Cytochem.* **25**, 295.

Horisberger, M., and Vauthey, M. (1984). Labelling of colloidal gold with protein. A quantitative study using β-lactoglobulin. *Histochemistry* **80**, 13.

Hütter, J. F., Piper, H. M., and Spieckermann, P. G. (1984). Myocardial fatty acid oxidation: Evidence for an albumin-receptor-mediated membrane transfer of fatty acids. *Basic Res. Cardiol.* **79**, 274.

Mason, J. C., Curry, F. E., and Michel, C. C. (1977). The effects of proteins upon the filtration coefficient of individually perfused frog mesenteric capillaries. *Microvasc. Res.* **13**, 185.

Mühlpfordt, H. (1982). The preparation of colloidal particles using tannic acid as an additional reducing agent. *Experientia* **38**, 1127.

Muller, W. E., and Wollert, U. (1979). Human serum albumin as a "silent receptor" for drugs and endogenous substances. *Pharmacology* **19**, 59.

Ockner, R. K., Weisiger, R. A., and Gollan, J. L. (1983). Hepatic uptake of albumin-bound substances: Albumin receptor concept. *Am. J. Physiol.* **245,** G13.

Peters, T., Jr. (1975). Serum albumin. In *The Plasma Proteins* (F. W. Putnam, ed.), 2nd ed., Vol. 1, pp. 133–181 Academic Press, New York.

Schneeberger, E. E., and Hamelin, M. (1984). Interaction of serum proteins with lung endothelial glycocalyx: Its effect on endothelial permeability. *Am. J. Physiol.* **247,** H206.

Villaschi, S., Johns, L., Cirigliano, M., and Pietra, G. G. (1986). Binding and uptake of native and glycosylated albumin–gold complexes in perfused rat lungs. *Microvasc. Res.* **32,** 190.

Warchol, J. B., Brelinska, R., and Herbert, D. C. (1982). Analysis of colloidal gold methods for labelling proteins. *Histochemistry* **76,** 567.

Weiser, H. B. (1949). *A Textbook of Colloid Chemistry*. Wiley, New York.

Weisiger, R. A., Gollan, J. L., and Ockner, R. K. (1981). Receptor for albumin on the liver cell surface may mediate uptake of fatty acids and other albumin-bound substances. *Science* **211,** 1048.

Weisiger, R. A., Gollan, J. L., and Ockner, R. K. (1982). The role of albumin in hepatic uptake process. *Prog. Liver Dis.* **7,** 71.

9

Label-Fracture Cytochemistry

FREDERICK W. K. KAN

Department of Anatomy
Faculty of Medicine
Université de Montréal
Montréal, Québec, Canada

and

PEDRO PINTO DA SILVA

Membrane Biology Section
Laboratory of Mathematical Biology
National Cancer Institute
Frederick Cancer Research Facility
Frederick, Maryland

INTRODUCTION
    Rationale and Development
METHODOLOGY
    Preparation of Cells for Label-Fracture
    Choice of Marker
    Fixation
    Cryoprotection, Mounting, and Freezing of Specimens
    Fracturing and Cleaning of Replicas
    Images of Label-Fracture Replicas
APPLICATIONS
    Molecular Demarcation of Surface Domains of Boar Spermatozoa
    Mapping of Antigens on Boar Sperm Cell Surface
    Other applications

ADVANTAGES AND LIMITATIONS
NEW DEVELOPMENTS AND OUTLOOK
  Replica-Staining Label-Fracture
  Fracture-Flip
REFERENCES

INTRODUCTION

The introduction of the freeze-fracture technique (Steere, 1957; Moor et al., 1961) in biological research some 30 years ago opened a new dimension in electron microscopic studies of biological specimens. Early images of membrane faces obtained with the freeze-fracture technique were interpreted as membrane surfaces (Moor and Mühlethaler, 1963; Weinstein and Bullivant, 1967; Koehler, 1968) or alternatively as inner hydrophobic regions within the membrane matrix (Branton, 1966). In 1970, Pinto da Silva and Branton performed freeze-fracture and etching experiments with erythrocyte ghosts covalently labeled with ferritin and proved that membranes were split during fracture. These results made possible the interpretation of images of freeze-fractured membranes and led to the synthesis of the globular and bilayer concepts of membrane structure: biological membranes were, therefore, hypothesized to consist of a bilayer continuum (a lipid bilayer plus adsorbed proteins) interrupted by proteins intercalated across the bilayer (Pinto da Silva and Branton, 1970). From 1970 to 1974 etching experiments with cytochemically labeled membranes proved this hypothesis (reviewed by Pinto da Silva, 1987). The following decade was a thriving period for freeze-fracture work. Electron micrographs of freeze-fracture preparations provided important morphological information on biological membranes in general and membrane specializations (e.g., junctions) and microdomains in particular.

Until 1980, however, the application of freeze-fracture was almost exclusively limited to the study of ultrastructural details in cell membranes. This was due to the limited applicability of freeze-etch cytochemistry. Thus, when used alone, freeze-fracture offered little and always indirect information on the chemical composition and function of membrane components. Over the past 10 years, we have developed various labeling techniques that allow *in situ* labeling of freeze-fractured plasma and intracellular membranes (for reviews, see Pinto da Silva *et al.*, 1984; Pinto da Silva, 1987). Taking advantage of the membrane splitting, one of the techniques—fracture-label—was introduced and is now used in certain favorable cases to determine the sidedness of membrane components and the presence and partition of transmembrane proteins and to follow intracellular traffic of membrane proteins. This is achieved by labeling both the protoplasmic and cytoplasmic halves of membranes after fracture. In some instances, the chemical composition of cytoplasmic components can also be revealed (Kan and Pinto da Silva, 1986; Chevalier *et al.*, 1987). The critical-point drying fracture-label technique (Pinto da Silva *et al.*, 1981a) has been combined with the backscattered electron imaging (BEI) mode of the scanning electron microscope (SEM) to visualize the *in vivo* distribution of lectin binding sites on freeze-fractured membranes in tissues and cells (Kan and Nanci, 1988). This BEI–fracture-label method provides additional information about the distribution of the labeling with respect to the three-dimensional organization of tissues and cells.

In fracture-label techniques, labeling of the tissue occurs after freezing, fracturing, and thawing of glutaraldehyde-fixed, glycerol-impregnated specimens. Clearly, fracture-label is not intended to be a substitute for surface labeling. Thawing of freeze-fractured specimens also leads to postfracture reorganization of various membrane components (Pinto da Silva *et al.*, 1981b,c; Barbosa and Pinto da Silva, 1983). As a result, intramembrane particles are not seen in fractured-labeled specimens. In order to obtain single coincident images of conventional replicas of freeze-fractured membranes and of the distribution of a surface antigen or receptor, we developed label-fracture (Pinto da Silva and Kan, 1984; Kan and Pinto da Silva, 1987). Fracture-label techniques have been reviewed (Pinto da Silva *et al.*, 1984; Pinto da Silva, 1987). Here we focus on the development, technical aspects, and applications of label-fracture as a new method for high-resolution labeling of cell surfaces.

Rationale and Development

The concept of label-fracture is based on the initial plan of some 20 years ago to prove the splitting of biological membranes during freeze-fracture (see review by Pinto da Silva, 1987). Initially, we labeled human and rat erythrocytes with lectin–gold complexes to examine whether colloidal gold particles could be pulled across the outer half of the membrane and be detected on protoplasmic faces after freeze-fracture. In this first experiment, we observed a persistent high density of labeling on the exoplasmic faces. The background (surrounding medium) was full of gold particles. Surprisingly, P faces were clear of contaminating gold spheres. The reason for this result was as follows: during thawing and cleaning of replicas on sodium hypochlorite and washing in distilled water, P faces were sheltered from gold particle contamination by the unfractured cell mass underneath and remained clean upon digestion by sodium hypochlorite and washing in distilled water. Because of these unexpected results, we repeated the labeling experiment, this time without the subsequent use of sodium hypochlorite solution. The first label-fracture replica showed lectin–gold-labeled E faces against a clean unlabeled background. P faces, on the other hand, were opaque because of the remaining unfractured and undigested cells attached to them.

METHODOLOGY

Preparation of Cells for Label-Fracture

Label-fracture is a cell surface-labeling technique. Although this technique has been applied to localize surface receptors of isolated epithelia (Chevalier *et al.*, 1987) and cells in culture (Boonstra *et al.*, 1985), it is best applied to cells in suspension or to isolated membrane fractions. The technique is relatively simple; the experimental sequence of label-fracture is schematically illustrated and summarized in Fig. 9.1.

Choice of Marker

Before starting the label-fracture experiment, one should decide on the most suitable electron-dense probe to be used. Much of the success in freeze-fracture cytochemistry is due to the availability and versatile nature of colloidal gold probes, which have been successfully conjugated to many lectins, enzymes, and antibodies. The electron opacity of colloidal gold and the well-documented methods of preparing particles of various sizes make it the marker of choice in label-fracture cytochemistry. Labeling of cell surfaces by colloidal gold also allows for quantitative analyses.

The size of colloidal gold is especially important since the gold size determines accessibility to membrane components, as well as resolution. In our experience of labeling surface glycoconjugates of boar spermatozoa, we found that not only is labeling intensity inversely proportional to the gold size but also large gold particles (>14 nm) may fail to reveal the specific topographical distribution of cell surface macromolecules, which can be demonstrated by the use of small gold particles (Figs. 9.2a and 9.3a). In our hands, best results are obtained with gold particles 6 to 10 nm in size (Figs. 9.2b and 9.3b). Due to steric hindrance, gold particles larger than 14 nm are generally not suitable for quantitative detection of surface molecules. On the other hand, gold particles smaller than 6 nm are difficult to visualize against the background of the Pt/C replica. This problem is particularly evident on a highly contrasted replica. In addition,

9. Label-Fracture Cytochemistry 179

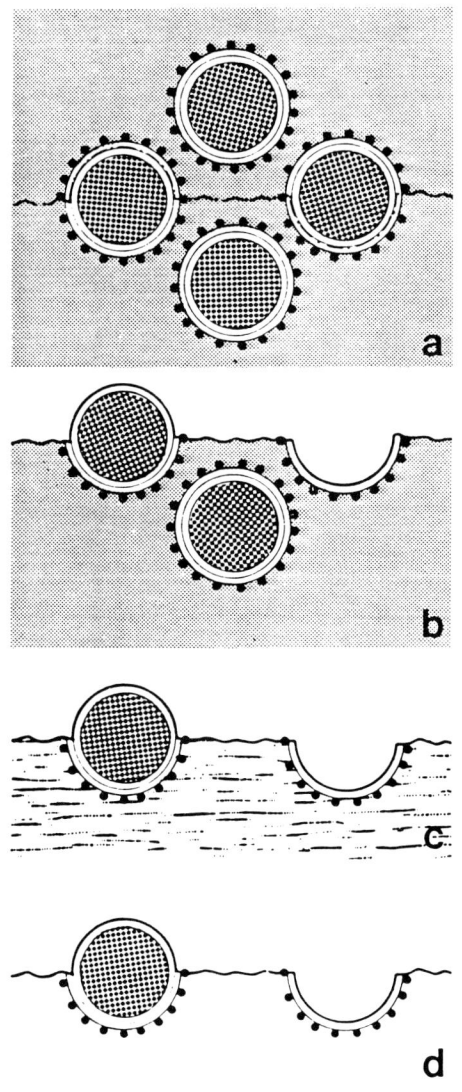

Fig. 9.1. Label-fracture. (a) Cells in suspension are labeled and frozen. (b) Freeze-fracture splits the plasma membranes of labeled cells into exoplasmic halves (with attached surface label; right) and protoplasmic halves (which remain attached to the cell body; left). Pt/C evaporation produces a high-resolution cast of the fractured cells (only label at the interface of fracture is exposed and shadowed). (c) Fractured, shadowed specimens are thawed and washed with distilled water, removing unfractured cells. Exoplasmic membrane halves remain attached to the replica. Coincident images of the Pt/C replica of the E face and the surface label are produced. Cells with fractured P faces (left) remain attached to the replica (but intrinsic electron density of the cell body prevents observation of the P face). (d) The replicas are then mounted on electron microscope grids, dried, and observed. (From Pinto da Silva and Kan, 1984.)

9.2a   b

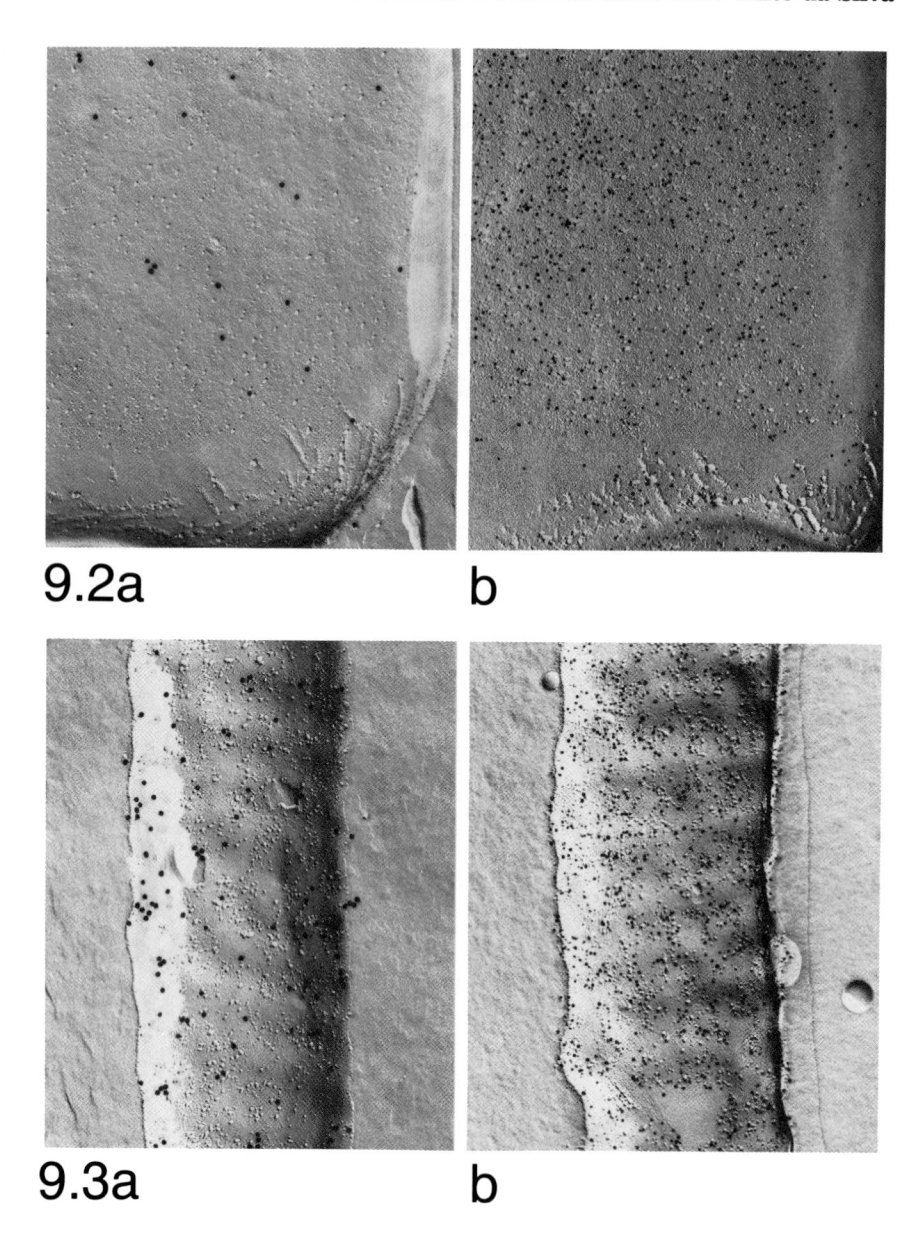

9.3a   b

Figs. 9.2 and 9.3. Con A label-fracture of boar sperm. Parts 2a and 3a show, respectively, portions of the head and tail region labeled with Con A followed by horseradish peroxidase–colloidal gold complex (20-nm gold). Gold particles are sparsely distributed over the exoplasmic membrane half of the head (2a) and tail (3a). Parts 2b and 3b show similar preparations of sperm cells labeled with colloidal gold particles of 10–13 nm in diameter; there is an increase of labeling density. Preferential association of the gold particles with intramembrane particles becomes apparent; ×42,000.

platinum-shadowed intramembrane particles can be confused with gold spheres smaller than 6 nm, making their identification on the replica difficult (note, however, that while the particles have a cone of shadow, the colloidal gold spheres do not).

Fixation

Isolated cells, cells in a monolayer, or epithelia are first fixed in a buffered fixative (normally 1 to 2.5% glutaraldehyde in phosphate-buffered saline, PBS, pH 7.4) for 1 hr at room temperature. The fixation step may be omitted or minimized if antigens are the target to be labeled, because different fixatives in various concentrations are known to have adverse effects on the retention of antigenicity (Hayat, 1986). After fixation, the cells are washed three times in PBS and then quenched by immersion in 0.1 M glycine for 1 hr at room temperature to remove aldehyde groups that might cause nonspecific binding. The cells are then washed three times by successive resuspension and centrifugation at 1500 rpm for 5 min each time. The cells are then directly labeled with a lectin–gold complex or with lectin/antibody followed by incubation in a protein–gold complex. For the detection of antigenic sites, cells are labeled with the antibody and then incubated in a protein A–gold solution. Methods of preparing various lectin–gold complexes and protein A–gold solutions at optimal conditions have been extensively described and well documented in the literature (see Volume 1).

Cryoprotection, Mounting, and Freezing of Specimens

Fixed and labeled cells are normally impregnated gradually with a cryoprotectant (e.g., 25–30% glycerol in PBS) prior to freezing in Freon 22. Cryoprotection may be omitted if the samples are to be frozen rapidly. For mounting the specimen, we have used the Balzers-type double-replica specimen table. We have found that the double replica method generates more fractured membrane faces, whereas freeze-fracturing with a precooled razor blade results in a great many cross-fractured profiles. Therefore, we recommend the use of the double-replica method in label-fracture.

To prepare cells for freeze-fracture, labeled cells in suspension are first concentrated into a soft pellet in a cuvette by centrifugation for 7 min at 1500 rpm. Most of the glycerol above the cell pellet in the cuvette is removed with a pipette and discarded. Using a pipette with a tapered end, the cells are stirred gently to loosen and disperse them. The density of this suspension should be about 1 : 3 (v/v). A small drop of this cell suspension is then transferred using a drawn Pasteur pipette onto a standard

Balzers-type specimen support plate (without central bore). The cells are spread to cover the entire surface of the top of the plate and resuspended, using a Pasteur pipette again, by pipetting so that the cells are loosely suspended in the glycerol medium and not aggregated. Too concentrated a sample of cells will not only yield few fractured faces but also result in piling of unfractured cells on top of one another, thus masking the details of the fractured membrane halves. Too high a concentration of cells is perhaps the most frequent source of failure in label-fracture. Following the transfer of cells, another specimen support plate (with a central bore) is placed on the top of the cells. In this way, the sample is sandwiched between two support disks. Excess cells in the suspension are squeezed out through the central bore and can be removed with a filter paper. After centering the support plates (which may be done with the help of a pair of centering tweezers), the cells are rapidly frozen by plunging into Freon 22 precooled with liquid nitrogen or into a solid/liquid nitrogen slush obtained by lowering the liquid nitrogen temperature at low pressure.

Fracturing and Cleaning of Replicas

Freeze-fracture by the double-replica method is performed in a Balzers-type freeze-fracture unit at $-130°C$ under a vacuum of 2×10^{-6} torr without etching and is followed by shadowing with platinum at a fixed angle of 45° and then coating with carbon at 90°. The thickness of the replica is ~2 nm platinum and 15 to 20 nm carbon as determined by a Balzers crystal thin-film monitor. With some training, the thickness of the film can also be judged by eye during the evaporation. After replication, instead of the conventional cleaning of replicas in acids or bases, replicas are thawed and then washed by successive floatings on distilled water (about 20 min per wash). To do this, immediately upon retrieval of the specimen from the chamber of the Balzers unit, the replica, while still frozen, is simply floated off the specimen support plate on a well of distilled water in a porcelain dish. During the washing, most of the unfractured cells will disperse. Normally, an intact replica with a surface area as large as the top of the specimen support plate can be obtained. Fragmentation of the replica sometimes occurs due to insufficient carbon coating. Breaking of the replica into a few pieces is not bad and may indicate a desirably thin carbon coat. A sample containing few cells can also result in fragmentation of the replica due to uneven distribution of cells on the top of the specimen support plate. To reveal protoplasmic fractured faces, replicas are cleaned with sodium hypochlorite solution for at least 1 hr to remove cell debris (as in conventional freeze-fracture). The replicas are then mounted on a (Formvar-coated) copper grid for electron microscopic examination.

Images of Label-Fracture Replicas

One of the expectations of the label-fracture technique was that it might be used for the chemical identification of intramembrane particles (i.e., correlation of a cytochemical marker with the distribution of intramembrane particles). To test this possibility, we carried out the first label-fracture experiment on human erythrocyte membrane labeled with two lectins, concanavalin A (Con A) (specific for mannose/glucose; Nicolson, 1974) and wheat germ agglutinin (WGA) (specific for N-acetylglucosamine/sialic acid; Bhavanandan and Katlic, 1979), followed by incubation in horseradish peroxidase–colloidal gold and ovomucoid–colloidal gold complexes, respectively (Geoghegan and Ackerman, 1977; Horisberger and Rosset, 1977).

The first label-fracture replicas revealed Con A and WGA labeling of the exoplasmic membrane half with high density (Figs. 9.4 and 9.5). The label was observed superimposed on the conventional replica image of the freeze-fracture exoplasmic face. Since the cleaning of replicas does not require digestion, the colloidal gold label apparently remains in place after repeated washing in distilled water. Our experiments indicated that thawing of the fractured specimens in glutaraldehyde does not result in denser labeling. Since the gold particles are under the fracture face and are attached to the exoplasmic membrane half, they are not replicated and therefore do not produce characteristic cones of shadow. However, at the edges of the fracture face where the membrane is cross-fractured, colloidal gold particles are frequently shadowed (Fig. 9.6). Exoplasmic fracture faces of labeled human leukocytes also showed specific labeling against a clear unlabeled background (Fig. 9.7). Colloidal gold particles were confined to within 10–20 nm of the edges of the fractured face. In human erythrocyte and leukocyte membranes, intramembrane particles are dense and evenly distributed and it is difficult to relate the gold labeling to the distribution of intramembrane particles. We then isolated rat thymocytes and induced aggregation of intramembrane particles by incubation of the cells at 37°C in PBS prior to glutaraldehyde fixation (Pinto da Silva and Martinez-Palomo, 1975). Label-fracture replica of thymocytes labeled with Con A–horseradish peroxidase–gold complex revealed a codistribution of Con A binding sites (indicated by gold labeling) and aggregates of intramembrane particles (Fig. 9.8). These early experimental results pointed to a novel surface-labeling method for chemical identification of intramembrane particles.

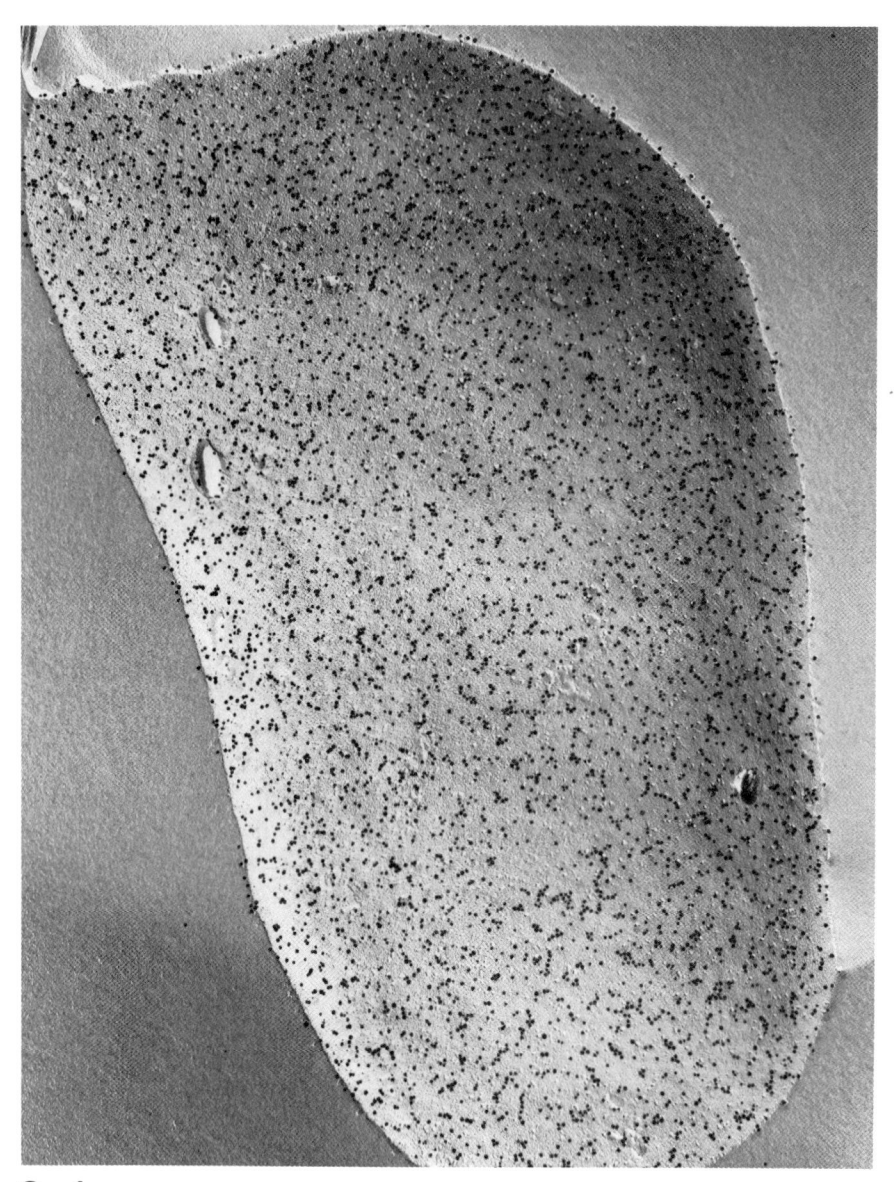

Figs. 9.4–9.6. Label-fracture of human erythrocyte membranes. Fig. 9.4: Con A; ×45,000. (From Pinto da Silva and Kan, 1984). Figs. 9.5 and 9.6: WGA. Colloidal gold particles are frequently shadowed (Fig. 9.6). Fig. 9.5, ×15,000; Fig. 9.6, ×62,000.

9. Label-Fracture Cytochemistry 185

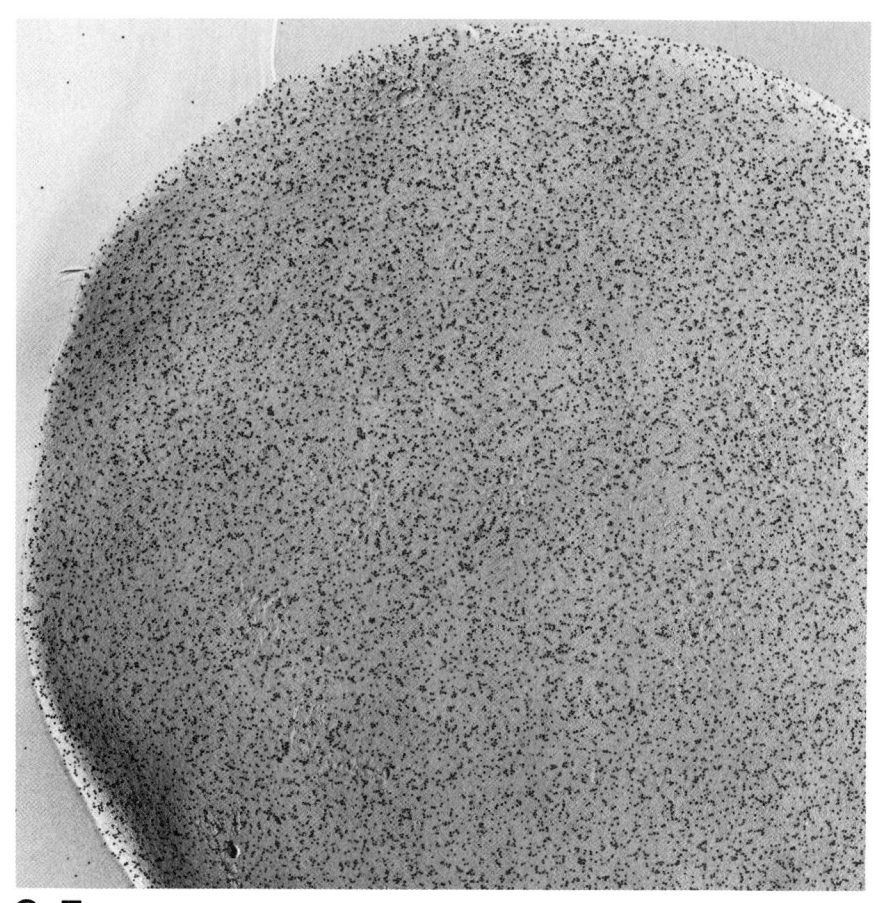

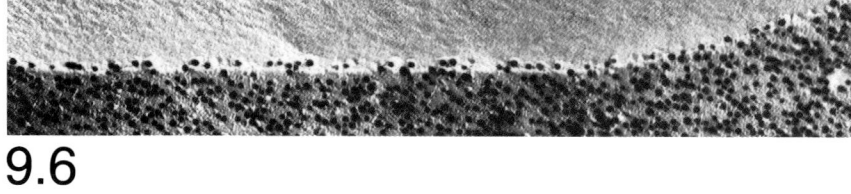

9.5

9.6

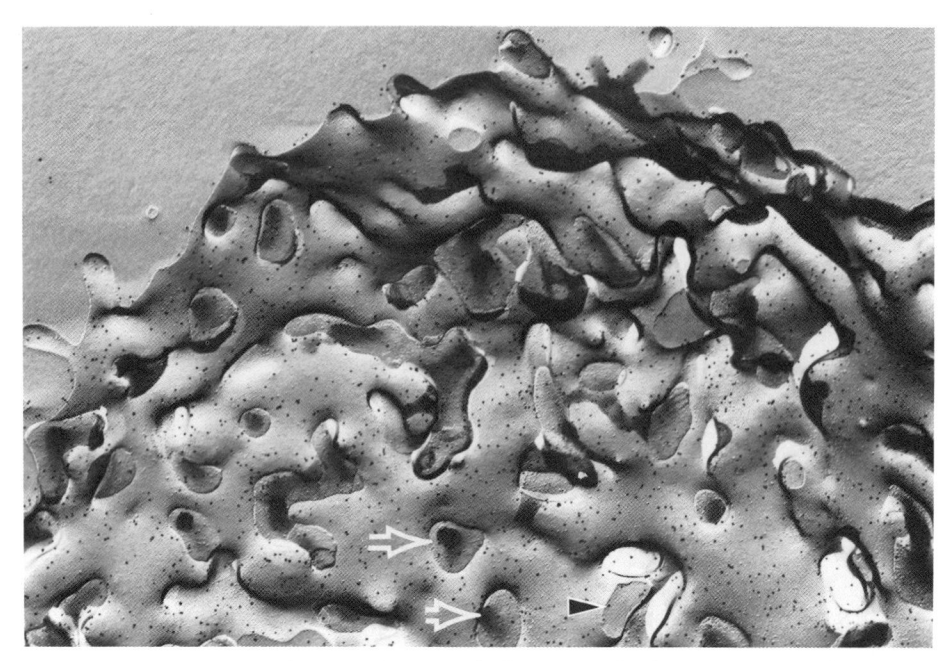

Fig. 9.7. WGA label-fracture of a human leukocyte. During freeze-fracture, cell projections and invaginations are cross-fractured. Cross-fractures of projections are generally labeled, with the labeling frequently obscured due to the electron density of the undigested, cross-fractured cytoplasm (white arrows). Cross-fractured invaginations are clean and, in general, labeled only at the perimeter (arrowhead); ×30,000. (From Pinto da Silva and Kan, 1984.)

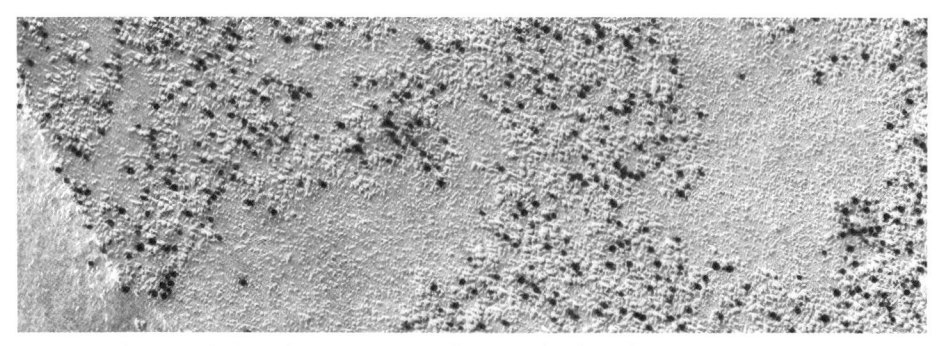

Fig. 9.8. Incubation of rat thymocyte cell suspension in PBS at 37°C induces aggregation of intramembrane particles. Con A binding sites in this cell are restricted to the particle aggregates; ×66,500. (From Pinto da Silva and Kan, 1984.)

APPLICATIONS

Molecular Demarcation of Surface Domains of Boar Spermatozoa

The surface of mammalian spermatozoa displays distinct surface domains (for review, see Fawcett, 1975). Most of the studies on the regionalization of surface sites have been carried out by immunofluorescence (Koo *et al.*, 1973; Toullet *et al.*, 1973; Fellous *et al.*, 1974; Koehler and Perkins, 1974; Myles *et al.*, 1981; Peterson *et al.*, 1981; Russell *et al.*, 1982, 1983). However, at higher resolution, cytochemical studies of the distribution of surface sites are best performed using methods that permit observation of large surface areas. We have used label-fracture to establish high-resolution maps of WGA and Con A receptor sites on the cell surface of boar spermatozoa and to investigate the possible association of these receptors to integral membrane components (Kan and Pinto da Silva, 1987). At a low magnification, a label-fracture replica of boar sperm cells consists of images of freeze-fractured exoplasmic membrane halves, protoplasmic halves with the remaining cell mass attached, and unfractured sperm cells (Fig. 9.9). The latter two appear as dark, opaque images on the replica.

At a high magnification, label-fracture reveals intense WGA labeling over the acrosomal region of the sperm plasma membrane with decreased WGA labeling from the postacrosomal area toward the base of the head (Fig. 9.10). The WGA labeling pattern in the sperm head appears to parallel the distribution of large intramembrane particles. At the base of the sperm head, where a distinct zone free of intramembrane particles is evident, there is an abrupt increase of WGA labeling density (Fig. 9.10). The particle-free zone, however, is not labeled by Con A, which is uniformly distributed on the sperm head (Fig. 9.11). Over the neck, aggregates of particles are frequently seen. These particle-rich areas are well labeled by Con A (Fig. 9.11). WGA, on the other hand, labels both particle-free and particle-rich areas (Fig. 9.10). The labeling pattern of the neck is seen also in the tail except that the annulus is not labeled by Con A (Figs. 9.12 and 9.13). Since Con A codistributes with the intramembrane particles, we take this as an indication of mannose-rich integral membrane proteins. The uniform distribution of WGA receptors over both smooth and particulate areas in the same regions suggests the presence of both glycoproteins and glycolipids. Thus, the application of label-fracture to study the surface domains of boar spermatozoa has demonstrated the power of the technique to provide cytochemical maps approaching the molecular level and to relate topochemistry of the cell surface to the freeze-fracture morphology of the plasma membrane.

Fig. 9.9. Low magnification of a label-fracture preparation of boar sperm cells labeled with WGA and ovomucoid–colloidal gold complex (10–13-nm gold). The replica consists of freeze-fractured exoplasmic membrane halves (asterisks) showing the topographical distribution of WGA receptor sites and an unfractured sperm cell (C). In label-fracture the replica is washed only with distilled water. Therefore, the tail (arrowheads) underneath a protoplasmic face remains attached to the replica; ×12,000.

9. Label-Fracture Cytochemistry 189

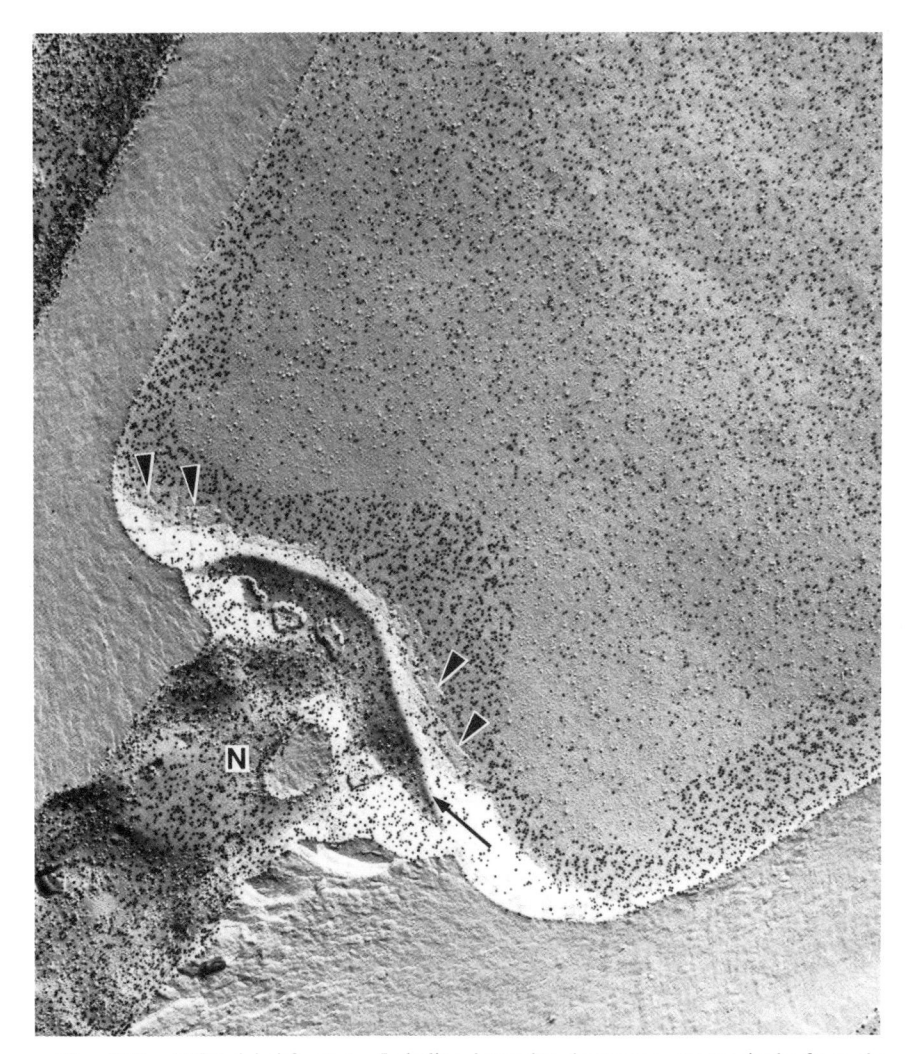

Fig. 9.10. WGA label-fracture. Labeling intensity decreases progressively from the equatorial to the postacrosomal region. A high concentration of WGA receptors is observed over the particle-free zone at the base of the head. The striated cords (arrowheads) below this zone are not labeled. However, gold particles are seen over the posterior ring (arrows). Over the neck (N), WGA labels uniformly both particle-free areas and particle-rich areas; ×33,000. (From Kan and Pinto da Silva, 1987.)

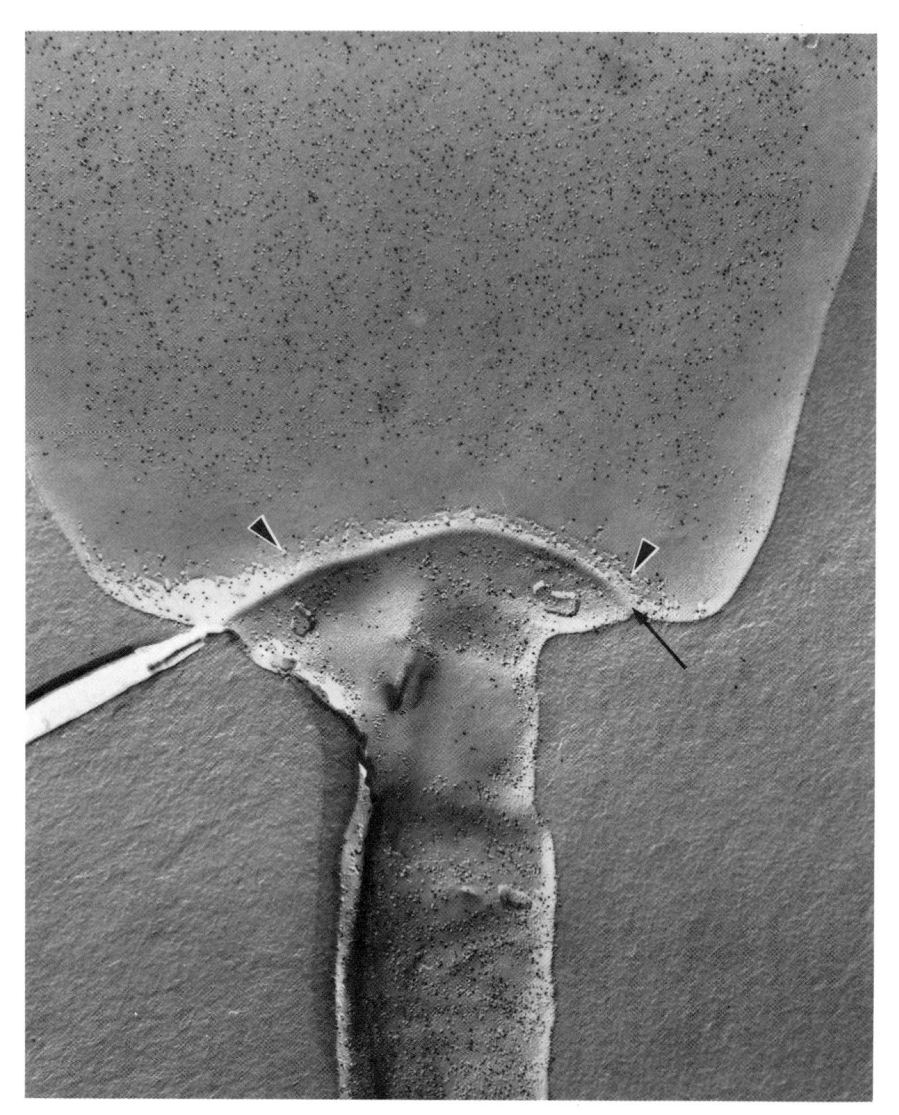

Fig. 9.11. Con A label-fracture of the boar sperm head. Con A labels both acrosomal (not shown in this figure except for basal portion of equatorial region) and postacrosomal regions. The particle-free zone at the base of the head is devoid of gold particles. A few gold particles are associated with the striated cords (arrowheads), but Con A label is absent over the posterior ring (arrow). Over the neck, Con A labels the particle-rich area, leaving the particle-free regions virtually unlabeled; ×33,000. (From Kan and Pinto da Silva, 1987.)

9. Label-Fracture Cytochemistry 191

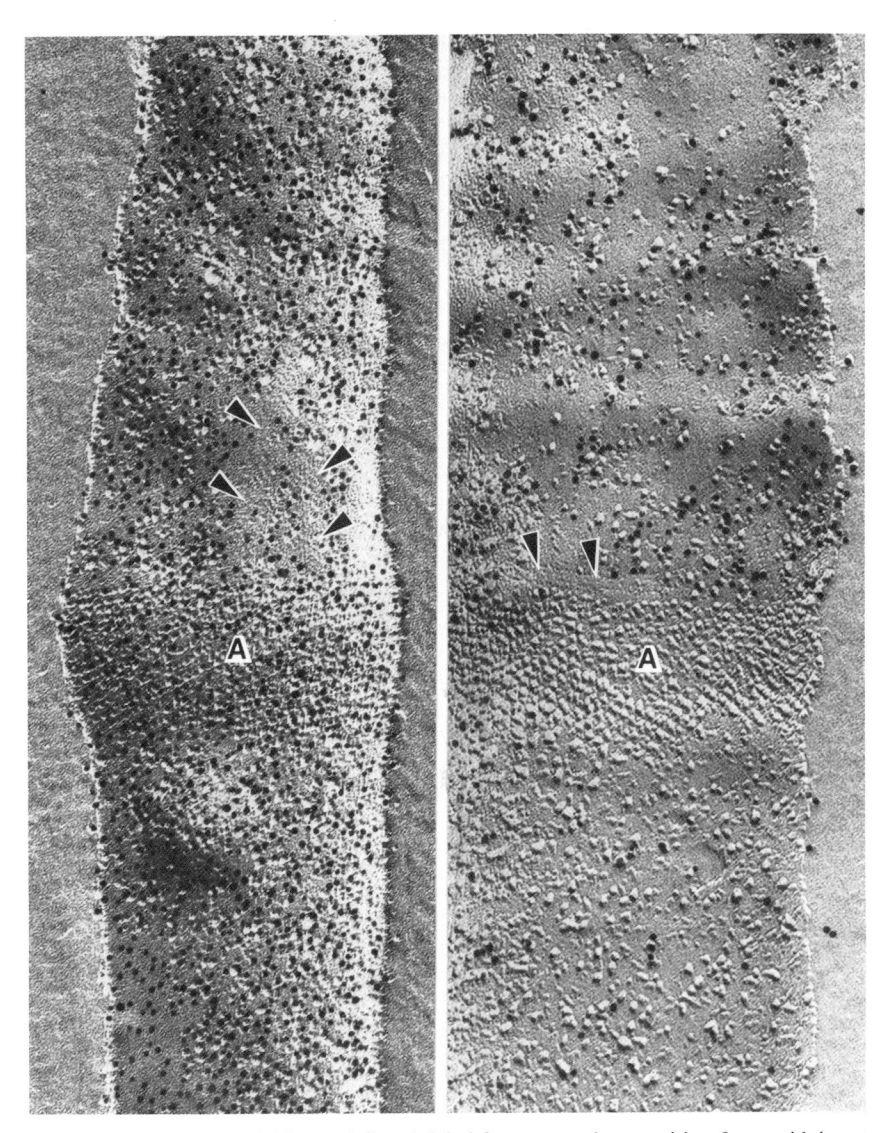

Figs. 9.12 and 9.13. WGA and Con A label-fracture at the transition from midpiece to principal piece. Fig. 9.12 (left): WGA; colloidal gold particles are distributed uniformly over midpiece (portion above annulus, A), annulus, and principal piece (below annulus), with the exception of the membrane microdomains composed of rectilinear arrays of pits (arrowheads) in the midpiece close to the annulus. Fig. 9.13 (right): Con A; colloidal gold particles in the midpiece appear to be associated with intramembrane particles. The annulus (A) and the rectilinear arrays of pits (arrowheads) are not labeled. ×56,000. (From Kan and Pinto da Silva, 1987.)

Mapping of Antigens on Boar Sperm Cell Surface

Advances in the characterization of membrane proteins and in the development of antibodies against various epitopes in many sperm cell plasma membrane proteins has allowed analysis of regional specificity in plasma membrane organization and function (Russell et al., 1982; Primakoff and Myles, 1983; Myles and Primakoff, 1984; Primakoff et al., 1985; Saxena et al., 1986; for a review, see Peterson and Russell, 1985). Label-fracture has the potential to provide cytochemical maps approaching molecular resolution on sperm surfaces. We have examined the distribution of antigenic sites on boar sperm cell surfaces labeled with anti-boar sperm plasma membrane IgG followed by incubation with protein A–gold complex. In label-fracture preparations of boar sperm ejaculates, the labeling is seen uniformly distributed throughout the entire sperm head with no apparent regionalization of the corresponding antigenic sites (Fig. 9.14). A weaker, but uniform, labeling is also observed over the midpiece and principal piece of the tail. Like Con A, anti-boar sperm plasma membrane IgG did not appear to label the annulus region. The results described here are only preliminary, but nevertheless they demonstrate the potential application of label-fracture for quantitative and high-resolution studies of antigenic sites on sperm cell surfaces.

Other Applications

Label-fracture has now been used to obtain marker density of polypeptides of the chloroplast envelope membrane of spinach (*Spinacia oleracea* L.) (Van Berkel et al., 1986); for analysis of freeze-fractured rod photoreceptor membranes using anti-opsin antibodies (De Foe and Besharse, 1985); and to investigate the distribution of wheat germ agglutinin receptor sites (Brown and Orci, 1986; Chevalier et al., 1986) and *Helix pomatia* lectin binding sites (Brown and Orci, 1986) on the luminal plasma membrane of freeze-fractured toad urinary bladder epithelial cells. The label-fracture technique has also been extended to the radioautographic localization of polypeptide ligands coupled to ^{125}I (Carpentier et al., 1985). The modification of the label-fracture method in combination with radioautography is relatively simple and reproducible. It may be used to localize plasma membrane-bound radioactive ligands on large membrane areas. It should also allow the direct correlation of radioactivity with the distribution of intramembrane particles.

ADVANTAGES AND LIMITATIONS

Label-fracture has two important advantages over the other commonly employed cell surface labeling techniques. First, colloidal gold marker

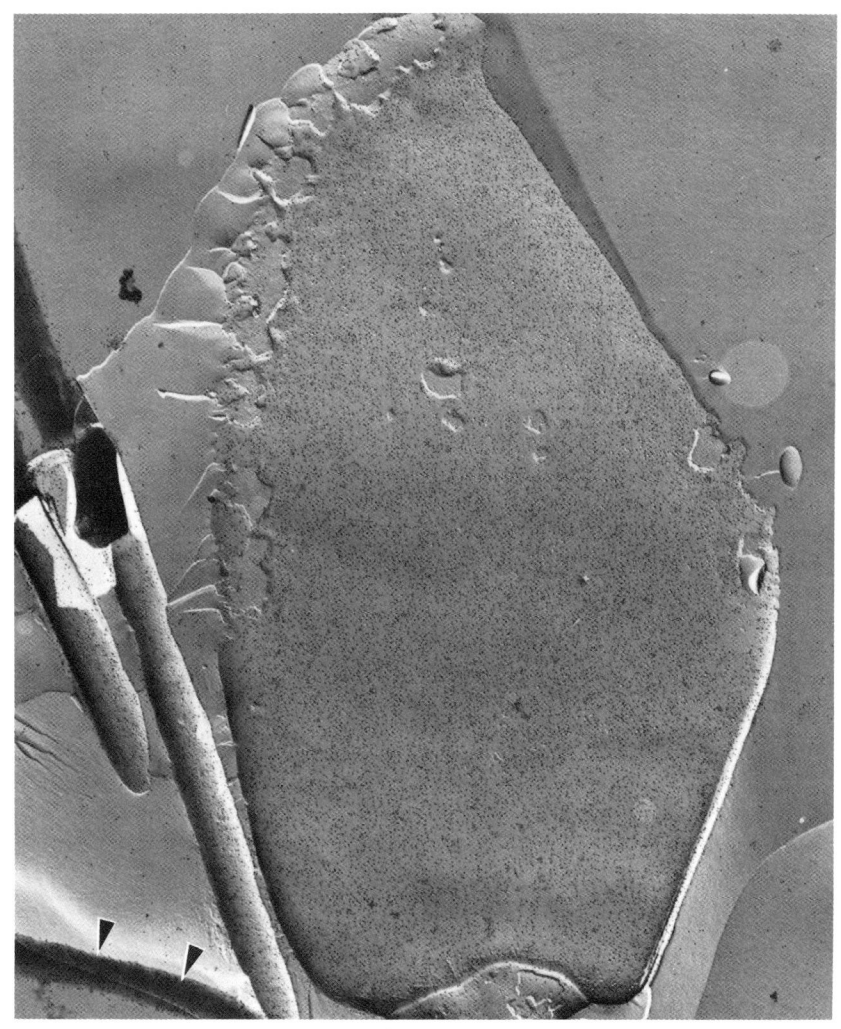

Fig. 9.14. Label-fracture preparation of a mature, ejaculated boar sperm previously labeled with anti-boar sperm plasma membrane IgG followed by protein A–gold. The labeling is uniformly distributed throughout the entire sperm head with no apparent regionalization; × 16,100.

can be visualized over large surface areas of fracture faces and can be quantitated. Second, the surface marker can be correlated with the distribution of intramembrane particles on the fractured faces. The technique is useful when studying surface domains in highly regionalized cells such as the spermatozoa described. Label-fracture allows chemical characterization of freeze-fractured plasma membranes and opens the way to direct chemical identification of intramembrane components.

Current cell surface labeling techniques involve the use of markers such as fluorescein isothiocyanate, peroxidase, ferritin, or colloidal gold. The former marker, having less resolution than the others, is normally used in light microscopy. The latter three are used in labeling cells which are subsequently embedded in a resin and sectioned for electron microscopic examination. Labeling by the postembedding method may also be done in this case. In these preparations, only the side view or cross section of the cell membrane is revealed. Consequently, this thin-section method is semiquantitative. Recently, the BEI mode of the SEM has been applied to localize gold-labeled cell surface antigens and glycoconjugates (Horisberger and Rosset, 1977; Horisberger, 1981; De Harven and Soligo, 1986; Choi and Siu, 1987; Namork *et al.*, 1987). Although the SEM has the advantage of providing information on the three-dimensional architecture of biological samples, it is also limited by its relatively poor resolution and cannot detect, without ambiquity, individual gold particles that are less than 14 nm in size.

Recent advances in freeze-fracture in combination with biochemistry, cytochemistry, and radioautography have opened new avenues for chemical identification and characterization of membrane components. For example, freeze-fracture combined with biochemistry has been used to analyze polypeptides of isolated outer membrane leaflet using gel electrophoresis (Nermut, 1983). Freeze-fracture can be combined with radioautography to localize radioactive glycoproteins, enzymes, and other membrane components to the E and P leaflets (Fisher and Branton, 1976; Rix *et al.*, 1976). This technique has been subject to various modifications (Schiller *et al.*, 1979; Nermut and Williams, 1980; Fisher, 1982; Kan *et al.*, 1984). The advantage of this technique is that it allows a time course study of the distribution of membrane components. Its limitation is that the replica has a shielding effect on radiation emitted from sources of radioactivity in cells or tissues. It is also a long and tedious procedure. Alternatively, the surface replica technique (Robenek and Severs, 1984) may be used for the visualization of cell surfaces labeled with gold-conjugated ligands. As this method does not require fracturing of the cells, one cannot correlate the distribution of intramembrane particles with some transmembrane proteins using colloidal gold probe. Recently, the sectioned, labeled-replica (SLR) method (Rash, 1979) has been proposed for the study of postreplication labeling of E-leaflet molecules. This technique has allowed the correlation of membrane immunoglobulins with the intramembrane particles in E-face replicas of murine B-lymphocyte plasma membranes (Dinchuk et al., 1987). In the SLR method, sectioning of the plastic-embedded labeled replica is required and only limited useful areas of replicas are obtained, as shown by previously published results. In highly regionalized cells such as mammalian spermatozoa, many sections have to be prepared in order to collect enough information on the microdomain distribution of a particular label. In view of the limitations inherent in other cell surface labeling techniques, label-fracture appears to be an easy, reproducible method for labeling cell surfaces with high resolution. When used in conjunction with complementary replicas, it is possible to relate the surface label directly to the distribution of intramembrane particles on protoplasmic faces.

The problems that remain to be solved in the future are related to whether changes in the distribution of the label occur during the thawing, washing, and drying steps of the label-fracture preparation. The freeze-fractured exoplasmic halves of the plasma membrane are stable since their apolar surface is stabilized by a Pt/C replica (2 nm thick) reinforced with a layer of carbon (10–20 nm thick). Therefore, postfracture reorganization into bilayered structures in fracture-label, where membranes are thawed directly in aqueous solution (Pinto da Silva et al., 1981a,b; Barbosa and Pinto da Silva, 1983), will not occur in label-fracture. Our previous experiments also indicated that thawing of the fractured specimens in glutaraldehyde solutions did not result in denser labeling. On the other hand, it has been established by fracture-label that transmembrane proteins may partition across the outer half of the membrane with the protoplasmic membrane half. Since colloidal gold particles do not appear to be pulled through the exoplasmic half of the membrane, it is possible that the gold particles may detach from the transmembrane proteins during thawing and washing. High-resolution labeling of cell surfaces in label-fracture requires the operation of a freeze-fracture unit, and therefore expertise in conventional freeze-fracture and knowledge of interpretation of images of label-fracture preparations are also important factors in obtaining good results. Despite these limitations, label-fracture offers high resolution and sensitivity that may be used to explore the ultrastructural patterns of the distribution of cell surface receptors and their relationship to membrane architecture.

NEW DEVELOPMENTS AND OUTLOOK

Replica-Staining Label-Fracture

A new variant of label-fracture called replica-staining label-fracture has just been developed (Forsman and Pinto da Silva, 1988a). The sequence of this new method (Fig. 9.15) follows from the demonstration that the exoplasmic half of the membrane remains attached to the Pt/C replica, thus allowing direct cytochemical labeling of freeze-fracture replicas. There are several advantages to this new method: (1) several fragments from a single replica can be labeled with different cytochemical reagents; (2) minimal amounts of biological materials are needed; (3) minute volumes (as low as 10 µl) of cytochemical reagents are required; (4) labeling of different specimens can be performed in one simple staining solution, an important consideration in quantitative immunocytochemistry. The price for this miniaturization is considerable: nonspecific adsorption of gold particles to the intervening Pt/C cast, leading to less pleasing images. This nonspecific adsorption is of no other consequence as specificity of the label can be judged only by comparison of the labeled (outer) surfaces in experiments and controls. Background can be reduced by high salt, wetting agents, and diluting colloidal gold solution.

Fracture-Flip

This is a new method to reveal the ultrastructure and topochemistry of membrane surfaces at macromolecular resolution (Forsman and Pinto da Silva, 1988b). Again, this is a corrollary of label-fracture; it was reasoned that, if the membrane halves are stabilized after fracture by the Pt/C replica, they might be stabilized by carbon evaporation alone. After thawing and washing, the replicas with their exoplasmic membranes attached are picked onto Formvar-coated grids and "flipped" (i.e., turned upside down). The membrane surfaces are now on top. High-resolution casts of the actual surfaces can now be obtained easily by Pt evaporation. The reader can best judge the new method, along with its potential for immunocytochemistry of membrane surfaces, in the papers that are now coming out (Forsman and Pinto da Silva, 1988b; Fujimoto and Pinto da Silva, 1988).

The combined promise of label-fracture (with postreplication label-fracture) and fracture-flip (with or without immunocytochemistry) is truly exciting. Not only are the methods for characterizing the fracture face in place but also a new method—fracture-flip—has multiple advantages over other existing methods for observing and studying the ultrastructure and topochemistry of cell and membrane surfaces. The resolution is far higher

9. Label-Fracture Cytochemistry

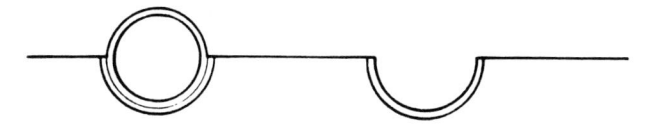

Cells are freeze-fractured and replicated with Pt/C

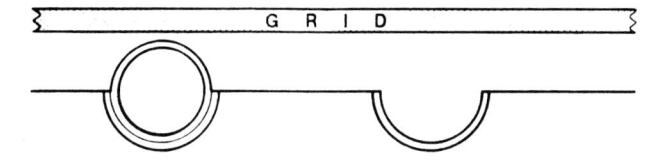

Pt/C replicas are washed and picked on grids

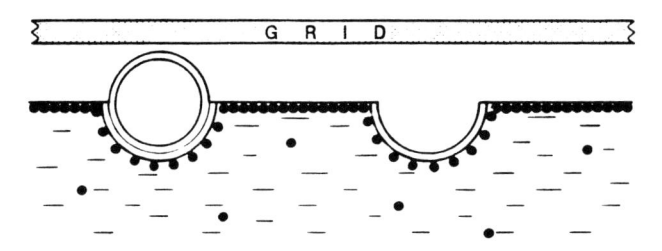

Cell surfaces are labeled

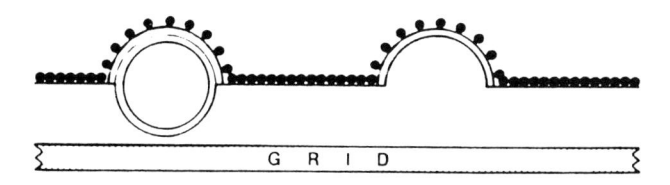

Labeled replicas are examined

Fig. 9.15. Schematic drawings showing the experimental procedure in replica-staining label-fracture. (From Forsman and Pinto da Silva, 1988a).

than that of scanning electron microscopy, and large areas of any membrane surface can easily be studied (in contrast to the limited exposure provided by freeze-etching and deep-etching). All this can be achieved with simple, readily available equipment (ultrarapid freezing can be used but is not required). There is much and exciting work ahead.

The authors wish to thank Ms. Johanne Chaîney for her secretarial assistance. This work was supported in part by the Medical Research Council of Canada (MA 9537) and the Fonds de la recherche en santé du Québec (870118) to F. W. K. Kan. F. W. K. Kan is the recipient of a scholarship from the MRC.

REFERENCES

Barbosa, M. L. F., and Pinto da Silva, P. (1983). Restriction of glycolipids to the outer half of a plasma membrane: Concanavalin A labeling of membrane halves in *Acanthamoeba castellani*. *Cell (Cambridge, Mass.)* **33,** 959.

Bhavanandan, V. P., and Katlic, A. W. (1979). The interaction of wheat germ agglutinin with sialoglycoproteins. The role of sialic acid. *J. Biol. Chem.* **254,** 4000.

Boonstra, J., Van Belzen, N., Van Maurik, P., Hage, W. J., Blok, F. J., Wiegant, F. A. C., and Verkleij, A. J. (1985). Immunocytochemical demonstrations of cytoplasmic and cell surface EGF receptors in A431 cells using cryoultramicrotomy, surface replication, freeze-etching and label-fracture. *J. Microsc. (Oxford)* **140,** 119.

Branton, D. (1966). Fracture faces of frozen membranes. *Proc. Natl. Acad. Sci. U.S.A.* **55,** 1048.

Brown, D., and Orci, L. (1986). Interactions of lectins with specific cell types in toad urinary bladder. Surface distribution revealed by colloidal gold probes and label fracture. *J. Histochem. Cytochem.* **34,** 1057.

Carpentier, J.-L., Brown, D., Iacopetta, B., and Orci, L. (1985). Detection of surface bound ligands by freeze-fracture autoradiography. *J. Cell Biol.* **101,** 887.

Chevalier, J., Pinto da Silva, P., Ripoche, P., Gobin, R., Wang, X. Y., Grossetete, J., and Bourguet, J. (1986). Structural and cytochemical differentiation of membrane elements of the apical membrane of amphibian bladder epithelial cells. A label-fracture study. *Biol. Cell.* **55,** 181.

Chevalier, J., Appay, M. D., Wang, X. Y., Bariety, J., and Pinto da Silva, P. (1987). Freeze-fracture cytochemistry of rat glomerular capillary tuft. Determination of wheat germ agglutinin binding sites and localization of anionic charges. *J. Histochem. Cytochem.* **35,** 1401.

Choi, A. H. C., and Siu, C. H. (1987). Filopodia are enriched in a cell cohesion molecule of Mr 80,000 and participate in cell–cell contact formation in *Dictyostelium discoideum*. *J. Cell Biol.* **104,** 1375.

De Foe, D. M., and Besharse, J. C. (1985). Membrane assembly in retinal photoreceptors. II. Immunocytochemical analysis of freeze-fractured rod photoreceptor membranes using anti-opsin antibodies. *J. Neurosci.* **5,** 1023.

De Harven, E., and Soligo, D. (1986). Scanning electron microscopy of cell surface antigens labeled with colloidal gold. *Am. J. Anat.* **175,** 277.

Dinchuk, F. E., Johnson, T. J. A., and Rash, J. E. (1987). Postreplication labeling of E-leaflet molecules: Membrane immunoglobulins localized in sectioned, labeled replicas examined by TEM and HVEM. *J. Electron Microsc. Tech.* **7,** 1.

9. Label-Fracture Cytochemistry 199

Fawcett, D. W. (1975). The mammalian spermatozoa. *Dev. Biol.* **44**, 394.
Fellous, M., Gachelin, G., Buc-Caron, M. H., Dubois, P., and Jacob, F. (1974). Similar location of an early embryonic antigen on mouse and human spermatozoa. *Dev. Biol.* **41**, 331.
Fisher, K. A., and Branton, D. (1976). Freeze-fracture autoradiography: Feasibility. *J. Cell Biol.* **70**, 453.
Fisher, K. A. (1982). Monolayer freeze-fracture autoradiography: Quantitative analysis of the transmembrane distribution of radioiodinated concanavalin A. *J. Cell Biol.* **93**, 155.
Forsman, C. A., and Pinto da Silva, P. (1988a). Label-fracture of cell surfaces by replica-staining. *J. Histochem. Cytochem.* **36**, 1413.
Forsman, C. A., and Pinto da Silva, P. (1988b). Fracture-flip: New high-resolution images of cell surfaces after carbon stabilization of freeze-fractured membranes. *J. Cell Sci.* **90**, 531.
Fujimoto, K., and Pinto da Silva, P. (1988). Macromolecular dynamics of the cell surface during the formation of coated pits is revealed by fracture-flip. *J. Cell Sci.* **91**, 161.
Geoghegan, W. D., and Ackerman, G. A. (1977). Adsorption of horseradish peroxidase, ovomucoid and anti-immunoglobulin to colloidal gold for the indirect detection of Con A, WGA, and goat anti-human IgG on cell surface at the electron microscope level: A new method, theory, and application. *J. Histochem. Cytochem.* **25**, 1187.
Hayat, M. A. (1986). Glutaraldehyde: Role in electron microscopy. *Micron Microsc. Acta* **17**, 115.
Horisberger, M., and Rosset, J. (1977). Colloidal gold: A useful marker for transmission and scanning electron microscopy. *J. Histochem. Cytochem.* **25**, 295.
Horisberger, M. (1981). Colloidal gold: A cytochemical marker for light and fluorescent microscopy and for transmission and scanning electron microscopy. *Scanning Electron Microsc.* **2**, 9.
Kan, F. W. K., Kopriwa, B. M., and Leblond, C. P. (1984). An improved method for freeze-fracture radioautography of tissues and cells, as applied to duodenal epithelium and thymic lymphocytes. *J. Histochem. Cytochem.* **32**, 17.
Kan, F. W. K., and Pinto da Silva, P. (1986). Preferential association of glycoproteins to the euchromatin regions of cross-fractured nuclei is revealed by fracture-label. *J. Cell Biol.* **102**, 576.
Kan, F. W. K., and Pinto da Silva, P. (1987). Molecular demarcation of surface domains as established by label-fracture cytochemistry of boar spermatozoa. *J. Histochem. Cytochem.* **35**, 1069.
Kan, F. W. K., and Nanci, A. (1988). Backscattered electron imaging of lectin binding sites in tissues following freeze-fracture cytochemistry. *J. Electron Microsc. Tech.* **8**, 363.
Koehler, J. K. (1968). Freeze-etching observations on nucleated erythrocytes with special references to the nuclear and plasma membrane. *Z. Zellforsch. Mikrosk. Anat.* **85**, 1.
Koehler, J. K., and Perkins, W. D. (1974). Fine structure observations on the distribution of antigenic sites on guinea pig spermatozoa. *J. Cell Biol.* **60**, 789.
Koo, G. C., Stackpde, C. W., Boyse, E. A., Hammerling, U., and Lardis, M. P. (1973). Topographical localization of H-Y antigen on mouse spermatozoa by immunoelectron-microscopy. *Proc. Natl. Acad. Sci. U.S.A.* **70**, 1502.
Moor, H., Mühlethaler, K., Waldner, H., and Frey-Wyssling, A. (1961). A new freezing-ultramicrotome. *J. Biophys. Biochem. Cytol.* **10**, 1.
Moor, H., and Mühlethaler, K. (1983). Fine structure of frozen-etched yeast cells. *J. Cell Biol.* **17**, 609.
Myles, D. G., Primakoff, P., and Bellvé, A. R. (1981). Surface domains of the guinea pig sperm defined with monoclonal antibodies. *Cell (Cambridge, Mass.)* **23**, 433.
Myles, D. G., and Primakoff, P. (1984). Localized surface antigens of guinea pig sperm migrate to new regions prior to fertilization. *J. Cell Biol.* **99**, 1631.

Namork, E., Heiter, H. E., and Falleth, E. (1987). Double labeling of cell surface antigens imaged with backscattered electrons. *J. Electron Microsc. Tech.* **6,** 87.

Nermut, M. V., and Williams, L. D. (1980). Freeze-fracture autoradiography of the red blood cell plasma membrane. *J. Microsc. (Oxford)* **118,** 453.

Nermut, M. V. (1983). The cell monolayer technique—an application of solid-phase biochemistry in ultrastructural research. *Trends Biochem. Sci.* **8,** 303.

Nicolson, G. L. (1974). The interaction of lectins with animal cell surfaces. *Int. Rev. Cytol.* **39,** 89.

Peterson, R. N., Russell, L. D., Spauling, G., Bundman, D., Buchanan, J., and Freund, M. (1981). Electrophoretic and chromatographic properties of boar sperm plasma membranes: antigens and polypeptides with affinity for isolated zonae pellucidae. *Int. J. Androl.* **2,** 300.

Peterson, R. N., and Russell, L. D. (1985). The mammalian spermatozoon: A model for the study of regional specificity in plasma membrane organization and function. *Tissue Cell* **17,** 769.

Pinto da Silva, P., and Branton, D. (1970). Membrane splitting in freeze-etching. Covalently bound ferritin as a membrane marker. *J. Cell Biol.* **45,** 598.

Pinto da Silva, P., and Martinez-Palomo, A. (1975). Distribution of membrane particles and gap junctions in normal and transformed 3T3 cells studied *in situ*, in suspension and treated with concanavalin A. *Proc. Natl. Acad. Sci. U.S.A.* **72,** 572.

Pinto da Silva, P., Kachar, B., Torrisi, M. R., Brown, C., and Parkinson, C. (1981a). Freeze-fracture cytochemistry: Replicas of critical point dried cells and tissues after "fracture-label." *Science* **213,** 230.

Pinto da Silva, P., Parkison, C., and Dwyer, N. (1981b). Fracture-label: Cytochemistry of freeze-fracture faces in the erythrocyte membrane. *Proc. Natl. Acad. Sci. U.S.A.* **78,** 343.

Pinto da Silva, P., Parkison, C., and Dwyer, N. (1981c). Freeze-fracture cytochemistry. II. Thin sections of cells and tissues after labeling of fracture faces. *J. Histochem. Cytochem.* **29,** 917.

Pinto da Silva, P., and Kan, F. W. K., (1984). Label-fracture: A method for high resolution labeling of cell surfaces. *J. Cell Biol.* **99,** 1156.

Pinto da Silva, P., Barbosa, M. L. F., and Aguas, A. P. (1984). A guide to fracture-label: Cytochemical labeling of freeze-fracture cells. In *Advanced Techniques in Biological Electron Microscopy* (J. K. Koehler, ed.), Vol. 3, p. 201. Springer-Verlag, Berlin and New York.

Pinto da Silva, P. (1987). Molecular cytochemistry of freeze-fractured cells: Freeze-etching, fracture-label, fracture-permeation, label-fracture. *Adv. Cell Biol.* **1,** 157.

Primakoff, P., and Myles, D. G. (1983). A map of the guinea-pig sperm surface constructed with monoclonal antibodies. *Dev. Biol.* **98,** 417.

Primakoff, P., Hyatt, H., and Myles, D. G. (1985). A role for the migrating sperm surface antigen PH-20 in guinea pig sperm binding to the egg zone pellucide. *J. Cell Biol.* **101,** 2239.

Rash, J. E. (1979). The sectioned-replica technique: Direct correlation of freeze-fracture replicas and conventional thin section images. In *Freeze-Fracture: Methods, Artifacts, and Interpretations* (J. E. Rash and C. S. Hudson, eds.), p. 153. Raven Press, New York.

Rix, E., Schiller, A., and Taugner, R. (1976). Freeze-fracture–autoradiography. *Histochemistry* **50,** 91.

Robenek, H., and Severs, N. J. (1984). Double labeling of lipoprotein receptors in fibroblast cell surface replicas. *J. Ultrastruct. Res.* **87,** 149.

Russell, L. D., Peterson, R. N., and Russell, T. A. (1982). Visualization of anti-sperm plasma membrane IgG and Fab as a method for localization of boar sperm membrane antigens. *J. Histochem. Cytochem.* **30,** 1217.

Russell, L. D., Peterson, R. N., Russell, T. A., and Hunt, W. (1983). Electrophoretic map of boar sperm plasma membrane polypeptides and localization and fractionation of specific polypeptide subclass. *Biol. Reprod.* **28,** 393.

Saxena, N. K., Russell, L. D., Saxena, N., and Peterson, R. N. (1986). Immunofluorescence antigen localization on boar sperm plasma membranes: Monoclonal antibodies reveal apparent new domains and apparent redistribution of surface antigens. *Anat. Rec.* **214,** 238.

Schiller, A., Rix, E., and Taugner, R. (1979). Freeze-fracture–autoradiography: The *in vacuo* coating technique. *Histochemistry* **59,** 9.

Steere, R. L. (1957). Electron microscopy of structural detail in frozen biological specimens. *J. Biophys. Biochem. Cytol.* **3,** 45.

Toullet, F., Voisior, G. A., and Hemirovsky, M. (1973). Histoimmunochemical localization of three guinea pig spermatozoal autoantigens. *Immunology* **24,** 635.

Van Berkel, J., Steup, M., Völker, W., Robenek, H., and Flügge, V. I. (1986). Polypeptides of the choroplast membranes as visualized by immunochemical techniques. *J. Histochem. Cytochem.* **34,** 577.

Weinstein, R. S., and Bullivant, S. (1967). The application of freeze-cleaving technics to studies on red cell fine structure. *Blood* **29,** 780.

10

Colloidal Gold Conjugates for Retrograde Neuronal Tracing

DANIEL MENÉTREY

Inserm, U-161
Physiopharmacologie du système nerveux
Paris, France

and

ALLAN I. BASBAUM

Departments of Anatomy and Physiology
University of California
San Francisco, California

INTRODUCTION
MATERIALS AND METHODS
  Preparation of Colloidal Gold
  Adsorption to Proteins
  Injection
  Survival Time
  Tissue Fixation
  Tissue Sectioning
  Tissue Processing
    Light Microscopy
    Electron Microscopy
APPLICATIONS
  Light Microscopy

Retrograde Labeling
  Use in Conjunction with Other Retrograde Tracers
  Use in Conjunction with Enzymatically (HRP) Localized Tracers
  Use in Conjunction with Immunohistochemistry
Electron Microscopy
  Single Retrograde Labeling
  Multiple Retrograde Labeling
  Use in Combination with Anterograde Tracers
  Use in Combination with HRP-Based Immunohistochemistry
CONCLUDING REMARKS AND PERSPECTIVES
REFERENCES

INTRODUCTION

Although Schwab and Thoenen introduced colloidal gold-labeled tetanus toxin as a retrograde tracer in 1978; it is only recently that the significant advantages of gold-labeled protein tracers have been emphasized. The general principle behind the procedure is to couple a protein that is readily endocytosed by nerve terminals to colloidal gold. The latter serves as the marker which is detected histochemically. Two tracers which we have been studying extensively are the topic of this review. They are colloidal gold-labeled wheat germ agglutinin–horseradish peroxidase (WGA-HRP-Au; Menétrey, 1985; Menétrey and Lee, 1985) and gold-labeled wheat germ agglutinin–apoHRP (WGA-apoHRP-Au; Basbaum and Menétrey, 1987). In the latter molecule, the heme portion of the HRP molecule is removed so that its peroxidase enzymatic activity is eliminated. The two complexes can be used in retrograde tracing studies, including studies of the axon collateralization of single neurons, and in combined immunocytochemical studies. The complexes can be effectively used at both the light and electron microscopic levels. In this review we describe the coupling procedure and illustrate the use of protein–gold tracers in a variety of tracing studies. We compare their advantages and disadvantages with those of more traditional tracers and discuss prospects for new developments in this technology.

MATERIALS AND METHODS:

Preparation of Colloidal Gold

Gold particles are prepared according to Mühlpfordi's procedure (1982). Briefly, 100 ml of tetrahydrochloroauric acid ($AuCl_3HCl \cdot 3H_2O$; MW 393.83) solution (0.01% w/v in distilled water) is boiled while vigorously stirring. Reducing agents [2 ml of trisodium citrate (Na$_3$C$_6$H$_5$O$_7$ · 2H$_2$O; MW 294.10) and 100 µl of tannic acid, both 1% w/v in distilled water] are mixed in a separate beaker and rapidly added to the boiling solution. The reaction is complete ($\approx$20 sec) when the solution quickly turns dark violet and then wine-red. Boiling and stirring are continued for 5 min, and then the solutions are cooled under running water. Solutions should have a final pH of $\approx$5.0. This can be checked on a 1-ml aliquot in the presence of 45 µl of 1% (w/v) polyethylene glycol (PEG; MW 20,000). Gold particles are $\approx$12 nm in diameter. The solutions can be kept at 4°C for several months if supplemented with 2% (w/v) sodium azide solution (0.2% v/v). Note that even minor alterations in the preparation procedure significantly affect the size of the gold particles. Different-size particles can be produced by changing the relative proportion of each reducing agent. If required, the gold particles can be adsorbed onto Formvar-coated grids for 15 min and then their size and dispersive range can be evaluated under the electron microscope. Borate buffer (0.1 M, pH 9.2, 20% v/v) should be added to populations of small particles to enhance their adsorption to the grid. This can be omitted for the particles of the largest size. Other protocols have been developed to get small particles with minor diameter dispersion (Baschong et al., 1985; Van Bergen en Henegouwen and Leunissen, 1986; Handley and Arbeeny, Chapter 7 in this volume). Narrow range gold solutions are also commercially available.

Adsorption to Proteins

The pH range and the amount of protein needed for gold–protein adsorption were determined according to Geoghegan and Ackerman (1977) and Goodman et al. (1981). Good coupling is achieved with 60 µg of WGA-HRP and 15 µg of WGA-apoHRP to 1 ml of gold solution at pH 8.4. These values were obtained for particles prepared with 2 ml of trisodium citrate and 100 µl of tannic acid as reducing agents per 100 ml of tetrahydrochloroauric acid solution. Small variations in the protein–gold ratios could be observed for particles of other sizes or particles obtained by other procedures. The difference in optimal protein–gold ratio for the two protein–gold complexes suggests that each molecule of the enzymatically inactive (apo) complex binds about four times as much gold as the active one. The pH of the gold solution is raised to $\approx$8.4 by adding small quantities ($\approx$4 µl/ml solution) of potassium carbonate (K$_2$CO$_3$, 0.2 M).

The gold solution is added to diluted protein (1 mg protein/ml distilled water) while vigorously stirring. After 5 min, filtered PEG (1% w/v) is added in a proportion of 1% (v/v) of the complete volume of the solution and stirred continuously for another 5 min. The PEG serves to coat the complex and prevents possible aggregation. The complex is centrifuged for 2 hr at 18,000 rpm. We use a Beckman J-21C centrifuge (JA-201 fixed-angle rotor). All but ≈1 ml of supernatant is aspirated and discarded. The soft, dark red pellet containing the protein–gold complex is gently aspirated, leaving the hard blue pellet stuck to the tube surface. The soft pellet is then resuspended in the remaining 1 ml of supernatant as stock solution. The stock solution is microcentrifuged in an Ependorf tube for 30 min at 18,000 rpm just prior to use. Injections are made from this final pellet. Consistent retrograde labeling is obtained when pellets have a final protein concentration of 2.5–5% or 0.5–1% for WGA-HRP or WGA-apoHRP, respectively. This protein percentage is expressed as the original weight of protein per volume obtained in the final pellet. If resuspended in supernatant to avoid desiccation, the pellet can be stored at 4°C for several months with no detectable loss of sensitivity. Care must be taken to avoid contamination of the tracer, since this could result in flocculation and probable inactivation. Some procedures for lyophilization of protein–gold complexes have been described (Baschong and Roth, 1985). Various lectin–gold complexes are now commercially available (E. Y. Laboratories, San Mateo, California; Sigma, St. Louis, Missouri).

Injection

Injections must be made by pressure. Iontophoresis has not been possible. We recommend the use of clean glass micropipettes with a tip diameter of ≈30 μm. This reduces tissue damage during penetration and allows high stereotaxic precision. The pipette is sealed to plastic tubing which is connected to a Hamilton microsyringe and the whole system is filled with mineral oil. The electrode is filled by aspiration through its tip. A little distilled water is then aspirated to avoid tip blocking. The micropipette is rapidly lowered to its final position in the brain. Injections are made over a period of minutes. Even with relatively large volumes there is minimal spread of the injection. Thus small injections can readily be made, a feature that is useful for mapping the afferent connections of discrete brain regions. The limited spread of the complex is probably due to its size and/or the presence of heavy gold particles. Increasing the amount of tracer injected does not substantially increase the size of the injection site. In fact, multiple injections must be made if larger injection sites are required.

Survival Time

The survival time between injection and perfusion steps is not critical. Typically we use survival times comparable to those which are used for WGA or WGA-HRP tracing methods (Schwab *et al.*, 1978; Gonatas *et al.*, 1979; Lechan *et al.*, 1981). For example, a 2-day survival is adequate for 8-cm transport in the rat. Survival times can be prolonged without change in the labeling pattern, which indicates that the gold particles are not subject to neuronal leaking, as are fluorescent dyes, or to transsynaptic transport or enzymatic degradation, as are protein tracers. In fact, we have kept animals up to 19 months and can still detect the tracer in retrogradely labeled cells. Shorter survival times, however, may be better for labeling distal parts of the cell body. The persistence of the tracer indicates that these tracers may be useful for some developmental or aging studies in which the fate of retrogradely labeled cell populations is studied.

Tissue Fixation

The choice of the fixative for tissue preparation is not as critical as it is for other retrograde tracing techniques. For example, fluorescent dyes require the use of monoaldhyde fixatives, such as paraformaldehyde, which are relatively nonfluorescent; HRP-based tracers work best in glutaraldehyde-fixed tissue. We have tried formol (10%), paraformaldehyde (4%), periodate–lysine–paraformaldehyde (4%), glutaraldehyde (3%), *p*-benzoquinone (0.5%) Bouin's, and acrolein (5%) as fixatives without any evidence that sensitivity of the technique is reduced. Postfixing the tissue for several weeks also does not affect the subsequent visualization of the gold. These properties are obviously related to the fact that what is eventually viewed is the stable gold particle, not the degradable protein to which it is conjugated. The wide range of possible fixatives is very advantageous since it permits this technique to be used in combination with many other neuroanatomical procedures (see below). We have not used the technique on unfixed tissue but believe that it will also work.

Tissue Sectioning

Any type of sectioning protocol can be used. To date, we have successfully used the protein–gold technique on frozen (cryostat or freezing microtome), paraffin, polyethylene glycol-embedded, or vibratome-sectioned tissue. The sections ranged from 6 to 50 μm in thickness. Sections are collected in phosphate buffer (PB; 0.1 *M*, pH 7.4). It is of particular value that the core of the injection site can be seen without further processing of the tissue.

Tissue Processing
Light Microscopy

The small size of gold particles is beyond the resolving power of the light microscope (LM) and thus it is only when the particles are aggregated in very heavily labeled cells that they can be detected. By intensifying the gold, however, it is possible to demonstrate that the number of gold-labeled cells detected at the LM level is only a small percentage of the total number of retrogradely labeled cells in the tissue. The intensification procedure we use was adapted from the silver intensification protocol of Danscher (1981), with buffer rinses replacing water rinses to improve tissue preservation. In this procedure the gold particles act as specific cores of nucleation for the deposition of metallic silver particles that are generated by reduction of a silver salt, such as silver lactate. The shell of silver that develops around the gold particle gives rise to a metallic speck directly visible under the light microscope. The silver intensification step can be done on either free-floating or slide-mounted sections with comparable success. It is performed in a dark box or under safe-light illumination and at room temperature (18–22°C). Keeping the temperature of the reaction in this range is essential for consistent results. Higher temperatures increase the reaction speed and can lead to an undesirable browning of the sections. It is essential to avoid any chloride salt in the reaction media and to keep materials as clean as possible to reduce background deposits. All reaction vessels are cleaned with bleach and rinsed well in distilled water.

After a brief rinse (5 min) in sodium citrate buffer (0.1 M, pH 3.8) sections are dipped in the physical developer for 1 hr. Although sections can remain in the developer for a longer period of time, they should be removed when myelinated fibers begin to turn brown. Each 100 ml of physical developer consists of 60 ml gum arabic solution (50% w/v in distilled water) to prevent autodevelopment of silver, 10 ml citrate buffer (1 M, pH 3.5), 15 ml hydroquinone (5.6% w/v in distilled water) as a reducing agent, and 15 ml silver lactate (0.7% w/v in distilled water) to supply silver ions. The silver lactate solution is carefully protected against light and added just before developer is used. After development, sections are washed twice (2 × 5 min) in PB (0.1 M, pH 7.4), put into sodium thiosulfate (2.5% w/v in PB) for 5 min, and finally rinsed for 5 min in PB. At this point, it is possible to identify labeled cells by screening the wet sections with bright-field illumination. After washing, the sections can be counterstained with neutral red or safranin. Silver deposits are readily detected with both bright- and dark-field illumination. Since it takes up to 2 days to dissolve gum arabic, we recommend that stock solutions be made up ahead of time and stored as aliquots at −20°C until needed. An alternative protocol for silver intensification is the neutral pH procedure developed by Janssen Life Sciences, Beerse, Belgium. Other intensification procedures that may give comparable results have been developed (Danscher, 1984; Danscher et al., 1987; Kopriwa, 1975); however, we have not tried them in retrograde transport studies.

Electron Microscopy

Since the fine gold particles are electron dense, they can be studied at the electron microscope (EM) level without silver enhancement. This is of particular value if gold particles of different sizes are to be used in multiple labeling studies. However, if only single labeling studies are planned, it is better to process the tissue with the silver intensification step. This makes detection of the labeled cells much easier. The osmication step that is required for EM, unfortunately, results in oxidation and significant loss of the silver precipitate. This can be prevented by gold toning the sections, i.e., substituting the silver deposits with metallic gold (Feigin and Naoumenko, 1976; Fairen et al., 1977) before osmication. The gold-toning steps are all performed at 4°C with continuous agitation in solutions made in distilled water. Sections are incubated for 10 min in the dark in 0.05% (w/v) tetrahydrochloroauric acid (in distilled water) and then washed three times in distilled water. The gold is then reduced with 0.05% (w/v) oxalic acid for 2 min. After three additional washes with distilled water, the residual, unreduced metal (silver and/or gold) is eliminated by washing the sections in 1% (w/v) sodium thiosulfate (3 × 20 min).

Although we have successfully used this approach for EM analysis (see below), it should be indicated that the ultrastructural quality of silver-intensified tissue is not ideal. Presumably the low pH of the silver reaction and possibly the gold-toning step are harmful to the tissue preservation. We are evaluating different silver enhancement procedures that we hope will overcome this problem.

APPLICATIONS

Although we checked for both anterograde and retrograde transport of the gold-labeled protein, we found only retrograde labeling, even after silver enhancement. All applications of the protein–gold markers are thus limited to retrograde tracing studies. The tracers can easily be combined with other techniques in various multiple-labeling studies. Except where indicated below, we recommend the use of the WGA-apoHRP-Au complex. The fixative of choice will vary with the requirements of the particular experiment.

Light Microscopy

Retrograde Labeling

Figure 10.1 is a bright-field photomicrograph of a silver-enhanced neuron in the cervical spinal cord that was retrogradely labeled after injection of protein–gold complex at the lumbar level. The cell body and proximal dendrites contain round black granules. It is significant that there is extensive dendritic labeling, a feature which is important for EM studies. Since there is almost no background silver labeling, detection of retrogradely labeled cells is very easy. When viewed with dark-field illumination, the

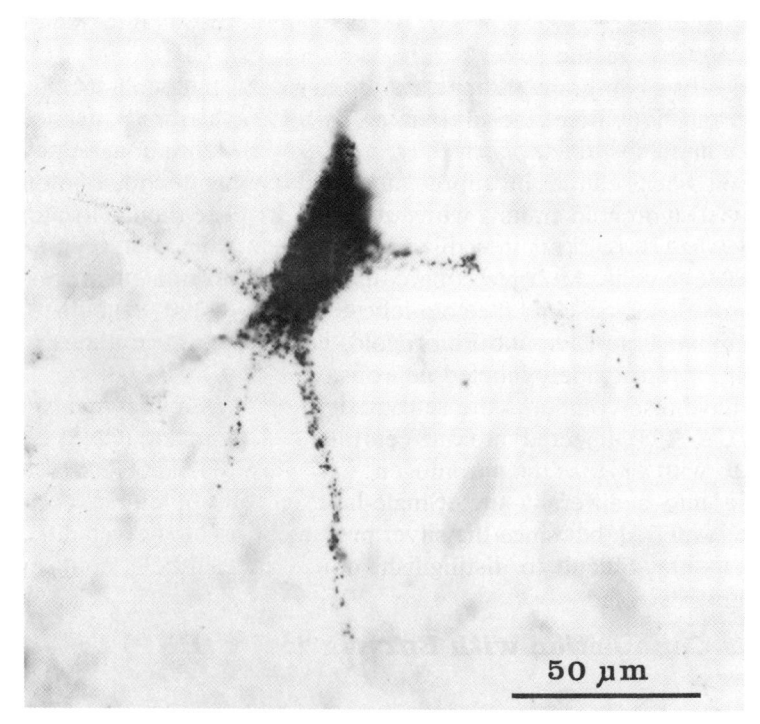

Fig. 10.1. Photomicrograph of a neuron in the cervical spinal cord that was retrogradely labeled after lumbar injection of WGA-HRP-Au. The section was silver enhanced and counterstained with neutral red; ×500.

silver particles appear as bright golden dots. There is some incorporation of the tracer after injection into bundles of fibers, which indicates that the protein–gold complex can be taken up by injured axons. However, the number of cells found after injection into white matter tract is much less than is seen when tetramethylbenzidine (TMB); Mesulam, 1982) is used as the chromogen to detect WGA-HRP injected into the same site. Control experiments established that the natural argyrophilia of neurons is not a problem and that the silver labeling is not related to degenerating axons or cell bodies.

Use in Conjunction with Other Retrograde Tracers

Silver intensification does not interfere with the detectability of the retrogradely transportable dyes, such as True Blue (Bentivoglio *et al.,* 1979) or Fluorogold (Schmued and Fallon, 1986), and thus studies of neurons with bifurcating axons can readily be performed. The protein–gold and fluorescent tracers are injected into different targets. Survival times are typically determined by the most appropriate time for the fluorescent tracer. A paraformaldehyde fixative is used. By simultaneously illuminating a section with epifluorescent and transmitted tungsten dark-field illumination, the dye and silver can be viewed at the same time. In single-labeled cells silver granules appear as golden dots. In double-labeled cells that contain True Blue, the silver takes on a pinkish granular cast against the blue fluorescent cytoplasm. Unfortunately, the ease of detecting double-labeled cells cannot be appreciated in a black-and-white photograph. Since the Fluorogold emits a white-gold fluorescence (hence its name), it is somewhat more difficult to distinguish the silver precipitate in Fluorogold-positive cells. Multiple collaterals could be examined with various combinations of dyes and the gold-labeled tracer. A disadvantage of some dyes, unlike True Blue and Fluorogold, is that they have a tendency to leak out of retrogradely labeled neurons.

We have also combined the retrograde transport of enzymatically active WGA-HRP, injected into one central nervous system (CNS) site and revealed with TMB as the chromogen, with the WGA-apoHRP-Au tracer injected into a different site. Single-labeled cells of either kind were readily identified, but since the silver precipitate and the crystalline TMB product were difficult to distinguish, double-labeled cells could not be accurately detected.

Use in Conjunction with Enzymatically (HRP) Localized Tracers

WGA-apoHRP-Au is of particular value when one wishes to evaluate the arborization pattern of afferents to an identified group of retrogradely labeled cells. It is possible to combine the anterograde transport of the lectin *Phaseolus vulgaris* leukoagglutinin (PHA-l; Gerfen and Sawchenko, 1984), localized with HRP-based immunocytochemistry and DAB as the chromogen, with the retrograde transport of WGA-apoHRP-Au. Conceivably, the anterograde transport of WGA-HRP could also be combined; however, since WGA-HRP is also transported retrogradely, confusion may arise if some cells are retrogradely labeled with both tracers. We believe that the PHA-l approach will prove most useful and may have its greatest utility at the EM level.

Use in Conjunction with Immunohistochemistry

Protein–gold tracers are particularly useful when combined with immunocytochemistry to demonstrate the cytochemistry of the projection neurons or the cytochemistry of afferents that project to retrogradely labeled cells. Studies of the latter type would best be performed at the EM level. Since gold-labeled tracers can be localized with almost any fixative, it is possible to do double-labeling studies with antisera against a variety of antigens, including peptides (the antigenicity of which is best preserved by paraformaldehyde fixation), and amino acids (the antigenicity of which is often dependent on glutaraldehyde being used). The gold-labeled tracer can be combined with HRP- or fluorescent-based immunocytochemical procedures. We have, in fact, found that retrogradely labeled peptide-immunoreactive cells can even be detected in rats that survived 19 months after injection of the WGA-apoHRP-Au.

With HRP-based Immunohistochemistry

Figure 10.2 illustrates retrogradely labeled neurons in lamina X of spinal cord that resulted from an injection of WGA-apoHRP-Au into the brain stem reticular formation. The tissue was silver intensified and then immunostained with an antiserum to cholecystokinin (CCK) using the avidin–biotin method (Hsu *et al.*, 1981) with DAB as a chromogen. To prevent darkening of the DAB immunoreaction product, it is essential that the silver intensification step be performed first. Cholecytostokinin-immunoreactive cells that project to the brain stem exhibit diffuse (reddish-brown) stain as well as distinct, round black granules; nonprojecting immunoreactive cells contain only diffuse cytoplasmic staining. The silver granules stand out clearly against the reddish-brown DAB immunoreaction product and thus double-labeled cells are readily detected. There is minimal background silver labeling, so there is little likelihood of false positive reactions.

With Fluorescent-based Immunocytochemistry

In these studies it is critical to use a nonfluorescent monoaldehyde fixative. Figure 10.3 shows examples of double-labeled cells in the locus

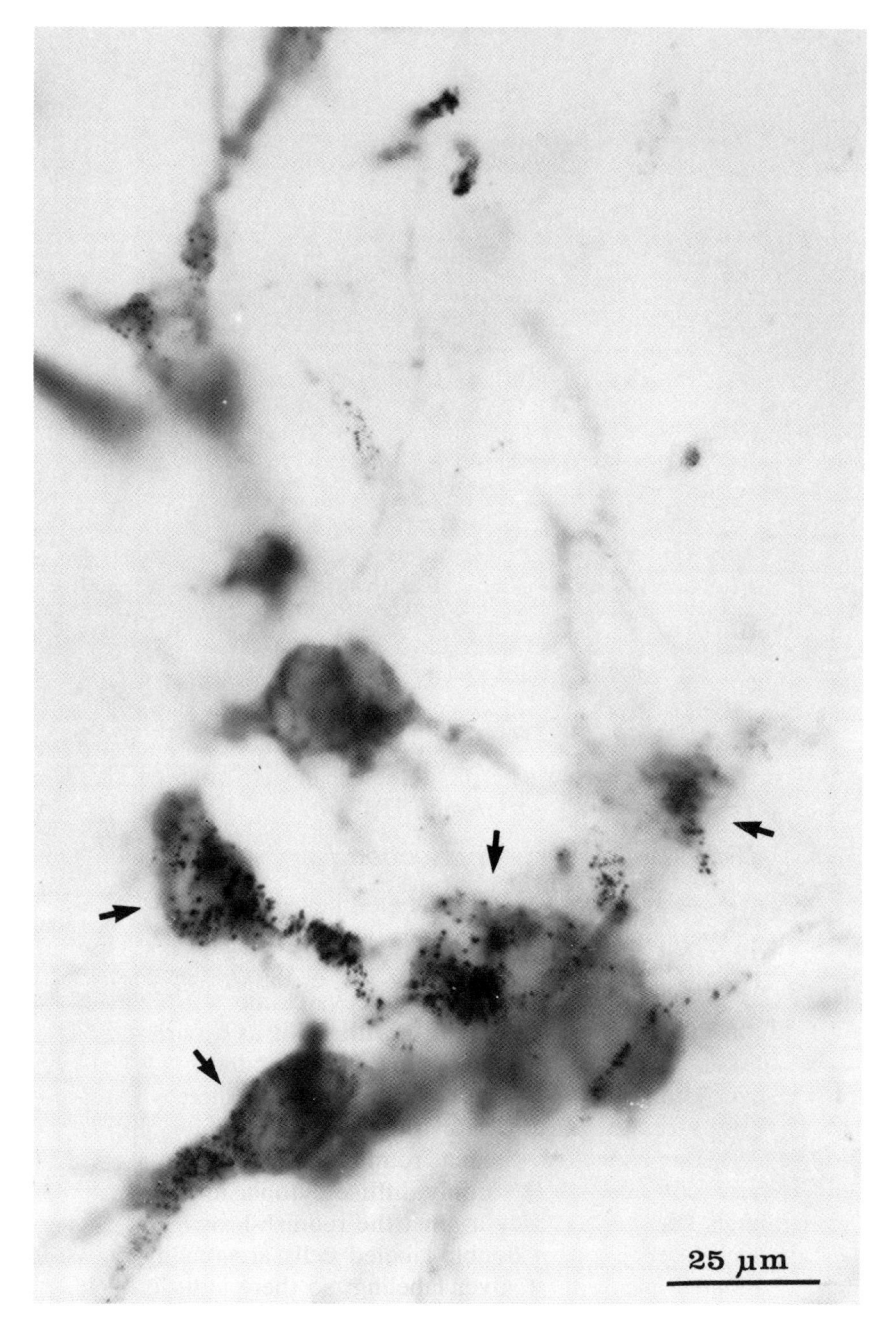

Fig. 10.2. Photomicrograph of neurons in lamina X of the lumbar spinal cord that were retrogradely labeled after injection of WGA-apoHRP-Au into the brain stem reticular formation. The section was silver enhanced and immunostained with an antiserum against cholecystokinin using DAB as the chromogen. Double-labeled cells (arrows) contain black dots within a diffuse, brown immunostained neuron; ×800.

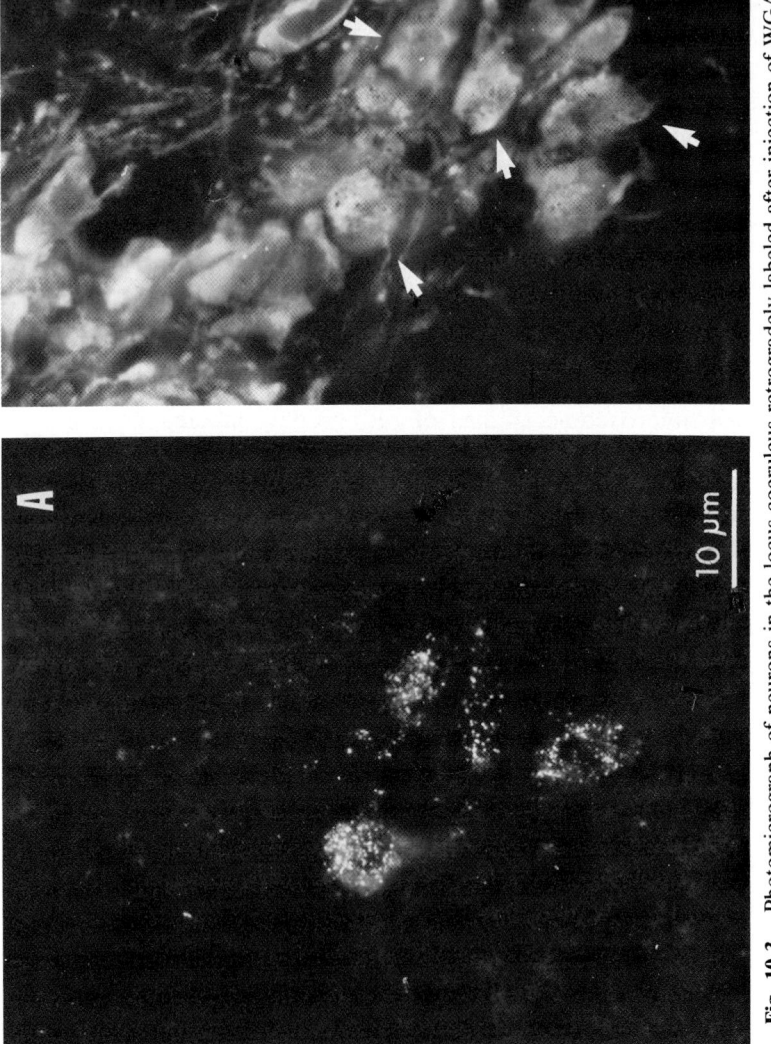

Fig. 10.3. Photomicrograph of neurons in the locus coeruleus retrogradely labeled after injection of WGA-apoHRP-Au into the lumbar spinal cord. The sections were silver enhanced and then immunostained with an antiserum against tyrosine hydroxylase using a fluorescein-conjugated secondary antibody. The sections were illuminated with transmitted tungsten light, with dark-field condenser (A) and with epifluorescent illumination (B; Leitz 12 filter cube). Double-labeled cells (arrows) contain bright, golden dots within the diffuse green cytoplasm; ×1500.

coeruleus immunostained with antisera directed against tyrosine hydroxylase (TH), using a fluorescein-conjugated second antibody. These neurons were retrogradely labeled after injection of protein–gold into the spinal cord. The sections were illuminated with epifluorescent light (Leitz, filter system I2) and transmitted tungsten (dark field). Tyrosine hydroxylase-positive neurons that project to the spinal cord contain golden dots (Fig. 10.3A) in a diffuse (green) cytoplasm generated by the fluorescein (Fig. 10.3B). Recognition of double-labeled cells is very simple; the typical silver-gold color of the retrogradely labeled cells "turns" a golden yellow when viewed in a fluorescein-positive cell. Again, these particular features and their simultaneous visualization cannot be appreciated in black-and-white photomicrographs.

Triple labeling of cells can, of course, be achieved by combining the retrograde transport of protein–gold complexes with that of various fluorescent tracers and using immunofluorescence to characterize the cytochemistry of neurons with collateralizing axons (Fig. 10.4)

Electron Microscopy

The properties of the protein–gold tracer are particularly well suited for electron microscopic studies of retrogradely labeled neurons. The tracers can be visualized by directly scanning colloidal gold particles (Fig. 10.5), or after silver enhancement and gold toning, which simplifies the identification of single-labeled cells. Both the colloidal gold particles and the silver deposits are easily distinguished from other electron-dense structures in the tissue.

Single Retrograde Labeling

The electron micrograph in Fig. 10.6 illustrates a retrogradely labeled cell body (A) and dendrite (B) in the nucleus raphe magnus of the medulla after an injection of WGA-apoHRP-Au into the cervical spinal cord. The tissue was silver enhanced and gold toned prior to osmication. Note that the background is very low. The label is restricted to the cytoplasm of the cell body and dendrites; it does not spread into surrounding neuropil. Although some nonlysosomal structures (Golgi apparatus) are labeled, the silver particles are predominantly located over lysosomes. The arrowheads point out very heavily labeled lysosomes. The latter have a characteristic distribution of the metal. There is a central, electron-dense core that is separated from another electron-dense core surrounded by a clean, thin annulus. The gold- or silver-enhanced particles are highly electron dense and thus there is no difficulty identifying label in lysosomes which are "normally" electron dense. At times the labeling is so intense that the underlying organelle is difficult to identify.

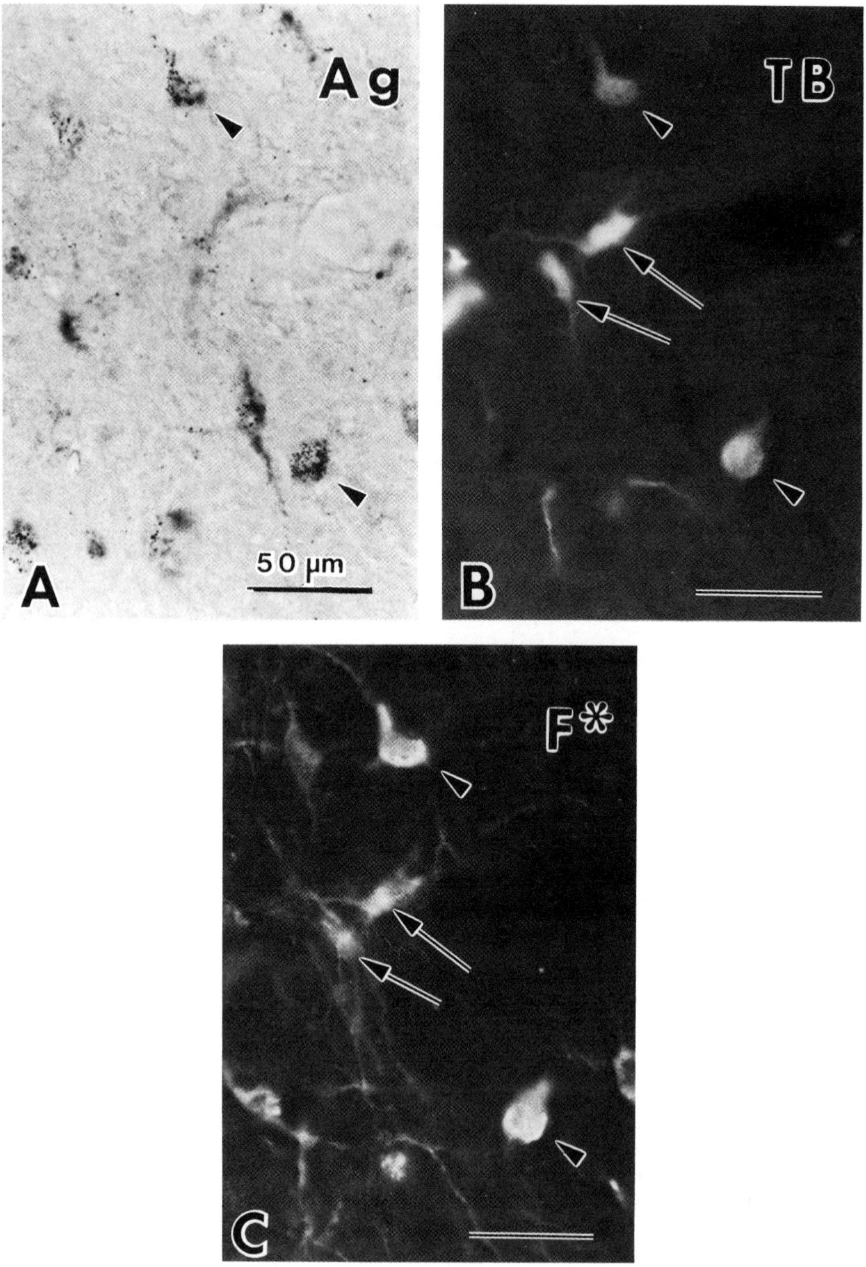
Ag
A
50 μm
TB
B
F*
C

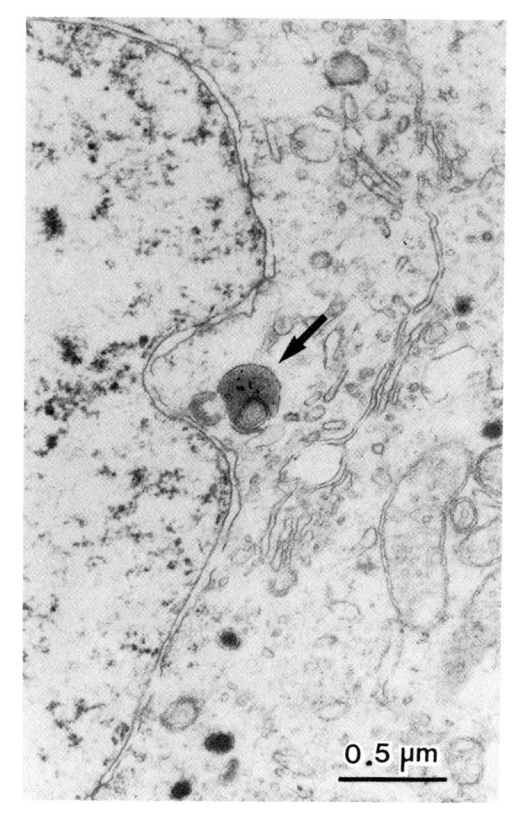

10. Conjugates for Retrograde Neuronal Tracing

Fig. 10.5. Electron micrograph illustrating colloidal gold particles in a lysosome (arrow) of a retrogradely labeled cell in the nucleus raphe magnus of the medulla after injection of WGA-apoHRP-Au into the spinal cord. The sections were neither silver enhanced nor gold toned, as they were for Fig. 10.6, to which they can be compared; ×28,750.

Fig. 10.4. Examples of triple-labeled neurons in the A5 cell group of the rostral medulla. WGA-apoHRP-Au was injected into the nucleus raphe magnus and the fluorescent retrograde tracer True Blue was injected into the spinal cord. Sections containing silver-enhanced cells were immunoreacted with an antiserum directed against tyrosine hydroxylase (TH), using a fluorescein-conjugated secondary antibody. Transmitted tungsten, bright-field illumination (A) reveals silver (Ag)-positive cells. Epifluorescent illumination of the same section with a different filter combination reveals True Blue (TB; Leitz A filter cube)-positive cells (B) and TH, i.e., fluorescein (F*; Leitz I2 filter cube)-positive cells (C). Arrowheads identify triple-labeled cells; arrows in B and C point to double-labeled cells that did not contain silver particles; ×340. (From Basbaum and Menétrey, 1987; reproduced with permission of the *Journal of Comparative Neurology*.)

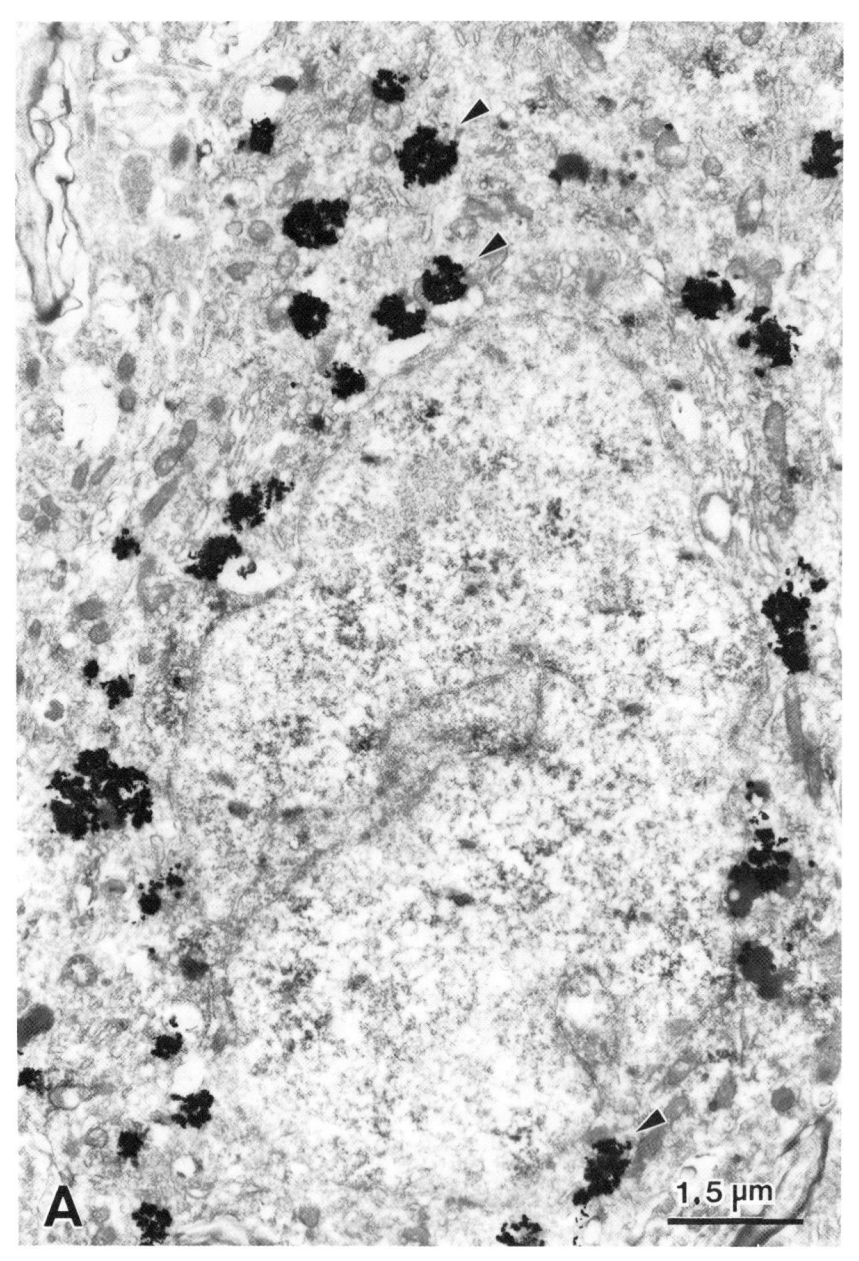

Fig. 10.6. Examples of retrograde labeling at the EM level after silver enhancement and gold toning. Cell body (A; ×12,000) and dendrite (B; ×25,000) contain heavily labeled lysosomes (arrowheads) and labeled nonlysosomal structures (arrows). (From Basbaum and Menétrey, 1987; reproduced with permission of the *Journal of Comparative Neurology*.)

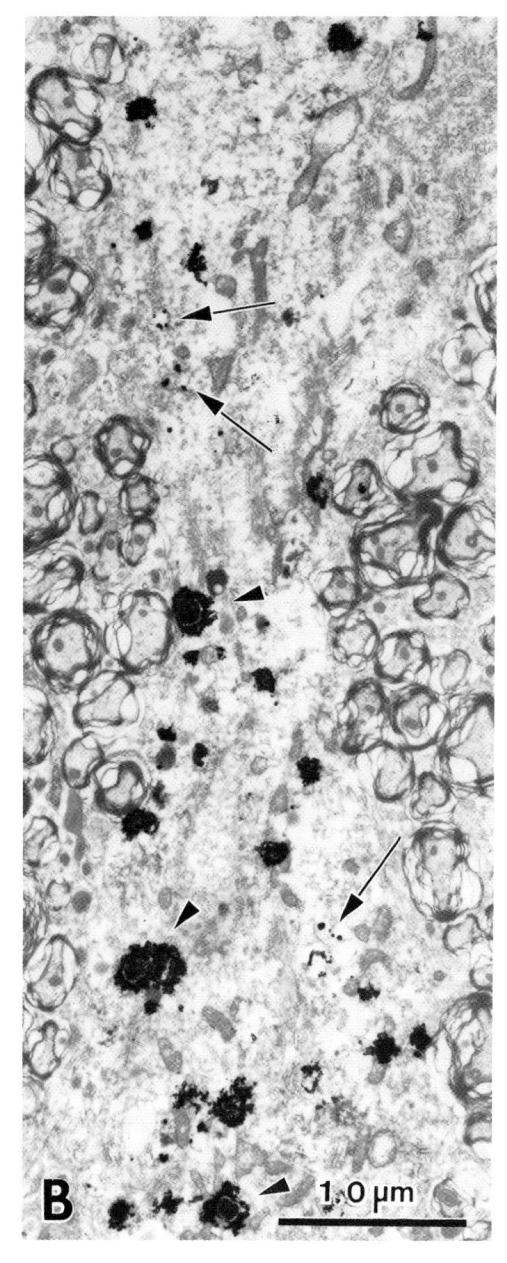

Fig. 10.6. *(Continued)*

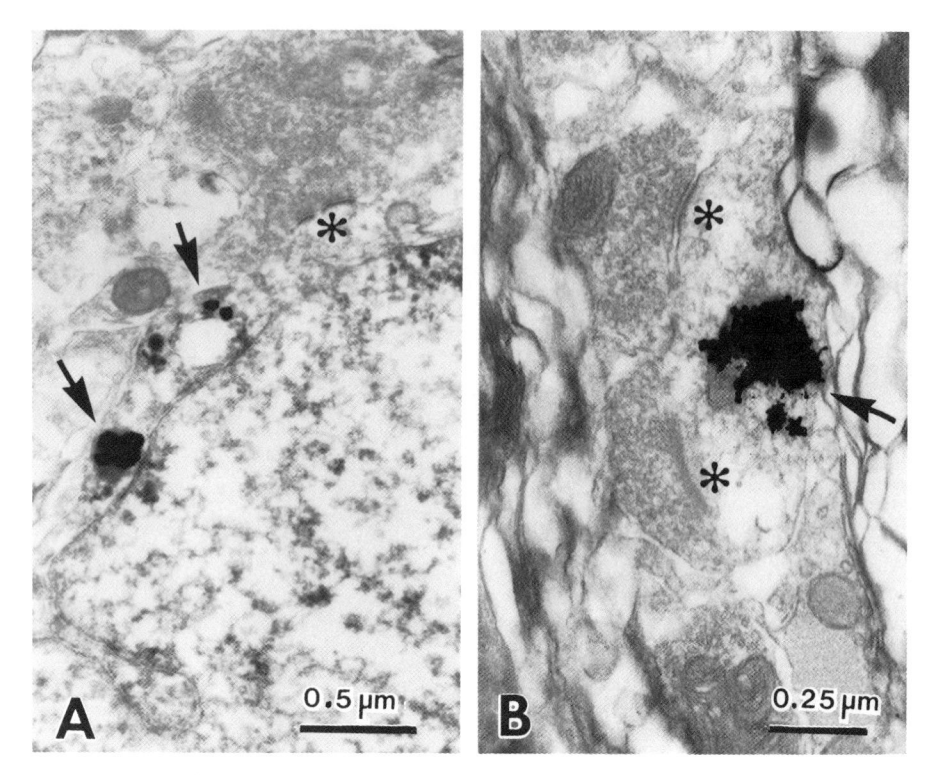

Fig. 10.7. Electron micrographs illustrating synaptic inputs (asterisk) onto a WGA-apoHRP-Au retrogradely labeled cell body (A; ×30,000) and distal dendrite (B; ×52,000) from silver-enhanced and gold-toned sections. The silver-enhanced reaction product is indicated by arrows in A and B. (From Basbaum and Menétrey, 1987; reproduced with permission of the *Journal of Comparative Neurology*.)

Figure 10.7A and B illustrate synaptic contacts onto WGA-apoHRP-Au retrogradely labeled cell body and distal dendrite, respectively. Both the very low background and the sensitivity of the procedure make detection of labeling of small dendritic profiles very reliable. Since the majority of synaptic inputs are found on distal dendrites, the ability to identify small dendrites that are retrogradely labeled is an important feature.

Multiple Retrograde Labeling

Gold particles of discrete, nonoverlapping sizes can be synthesized and coupled to the protein. These can be used as direct, specific markers for EM detection in multiple retrograde labeling studies in which the different tracers are injected into different brain regions. Since the gold cannot be visualized at the LM level, it is difficult to define the region of tissue which contains the labeled cells and from which the EM study is to be performed. To overcome this problem we use the active HRP conjugate. The distribution of labeled cells is first identified at the LM level by HRP histochemistry. Next the area containing labeled cells is embedded and processed for EM.

Use in Combination with Anterograde Tracers

As indicated above, protein–gold complexes may be of particular value for studying afferent inputs to retrogradely labeled cells. To this end, one of the following approaches can be used. It is possible to combine WGA-apoHRP-Au with the anterograde transport of WGA-HRP localized with TMB (Mesulam, 1982) or benzidine dihydrochloride (BDHC) (Lakos and Basbaum, 1986) as the chromogen. This approach offers high sensitivity of the anterograde label; however, the fact that there may be simultaneous retrograde labeling of cells with WGA-HRP must be considered. Alternatively, anterogradely transported PHA-1 (localized by HRP-based immunocytochemistry with DAB as the chromogen) can be used. Finally, anterograde tracing with tritiated amino acids could be used. EM autoradiography after silver intensification, although time consuming, is feasible.

Use in Combination with HRP-Based Immunocytochemistry

Cytochemistry of Projection Neurons

Figure 10.8 illustrates an example of a double-labeled raphe-spinal neuron after injection of WGA-apoHRP-Au into the cervical spinal cord and use of immunocytochemistry with an antiserotonin antibody to identify the cytochemistry of retrogradely labeled raphe-spinal neurons. The DAB immunoreaction product is dark and flocculent and is readily distinguished from the very dense gold precipitate that denotes the presence of the WGA-apoHRP-Au. Note that there is also a serotonin immunoreaction product in the nucleus of the cell, a characteristic feature of serotonin-associated neurons.

Cytochemistry of Afferents to Projection Neurons

Figure 10.9 illustrates a neuron of the midbrain periaqueductal gray (PAG) that was retrogradely labeled with WGA-apoHRP-Au from the medullary nucleus raphe magnus. Sections containing labeled cells were immunoreacted with an antiserum directed against γ-aminobutyric acid (GABA). This figure illustrates that there are axosomatic GABA-immunoreactive contacts that are presynaptic to the retrogradely labeled PAG cells.

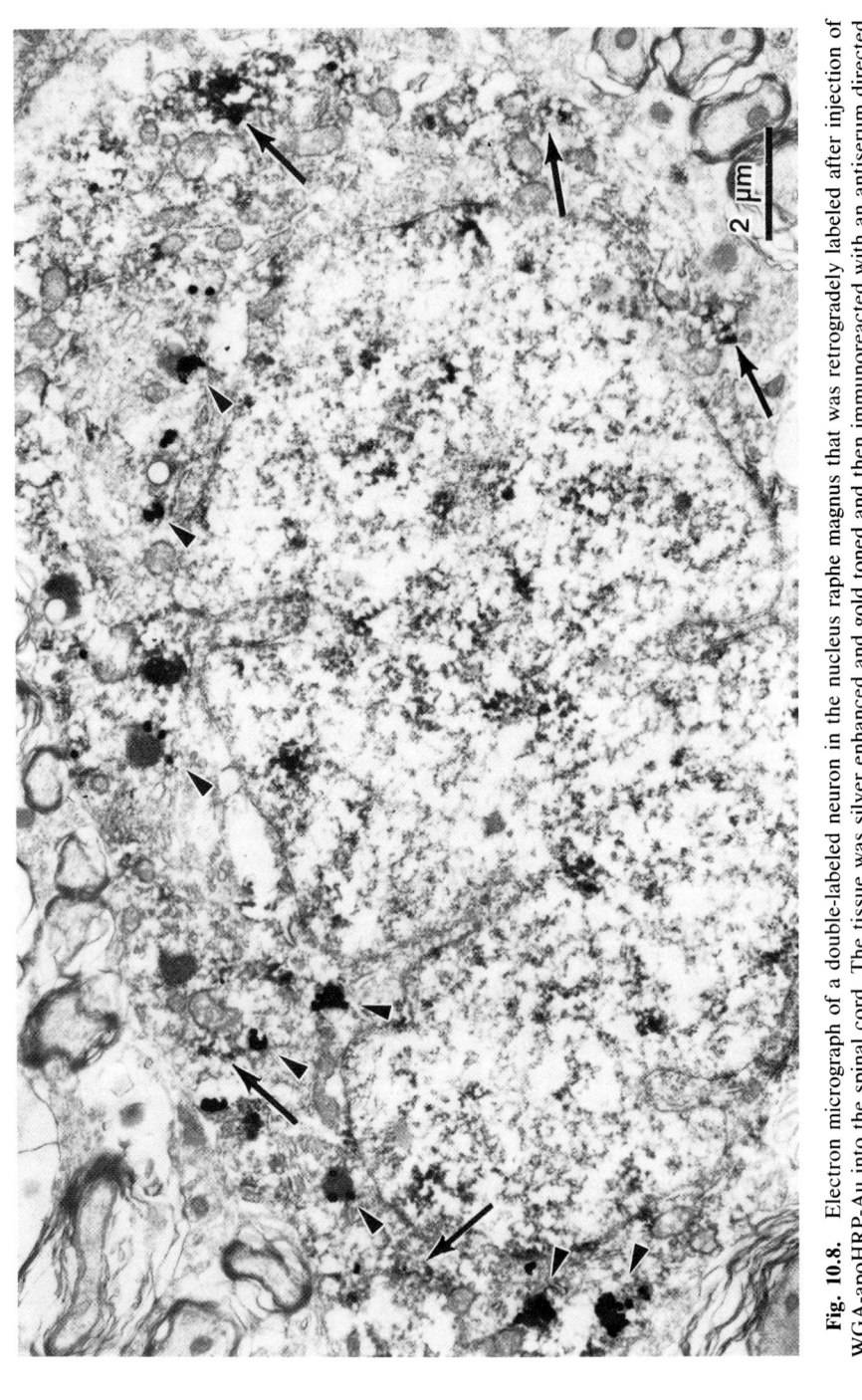

Fig. 10.8. Electron micrograph of a double-labeled neuron in the nucleus raphe magnus that was retrogradely labeled after injection of WGA-apoHRP-Au into the spinal cord. The tissue was silver enhanced and gold toned and then immunoreacted with an antiserum directed against serotonin using DAB as the chromogen. Arrowheads point to the gold deposits in the cytoplasm; arrows point to the flocculent, DAB reaction immunoreaction product; × 7500. (From Basbaum and Menétrey, 1987; reproduced with permission of the *Journal of Comparative Neurology*.)

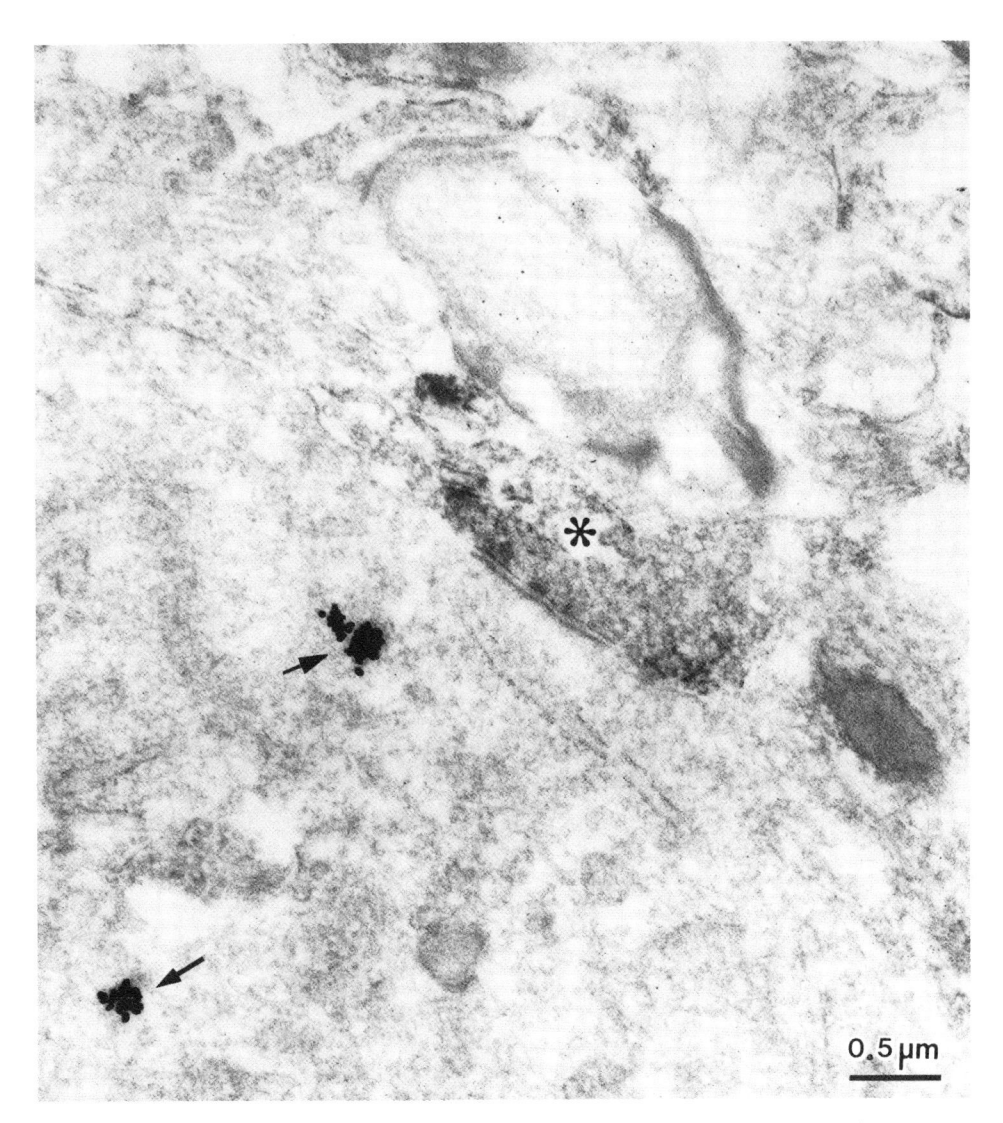

Fig. 10.9. Electron micrograph of a neuron in the midbrain periaqueductal gray that was retrogradely labeled after injection of WGA-apoHRP-Au into the nucleus raphe magnus of the medulla. The section was silver enhanced and gold toned and then immunoreacted with an antiserum directed against γ-aminobutyric acid. Immunoreactive GABA terminals (asterisk) are located presynaptic to the retrogradely labeled neuron. The silver-enhanced reaction product is indicated by arrows; ×45,000.

CONCLUDING REMARKS AND PERSPECTIVES

We believe that this review establishes that protein–gold complexes can be extremely useful for retrogradely tracing neural connections at either the LM or EM level. The ease with which these tracers can be used in multiple-labeling studies is a great advantage. They can be used with various fixative solutions and can be combined with a variety of HRP- or fluorescent-based tracing and immunocytochemical procedures. It is highly probable that the high sensitivity of the protein–gold procedure is related to an adsorptive capture of the complex through binding of WGA to surface receptors. A significant although unexplained fact is that the protein–gold complexes do not appear to be transported in the anterograde direction. One possibility is that the gold is dissociated from the complex in the cell body. Although the protein may be transported anterogradely, the loss of the gold would make the tracer undetectable, even after silver intensification. Alternatively, the very large size of the complex may hinder anterograde transport. Note that PEG-coated colloidal gold particles, by themselves, can be taken up and transported to the cell body if injected in large amounts in the central nervous system. The sensitivity in this case is, however, far below that seen for the protein–gold procedure. The fact that colloidal gold alone is transported suggests that altering the size of the complex may be an effective approach to the development of new, more efficient tracers.

Since colloidal gold can be efficiently coupled to a wide range of proteins, it should not be difficult to create and test new complexes that are retrogradely transported. Free lectins, including WGA, have, in fact, been gold labeled. With WGA, however, the gold–protein ratio is much lower than with WGA-apoHRP-Au. Presumably, the larger the protein "carrier" molecule, the greater the amount of gold that can be complexed (De Roe *et al.*, 1987). An interesting alternative may be to couple gold to a WGA dimer. This might enhance retrograde transport by increasing the adsorptive capture of the complex to surface sugars which are bound by WGA. Protein molecules which are transported with greatest efficiency and sensitivity, such as cholera toxin (Shapiro and Miselis, 1985), may also prove useful for gold labeling.

Since the complexes can be taken up and stored for long periods by cultured cells, they can be used in development or grafting studies. Smits-Van Prooije *et al.* (1987) have injected a WGA-Au complex into the amniotic cavity and used it to follow the migration of embryonic cells in culture. Seeley and Field (1988) have labeled embryonic hippocampal cells in suspension culture before transplanting into the brains of adult host animals to follow the fate of transplanted neurons. Other histochemical uses of these complexes can be envisaged, including the combination of retrograde tracing with *in situ* hybridization techniques or with immunogold histochemistry. Finally, the development of gold-labeled anterograde tracers and new silver intensification procedures that are compatible with good tissue preservation will prove extremely beneficial in neuroanatomical studies.

We thank David Reichling for permission to reproduce Fig. 10.9. This work was supported by NIH grants NS21445 and 14627 and by the CNRS and INSERM, France.

REFERENCES

Basbaum, A. I., and Menétrey, D. (1987). Wheatgerm agglutinin-apoHRP gold: A new retrograde tracer for light- and electron-microscopic single- and double-label studies. *J. Comp. Neurol.* **261,** 306.

Baschong, W., and Roth, J. (1985). Lyophilization of protein–gold complexes. *Histochemistry* **17,** 1147.

Baschong, W., Lucocq, J. M., and Roth, J. (1985). "Thiocyanate gold": Small (2–3 nm) colloidal gold for affinity cytochemical labeling in electron microscopy. *Histochemistry* **83,** 409.

Bentivoglio, M., Kuypers, H. G. J. M., Catsman-Berrevoets, C. E., and Dann, O. (1979). Fluorescent retrograde neuronal labeling in rat by means of substances binding specifically to adenine–thymine rich DNA. *Neurosci. Lett.* **12,** 235.

Danscher, G. (1981). Localization of gold in biological tissue. A photochemical method for light and electronmicroscopy. *Histochemistry* **71,** 81.

Danscher, G. (1984). Autometallography. A new technique for light and electron microscopic visualization of metals in biological tissues (gold, silver, metal sulphides and metal selenides). *Histochemistry* **81,** 331.

Danscher, G., Rytter Norgaard, J. O., and Baatrup, E. (1987). Autometallography: Tissue metals demonstrated by a silver enhancement kit. *Histochemistry* **86,** 465.

De Roe, C., Courtoy, P. J. and Baudhuin, P. (1987). A model of protein–colloidal gold interactions. *J. Histochem. Cytochem.* **35,** 1191.

Fairen, A., Peters, A., and Saldanha, J. (1977). A new procedure for examining Golgi impregnated neurons by light and electron microscopy. *J. Neurocytol.* **6,** 311.

Feigin, I., and Naoumenko, J. (1976). Some chemical principles applicable to some silver and gold staining methods for neuropathological studies. *J. Neuropathol. Exp. Neurol.* **35,** 495.

Geoghegan, W. D., and Ackerman, G. A. (1977). Adsorption of horseradish peroxidase, ovomucoid and anti-immunoglobulin to colloidal gold for the indirect detection of concanavalin A, wheat germ agglutinin and goat anti-human immunoglobulin G on cell surfaces at the electron microscopic level: A new method, theory and application. *J. Histochem. Cytochem.* **25,** 1187.

Gerfen, C. R., and Sawchenko, P. E. (1984). An anterograde neuroanatomical tracing method that shows the detailed morphology of neurons, their axons and terminals:

Immunohistochemical localization of an axonally transported plant lectin, *Phaseolus vulgaris* leucoagglutinin (PHA-1). *Brain Res.* **290**, 219.

Gonatas, N. K., Harper, C., Mizutani, T., and Gonatas, J. O. (1979). Superior sensitivity of conjugates of horseradish peroxidase with wheat germ agglutinin for studies of retrograde axonal transport. *J. Histochem. Cytochem.* **27**, 728.

Goodman, S. L., Hodges, G. M., Trejdosiewicz, L. K., and Livingston, D. C. (1981). Colloidal gold markers and probes for routine application in microscopy. *J. Microsc. (Oxford)* **123**, 201.

Hsu, S., Raine, L., and Fanger, H. (1981). A comparative study of the peroxidase–antiperoxidase method and an avidin–biotin complex method for studying polypeptide hormones with radioimmunoassay antibodies. *Am. J. Clin. Pathol.* **75**, 734.

Kopriwa, B. M. (1975). A comparison of various procedures for fine grain development in electron microscopy radioautography. *Histochemistry* **44**, 201.

Lakos, S., and Basbaum, A. I. (1986). Benzidine dihydrochloride as a chromogen for single- and double-label light and electron microscopic immunocytochemical studies. *J. Histochem. Cytochem.* **34**, 1047.

Lechan, R. M., Nestler, J. L., and Jacobson, S. (1981). Immunohistochemical localization of retrogradely and anterogradely transported wheat germ agglutinin (WGA) within the central nervous system of the rat: Application to immunostaining of a second antigen within the same neuron. *J. Histochem. Cytochem.* **29**, 1255.

Menétrey, D. (1985). Retrograde tracing of neural pathways with a protein–gold complex. I. Light microscopic detection after silver intensification. *Histochemistry* **83**, 391.

Menétrey, D., and Lee, C. L. (1985). Retrograde tracing of neural pathways with a protein–gold complex. II. Electron microscopic demonstration of projections and collaterals. *Histochemistry* **83**, 525.

Mesulam, M. M. (1982). *Tracing Neural Connection with Horseradish Peroxidase.* Wiley, New York.

Mühlpfordi, H. (1982). The preparation of colloidal gold particles using tannic acid as an additional reducing agent. *Experientia* **38**, 1127.

Schmued, L. C., and Fallon, J. H. (1986). Fluoro-gold: A new fluorescent retrograde axonal tracer with numerous unique properties. *Brain Res.* **377**, 147.

Schwab, M. E., and Thoenen, H. (1978). Selective binding, uptake, and retrograde transport of tetanus toxin by nerve terminals in the rat iris. An electron microscope study using colloidal gold as a tracer. *J. Cell Biol.* **77**, 1.

Schwab, M. E., Javoy-Agid, F., and Agid, Y. (1978). Labeled wheatgerm agglutinin (WGA) as a new, highly sensitive retrograde tracer in the rat brain hippocampal system. *Brain Res.* **152**, 145.

Seeley, P. J., and Field, P. M. (1988). Use of colloidal gold complexes of wheat germ agglutinin as a label for neural cells. *Brain Res.* **449**, 177.

Shapiro, R. E., and Miselis, R. R. (1985). The central organization of the vagus nerve innervating the stomach of the rat. *J. Comp. Neurol.* **238**, 473.

Smits-Van Prooije, A. E., Vermeij-Keers, C., Dubbeldam, J. A., Mentink, M. M. T., and Poelmann, R. E. (1987). The formation of mesoderm and mesectoderm in presomite rat embryos cultured in vitro, using WGA-Au as a marker. *Anat. Embryol.* **176**, 71.

Van Bergen en Henegouwen, P. M. P., and Leunissen, J. L. M. (1986). Controlled growth of colloidal gold particles and implications for labeling efficiency. *Histochemistry* **85**, 81.

11

Colloidal Gold Labeling of Microtubules in Cleaved Whole Mounts of Cells

JAN A. TRAAS

I.N.R.A.
Institut National de la Recherche Agronomique
Dijon, France

INTRODUCTION
METHODOLOGY
  Specimen Preparation
    Preparation of Cells
    Preparation of Grids and Coverslips
  Fixation
    Plant Cells in Tissues (Roots of Seedlings and Young Plants)
    Protoplasts from Plant Cells
    Animal Cells Grown in Culture
    The Use of Formaldehyde
  Cleaving
    Plant Cells in Tissues (Roots of Seedlings)
    Protoplasts of Plant Cells
    Animal Cells in Culture
  Immunogold Labeling
    Blocking of Nonspecific Binding
    Immunolabeling

Postfixation and Critical-Point Drying
  Postfixation
  Dehydration and Critical-Point Drying
Examining the Specimen in the Electron Microscope
Controls
  Negative Staining
  Immunofluorescence
CONCLUDING REMARKS
REFERENCES

INTRODUCTION

Immunocytochemical methods have made a major impact on cell biology, perhaps most clearly where our understanding of the cytoskeleton is concerned. Immunofluorescence microscopy has been essential in analyzing the distribution, dynamics, and function of specific cytoskeletal elements in a wide range of both animal and plant cells (e.g., Osborn and Weber, 1982; Lloyd, 1987). With the introduction of immunolabeling techniques for electron microscopy, it also became possible to study the fine distribution of specific cytoskeletal elements at the ultrastructural level. The development of these techniques for electron microscopy has encountered several problems such as the establishment of optimal conditions for fixation, permeabilization of the cell membrane for antibodies, and the choice of the suitable electron-dense label. Colloidal gold particles are now most commonly used as markers for immunoelectron microscopy (De Brabander *et al.*, 1977; De Mey *et al.*, 1981, 1986; Geuze *et al.*, 1981, and references therein). As antibodies and gold particles do not readily penetrate into the cytoplasm, they are usually applied to the surface of thin sections of frozen or (low temperature) embedded material.

For the labeling of cytoskeletal elements a different technique has been developed and this involves the application of antibodies to fixed, extracted cells prior to embedding (Langanger *et al.*, 1984, 1986; De Mey *et al.*, 1986). Although this method has yielded excellent results, it still has the disadvantage of giving two-dimensional information about a three-dimensional network. However, sectioning can be avoided by preparing fixed/extracted and gold-labeled cells for whole-mount microscopy or for replication with carbon/platinum using critical-point drying or freeze drying (De Mey *et al.*, 1981; see also Heuser and Kirschner, 1980; Hartwig and Shevlin, 1986). Even so, this method still has the major drawback that detergents affect the fine structure of cytoplasm and interactions between membranes and the cytoskeleton.

11. Labeling of Microtubules in Cleaved Whole Mounts of Cells

Different techniques have been developed allowing the visualization of the membrane-bound cytoskeleton, avoiding both sectioning and extraction. These methods usually require the cleaving of cells in order to obtain samples of exposed membrane with adherent structures (Boyles and Bainton, 1979; Mesland et al., 1981; Aggeler et al., 1983; for review, see Nermut, 1982). More recently, a technique (based on the dry cleaving method; see Mesland et al., 1981; Mesland and Spiele, 1984; Traas, 1984) was developed for plant cells: fixed cells are attached to poly-L-lysine-coated grids and broken open while still in buffer. Thus the cytoplasm becomes accessible to antibodies. After incubation with the antibodies, the cleaved cells are postfixed and critical-point dried (Traas and Kengen, 1986). A similar technique for the labeling of cytoskeletal elements has also been introduced for animal cells by Nicol et al. (1987), who used the "lysis squirting" technique to expose the plasma membrane-adherent cytoplasm (see also Nermut, 1982). In this chapter the optimal conditions for the immunolabeling of microtubules in cleaved whole mounts are described for a number of cells. The technique can easily be extended to other cytoskeletal systems.

METHODOLOGY

Specimen Preparation

Preparation of Cells

The procedure will be described for the following cell types: plant cells in tissues (specifically root cells), protoplasts from plant cells, and animal cells grown in culture.

Plant Cells in Tissues

No specific treatment is necessary prior to fixation.

Protoplasts from Plant Cells

Protoplasts are prepared under standard conditions. After removal of the enzymes, the protoplasts are first attached to poly-L-lysine-coated grids (for the coating of grids see below). Drops of a protoplast suspension are pipetted on grids in a petri dish. The cells are allowed to settle for 20 min. The nonadhering cells are then removed by a quick wash in the protoplast medium prior to fixation. Some types of protoplasts do not adhere to the poly-L-lysine surface and they have to be attached to the grids after fixation.

Animal Cells in Culture

Cells are seeded and cultured under standard conditions in the appropriate petri dishes. The cells are used when they have reached a confluency of ≈80%. No further pretreatment is necessary.

Preparation of Grids and Coverslips

In general, nickel grids (100–200 mesh) are used because they bend less easily than copper grids. Grids are first coated with Formvar (0.5% Formvar in chloroform) and subsequently with carbon. Grids are then coated with poly-L-lysine (for references on poly-L-lysine coating, see Mazia et al., 1975; Nermut, 1982). We have obtained optimal results using poly-L-lysine with a high molecular weight (>300 000; e.g., Sigma, P1524) at a concentration of 1 mg/ml in water. For plant cells 5 mg/ml gives better results. Grids are inverted and floated on top of drops of poly-L-lysine for at least 30 min. They are washed thoroughly in a large volume of water and used immediately without drying. Glass coverslips are first cleaned in concentrated nitric acid. They are washed in water, cleaned in acetone, and wiped dry. The coverslips are then coated with poly-L-lysine, just before use, as described for grids.

Fixation

Fixation is perhaps the most critical step in the procedure as it is the most difficult one to control. First, the choice of the fixative is influenced by two requirements which are not necessarily compatible: the need for ultrastructural preservation and the need to retain antigenicity. Glutaraldehyde usually gives an excellent structural stabilization but it can severely affect immunoreactivity. Alternatively, formaldehyde tends to have less effect upon antigenicity, but it is also less suitable as a fixative for electron microscopy. This can change from antibody to antibody, cell to cell, and even cell element to cell element. In general, low concentrations of glutaraldehyde (0.1–0.5%) in combination with short fixation times (usually 10–30 min, for tissues up to 1 hr) form a good compromise. Only when glutaraldehyde fails should formaldehyde be used. It is beyond the scope of this chapter to discuss in detail the problems which can arise during fixation. For extensive reviews, see Hayat (1981, 1986) and Robards (1985) (for the plant cell cytoskeleton, see also Traas et al., 1987).

Plant Cells in Tissues (Roots of Seedlings and Young Plants)

Roots are cut from the seedlings and immersed immediately in the fixative containing 100 mM PIPES (pH 6.9) (piperazine-N,N'-bis-2-ethanesulfonic acid), 10 mM MgSO$_4$, 10 mM EGTA (ethyleneglycol-bis(β-aminoethyl ether)-N,N,N',N'-tetraacetic acid), and 0.5 % glutaraldehyde. This buffer with Mg and EGTA will be referred to as PME. Instead of PIPES buffer, phosphate buffer can be used (100 mM, pH 6.9). However, phosphate has been reported to induce distortion of microtubules (Lloyd and Wells, 1985). The tissues are fixed for 2 hr. When only the epidermal cells need to be studied, 30 min will be sufficient. The unbound fixative is removed from the tissue with five washes in buffer of 10 min each.

Protoplasts from Plant Cells

Living protoplasts adhering to grids are immersed in the fixative containing 250 mM mannitol, PME, and 0.1% glutaraldehyde for 30–60 min. This fixative has given good results for protoplasts of three different plant species. It is advisable, however, to optimize the concentration of mannitol for each new species. The protoplasts are washed in PME for 1 hr with five changes of buffer of 10 min each.

In some cases better results are obtained when the protoplasts are fixed in suspension. About 5 ml of protoplasts in protoplast medium is pipetted into a test tube. About 5 ml of a solution containing 1% glutaraldehyde in PME buffer with mannitol is added dropwise to the suspension. In this way the concentration of glutaraldehyde is increased gradually to 0.5% over a period of 10 min.

The cells are allowed to settle (20 min) and the supernatant is replaced with fresh buffer. This is repeated four times. As the cells settle very slowly, this is a very long procedure. It can be speeded up by centrifugation (300 rpm), but this should be kept to a minimum as the cells are very fragile. The cells are then washed in water. The protoplast suspension is pipetted into a petri dish and poly-L-lysine-coated grids are immersed in this suspension. The fixed cells are allowed to settle for 20 min.

Animal Cells Grown in Culture

The cells are briefly rinsed with PBS (37°C) just before fixation. The PBS is removed and the fixative is added, which contains 0.1% glutaraldehyde. In our hands PME buffer has given good results for bovine lens cells and C6 (rat glioma) cells but optimal fixation conditions (osmolarity, pH, buffer, etc.) should be determined first for each cell type. After 30 min, the fixative is removed and the cells are washed for 1 hr with five changes of buffer.

The Use of Formaldehyde

For some antibodies it might be necessary to use formaldehyde as a primary fixative. In that case the glutaraldehyde is replaced by 4% (w/v) formaldehyde. Formaldehyde is always prepared freshly from paraformaldehyde powder. The powder is dissolved in buffer at high pH (>9) at a high temperature (60–70°C). After the powder has dissolved, the pH is adjusted to 6.9, avoiding chloride ions, which can make a carcinogenic product. Cell suspensions and monolayers are fixed for 30 min, tissues for 1 hr.

Cleaving

Plant Cells in Tissues (Roots of Seedlings)

Roots are immersed in a solution containing 2–5% (w/v) cellulase (Onozuka R10) and PME. This treatment weakens the cell wall and facilitates cleaving. After 30 min, the cellulase is removed and the roots are washed for 1 hr with five changes of buffer. The roots are cut into segments of ~1 cm, which are split in two using fine tweezers. The halves are attached to the poly-L-lysine-coated grids, in distilled water, as shown in Fig. 11.1. Note that the cells will only adhere well to the grids when in water. The halves are allowed to adhere to the grids for 10 min; then they are removed from the grids. At least 10–15 cleaved cells per half root will remain on the grids.

Protoplasts of Plant Cells

Grids with adhering protoplasts are inverted on poly-L-lysine-coated coverslips in water and the upper surface of the protoplasts is allowed to

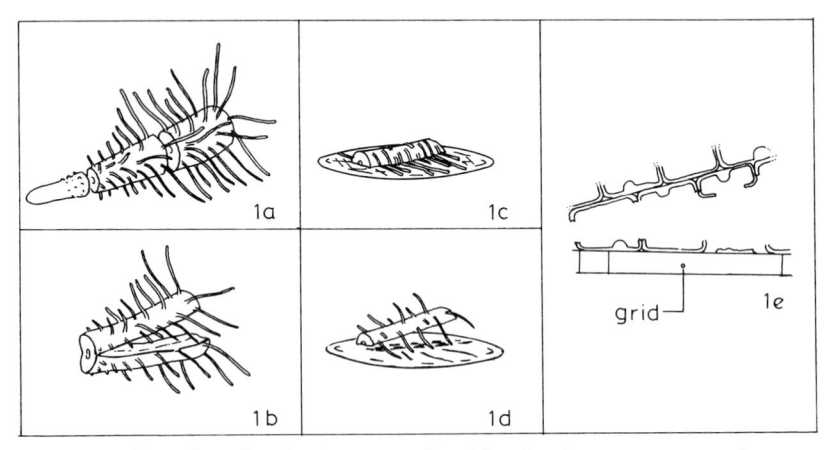

Fig. 11.1. Procedure for cleaving root cells. After fixation, roots are cut in segments (a). The segments are split in two (b) and allowed to adhere to the surface of a grid (c). The cells are cleaved by removing the half root from the grid (d, e).

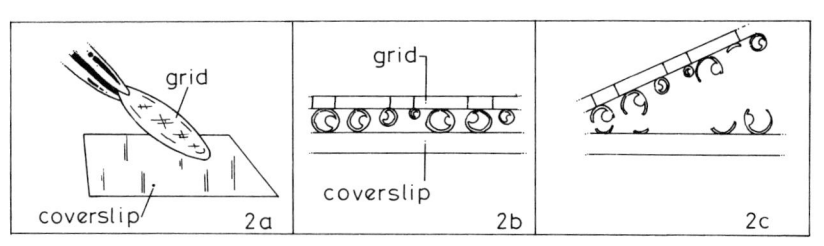

11. Labeling of Microtubules in Cleaved Whole Mounts of Cells

Fig. 11.2. Procedure for cleaving plant protoplasts. Protoplasts on grids are allowed to adhere to the surface of a coverslip (a, b). The cells are cleaved by removing the grid from the coverslip (c).

adhere to the glass surface for 20 min. Thus the cells will adhere to both the grid and the coverslip. The grids are then removed, causing a number of cells to break open, a proportion of which will remain on the grids. In general, 50% of the cells will cleave (Fig. 11.2).

Animal Cells in Culture

The cells on the bottom of petri dishes are washed in water. Poly-L-lysine-coated grids are inverted onto the monolayers. Cells are allowed to adhere to the grids for 20 min. The grids are removed and a number of cleaved cells will remain on the grids. In general, at least 20% of the cells will cleave (Fig. 11.3).

At this stage there are two potential problems. First, the cells may not adhere well enough to be cleaved. This is especially true for plant cells with thick walls such as pollen grains, which cannot be cleaved. Although we have not encountered this problem with animal cells, it is possible that, in particular, cells with an irregular surface do not adhere well to the poly-L-lysine coat. In that case it might be useful to spin the grids down on the cell surface or to use other methods to attach the cells to the grids (reviewed in Nermut, 1982). A second problem might be the irregu-

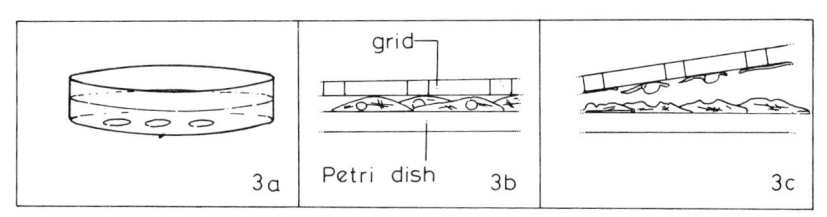

Fig. 11.3. Procedure for cleaving animal cells in culture. Grids are inverted on top of the fixed cells in a petri dish (a) and the cells are allowed to adhere to the surface of the grid (b). Cells are cleaved by removing the grids from the monolayer of cells (c).

lar plane of cleavage. In our experience a large number of cells cleave sufficiently close to the plasma membrane (see also Mesland et al., 1981). However, an excess of cytoplasm associated with the membrane can easily be removed by repeating the cleaving procedure. In that case broken-open cells on grids are inverted on poly-L-lysine-coated coverslips.

Immunogold Labeling
Blocking of Nonspecific Binding

There are different methods for blocking nonspecific binding (mainly due to free aldehyde groups):

1. The grids with the cleaved cells are immersed in a solution of 5% (w/v) bovine serum albumin (BSA) in Tris-buffered saline (TBS) for 10 min. This has given sufficient blocking for at least four antibodies that we have used.

2. The grids with the cells are treated twice with $NaBH_4$ (1 mg/ml) in PME for 10 min in order to reduce free aldehydes. This treatment is followed by an incubation of 10 min in BSA/TBS. In our hands this method caused a distortion of cellular fine structure in plant cells (Traas and Kengen, 1986). It has, however, been applied to animal cells with excellent results (e.g., Osborn and Weber, 1982; De Mey et al., 1986; Nicol et al., 1987). After blocking, the cells are washed in TBS for 5 min.

Immunolabeling

The primary antibody is diluted in TBS/BSA (1% w/v) to the appropriate concentration (MAS 077B, Sera Lab for antitubulin). Drops of the first antibody are pipetted on multiwell slides (Flow Laboratories). The grids with the cleaved cells are inverted onto the drops. The multiwells are put in sandwich boxes at 100% relative humidity on a shaker at room temperature. After 1–3 hr (depending on the antibody) the grids are removed from the multiwells and immersed in petri dishes with TBS. They are washed thoroughly for 30 min with five changes of buffer.

The cells are incubated on multiwell slides with a gold-conjugated second antibody (Janssen Life Sciences Products). Antibodies labeled with 5- or 10-nm particles should be used. Larger gold particles will not easily penetrate the layer of membrane-associated cytoplasm. The second antibodies should be used at the lowest possible concentration (1/10–1/100) in order to reduce background labeling to a minimum. Incubations are at room temperature for 2–4 hr or at 4°C overnight. After immunolabeling, the cleaved cells are washed thoroughly for 1 hr with several changes of buffer. This step is important as it reduces the background considerably.

Postfixation and Critical-Point Drying

Postfixation

Cells are fixed for 30 min in a solution containing PME, 0.1% (w/v) tannic acid, and 1.0% (w/v) glutaraldehyde. Tannic acid treatment results in better preservation of cytoplasmic structures and especially of cytoskeletal elements (e.g., Maupin and Pollard, 1983; Hayat, 1989). We have used tannic acid from Mallinckrodt, as tannic acid from other sources can produce precipitates in the cytoplasm (Traas and Kengen, 1986). Cells are washed thoroughly in distilled water after postfixation with glutaraldehyde/tannic acid. The washing step is important, as tannic acid can precipitate with osmium. Cells are postfixed in 0.5% (w/v) OsO_4 in distilled water for 20 min. Cells are stained in uranyl acetate (0.5% w/v in distilled water) for 20 min.

Dehydration and Critical-Point Drying

To facilitate further handling, the grids are transferred to a grid holder. The cells are dehydrated in a series of ethanol:

50% ethanol	(5 min)
70% ethanol	(5 min)
90% ethanol	(10 min)
100% ethanol	(5 min)
100% ethanol	(10 min)
100% ethanol	(dried with molecular sieve) (10 min)
100% ethanol	(dried with molecular sieve) (10 min)

It is important to remove all traces of water, which can cause severe distortion of the cytoplasm during critical-point drying.

Critical-Point Drying

The grids are transferred to the specimen container of the critical-point drier, which contains dried ethanol. The cells are critical-point dried using CO_2. It is important to remove all traces of ethanol with at least 15 washes of liquid CO_2. The first washes can be at intervals of 1–2 min. After 15 min when most of the ethanol has been removed the intervals can be longer (5–10 min). In addition, it may be necessary to dry the CO_2 as well by passing the liquid through a molecular sieve. After ±90 min, the CO_2 is heated slowly (10 min) to 40°C (pressure should not exceed 100 bar). The pressure is released very slowly (20 min) at constant temperature.

Examining the Specimen in the Electron Microscope

The preparations can be examined immediately in the electron microscope. In order to prevent beam damage and contamination it is essential to use the anticontamination device, commonly a liquid nitrogen trap cooling the space around the specimen. It is also important to use the highest voltage possible (in any event >80 kV) with low light intensity. If the preparations cannot be examined directly, they must be kept with silica gel under vacuum. Under these conditions the fine structure of the cytoplasm will remain intact for at least 2 weeks.

Controls

Besides the obvious controls for nonspecific binding of the antibody, the following methods may help to solve problems that arise during the procedure.

Negative Staining

Negative staining can be helpful at any stage of the procedure to check for the quality of fixation, the proportion of cleaved cells, the degree of labeling, and background. Cells can be stained using standard procedures with 1% uranyl acetate in distilled water (Hayat, 1989; Hayat and Miller, 1989). It is always useful to prepare a number of grids for negative staining during different stages of the procedure. In this way time, antibodies, and chemicals can be saved. Moreover, the technique can be, in some cases, a good alternative for critical-point drying (e.g., Claviez *et al.*, 1986; Nermut, 1982, for review).

Immunofluorescence

Especially when gold labeling is poor or when high backgrounds are observed, it can be helpful to use the immunofluorescence method. This is especially true when there are doubts about the suitable conditions for fixation and about the binding of the first antibody. For immunofluorescence, the cells on grids (or coverslips) can be fixed, cleaved, and incubated with the first antibody as described above. Instead of the gold-labeled second antibody, an appropriate fluorescent antibody is used. After labeling, the cells are mounted on slides in glycerol/buffer with an antifading agent (e.g., Citifluor; City University, London) and examined in the fluorescence microscope. When glutaraldehyde is used as a fixative the cells have to be treated first with $NaBH_4$ to reduce background fluorescence. This is not necessary when formaldehyde is used. For more details about the immunofluorescence procedure, see Lloyd *et al.* (1979), Osborn and Weber (1982), and Wick *et al.* (1981).

CONCLUDING REMARKS

The method described here for the immunolabeling of cleaved cells is a useful technique for studying the cortical cytoplasm free of embedding materials. It not only allows the identification of cytoskeletal elements in the cell, but also can be combined with quantitative analyses since large surface areas of a great number of cells are usually left on the grid (see also Mesland and Spiele, 1984; Traas *et al.*, 1984). We have used the technique successfully on different cell types to label microtubules (Fig. 11.4) (Traas and Kengen, 1986) and on protoplasts of *Daucus carota* to label fibrillar bundles (Fig. 11.4d,e) (J. Traas, unpublished data; see also Hargreaves, *et al.*, 1989). For animal cells grown in culture the technique can be used to study the dorsal (medium-facing) cytoplasm and (if cells are grown on the grids) substrate-facing cortical cytoplasm. Moreover, it can be combined with the labeling of cell surface proteins to study membrane–cytoskeleton interactions (Roos *et al.*, 1985). There are, however,

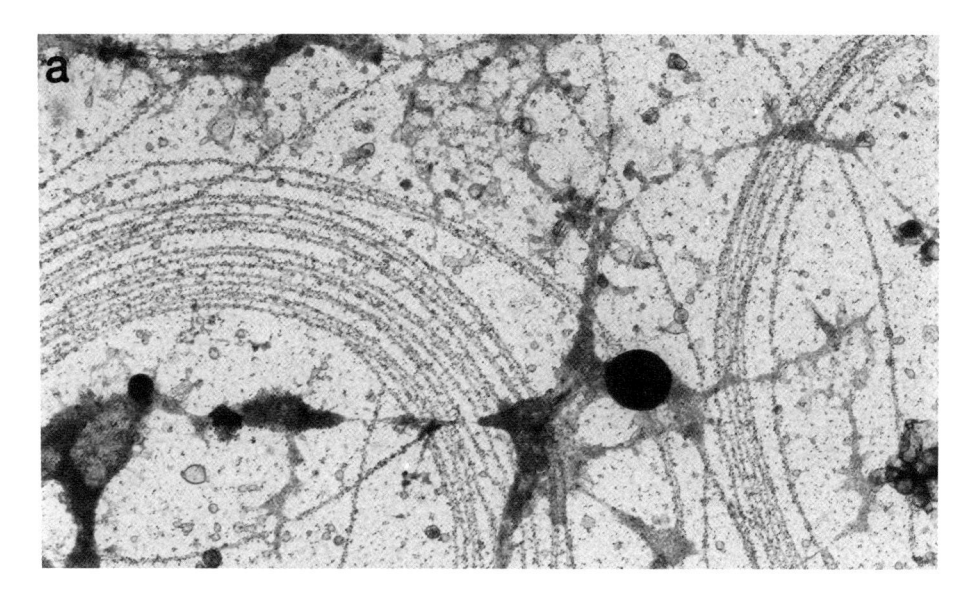

Fig. 11.4. Gold-labeled microtubules in cleaved protoplasts of cell suspension culture cells of *Daucus carota* (a, b) and in cortical root cells of *Equisetum hyemale* (c). (d) Low magnification of a fibrillar bundle in a protoplast of *Daucus carota*. These bundles are cytoskeletal elements which have antigenic sites in common with intermediate filaments (Dawson *et al.*, 1985; Hargreaves *et al.*, 1989). In (e) a detail is shown of a fibrillar bundle labeled with a monoclonal antibody against fibrillar bundle proteins. (a) ×19,000; (b) ×75,000; (c) ×75,000; (d) ×14,500; (e) ×110,000. (Figure continued on next page).

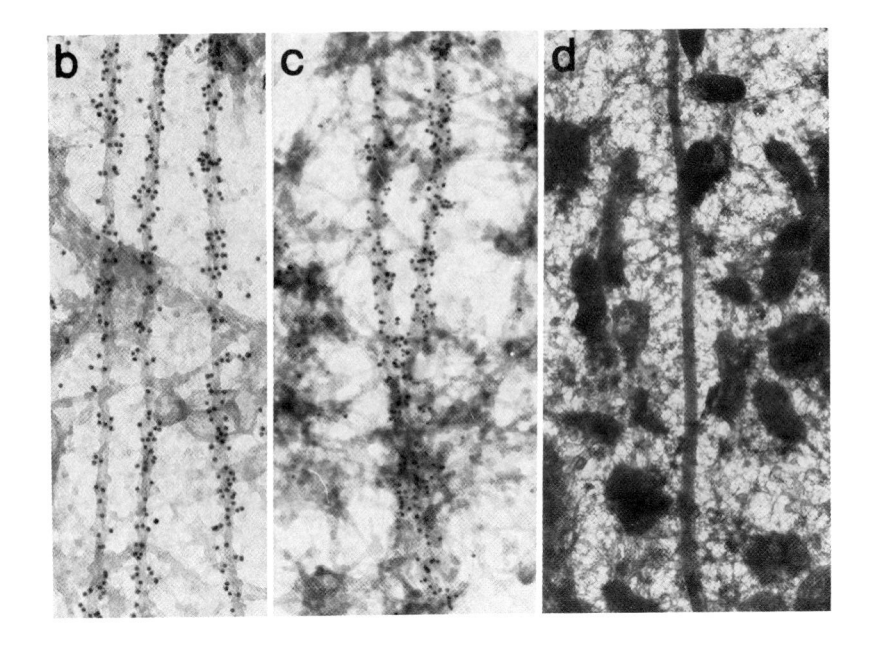

Fig. 11.4. *(Continued)*

11. Labeling of Microtubules in Cleaved Whole Mounts of Cells

some potential problems and restrictions which have to be considered before one decides to apply the method.

Although most difficulties have been discussed in the different sections of this chapter, it is important to stress some of the limitations: the modification of antigenic sites by the fixative, the accessibility of the antigen, and the impossibility to cleave some cell types. Especially for tissue cells, preservation of fine structure is far superior with glutaraldehyde (Hayat, 1981; De Mey *et al.*, 1986; Traas and Kengen, 1986), which can severely affect antigenicity. For the antitubulin that we have used, but also for a number of other antibodies, this poses no problem. For some antibodies formaldehyde prefixation must be used (e.g., De Mey *et al.*, 1986), which can result in poor preservation of ultrastructure. Another problem might be the accessibility of the antigen to the antibody. The method avoids extraction and so dense structures and regions of the cytoplasm (such as filament bundles and dense networks) may not be opened up to the gold-labeled antibodies.

Finally, some cell types cannot be cleaved. This is especially true for plant cells with thick walls but it may also be a problem for animal cells with an irregular surface which does not adhere to the poly-L-lysine coat. In our experience, however, these cell types form only a minority and the technique is applicable to a wide range of cells. Therefore, the method is an important complement to other immunocytochemical techniques.

This work was partly supported by an EMBO long-term fellowship. I wish to thank Dr. C. W. Lloyd for helpful comments. J. Traas is a visiting scientist from the I. V. T., Wageningen (NL).

REFERENCES

Aggeler, J., Takemura, R., and Werb, Z. (1983). High-resolution three dimensional views of membrane associated clathrin and cytoskeleton in critical point dried macrophages. *J. Cell Biol.* **97**, 1452.

Boyles, J., and Bainton, D. (1979). Changing patterns of plasma membrane-associated filaments during the initial phases of polymorphonuclear adherence. *J. Cell Biol.* **82**, 347.

Claviez, M., Brink, M., and Gerisch, G. (1986). Cytoskeletons from a mutant of *Dictyostelium discoideum* with flattened cells. *J. Cell Sci.* **86**, 69.

Dawson, P. J., Hulme, J. S., and Lloyd, C. W. (1985). Monoclonal antibody to intermediate filament antigen crossreacts with higher plant cells. *J. Cell Biol.* **100**, 1793.

De Brabander, M., De Mey, J., Joniau, M., and Geuens, G. (1977). Immunocytochemical visualization of micro-tubules and tubulin at the light and electron microscopical level. *J. Cell Sci.* **28**, 283.

De Mey, J., Moeremans, M., Geuens, G., Nuydens, R., and De Brabander, M. (1981). High resolution light and electron microscopic localization of tubulin with the IGS (immuno gold staining) method. *Cell Biol. Int. Rep.* **5**, 889.

De Mey, J., Langanger, G., Geuens, G., Nuydens, R., and De Brabander, M. (1986). Preembedding for localization by electron microscopy of cytoskeletal antigens in cultured cell monolayers using gold labeled antibodies. In *Methods in Enzymology* (R. B. Vallee, ed.), vol. 134, p. 592. Academic Press, Orlando, Florida.

Geuze, H., Slot, J., Van der Ley, P., Scheffer, R., and Griffith, J. (1981). Use of colloidal gold particles in double labeling immunoelectron microscopy of ultrathin frozen tissue sections. *J. Cell Biol.* **89**, 653.

Hargreaves, A. J., Dawson, P. J., Butcher, G. W., Larkins, A., Goodbody, K. C., and Lloyd, C. W. (1989). A monoclonal antibody against cytoplasmic fibrillar bundles from carrot cells, and its cross-reaction with animal intermediate filaments *J. Cell Sci.* **92**, 371–378.

Hartwig, J. H., and Shevlin, P. (1986). The architecture of actin filaments and the ultrastructural location of actin binding protein in the periphery of lung macrophages. *J. Cell Biol.* **103**, 1007.

Hayat, M. A. (1981). *Fixation for Electron Microscopy*. Academic Press, New York and London.

Hayat, M. A. (1986). Glutaraldehyde: Role in electron microscopy. *Micron Microsc. Acta* **17**, 115.

Hayat, M. A. (1989). *Principles and Techniques of Electron Microscopy*, 3rd ed. Macmillan, London and CRC Press, Boca Raton, Florida.

Hayat, M. A., and Miller, S. (1989). *Negative Staining*. McGraw-Hill, New York.

Heuser, J. E., and Kirschner, M. W. (1980). Filament organization revealed in platinum replicas of freeze dried cytoskeletons. *J. Cell Biol.* **86**, 212.

Langanger, G., De Mey, J., Moeremans, M., Daneels, G., De Brabander, M., and Small, V. (1984). Ultrastructural localization of alpha actinin and filamin in cultured cells with the immunogold staining method. *J. Cell Biol.* **99**, 1324.

Langanger, G., Moeremans, M., Daneels, G., Sobieszek, A., and De Brabander, M. (1986). The molecular organization of myosin in stress fibres of cultured cells. *J. Cell Biol.* **102**, 200.

Lloyd, C. W., Slabas, A., Powell, A., Macdonald, G., and Badley, R. (1979). Cytoplasmic microtubules of higher plant cells visualized with anti-tubulin antibodies. *Nature (London)* **279**, 239.

Lloyd, C. W., and Wells, B. (1985). Microtubules are at the tips of root hairs and form helical patterns corresponding to inner wall fibrils. *J. Cell Sci.* **75**, 225.

Lloyd, C. W. (1987). The plant cytoskeleton: The impact of fluorescence microscopy. *Annu. Rev. Plant Physiol.* **38**, 119.

Maupin, P., and Pollard, T. D. (1983). Improved preservation and staining of HeLa cell actin filaments, clathrin coated membranes and other cytoplasmic structures by tannic acid–glutaraldehyde–saponin fixation. *J. Cell Biol.* **96**, 51.

Mazia, D., Sale, W. S., and Schatten, G. (1975). Adhesion of cells to surfaces coated with polylysine. Applications to electron microscopy. *J. Cell Biol.* **66**, 198.

Mesland, D., Spiele, H., and Roos, E. (1981). Membrane associated cytoskeleton and coated vesicles in cultured hepatocytes visualized by dry cleaving. *Exp. Cell Res.* **132**, 169.

Mesland, D., and Spiele, H. (1984). Brief extraction with detergent induces the appearance of many plasma membrane associated microtubules in hepatocytic cells. *J. Cell Sci.* **68**, 113.

11. Labeling of Microtubules in Cleaved Whole Mounts of Cells 241

Nermut, M. V. (1982). The cell monolayer technique in membrane research. *Eur. J. Cell Biol.* **28,** 160.

Nicol, A., Nermut, M., Doeinck, A., Robenek, H., Wiegand, C., and Jokusch, B. M. (1987). Labeling of structural elements at the ventral plasma membrane of fibroblasts with the immunogold technique. *J. Histochem. Cytochem.* **35,** 499.

Osborn, M., and Weber, K. (1982). Immunofluorescence and immunocytochemical procedures with affinity purified antibodies: Tubulin containing structures. *Methods Cell Biol.* **24A,** 97.

Robards, A. W., ed. (1985). *Botanical Microscopy.* Oxford Univ. Press, London and New York.

Roos, E., Spiele, H., Feltkamp, C. A., Huisman, H., Wiegant, F. A., Traas, J. A., and Mesland, D. A. M. (1985). Localization of cell surface glycoproteins in membrane domains associated with the underlying filament network. *J. Cell Biol.* **101,** 1817.

Traas, J. A. (1984). Visualization of the membrane bound cytoskeleton and coated pits of plant cells by means of dry cleaving. *Protoplasma* **119,** 212.

Traas, J. A., Braat, P., and Derksen, J. (1984). Changes in microtubule arrays during the differentiation of cortical root cells of *Raphanus sativus. Eur. J. Cell Biol.* **34,** 229.

Traas, J. A., and Kengen, H. (1986). Gold labeling of microtubules in cleaved whole mounts of cortical root cells. *J. Histochem. Cytochem.* **34,** 1501.

Traas, J. A., Doonan, J., Rawlins, D., Shaw, P., Watts, J., and Lloyd, C. (1987). An actin network is present in the cytoplasm throughout the cell cycle of carrot cells and associates with the dividing nucleus. *J. Cell Biol.* **105,** 387.

Wick, S. M., Seagull, R. W., Osborn, M., Weber, K., and Gunning, B. E. S. (1981). Immunofluorescence microscopy of organized microtubule arrays in structurally stabilized meristematic plant cells. *J. Cell Biol.* **97,** 234.

12

Colloidal Gold: Immunonegative Staining Method

JULIAN E. BEESLEY

Wellcome Research Laboratories
Beckenham, Kent, England

INTRODUCTION
METHODOLOGY
   Immunolabeling Schedule
VARIATION OF THE TECHNIQUE
DOUBLE LABELING EXPERIMENTS
QUANTITATION
ASSESSMENT
APPLICATIONS
   Virology
   Bacteriology
CONCLUSIONS
REFERENCES

INTRODUCTION

The immunological study of small structures such as viruses and bacteria with the electron microscope demands a method which can attain good contrast and resolution to observe the specimen. Furthermore, small structures possess a very low antigenic mass, and for successful immunolabeling the method must possess a high immunolabeling efficiency and the antigen should not be blocked by embedding media or damaged by fixation and dehydration. In addition, examination of suspect viral and bacterial specimens for diagnostic purposes requires that specimen preparation and immunolabeling should use a minimal amount of sample and that the process should be as short as possible.

Immunoelectron microscope methods have existed for many years. The aggregation method, for instance, relies upon specific antibodies to decorate virus particles and form clumps (Almeida and Waterson, 1969). Alternatively, immunosorbent methods which selectively trap viruses on electron microscope grids previously coated with specific antibodies are popular. These trapped viruses may be further decorated with antibodies to enhance the immunolabeling and for double-labeling experiments (Derrick, 1973; Milne and Luisoni, 1977; Shukla and Gough, 1979; Nicolaieff et al., 1980).

Colloidal gold probes, reported initially by Faulk and Taylor (1971), are now widely used in electron immunocytochemistry. The immunonegative stain method in conjunction with colloidal gold probes is used for the immunological study of small particles.

METHODOLOGY

The method was developed in this laboratory for the examination of viruses and bacterial pili (Beesley and Betts, 1984) and named the immunonegative stain method but it is, however, the method of choice for immunolabeling any structure which can be dried onto a grid, immunolabeled *in situ,* and then visualized by negative staining (Fig. 12.1).

There is no limitation as to which protein–gold complex is used in the technique. Furthermore, either monoclonal or polyclonal antibodies can be used. The primary antibody is naturally central to the technique and a high-titer specific antibody is desirable. The immunonegative stain method examines whole organisms dried onto the grid. Internal antigens are therefore not usually exposed and will not be labeled, even by high-titer specific antisera. Antisera raised against external antigens are therefore a necessity. Care must be taken when choosing these antisera for virus diagnosis since some virus particles such as herpes may disintegrate, releasing internal antigens. The antisera may be used in conjunction with any of the gold probes currently described.

12. Colloidal Gold: Immunonegative Staining Method 245

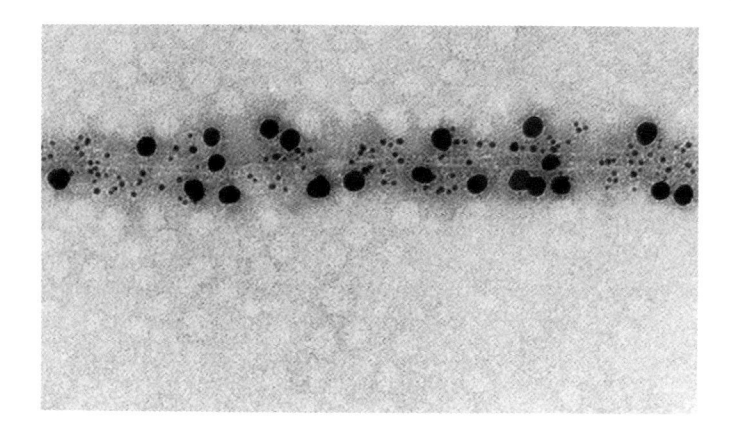

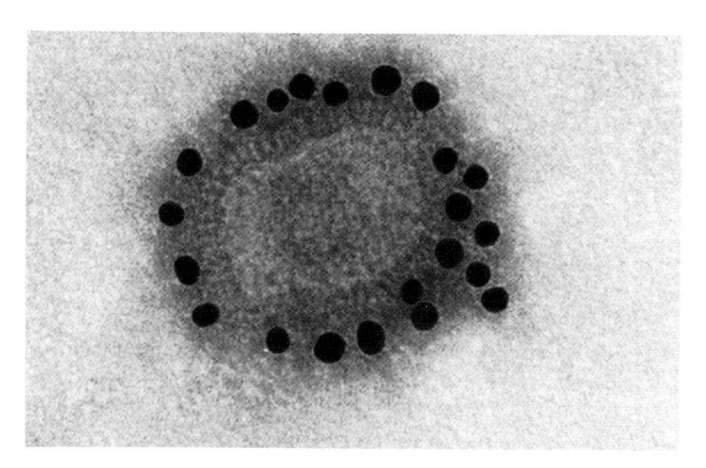

Fig. 12.1. (Top) *Bacteroides nodus* pili preparation sequentially immunolabeled with heterologous antibody and the 5-nm immunogold probe followed by the homologous antibody and finally the 20-nm immunogold probe; ×146,000. (Bottom) Influenza virus incubated with antibody against viral hemagglutinin and then the 20-nm immunogold probe; ×220,000. (From Beesley and Betts, 1984).

The technique is relatively straightforward and does not require complex preparatory steps. The gold probes may be stored at 4°C and small aliquots of the antiserum may be stored at −20°C. When immunological examination of the sample is required many grids can be made from a small aliquot of starting material and the specimen can be immunolabeled with several antisera and viewed within the hour. The reliability of the technique is extremely high.

If the sample is thought to contain a vigorous pathogen, this should be killed before examination in the microscope. This may be effected by floating the grid on 3% glutaraldehyde in phosphate buffer immediately before contrasting (step 7 in the following schedule). A further rinse in distilled water before application of the stain will prevent contamination of the grid by crystals from the buffer. Glutaraldehyde treatment of the pathogen before immunolabeling will severely reduce immunolabeling, especially if monoclonal antibodies are being used.

Immunolabeling Schedule

1. Prepare concentrated specimen suspension in water.
2. Dry suitable aliquots of sample onto Butvar/carbon-coated 400-mesh gold grids.
3. Float grid, specimen side down, on 15-µl aliquots of antibody diluted with phosphate-buffered saline (pH 7.2) containing 1% bovine serum albumin* (PBS-BSA) for 15 min.
4. Rinse by floating grids on four droplets of PBS-BSA for 1 min each.
5. Float grid, specimen side down, on 15-µl aliquots of gold probe suitably diluted with PBS-BSA for 15 min.
6. Rinse by floating grids three times for 1 min each on droplets of distilled water.
7. Contrast before examination with desired negative stain, e.g., 1.5% ammonium molybdate (pH 6.8) or 1% sodium phosphotungstic acid (pH 7.5).

VARIATIONS OF THE TECHNIQUE

Stannard et al. (1982, 1987) prepare IgG–gold complexes with both monoclonal (Stannard et al., 1987) and polyclonal (Stannard et al., 1982) antibodies. Virus particles are concentrated from culture fluids by centrifugation at 48,000 g and resuspended in 300 µl of phosphate buffer (pH 7.2). About 20 to 30 µl of the selected gold probe is added, mixed, and

---

*Phosphate-buffered saline (pH 7.2) containing 1% BSA is a suitable buffer for diluting the antisera and gold. The concentrations of antibody and gold probe are found empirically by diluting the reagents and testing on known positive antigens until specific labeling with low background is achieved. Care must be taken when using buffers containing detergents such as Tween 20, which will destroy the structure of the unfixed organisms.

left overnight at room temperature. The volume of each sample is increased to 5 ml with phosphate buffer, and the immunolabeled virus particles are concentrated by centrifugation at 27,000 g for 20 min. The pellets are then negatively stained with 1% sodium phosphotungstic acid. The technique produces excellent immunolabeling but, in contrast to the previous technique, it is lengthy and relatively large volumes of virus must be committed to each antiserum tested.

DOUBLE-LABELING EXPERIMENTS

An overwhelming advantage of gold probes over other immunoelectron microscope techniques is that gold probes of different sizes can be prepared and used for double-labeling experiments. Experience has shown that there appears to be an extremely efficient attachment, presumably an electrostatic binding, between the specimen and the carbon coating of the grid. This bond resists long flotation on the various immunological reagents necessary for double-labeling experiments. The only evidence of specimen movement is that bacterial pili appear to be aggregated by specific antisera. The bacteria do not aggregate and it is therefore felt that the pili probably float free in the medium during immunolabeling. Viruses are not aggregated in this way.

There have been few examples reported of multiple labeling using the immunonegative stain method, although it is theoretically possible to carry out most of the double-labeling techniques so far described using colloidal gold. The double-sided labeling method (Beesley et al., 1984; Bendayan and Stevens, 1984) is an exception since small particles need a supporting film on the grid and are therefore not exposed on both sides. This method is, however, irrelevant for the immunonegative stain method since small particles do not usually possess two distinct faces.

A simple double-labeling technique would be to coat gold particles of different sizes with different antisera and use mixtures of these for simultaneous immunolabeling of both antigens. A variation of this method has been carried out by incubating protein A–gold probes of different sizes with different specific antibodies before immunolabeling the antigen (Robinson et al., 1984).

A highly specific technique has been reported by Larsson (1979), who produced antigen–gold complexes. The specimen is incubated with an excess of primary antibody so that one Fab site on the antibody molecule remains free to bind with the antigen–gold complex. This may be repeated for the localization of the second antigenic site. There is no risk of cross-contamination but the availability of pure antigen limits the method.

A practical double-labeling method was designed by Tapia et al. (1983). Specific antibodies are raised in different hosts. These host species are selected for their lack of cross-reactions. The sample is incubated with a mixture of both antibodies, followed, after appropriate washes, by an incubation with two different gold probes, each coated with an antiserum raised against the different primary antibody species.

The protein A–gold method has been successfully employed for double-labeling experiments (Roth, 1982; Slot and Geuze, 1984). The sample is incubated, first with one antibody, then with the smallest gold probe (usually 2–4 nm in diameter). The sample is incubated with free protein A (1 mg/ml) for 10 min to saturate any Fc sites of the primary antibody not bound with the gold probe. This step inhibits cross-reactions when the second antigen is localized with the second antibody and the second protein A–gold probe.

Incubating the antigen with antibody and gold probe does not usually saturate all the available antibody complexed with the antigen. If this first incubation were followed by a further incubation with a second antibody and gold probe, there would then be a high degree of cross-reaction of the second gold probe on the first antibody. It is possible after the first antibody incubation to carry out successive incubations with gold probe only in order to saturate almost completely all the available antibody sites. Arrival at this stage is determined by one further incubation with a probe of different size. Absence of labeling with the probe of a second size proves the absence of free antibody in the sample. Once this point has been reached, double labeling may be effected by further incubation with second antibody and another gold probe (Beesley et al., 1984).

Should fortune prevail, double-labeling immunonegative stain preparations can be effected after judicious selection of antibody and gold concentrations. The antigen is incubated with the first antibody, then with the first gold probe, followed with no blocking treatment by the second antibody and then the second-sized gold particle (Beesley, 1987). It is imperative, when carrying out double-labeling experiments, to perform adequate control experiments.

QUANTITATION

Quantitation using colloidal gold probes is still in its infancy. Although there have been many attempts at relative quantitation, that is, comparing the amount of immunolabeling on test specimens with known standards, only one preliminary attempt has been made to quantify absolute numbers of antigens in a preparation (Griffiths and Hoppeler, 1986). Quantitation using the immunonegative stain method has rarely been attempted. Beesley and Betts (1985) distinguished the three poliovirus types by quantifying the immunolabeling on each virus type after immunolabeling with antisera raised against each of the three different types of viruses.

ASSESSMENT

The immunonegative stain method is a technique of high immunolabeling efficiency that produces high contrast and high resolution in the sample. The antigen is well preserved because there is no prefixation before immunolabeling. The method is therefore highly sensitive (Beesley and Betts, 1984). Immunolabeling depends upon the choice of the correct antibody in conjunction with the gold probes. Antibodies raised against external antigens of the specimen under examination must be used. This labeling is easily detected, even on dark samples containing debris, because the gold probes are very dense. Therefore, even relatively little labeling can be detected. The method uses extremely small amounts of sample, is quick, and necessitates the minimum of preparation before immunolabeling. There is very little methodological nonspecific labeling (Beesley and Betts, 1987). The method is applicable to any structure which may be dried down onto an electron microscope grid. It could be used either to identify an unknown antigen with a known antibody or to identify an unknown antibody by using a known antigen (Muller and Baigent, 1980). The method is applicable also to multiple-labeling studies as well as for quantitative assessment of antigenic sites. It should therefore be a valuable method for both research and diagnosis.

APPLICATIONS

Virology

Electron microscopic examination continues to be the method of choice for the diagnosis of viral particles. Identification of viral particles after centrifugation is relatively straightforward if the viruses are presented as a concentrated suspension and the viruses possess characteristic, easily recognizable features. If these parameters are not fulfilled, then an immunological test is necessary. It is not therefore surprising that a major emphasis on the immunonegative stain method has been directed toward virus diagnosis.

Beesley and Betts (1984) described some of the pitfalls associated with this method. This was followed in 1985 (Beesley and Betts 1985) and 1987 (Beesley and Betts 1987) with a more detailed account of the method for virus diagnosis.

Louro and Lesemann (1984) described the use of the protein A–gold complex for specific labeling of antibodies bound to 12 different plant viruses. They described the conditions necessary for highly specific labeling of antigens combined with a low nonspecific background. They found that protein A–gold labeling increased the sensitivity by up to four twofold dilutions of antibody for detecting low amounts of antibodies on virus particles compared with immunosorbent electron microscope techniques. In other studies, these authors found that applications of the protein A–gold complex allowed detection of small antigens having no distinct morphology and that the combination of dense antibody coating and protein A–gold labeling significantly increased virus particle diameter and contrast. They also described double antibody coating of the particles using rabbit anti-virus IgG followed by goat anti-rabbit IgG prior to protein A–gold labeling, which facilitated screening for elongate virus particles at a low concentration in crude sap. Colloidal gold probes have also been used by Pares and Whitecross (1982) for identifying viruses in plant sap. They explore the possibility of using colloidal gold–protein A–antibody complexes for the detection of several strains of tobacco mosaic virus.

Kjeldsberg (1985) described specific immunolabeling of human rotaviruses and adenoviruses using rabbit primary antibodies in conjunction with 20-nm gold probes coated with goat anti-rabbit IgG. She found that precoating of grids with bovine rotavirus antibody decreased nonspecific background staining and at the same time slightly increased the number of rotavirus particles adhering to the grid.

These studies were closely followed by Hopley and Doane (1985), who described the protein A–gold method for immunoelectron microscopy of viruses in fluid samples, and by Murti and Webster (1986), who reported the distribution of hemagglutinin and neuraminidase on influenza viruses. Manuelidis and Manuelidis (1986) also used the technique to localize scrapie viral antigens and lectins in preparations of Creutzfeldt–Jakob disease. Although antibodies and also specific lectins label appropriate proteins in Western blots, in immunonegatively stained preparations the majority of label adheres to fluffy small proteinaceous aggregates and not on visible scrapie-associated fibrils. More research is therefore needed to clarify this.

Muller and Baigent (1980) used the immunonegative stain method in conjunction with protein A–gold probes as a test for the presence of antibodies in sera. Immunolabeling of a morphologically identifiable antigen serves as the criterion for specificity. It was applied to antisera containing antibodies against T4 bacteriophage, vaccinia virus, and *Enterobacterium yersinia*. An interesting development of this is the possibility of identifying antibodies in "unknown" sera by testing the sera against multiple morphologically distinguishable antigens on the same grid. Clearly, this could be of use in testing specificity of antisera.

Finally, Stannard *et al.* (1982) used their variation of the immunonegative stain method to provide the first reported visual evidence of circulating complexes of hepatitis B e-antigen. A later study by this group (Stannard *et al.,* 1987) detected spikes of different kinds, distinct in size and appearance, on the surface of herpes simplex virions by the standard negative staining method. The use of monoclonal antibodies complexed to colloidal gold permitted identification of the viral glycoproteins gB, gC, and gD present in different structures projecting from the viral envelope.

Bacteriology

Beesley *et al.* (1984) used the immunonegative stain method to demonstrate multiple antigenicity in the pili of *Bacteroides nodosus*. At least 17 types of *B. nodosus* are responsible for foot rot in sheep. A vaccine including more than 10 serotypes would be difficult to manufacture and furthermore, the antigenic composition would lead to a reduced response to the individual antigens (Day *et al.,* 1986). Three antisera, raised against three distinct pili serotypes, exhibited considerable cross-reactions. It now appears likely that each of these pilus types may contain at least four different antigenic sites (Beesley *et al.,* 1984). The selection of bacterial strains shown by immunocytochemistry to possess appropriate cross-reactions could lead to a vaccine with a wide cover in the field.

Levine *et al.* (1984) used the technique to investigate the characteristics of coli surface antigen CS3 on bacterial cells. After purification, CS3 consisted of thin, flexible fimbriae, which were also visible on the bacteria. These contrasted with CS1, which by use of appropriate antibodies was shown to be immunologically distinct.

Robinson *et al.* (1984) used gold–protein A–antibody complexes to localize specific gonococcal macromolecules on *Neisseria gonorrhoeae*. These probes allowed the demonstration of antibody binding to pilus structures of the same gonococcal stain whose pili were used to raise the antibody and demonstrated the lack of antibody recognition of pilus structures on two other gonococcal stains. The failure to achieve labeling on isogenic nonpiliated clones of the homologous gonoccoccus indicated the absence of pilus antigens on the surface of these organisms. Further double-label experiments allowed the simultaneous localization of two different gonococcal antigens.

The location of antigenic determinants of conjugative F-like pili has been determined on genetically engineered *Escherichia coli* (Worobec *et al.*, 1986). The amino terminus-specific antibodies did not bind to the sides of the pili but appeared to be associated with the pilus tip. These antibodies also bound to the vesiclelike structure at the base of the pilus. Antipilus antibodies, not specific for the amino terminus (unbound immunoglobulin G), bound to the sides of the pili. Anti-F and anti-Col B2 pilus antibodies bound to the sides of F, Col B2, and R1–19 pili, which have only their secondary epitope in common. The authors assume that the carboxy-terminal lysine of R1–19 pilin prevents the absorption of anti-F pilus antiserum but not anti-Col B2 pilus antiserum to the sides of the pilus, presumably by interfering with the recognition of the secondary epitope.

Hiemstra *et al.* (1986) preincubated OsO_4-fixed cells with antibody and protein A–gold probes before drying the cells onto Formvar–carbon-coated grids for examination. These experiments detected lipoproteins accessible at the cell surface after various pretreatments. A further study using the immunonegative stain method (McLaughlin *et al.*, 1986) demonstrated five patterns of spatial arrangements of antigens on the cell surface of *Listeria monocytogenes*.

CONCLUSIONS

The immunonegative stain method is ideal for the study of external antigens on any small structure which may be dried onto a coated grid and immunolabeled in suspension. It is a method which attains high morphological resolution and impressive, clear immunolabeling. It is not a concentrating method and therefore a limitation of the technique is that any particles presented for examination should be in a sufficiently high concentration to enable them to be readily apparent after drying onto a coated grid. Once this limitation has been overcome, the method can be used for almost all multiple-labeling experiments and quantitative studies. Reports of the method have been appearing for the last 3 years. It is not at present widely used but nonetheless it possesses considerable potential for wide application to the study of small particulate antigens.

REFERENCES

Almeida, J. D., and Waterson, A. P. (1969). The morphology of virus–antibody interaction. *Adv. Virus Res.* **15**, 307.

Beesley, J. E., and Betts, M. P. (1984). Applications of the immunonegative stain technique to the study of viral and bacterial antigens. *Proc. Eur. Congr. Electron Microsco.*, *8th, 1984* Vol. 3, p. 1595.

12. Colloidal Gold: Immunonegative Staining Method

Beesley, J. E., Day, S. E. J., Betts, M. P., and Thorley, C. M. (1984). Immunocytochemical labeling of *Bacteroides nodosus* pili using an immunogold technique. *J. Gen. Microbiol.* **130,** 1487.

Beesley, J. E., and Betts, M. P. (1985). Virus diagnosis, a novel use for the protein A–gold probe. *Med. Lab. Sci.* **42,** 161.

Beesley, J. E. (1987). Colloidal gold electron immunocytochemistry: Its potential in medical microbiology. *Serdiagn. Immunother.* **1,** 239.

Beesley, J. E., and Betts, M. P. (1987). Colloidal gold probes for the identification of virus particles: An appraisal. *Micron Microsc. Acta* **18,** 299.

Bendayan, M., and Stevens, H. (1984). Double labeling cytochemistry applying the protein A–gold technique. In *Immunolabelling for Electron Microscopy* (J. M. Polak and I. M. Varndell, eds.), pp. 143–154. Elsevier, Amsterdam.

Day, S. E. J., Thorley, C. M., and Beesley, J. E. (1986). Serotyping of *Bacteroides nodosus*: Proposal for a further 9 serotypes (J–R) and a study of the antigenic complexity of *B. nodosus* pili. In *Foot Rot in Ruminants* (D. J. Stewart, J. E. Petersen, N. M. McKen, and D. L. Emery, eds.), pp. 147–159. CSIRO Div. Anim. Health, Australia Wool Corporation, NSW.

Derrick, K. S. (1973). Detection and identification of plant viruses by serologically specific electron microscopy. *Phytopathology* **63,** 441.

Faulk, W. P., and Taylor, G. M. (1971). An immunocolloid method for the electron microscope. *Immunochemistry* **8,** 1087.

Griffiths, G., and Hoppeler, H. (1986). Quantitation in immunocytochemistry; correlation of immunogold labeling to absolute numbers of membrane antigens. *J. Histochem.Cytochem.* **34,** 1389.

Hiemstra, H., de Hoop, M. H., Inouge, M., and Witholt, B. (1986). Induction kinetics and cell surface distribution of *Escherichia coli* lipoprotein under *Lac* promoter control. *J. Bacteriol.* **168,** 140.

Hopley, J. F. A., and Doane, F. W. (1985). Development of a sensitive protein-A gold immunoelectron microscopy method for detecting viral antigens in fluid specimens. *J. Virol. Methods* **12,** 135.

Kjeldsberg, E. (1985). Specific labeling of human rotavirus and adenovirus with gold–IgG complexes. *J. Virol. Methods* **12,** 47.

Kjeldsberg, E. (1989). Immunogold labeling of viruses in suspension. In *Colloidal Gold: Principles, Methods, and Applications, Vol. I.* (M. A. Hayat, ed.), pp. 433–449. Academic Press, San Diego, California.

Larsson, L. I. (1979). Simultaneous ultrastructural demonstration of multiple peptides in endocrine cells by a novel immunocytochemical method. *Nature (London)* **282,** 743.

Levine, M. M., Ristaino, P., Marley, G., Smith, C., Knutton, S., Boedeker, E., Black, R., Young, C., Clements, M. L., Cherey, D., and Patnak, R. (1984). Coli surface antigen 1 and 2 of colonisation factor antigen 11 positive enterogenic *Escherichia coli*: Morphology, purification and immune responses in humans. *Infect. Immun.* **44,** 409.

Louro, D., and Lesemann, D. E. (1984). Use of protein A–gold complex for specific labeling of antibodies bound to plant viruses. I. Viral antigens in suspensions. *J. Virol. Methods* **9,** 107.

Manuelidis, L., and Manuelidis, E. E. (1986). Recent developments in scrapie and Creutzfeldt–Jakob disease. *Prog. Med. Virol.* **33,** 78.

McLaughlin J., Beesley, J. E., and Betts, M. P. (1986). Monoclonal antibodies against *Listeria monocytogenes* serogroup 4 surface antigens. *Rev. Inst. Pasteur Lyon, Proc. Int. Symp. Probl. Listeriosis 9th* pp. 102–105.

Milne, R. G., and Luisoni, E. (1977). Rapid immune electron microscopy of virus preparations. *Methods Virol.* **6,** 265–281.

Muller, G., and Baigent, D. L. (1980). Antigen controlled immunodiagnosis—"acid test." *J. Immunol. Methods* **37**, 185.

Murti, K. G., and Webster, R. G. (1986). Distribution of hemagglutinin and neuraminidase on influenza virus as revealed by immunoelectron microscopy. *Virology* **149**, 36.

Nicolaieff, A., Obert, G., and Van Regenmortel, M. H. V. (1980). Detection of rotavirus by serological trapping on antibody-coated electron microscope grids. *J. Clin. Microbiol.* **12**, 101.

Pares, R. D., and Whitecross, M. I. (1982). Gold-labeled antibody decoration (GLAD) in the diagnosis of plant viruses by immuno-electron microscopy. *J. Immunol Methods* **51**, 23–28.

Robinson, E. N., McGee, Z. A., Kaplan, J., Hammond E., Larson, J. K., Buchanan, T. M., and Schoolnik, G. K. (1984). Ultrastructural localisation of specific gonococcal macromolecules with antibody–gold sphere immunological probes. *Infect. Immun.* **46**, 361.

Roth, H. (1982). The preparation of protein A–gold complexes with 3nm and 15nm gold particles and their use in labeling multiple antigens on ultrathin sections. *Histochem. J.* **14**, 791.

Shukla, D. D., and Gough, K. H. (1979). The use of protein A from *Staphylococcus aureus* in immune electron microscopy for detecting plant virus particles. *J. Gen. Virol.* **45**, 533.

Slot, J. W., and Gueze, H. J. (1984). Gold markers for single and double immunolabeling of ultrathin cryosections. In *Immunolabelling for Electron Microscopy* (J. M. Polak and I. M. Varndell, eds.), pp. 129–142 Elsevier, Amsterdam.

Stannard, L. M., Lennon, M., Hodgkiss, M., and Smuts, H. (1982). An electron microscopic demonstration of immune complexes of hepatitis B e-antigen using colloidal gold as a marker. *J. Med. Virol.* **9**, 165.

Stannard, L. M., Fuller, A. O., and Spear, P. G. (1987). Herpes simplex virus glycoproteins associated with different morphological entities projecting from the virus envelope. *J. Gen. Virol.* **68**, 715.

Tapia, F. J., Varndell, I. M., Probert, L., DeMey, J., and Polak, J. M. (1983). Double immunogold staining method for the simultaneous ultrastructural localisation of regulatory peptides. *J. Histochem. Cytochem.* **31**, 977.

Worobec, E. A., Frost, L. S., Pieroni, P., Armstrong, G. D., Hodges, R. S., Parker, J. M. R., Finlay, B. B., and Paranchych, W. (1986). Location of the antigenic determinants of conjugative F-like pili. *J. Bacteriol.* **167**, 660.

13

Immunogold Labeling of Viruses *in Situ*

SYLVIA M. PIETSCHMANN, ELDA H. S. HAUSMANN, and HANS R. GELDERBLOM

Robert Koch-Institut des Bundesgesundheitsamtes
Berlin, Federal Republic of Germany

INTRODUCTION
PREEMBEDDING IMMUNOELECTRON MICROSCOPY
   VERSUS POSTEMBEDDING LABELING OF THIN
   SECTIONS FOR IMMUNOELECTRON MICROSCOPY
PREEMBEDDING IMMUNOELECTRON MICROSCOPY
   OF VIRUS-INFECTED CELLS.
  Preparation of Cells for Preembedding Labeling
  Coating of Plastic and Glass Surfaces
  Processing of Cell Monolayers
  Embedding and Evaluation of Microtest Cultures
IMMUNOCRYOULTRAMICROTOMY
  Preparation of Cells
  Incubation and Stabilization of Thin Frozen Sections

USE OF SEMITHIN CRYOSECTIONS IN
   IMMUNOCYTOCHEMISTRY
CHOICE OF IMMUNOELECTRON MICROSCOPIC
   MARKERS AND LABELING TECHNIQUES
   Electron-Dense Markers
TACTICS OF IMMUNOLABELING: DIRECT VERSUS IN-
   DIRECT TECHNIQUES
CONCLUDING REMARKS
REFERENCES

INTRODUCTION

The increasing availability of monoclonal antibodies and the preparatory and labeling techniques have furthered ultrastructural studies in virology in the last 10 years. Progress in methodology has allowed the widespread application of immunogold negative staining (see Chapter 16 by Kjeldsberg in Volume 1 of this series) to both diagnostic virology and basic research, permitting rapid typing of virus isolates and at higher resolution the mapping of antigenic determinants.

By studying virus-infected cells, however, additional questions may be answered regarding, for instance, the processing of viral information and intricate virus–cell interactions. To learn more about the complex system of the virus-infected cell, a variety of preparative techniques are used. While classical thin sectioning based on reliable glutaraldehyde–OsO_4 double fixation and epoxy resin embedding guarantees high-resolution information on structural aspects, the utilization of immunoelectron microscopy (IEM) reveals the antigenic makeup of viruses and cells. Antigenicity can be used as an additional marker to describe and discriminate biological properties. Thus the application of conventional electron microscopy and IEM techniques will help to elucidate the fine structure and antigenic makeup of a particular new virus. It also permits an insight into functional relationships and practically important aspects of virus pathogenicity in addition to morphology.

In this chapter we describe some recent experiences in studying antigenic properties of three different virus systems, i.e., yellow fever virus (Gelderblom et al., 1985a), human immunodeficiency virus (HIV) (Gelderblom et al., 1985b, 1987, 1988; Hausmann et al., 1987), and mammalian herpesviruses (Gelderblom et al., 1982; Schenk et al., 1988). From our observations certain strategies may be deduced regarding the efficient and straightforward application of the different IEM preparation and labeling techniques.

PREEMBEDDING IMMUNOELECTRON MICROSCOPY VERSUS POSTEMBEDDING LABELING OF THIN SECTIONS FOR IMMUNOELECTRON MICROSCOPY

Some general considerations might be helpful for deciding a particular IEM technique to be used. It is generally accepted that virus surface determinants are essential for virus–cell interactions and for the host's antiviral immune response. Preembedding labeling detects primarily external determinants such as envelope antigens of budding viruses. Hence, where the expression of viral antigens on the cell surface is of interest, preembedding labeling should be performed because of the better preservation of the fine structure and the inherent simplicity and sensitivity of this technique.

Postembedding labeling of thin sections in principle allows access also to determinants hidden inside the cells and viruses, as they become exposed by the sectioning process. Initially, water-soluble media such as serum albumin after cross-linking (McLean and Singer, 1970; Griffiths and Jokusch, 1980) or glycol methacrylates (Leduc and Bernhard, 1967; Shahrabadi and Yamamoto, 1971) were used for embedding. In the last decade, however, new water-miscible resins (LR White; Newman *et al.*, 1983; Newman, Chapter 4, this volume; Carlemalm *et al.*, 1982; Hobot, Chapter 5, this volume) as well as the use of thin frozen sections have been developed for IEM. Based on the pioneering work of W. Bernhard in the 1960s, immunocryoultramicrotomy has become a reliable technique in cell biology and virology mainly through the efforts of two groups (Tokuyasu, 1980, 1983; Griffiths *et al.*, 1983, 1984; Singer *et al.*, 1987). The advantages and limitations of pre- and postembedding IEM techniques have been extensively discussed by Boonstra *et al.* (1987), Gelderblom *et al.* (1985a), and Nermut *et al.* (1987).

There are several reasons why preembedding IEM is more sensitive than postembedding labeling of thin resin or cryosections: (1) In preembedding incubations, determinants are freely accessible for the antibody, whereas in postembedding IEM, only determinants on the surface of the sections can be detected (Stierhof *et al.*, 1986). (2) Resin embedding causes cross-linking and/or steric hindrance of determinants in the specimen, which inevitably decreases antibody binding capacity. (3) Chemical prefixation usually diminishes antigenicity. (4) Dehydration of the specimen during embedding results in partial denaturation, i.e., loss of antibody binding sites (Causton, 1984). On the other hand, the newly developed resins for low-temperature embedding represent remarkable progress in postembedding IEM of virus cell specimens (Carrascosa *et* al., 1986; Richardson et al., 1986), revealing a better preservation of fine structure. Resin-embedded specimens have an additional advantage as they can be stored for later IEM investigations.

If resin sections fail to yield a well-defined IEM signal, the technically more delicate immunocryoultramicrotomy should be applied. This technique has the advantage of rapidity and, as there is no cross-linking due to resin polymerization involved, more antigenic determinants are available for detection.

PREEMBEDDING IMMUNOELECTRON MICROSCOPY OF VIRUS-INFECTED CELLS

Both diagnostic and basic virology make ample use of cell cultures. A thorough characterization of most of the existing cell strains regarding their biology and virus susceptibility is given in the catalogue of the American Type Culture Collection (ATCC, Rockville, MD). To perform IEM in an effective way, only high-producer virus–cell systems guaranteeing high concentrations of virus, of at least one particle/cell section profile or 1000/cell around and/or within the host cell, are used. Contamination of cell cultures or the seed virus with mycoplasma or other agents may interfere with effective virus production. Thus, it is generally advisable to control the virus–cell system by thin-section electron microscopy and, if the result is unsatisfactory, to look for alternative cell sources which might support virus growth more efficiently.

Preparation of Cells for Preembedding Labeling

Many cell strains can be grown as monolayer cultures adhering to the bottom of the tissue culture vessel. To prepare specimens suitable for *in situ* IEM, cells are seeded into 5- or 9-cm-diameter cell culture dishes containing cover glasses or special plastic supports (Biofolie as the bottom in a Petriperm dish, Heraeus, Hanau, FRG; Thermanox dishes, Lab-Tek Miles). We have adopted a microtechnique, described initially by Schwarz et al. (1976), using the 60-well Microtest No. 1 plate (Fig. 13.1 Falcon Plastics, Oxnard, CA). In the flat-bottomed wells, having a diameter of 1.1 mm, between 2000 and 4000 cells per well are seeded. To achieve optimal cell propagation, it is advisable to use a relatively rich cell culture medium supplemented with 5 or 10% newborn or fetal calf serum and a CO_2-enriched atmosphere. If grown near confluence, the monolayers are infected with virus. The use of dense monolayers is not advisable, since with many virus systems they do not give rise to a highly

13. Immunogold Labeling of Viruses *in Situ* 259

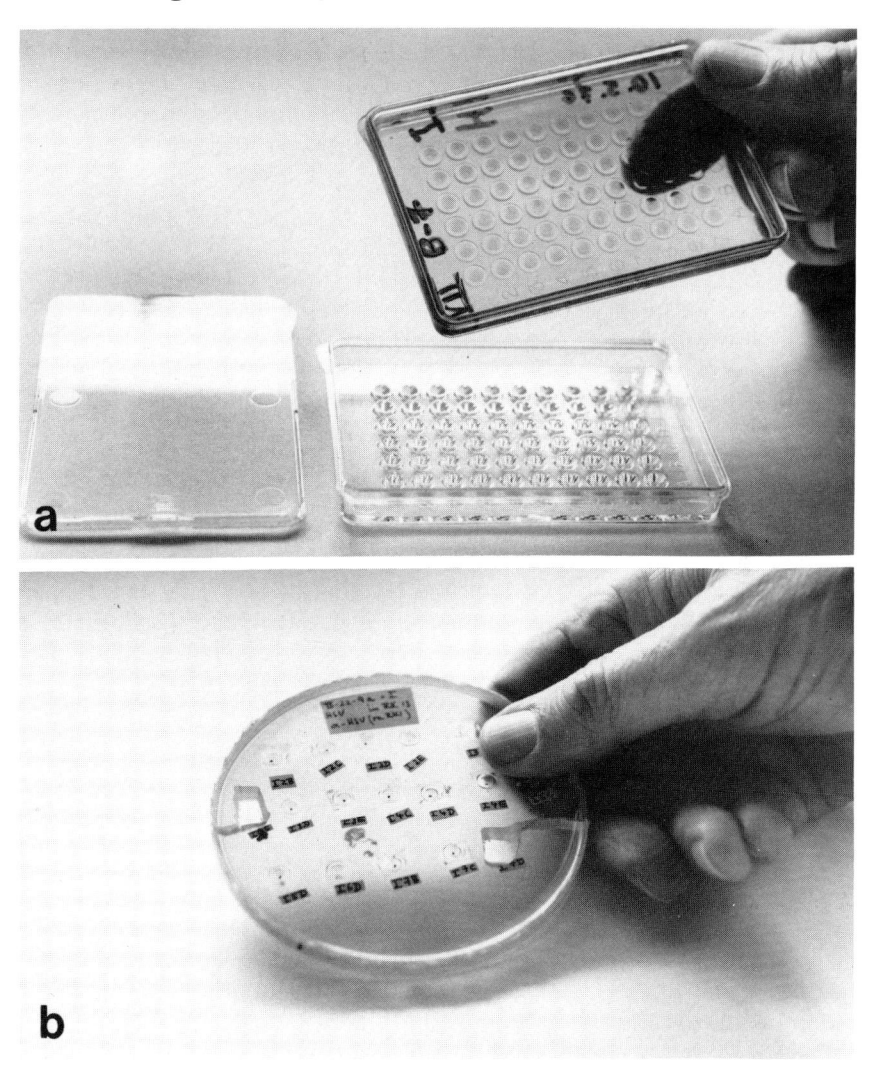

Fig. 13.1. Use of Microtest No. 1 plates in preembedding IEM. This 60-well tissue culture plate enables *in situ* labeling, embedding, and electron microscopic evaluation of virus-infected cell cultures. In (a) both an empty and an Epon-embedded plate are shown, while in (b) the reembedding is demonstrated. Small labels placed in the plate during first or second embedding allow accurate identification of the specimens.

effective infection in a one-step growth curve. The appropriate virus dose and incubation time for a given system may be evaluated before use in IEM by light microscopy (cytopathic effect), immunofluorescence (in noncytolytic systems), or thin-section electron microscopy. For preembedding labeling of budding viruses, like orthomyxo-, rhabdo-, retro-, and togaviruses, cells should be used before a cytopathic effect has fully developed; otherwise the preservation of cell structure becomes unsatisfactory. The severe cytopathic effect of some cytolytic viruses, like adeno-, papova-, reo-, and rotaviruses, might also permit preembedding labeling of intracellular virus particles (Oshiro et al., 1967; Tao et al., 1987). Such a procedure, however, does not preserve the cell fine structure and therefore cannot give detailed morphological information.

Lymphoblastoid cells, which are routinely propagated in suspension, may be immunolabeled in suspension; however, they can also be immune-incubated *in situ,* when "glued" to an appropriate support. Immobilization of suspension cells may be performed in the wells of a Microtest No. 1 plate. Two chemicals with a cationic charge, poly-L-lysine (Fisher, 1975) and the dye alcian blue (Sommer, 1977), have proved successful in coating cell supports, thus allowing reliable cell adhesion (see also Chapter 12 by Nermut and Nicole in volume 1 of this series).

Coating of Plastic and Glass Surfaces

Poly-L-lysine (mol. wt. 70,000; Sigma Chemical) is dissolved in distilled water at 50 µg/ml. Glass slides or coverslips are carefully cleaned with 70% ethanol acidified with a few drops of 1 M HCl and dried using a cotton cloth. Slides are incubated for 30 min in polylysine. Excess polylysine is removed by several washes in distilled water or phosphate-buffered saline (PBS). A dense cell suspension in a well-buffered medium containing no serum protein is applied to the wet, coated support and left undisturbed for 15 min to settle at room temperature or at 37°C. High protein concentrations in the medium should be avoided, since they will interfere with cell attachment. Nonadhering cells are washed away by floating the plate or slide with PBS, thus giving rise to a uniform and tightly adhering cell monolayer.

Alternatively, alcian blue, a cationic dye, can be used. It has the advantage of being cheaper than polylysine and is applied at 10 mg/ml in 1% acetic acid (Sommer, 1977). The dye solution is stable for several months. Coating of supports and cell attachment are performed essentially as described for polylysine.

Processing of Cell Monolayers

Prefixation of virus-producing cells contributes to a good structural preservation of the specimen, avoids redistribution of membrane antigens on live cells during incubation with the antibodies, and prevents laboratory infection in working with infectious viruses.

Usually, chemical fixation is performed with ultrapure monomeric glutaraldehyde stored under nitrogen (Sigma Chemical). Glutaraldehyde is used for prefixation at concentrations of 0.05–0.1% in PBS for 15 min at 20°C. Where glutaraldehyde prefixation fails in retaining sufficient antigenicity, milder fixatives may be used, such as formaldehyde (freshly prepared according to Karnovsky, 1965; 1–8% in PBS), dimethylsuberimidate (Hassell and Hand, 1974), or mixtures of fixatives comprising periodate–lysine–paraformaldehye (McLean and Nakane, 1974) or containing acrolein (Boonstra et al., 1987). In certain cases one could try to omit any prefixation and perform the immune incubations on ice. For detailed information on alternative fixatives the reader should refer to the comprehensive reviews by Bullock (1984) and Hayat (1981, 1986, 1989).

Subsequently, the processing of virus-producing cells for IEM is described (Table 13.1) utilizing Microtest No. 1 plates (Fig. 13.1). It allows the incubation of a large number of identical virus–cell specimens with 10 or 20 different antibodies at a time. To avoid viral cross-contamination, only one particular type of infected or uninfected control cells should be grown on a single plate. It is convenient to have three or even six wells per individual specimen. Also advantageous are the small amount of reagents, 7–10 µl per well, necessary for the incubations and the possibility of evaluating virus–cell interactions *in situ* in an undisturbed and defined orientation (Fig. 13.2 Gelderblom and Schwarz, 1976; Schwarz et al., 1976; Gelderblom et al., 1985a). Before fixation, virus-producing cells must be handled under appropriate biohazard conditions in a laminar airflow hood using gloves and 70% isopropyl alcohol or 1% sodium hypochlorite as a disinfectant.

It is better that two persons work together to perform the labeling procedure. After light microscopic control, the plates are labeled with the pertinent protocol data and washed by flooding with 15 ml of PBS and shaking gently for 1 min at 20°C in order to remove excess proteins that would interfere with prefixation. The buffer is poured over one corner of the plate, keeping it vertical for a few seconds. Under no circumstances during the entire processing should the specimens be allowed to dry. Prefixation is performed by flooding the plate up to the rim with glutaraldehyde and leaving it for 10 to 15 min at room temperature with occasional shaking. The fixative is then removed and the plate is treated (Table 13.1)

TABLE 13.1
Processing of Virus-Producing Cell Cultures for *in Situ* Preembedding IEM Using Microtest No. 1 Plates

Procedure[a]	Temperature (°C)	Time
2 washings with PBS	20	30 sec each
Prefixation with 0.05 or 0.1% GA in PBS	20	15 min
2 washings with PBS	20	2 min each
1 washing with conditioned PBS	20	15 min
Incubation with the primary antibody	37	30 min
2 washings with conditioned PBS	20	3 min each
Incubation with the labeled antibody	37	30 min
3 washings with conditioned PBS	20	3 min each
1 washing with PBS	20	3 min
Fixation with 2.5% GA in PBS	20	15 min
1 washing with distilled water	20	5 min
Postfixation with 1% OsO_4 in distilled water	30	60 min
1 washing with distilled water	20	5 min
Staining with 1 to 2.5% uranyl acetate in distilled water in the dark	20	60 min
Dehydration with graded ethanol: 30, 50, 70, 90, and 95%	20	5 min each
Dehydration with three changes of 100% ehtanol	20	10 min
Infiltration with two changes of a 1 : 3 mixture of Epon and ethanol	20	30 min each
Infiltration with two changes of a 2 : 3 mixture of Epon and ethanol	20	30 min each
Infiltration with Epon containing accelerator	20	Overnight
Embedding in Epon with accelerator, polymerization	60	24–48 hr
Cutting out and oriented reembedding	60	48 hr
Thin sectioning and mounting on 360-mesh grids		
Poststaining with 0.4% lead citrate and carbon reinforcement		

[a] Conditioned PBS is PBS containing 0.1% gelatin and 0.02 M lysine; GA, ultrapure monomeric glutaraldehyde.

consecutively with PBS and conditioned PBS (PBS containing 0.1% bovine serum albumin or 0.1% gelatin and 10 mM lysine) to remove the fixative and to block any remaining free aldehyde groups. Conditioned PBS also serves as a diluent for primary and secondary antibodies and as a washing solution, since due to its protein content it reduces nonspecific binding of the immunoreagents.

The wells to be treated with a given antibody are freed from buffer using a Pasteur pipette connected to a vacuum line. The pipette is held tangentially in order to avoid destruction of the monolayer at the bottom of the well. According to the protocol determined, antibodies at pretested

13. Immunogold Labeling of Viruses in Situ

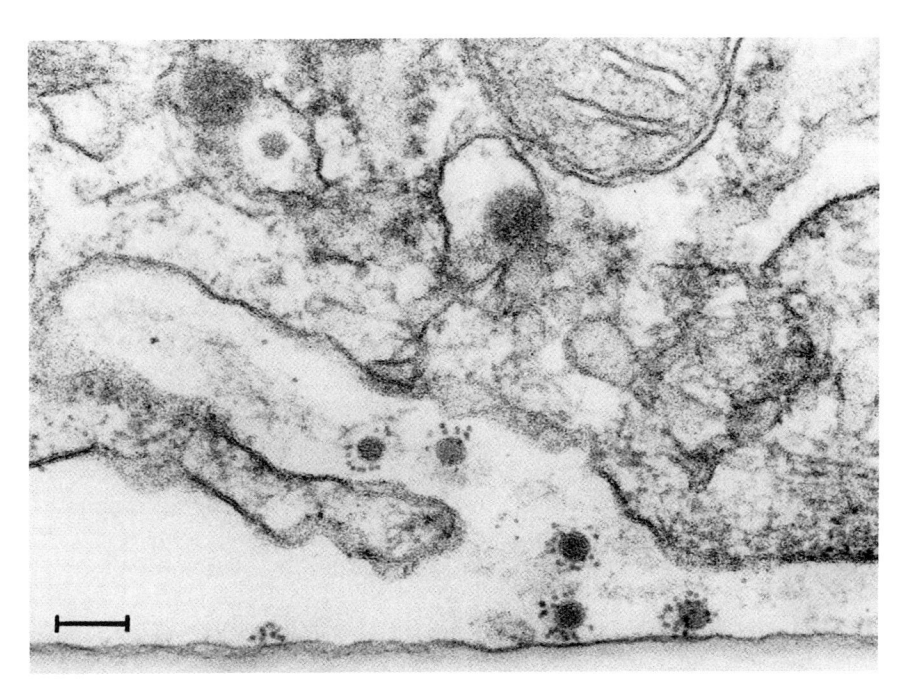

Fig. 13.2. Preembedding IEM of yellow fever virus-infected cells using virus-specific monoclonal antibody at a concentration of 100 μg/ml followed by anti-mouse IgG–ferritin conjugate (1 : 10 dilution). Sections were cut perpendicular to the bottom of the well of the Microtest No. 1 plate. (From Gelderblom *et al.*, 1985a, with permission of Elsevier Science Publishers, Amsterdam). Bar marker represents 100 nm; ×100,000.

dilutions are added to the individual wells. Finally, PBS droplets are placed in the corners of the plate, and the lid is closed to keep a humid atmosphere during incubation at 37°C for 30 min. Primary antibodies are sucked out of the wells with a pipette, which is flushed between individual antibodies with plain PBS. The plates are washed by flooding them twice with 10 ml of conditioned PBS. After the addition of appropriately diluted second, ferritin-, or gold-labeled antibody, plates are incubated at the same conditions as for the primary antibody. The marker antibody conjugate is removed again individually from each well. The plates are washed and fixed for 15 min with 2.5% glutaraldehyde in PBS at room temperature. The plates can be processed further for embedding without delay, but for convenience they may also be stored for several days without appreciable damage to the ultrastructure if kept in glutaraldehyde at 4°C. The proper choice and the dilution of the detection system, as well as the advantages and limitations of some of the most widely used incubation schemes and marker systems, will be discussed below.

Embedding and Evaluation of Microtest Cultures

The embedding protocol follows generally established routines for epoxy resin embedding (Hayat, 1989). However, dehydration and infiltration of monolayer specimens can be abridged considerably. Glutaraldehyde is poured off the plates, and after briefly rinsing with PBS followed by distilled water, the plates are filled with OsO_4 for postfixation (1% OsO_4 in distilled water) for 1 hr at room temperature. This treatment causes slight shrinkage of the specimen, which results in improved delineation of cells concomitant with easier detectability of the virions. Based on comparative studies, using several hypo- and hypertonic solutions, no changes in the fine structure of the labile retroviruses could be detected with the above treatment (Gelderblom et al., 1974). After postfixation and rinsing in distilled water, the plates are treated with uranyl acetate (1–2.5% in distilled water) for 1 hr in the dark. Incident light and/or the presence of phosphate ions may result in severe nonspecific precipitates. Uranyl ions bind mainly to phosphate groups of nucleic acids and lipids and therefore help to differentiate structural details, especially after lead poststaining.

Dehydration is performed rapidly with five increasing concentrations of ethanol in distilled water (30, 50, 70, 90, and 95% each for 5 min) at room temperature, followed by three changes with 100% ethanol (10 min each), with the plates being closed with the respective lids. Ethanol, like some Epon constituents, readily absorbs water, which may cause problems during polymerization. Infiltration of the monolayer is carried out in three steps using a 1 : 3 mixture of Epon in ethanol for 1 hr, followed by a 2 : 3 mixture for 1 hr at room temperature, and finally Epon (a 7 : 3 mixture of Epon I and Epon II is used with or without the addition of DMP-30 accelerator). Since the thorough exchange of these mixtures is essential for proper embedding, each solution is changed twice.

Filled with Epon plus the accelerator, the plates are left overnight. After the final change, Epon should cover the plate by not more than 1 mm. The specimens are polymerized at 60°C. Thicker Epon layers will cause incomplete polymerization. The amount of the resin as well as the horizontal position of the plates in the oven should be controlled. After polymerization for at least 24 hr at 60°C, individual wells are cut out using a small hand saw. By immersion in liquid nitrogen or with the help of a scalpel and a small hammer, the tissue culture plastic is split from the Epon block, which now shows the embedded monolayer freely exposed.

To stabilize the specimens in the electron beam, we routinely reembed them in a plastic dish (13 cm in diameter) together with small labels. For safer identification the label may also be included in the first Epon layer. The cells are covered by 2 mm of Epon (Fig. 13.1). Individual blocks are sawed out from this plate, clamped in flat specimen holders, and trimmed for thin sectioning using either a Reichert TM 60 Specimen Trimmer or an LKB Pyramitome. Serial sections are cut perpendicular to the bottom of the culture, permitting easy orientation, and are collected on bare 200-mesh copper grids. After poststaining with lead citrate (Venable and Coggeshall, 1965) and stabilization with a 15-nm layer of carbon, the specimens are evaluated, preferably at 60 kV. Micrographs of viruses are recorded routinely at magnifications between 20,000 and 40,000.

Special advantages of this procedure become readily apparent. Besides using small amounts of reagents and a large number of specimens (which can be handled simultaneously), the *in situ* processing permits assessing not only the virus–cell relationship but also cell-to-cell interactions in an oriented way. The handling of this microtechnique does not pose any serious obstacles. Nevertheless, for preembedding immunolabeling larger amounts of cells grown on cover glasses or in suspension may also be processed (Oshiro *et al.*, 1967; Calafat *et al.*, 1983; Evans and Webb, 1984; Carrascosa *et al.*, 1986).

IMMUNOCRYOULTRAMICROTOMY

In our hands, immunolabeling of thin frozen sections of virus-infected cells has proved consistently more specific and sensitive than postembedding labeling of thin methacrylate sections. Using monoclonal antibodies, we were unable to get clear-cut results with the latter technique. We therefore confine ourselves in this chapter to the description of the immunocryoultramicrotomy. The reader interested in postembedding labeling of thin resin sections, which may be the technique of choice in many other systems, is referred to some other chapters in this volume and volume 1. Immunocryoultramicrotomy can be learned efficiently only in an experienced laboratory rather than from a written source, though the matter is dealt with in a number of detailed papers (Griffiths *et al.*, 1983, 1984; Tokuyasu, 1983, 1986; Gelderblom *et al.*, 1985a; Boonstra *et al.*, 1987; Singer *et al.*, 1987; and chapters in volume 1). Here we will outline only the general principles of this technique (Fig. 13.3) and, in addition, provide hints for successful preparation of virus-infected cells.

Preparation of Cells

Cells are grown and infected following the strategies already described for preembedding IEM. However, larger numbers of cells ($5–10 \times 10^6$) are required to prepare the densely packed blocks from which the sec-

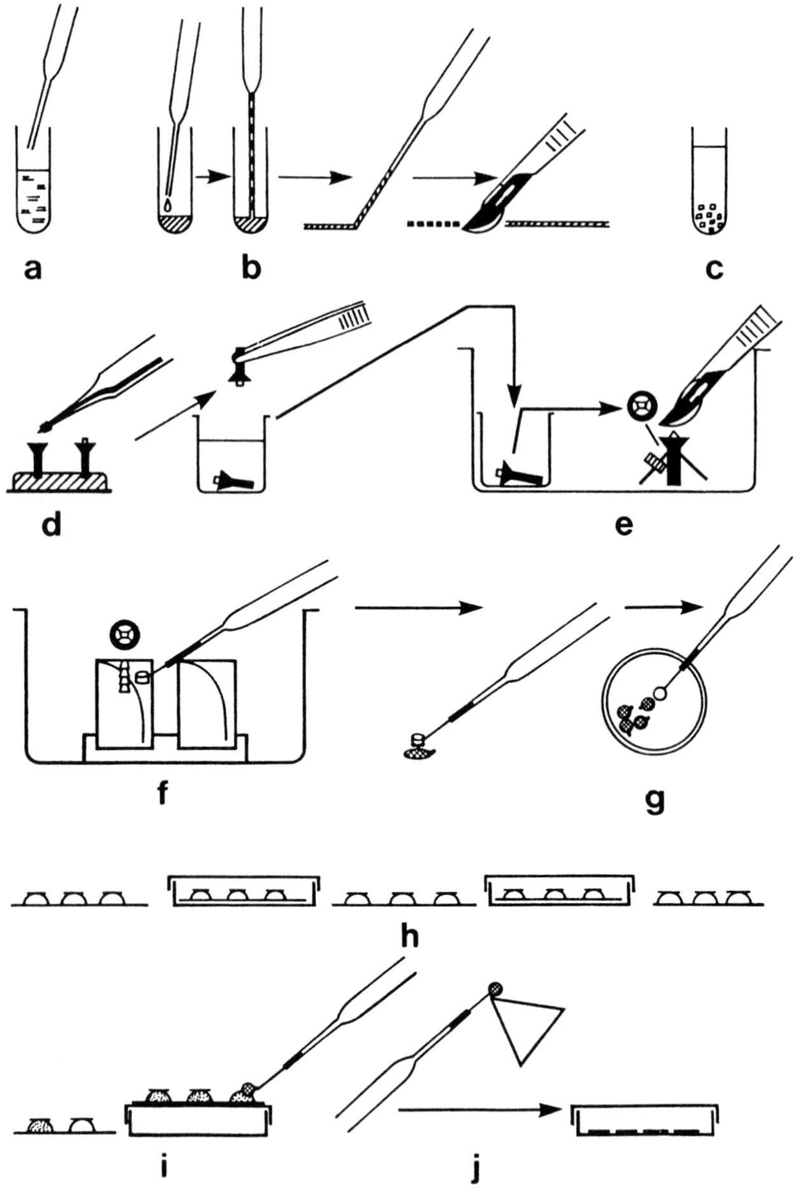
a
b
c
d
e
f
g
h
i
j

13. Immunogold Labeling of Viruses *in Situ*   267

tions are cut. Cells, grown as monolayer cultures or in suspension, should be prefixed (see below) when they show a moderate cytopathic effect or other signs of virus production. This step is necessary for immunocryoultramicrotomy, since we are interested in tracing antigenic determinants inside the cells, which suffer from the ongoing cytopathic effect. This effect is accompanied by massive swelling of cells by water uptake, which finally may lead to dislocation of cytoplasmic constituents during handling of the sections. Therefore, prefixation for immunocryoultramicrotomy also helps to keep antigens *in situ* during the extensive labeling procedure.

After washing in PBS, cells are fixed with a mixture of 0.1% glutaraldehyde and 4% freshly prepared formaldehyde in PBS containing 5% sucrose for 15 min at 20°C; 0.05% glutaraldehyde often gives insufficient structural preservation. Monolayers are then gently scraped off the bottom of the dish with the help of a rubber policeman by a few contiguous movements, resulting in coherent cell sheets. Cells are sedimented for 10 min at 200–300 *g*. Depending on the particular cell type, the resulting pellet will be sufficiently stable for further manipulation. This is also true for many epithelioid cell strains such as HeLa, Vero, and PS. Other cells may be optimally held together when embedded in gelatin or agarose blocks. Indeed, this step may also be essential to keep macromolecules *in situ* during immunolabeling, as shown for the loosely anchored envelope glycoprotein knobs of the human immunodeficiency virus type 1 (HIV-1) (Gelderblom *et al.*, 1985b, 1987; Hausmann *et al.*, 1987) (see Figs. 13.4 and 13.5). This block enclosure, however, was obsolete when studying herpesvirus antigens due to the more stable anchorage of herpesvirus envelope projections (see Figs. 13.6, 13.7, and 13.8).

To preserve both the fine structure and the antigenicity of the cells, the use of a low-melting-point agarose (Bethesda Research Laboratory, Bethesda, MD) is recommended (H. Schwarz, personal communication). To the loosely packed cell pellet an equal volume of agarose (3% in PBS) is added (Fig. 13.3b) and gently mixed with a thin glass rod at 35–40°C.

Fig. 13.3. Schematic representation of immunocryoultramicrotomy. After prefixation, cells are cryoprotected by infiltration with sucrose (a), embedded in agarose (b), and dissected into small blocks (c). Specimens are mounted on aluminum specimen stubs (d), snap-frozen by rapid immersion in undercooled liquid nitrogen, and transferred into the cryochamber, where the blocks are trimmed (e). Thin, frozen sections are retrieved with a platinum loop containing 2.3 *M* sucrose (f). The thawed sections are placed on plastic and carbon-coated hexagonal grids, which are stored (sections down) on PBS (g). Following immunolabeling (h), sections are stained and stabilized in drops of uranyl acetate and Methocel (i). Finally, excess fluid is removed with a filter paper, and the grid is air dried while held in the loop (j).

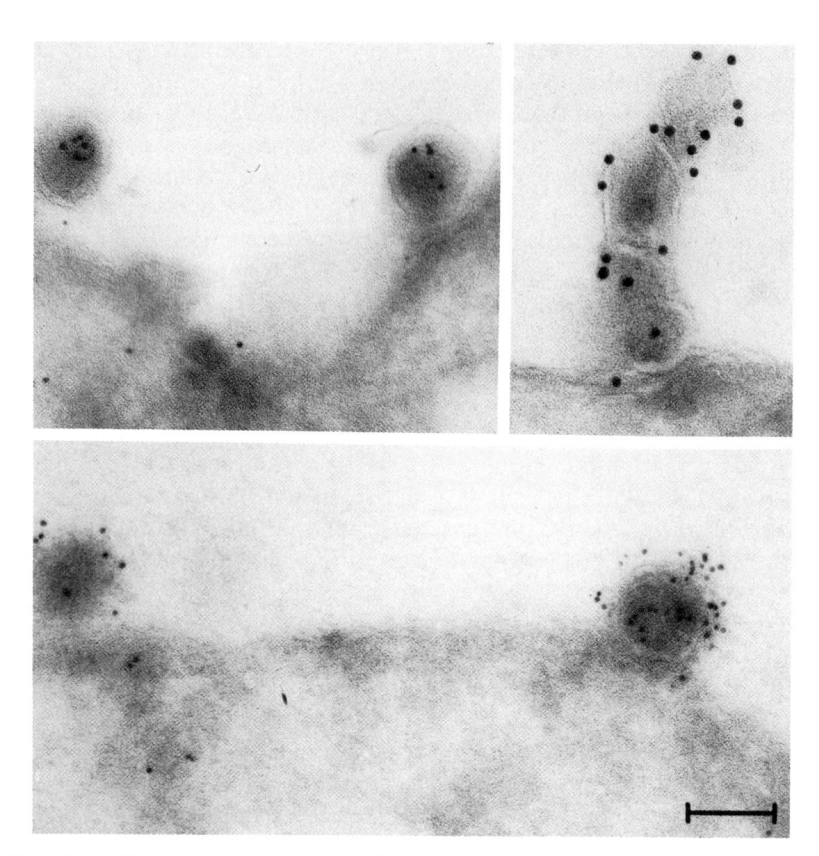

Fig. 13.4. Comparative immunocryoultramicrotomy of HIV-1 strain HTLV-IIIB-producing H9 cells embedded in agarose (upper right, bottom) or without prior agarose protection (upper left). All sections were incubated with the same serum from a human ARC patient, followed by anti-human IgG gold particles (5 nm) (upper left, bottom) or gold particles (10 nm) (upper right). Envelope determinants are revealed only with agarose-embedded specimens (upper right, bottom), while internal HIV antigens are detected in all three preparations. The IgG–gold (5-nm) probe leads to a two- to threefold higher labeling density than the IgG–gold (10-nm) probe. (From Hausmann *et al.*, 1987, with permission of Elsevier Science Publishers.) Bar marker represents 100 nm; ×120,000.

The mixture is drawn up 1 or 2 cm into the tip of a Pasteur pipette. After gelling in an ice bath, the rod-shaped block is removed from the pipette and divided under a binocular light microscope with a scalpel into blocks 0.5 to 0.8 mm in length. In an Eppendorf vial they are infiltrated for cryoprotection at 4°C with increasing concentrations of 0.3, 1.2, and 2.3 *M* sucrose in PBS. When they have sunk to the bottom of the vial, they are sufficiently infiltrated. They can be stored in 2.3 *M* sucrose without loss

13. Immunogold Labeling of Viruses *in Situ*

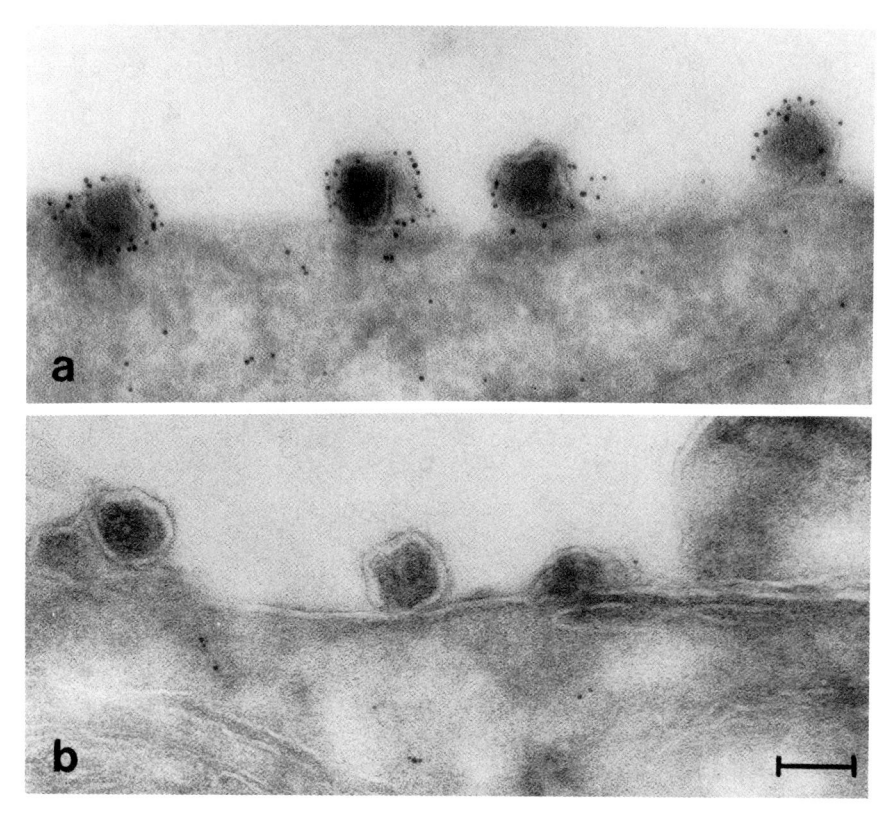

Fig. 13.5. Thin frozen sections of HIV-1-producing cells after incubation with a serum from an ARC patient and detection of bound antibodies by IgG–gold (5 nm) (a). This serum labels envelope as well as interior determinants of HIV, while human control sera (b) are negative. Bar marker represents 100 nm; ×100,000.

of antigenicity at 4°C for several weeks with the addition of 0.02% NaN_3 (Gelderblom *et al.*, 1985a; Griffiths *et al.*, 1984). The blocks are mounted on aluminum specimen stubs of the Reichert-Jung FC4 cryoattachment system (Fig. 13.3d). The preparation and storage of at least 10 samples will allow processing of identical specimens for longer periods of time. During mounting under the binocular light microscope, drying artifacts must be avoided.

The specimens are then shock-frozen by quickly dropping the stubs with the blocks into undercooled liquid nitrogen contained in a Styrofoam box. This so-called nitrogen slush ($-210°C$, 63 K) is generated from liquid nitrogen ($-196°C$, 77 K) by applying a vacuum of 26 Pa (0.2 torr) for 5 min. It does not show the insulating effect of the Leydenfrost boiling

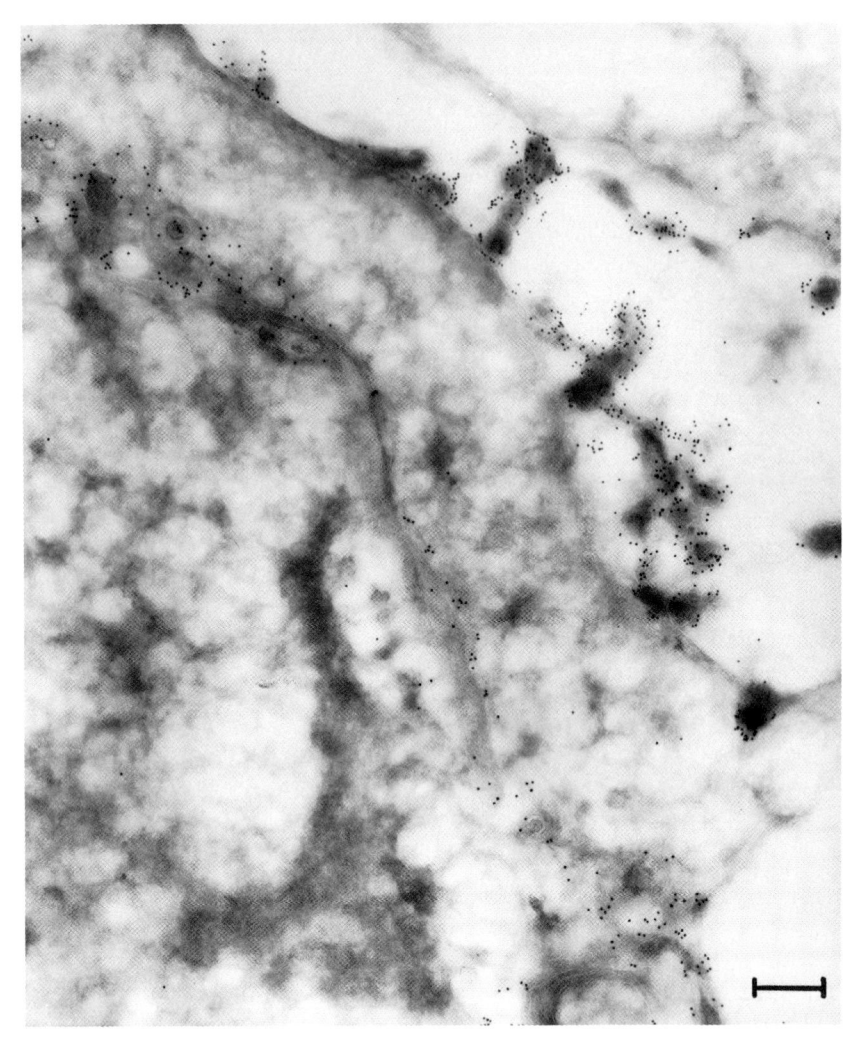

Fig. 13.6. Immunocryoultramicrotomy of herpes simplex virus type 1-producing Vero cells. The virus glycoprotein-specific monoclonal antibody revealed by the IgG–gold (10-nm) probe is detecting antigens associated with the envelope of mature cell-released virions and with the outer nuclear envelope. Bar marker represents 300 nm; ×30,000.

phenomenon and thus allows rapid freezing (Robards and Sleytr, 1985). One or preferably two of the frozen specimens are transferred for sectioning into the precooled FC4 cryoattachment of the Reichert Ultracut microtome, while the remaining samples are stored for future use in liquid nitrogen. When combined with rapid freezing, sucrose prevents ice crystal formation (Griffiths *et al.*, 1984). The high sucrose concentration ex-

13. Immunogold Labeling of Viruses in Situ 271

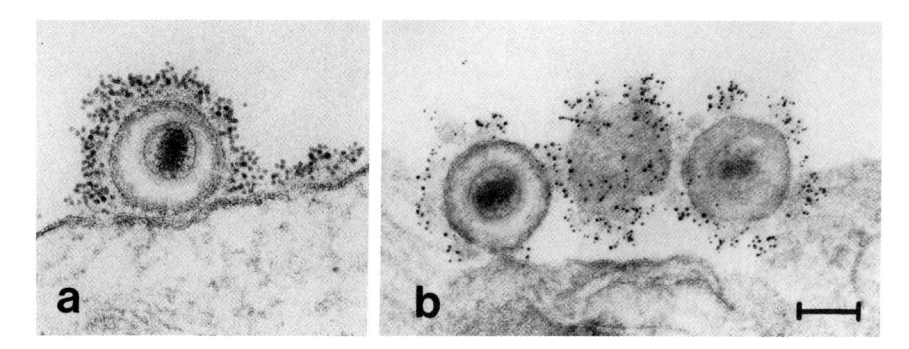

Fig. 13.7. In preembedding IEM, extracellular herpesviruses reveal a fringe of marker molecules after incubation with a polyclonal hyperimmune serum. The ferritin probe displays compact tagging (a), while the smaller, more electron-dense marker system IgG–biotin/streptavidin–gold (4 nm) shows distinct clusters of gold particles (b). Bar marker represents 100 nm; ×80,000.

erts a further beneficial effect. The cryoprotected material is relatively soft, and therefore it can be sectioned even at $-140°C$ using glass knives. Glass knives are more advantageous for cryosectioning than diamond knives; the former are prepared from conventional glass strips used in ultramicrotomy (LKB, Bromma, Sweden; 400 × 25 × 6.3 mm). When used for cryosectioning, the edges of the inexpensive glass knife are stable, allowing hundreds of sections. In our hands, the reinforcement of glass knifes by a thinly evaporated tungsten layer as recommended by Roberts (1975) does not improve knife edge quality.

Reaching temperature equilibrium of about $-100°C$ (173 K) in the cryochamber, with two knives already in position, takes 40 min. At equilibrium the specimen stub is trimmed to appropriate size and pyramidal shape using a precooled scalpel. For thin sections, cutting is performed at -100 to $-90°C$ using a low cutting speed. Fine sections of a thickness between 50 and 80 nm have been obtained routinely using hand-driven or automatic cutting. With the help of an eyelash, sections are grouped on the back of the dry knife. Finally, they are transferred to grids using a droplet of 2.3 M sucrose, held in a platinum wire loop. During the transport the frozen sections will thaw and attach tightly to the grids. High-transparency hexagonal grids (300 or 400 mesh) with thin bars of 5 μm (Fig. 13.4) are used. These grids are more stable than those with rectangular or square holes when coated with a Pioloform plastic film (Stockem, 1970; Wacker Chemie, München, FRG) and reinforced with a 20-nm layer of carbon. Before use, the carbon surface of the grids is made hydrophilic by glow discharge for 0.5–3 min at 26–67 Pa (0.2–0.5 torr). The sections

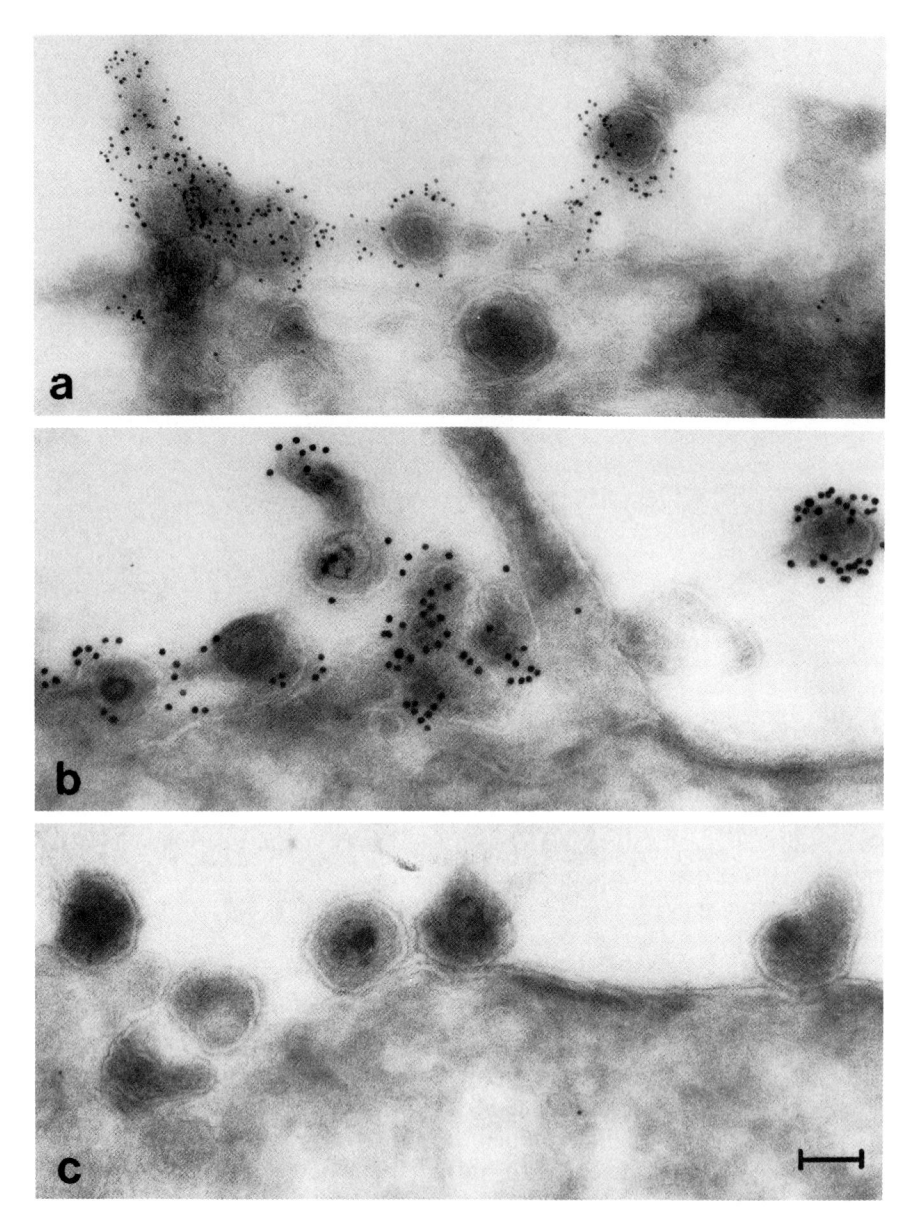

Fig. 13.8. Immunocryoultramicrotomy of Vero cells infected with different herpesviruses, BHV-2 (a) and HSV-1 (b, c). The binding of a cross-reactive monoclonal antibody was visualized by IgG–gold (5 nm) (a) or IgG–gold (10 nm) (b), while an unrelated monoclonal antibody (c) was completely unreactive. Bar marker represents 100 nm; ×80,000.

13. Immunogold Labeling of Viruses in Situ 273

Fig. 13.9. High-transparency 300-mesh copper support grid with hexagonal holes and thin bars. These grids, after coating with a 20-nm thin plastic film, carbon reinforcement, and glow discharge pretreatment, allow a good compromise between specimen stability and ease of evaluation of thin frozen sections. Bar marker represents 12.5 µm; ×600.

on the grids can be stored until labeling for up to several hours in a small plastic dish with the sections floating face down on PBS.

Incubation and Stabilization of Thin Frozen Sections

Labeling is performed by floating the grids on solutions placed as droplets onto Parafilm sheets (Griffiths *et al.*, 1983). According to the flow diagram in Table 13.2, a large number of individual specimens can be conveniently treated simultaneously. The general principles discussed for preembedding IEM labeling are also true for the incubation of thin cryosections. However, the search for intracellular determinants poses additional problems due to the inherent stickiness of intracellular constituents. There are two usually successful ways to block high nonspecific background: (1) preincubation of sections with the preimmune or a "normal," nonimmune serum of the animal species which supplied the anti-IgG antibody for the second labeled antibody and (2) dilution of the primary antibody in 10% normal serum derived from the same species.

TABLE 13.2
Indirect Antibody Labeling Procedure for Thin Cryosections[a]

Number and Volume of Droplets	Procedure	Temperature (°C)	Time
1 × 100 μl	Conditioning with 0.02 M lysine in PBS	20	5 min
2 × 100 μl	Conditioning with 0.1% gelatin in PBS	20	2 min each
1 × 10 μl	Incubation with the primary antibody	37	30 min
5 × 100 μl	Washings in 0.1% gelatin in PBS	20	4 min each
1 × 10 μl	Incubation with labeled antibody	37	30 min
5 × 100 μl	Washings in 0.1% gelatin in PBS	20	4 min each
3 × 100 μl	Washings in PBS	20	2 min each
1 × 50	Postfixation with 2.5% GA in PBS	20	10 min
3 × 100 μl	Washings in PBS	20	4 min each
5 × 100 μl	Washings in distilled water	20	2 min each
1 × 50 μl	Staining with 2% uranyl acetate/oxalate (pH 7) in the dark	20	5 min
1 × 100 μl	Washing in distilled water	20	1 min
1 × 100 μl	Staining and stabilizing with 1% Methocel containing 0.2% uranyl acetate in distilled water	0 (ice)	10 min

[a] The whole labeling sequence is carried out by floating the grids on the surface of drops placed on Parafilm sheets (modified from Griffiths et al., 1984).

To visualize the fine structure, sections need to be stabilized and stained using 2% uranyl acetate/oxalate (pH 7) (mix equal volumes of 0.3 M oxalic acid with 4% aqueous uranyl acetate and adjust pH with 5% NH_4OH; Griffiths et al., 1984). This staining solution introduces a mixture of positive and negative contrast. The labile sections are further stained and protected against collapse during air drying by infiltration and coating in a drop of methyl cellulose (Methocel, 25 cP Fluka, Buchs, Switzerland; 1% methyl cellulose in distilled water containing 0.2% aqueous uranyl acetate). Methocel, prepared according to Griffiths et al. (1984), is applied to the sections for 10 min on ice (Fig. 13.3i). Subsequently, excess fluid is removed from the grid by touching it with a filter paper, and the specimen is air dried while held in a platinum wire loop. The fine structure may be further preserved by finally coating the sections with a thin film of a resin (e.g., Epon; Keller *et al.*, 1984) and by another fixation step applied before the final Methocel treatment (2.5% glutaraldehyde in PBS). Because of the inherent high contrast, frozen sections are evaluated preferably at 80 kV.

USE OF SEMITHIN CRYOSECTIONS IN IMMUNOCYTOCHEMISTRY

Semithin sections (0.5 μm thick) may be cut for light microscopic immunocytochemistry from the same specimen block, but at a higher temperature ($-60°C$) using a considerably lower cutting speed. Such sections are collected on glass slides and processed as detailed by Griffiths *et al.* (1984). The evaluation of semithin sections of virus-infected cells by immunofluorescence or the alkaline phosphatase–anti-alkaline phosphatase (APAAP) technique (Cordell *et al.*, 1984) presents an attractive alternative to the hitherto widely applied technique of studying whole cells after permeabilization. When a new virus–cell system needs to be investigated by IEM, light microscopic or dot blot techniques, due to their comparative sensitivity and speed, help to assess rapidly the feasibility of a given system and to establish preparation parameters (e.g., prefixation) for successful IEM.

CHOICE OF IMMUNOELECTRON MICROSCOPIC MARKERS AND LABELING TECHNIQUES

For screening purposes for light and electron microscopic cytochemistry, polyclonal antibodies, as present in hyperimmune sera, are preferable to monoclonal antibodies, since mono- or polyspecific antisera yield a much higher signal due to the detection of a variety of determinants. Furthermore, hyperimmune sera usually contain within their spectrum of polyclonal antibodies a fraction of high-affinity antibodies, leading to reliable labeling. Monoclonal antibodies binding exclusively to one particular antigenic determinant may be well suited to describe the antigenic makeup of virus particles (see Chapter 15 and 16 by Beesley and Kjeldsberg, respectively, in Volume 1 of this series) and to look for virus–cell interactions by *in situ* IEM, provided they exert high-affinity binding.

Because of their low electron-scattering power, native antibodies are scarcely detectable when applied as a probe in virus-producing cell cultures. However, the use of unlabeled antibodies should be considered as an alternative to the other, well-established preembedding IEM techniques, as was shown in a study of mouse mammary tumor virus-producing cells by Lasfargues et al. (1974). Using polyclonal antisera and indirect immunonegative staining, a heavy fringe of antibodies was detected around budding and immature virus particles. Immunonegative staining is applicable to virus-producing cells only when large amounts of virus are available on the cell surface. This technique, however, fails to provide insight into cytoplasmic or nuclear compartments. Therefore, this application apparently represents a rare exception to the rule; i.e., immunoelectron microscopy depends on the use of electron-dense labeled antibodies.

Electron-Dense Markers

Since the introduction in 1959 of ferritin as an electron-dense marker (Singer, 1959), two valuable additions to the repertoire of markers can be noted (Beesley et al., 1982): the immunoenzyme techniques (Nakane and Pierce, 1966; Avrameas and Uriel, 1968; Nakane and Kawaoi, 1974) and the use of metallic colloidal gold (Romano and Romano, 1977; Roth, 1982, 1984; for a comprehensive review, see Horisberger, 1984). These three markers have been successfully applied in virological IEM. However, depending on the particular problem and the techniques applied, each marker system shows specific advantages. Ferritin, being a ubiquitous iron storage protein, shows an amorphous protein shell 11 nm in diameter surrounding a 7-nm, moderately electron-dense core composed of up to 4000 iron atoms. Ferritin can easily be isolated from horse spleens and is covalently conjugated to affinity-purified IgG by bifunctional reagents such as diisocyanates (Singer, 1959) and glutaraldehyde (Takamiya et al., 1975).

Commercially available ferritin conjugates (Cappel Laboratories, Cochranville, PA) have been used routinely with success and were compared for labeling efficiency with IgG–gold conjugates of different sizes (Janssen Life Sciences, Beerse, Belgium) or with protein A–gold conjugates commercially purchased or prepared in one's own laboratory (Beesley et al., 1982; Evans and Webb, 1984; Van Bergen en Henegouwen and Leunissen, 1986).

IgG–gold probes are made by adsorption of affinity-purified IgG to the highly negatively charged surface of colloidal gold particles. The preparation of colloidal gold particles has been explained by Handley in volume 1 of this series. Colloidal gold particles 5–10 nm in diameter are used most often in virological IEM. Coupling efficiency and the reactivity of the

13. Immunogold Labeling of Viruses *in Situ* 277

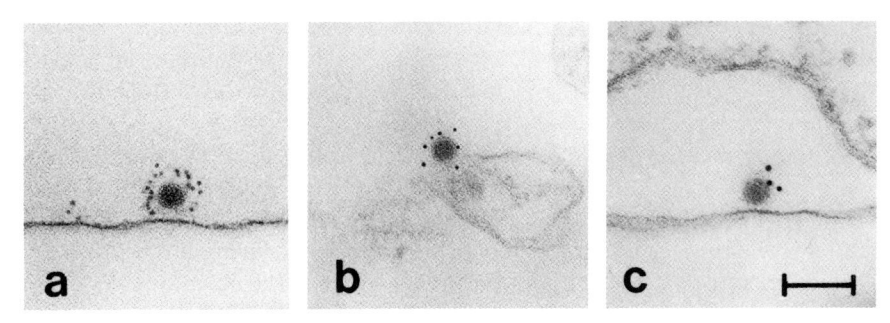

Fig. 13.10. Comparison of IgG–ferritin (a), IgG–gold (5-nm) (b), and protein A–gold (10-nm) (c) probes in preembedding IEM of yellow fever virus-producing cells. Using the same primary monoclonal antibody at 100 μg/ml, the ferritin marker, in contrast to the gold probes, forms a dense corona. Bar marker represents 100 nm; ×100,000.

bound IgG or protein A are dependent on the pH used during adsorption. A value slightly above the pI of the protein will give the most reactive probes. Attempts to prepare conjugates are often hampered by the lack of the appropriate high-affinity IgG fractions.

The marker systems are used at pretested dilutions in conditioned PBS. For preembedding IEM, ferritin conjugates usually are diluted 1 : 5 to 1 : 50, while the IgG–gold probe in immunocryoultramicrotomy must be diluted 1 : 60 or 1 : 80 (OD 0.3 at 550 nm) to avoid nonspecific binding (De Mey, 1986). Before use, all immunoreagents are centrifuged at 8000 g for 4 min to remove aggregates.

Remarkable differences become apparent when gold and ferritin conjugates are compared with regard to labeling efficiency. Assessing the number of marker molecules deposited, IgG–ferritin conjugates are 2 to 5 times more sensitive than gold conjugates (Fig. 13.10) Tokuyasu, 1980, 1983; Evans and Webb, 1984; Gelderblom *et al.*, 1985a). Since the relatively low scattering power of ferritin does not interfere with its application to preembedding IEM, ferritin conjugates represent ideal markers for labeling cell surface determinants or cell free virus in sections.

To investigate intracellular compartments, the cells need to be permeabilized using freezing and thawing or combinations of detergents and fixatives (Bohn, 1978; Willingham *et al.*, 1978; Palmer *et al.*, 1985). This treatment, however, adversely affects the fine structure. Furthermore, due to its low contrast, the ferritin marker is barely detectable in the electron-opaque cytoplasm and, in addition, shows a comparatively high degree of nonspecific sticking to intracellular constituents. For these reasons, we prefer IgG–gold labeling of thin cryosections when intracellular determinants need to be located. As observed in several studies, the diameter of the gold probe has a drastic influence on labeling density. Gold probes 5 nm in diameter are two times more effective in labeling than 10-nm probes (Fig. 13.4 and 13.8). However, because of their easier detectability, gold particles of 10-nm diameter are used with advantage in low-resolution IEM (Fig. 13.6).

Immunoenzyme conjugates represent the third marker system. They are considerably smaller than ferritin conjugates and therefore do not meet the same steric difficulties in the intracellular compartments. The bound conjugate is detected by an insoluble, electron-dense reaction product of the enzyme, usually peroxidase. Since the resulting insoluble dye forms chelates with osmium, it can be detected by both light and electron microscopy. Furthermore, depending on incubation time, temperature, and concentration of the substrate, the signal may be amplified, thus rendering the immunoenzyme technique more sensitive than immunoferritin labeling. Sensitivity and specificity of this technique can be further increased by using soluble complexes of peroxidase and antiperoxidase antibodies (PAP technique; Sternberger, 1979). Compared to the particulate ferritin or gold markers, however, the immunoenzyme technique shows only low contrast and a reduced spatial resolution in IEM (Dourmashkin in et al., 1982; Palmer et al., 1985). Therefore immunoenzyme conjugates are more suitable for low-resolution cytochemistry.

TACTICS OF IMMUNOLABELING: DIRECT VERSUS INDIRECT TECHNIQUES

Since covalent coupling to ligands diminishes considerably the binding ability of the conjugated antibody because of steric hindrance, direct IEM using tagged primary antibody is possible only in rare cases where high-affinity antibodies are supplied in large amounts. In addition, since the individual coupling of a series of antibodies is cumbersome, indirect two- or three-step procedures are preferable in most instances (Fig. 13.11). They are more effective, as with the same anti-mouse Ig marker conjugate, all mouse monoclonal antibodies can be detected. As an additional advantage, the primary unlabeled antibody may be detected by more than one marker antibody, leading to an enhanced signal. Alternatively, the gap between the antigenic determinant detected by the primary antibody and the prospective marker may be bridged solely by immunological means. This is exemplified by the use of hybrid antibodies, which are heterodimeric $F(ab')_2$ molecules composed of Fab' with differing specificity (Hämmerling et al., 1968; Gelderblom et al., 1972; Gelderblom, 1975), and the unlabeled antibody bridging technique (Mason et al., 1969). Due to the exclusive immunological binding, both techniques show a very low background.

13. Immunogold Labeling of Viruses *in Situ*

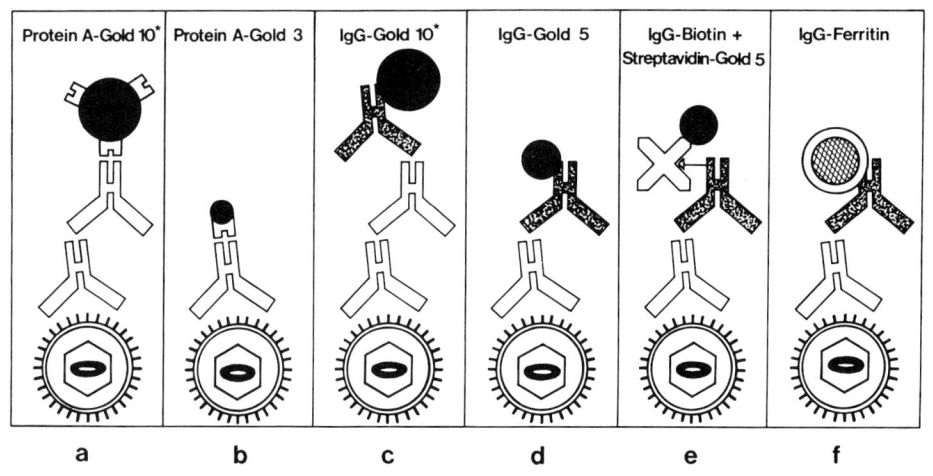

Fig. 13.11. Diagram illustrating different indirect detection systems for visualization with the electron microscope. Protein A–gold, binding specifically to the Fc fragment of immunoglobulins, enables the detection of bound antibody (a, b). Colloidal gold-labeled secondary antibodies react with bound primary antibodies (c, d). Biotin-labeled IgG antibodies directed against the primary antibody are revealed by the high-affinity interaction of biotin with streptavidin tagged with colloidal gold (e). In contrast to gold marker systems, ferritin linked to IgG molecules is less electron dense (f). The lower labeling density of markers with gold colloids of 10-nm diameter or more is enhanced by sequentially incubating the antigen-bearing specimen with the primary antibody, an unlabeled bridge antibody, and the gold-conjugated marker (a, c).

CONCLUDING REMARKS

Most critical for the outcome of IEM labeling is the qualitiy of the primary antibodies used. High-affinity antibodies with titers in Western blot and conventional enzyme-linked immunosorbent assay (ELISA) higher than 1 : 50 and 1 : 1000, respectively, are a prerequisite for reliable labeling in IEM. Monospecific polyclonal hyperimmune sera usually fulfill these demands; however, since the amount of specific antibodies in hyperimmune serum rarely exeeds 10% of the total immunoglobulins, "nonspecific" reactions might occur. Absorption of the antisera to prefixed noninfected cells will diminish these undesired reactions by removal of the antibodies directed against normal cellular constituents.

As monoclonal antibodies are homogeneous, exquisitely specific antibodies with identical affinity, the labeling density of hybridoma culture supernatants (1–50 µg/ml hybridoma protein) in most cases is sparse. Therefore, culture fluid has to be concentrated for IEM to 10–100 µg/ml, or used as diluted ascites, which usually contains up to 10 mg/ml of specific antibody. With highly reactive monoclonal antibodies a concentration as low as 1 ng/ml has led to a distinct labeling signal in preembedding immunoferritin IEM (Gelderblom et al., 1985a). A weak signal observed after incubation with an individual monoclonal antibody can be considerably enhanced when a mixture of homologous monoclonal antibody is applied.

Chemical fixation is another major problem in IEM. As pointed out earlier, prefixation is useful in preembedding IEM mainly to immobilize surface antigens and avoid an antibody-induced redistribution. Without prefixation, cell surface antigens will appear as dense patches (Hämmerling et al., 1968; Calafat et al., 1983). In immunocryoultramicrotomy chemical prefixation represents an obligatory prerequisite to retain ultrastructure in the thawed section. Chemical fixation by low glutaraldehyde concentrations (0.1%) or mixtures of glutaraldehyde and paraformaldehyde (0.1% and 4 to 8%) is efficient and unsurpassed in preservation of ultrastructure. However, because of the extensive cross-linking effect of glutaraldehyde (for discussion, see Hayat, 1981, 1986; Bullock, 1984), the reactivity of a particular antigen may be abolished, whereas other antigens such as surface determinants of parasites withstand even fixation with 8% glutaraldehyde (H. R. Gelderblom, unpublished data).

In special cases more selective fixatives have been applied with success: acrolein (0.1% in a solution of 2% formaldehyde; Boonstra et al., 1987), carbodiimides (Willingham et al., 1978), and dimethylsuberimidate (Hassell and Hand, 1974). Because of the detrimental effect of chemical fixation on the antigenicity, several alternatives using physical fixation, such as the labeling of sections prepared from freeze-substituted specimens (M. Müller, personal communication; Humbel and Schwarz, 1989) are currently under investigation.

In spite of the problems mentioned above, IEM of virus-infected cells has become a powerful tool in virology, due to the improvement of labeling and preparation techniques and the ever-increasing number of monoclonal antibodies. The combination of conventional thin-section electron microscopy and IEM with a number of techniques, such as freeze substitution and pre- and postembedding labeling, may ultimately help to elucidate intricate virus–cell relationships.

We thank Dr. Heinz Schwarz for sharing many of his constructive ideas and for critically reading the manuscript and Dr. Georg Pauli for his interest in these investigations, the reliable supply of virus-infected cultures, and many helpful comments. We are grateful to Mrs Bärbel Jungnickl and Mr Hilmar Reupke for patient and diligent assistance with the photographic work and electron microscopy, respectively.

REFERENCES

Avrameas, S., and Uriel, J. (1968). Méthode de marguage d'anticorps avec des enzymes et son application en immunodiffusion. *C. R. Hebd. Seances Acad. Sci.* **262**, 2543.

Beesley, J. E., Orpin, A., and Adlam, C. A. (1982). A comparison of immunoferritin, immunoenzyme and gold-labelled protein A methods for the localization of capsular antigen on frozen thin sections of *Pasteurella haemolytica*. *Histochem. J.* **14**, 803.

Bohn, W. A. (1978). A fixation method for improved antibody penetration in electron microscopical immune-peroxidase studies. *J. Histochem. Cytochem.* **26**, 293.

Boonstra, J., van Maurik, P., and Verkleij, A. J. (1987). Immunogold labelling of cryosections and cryofractures. In *Cryotechniques in Biological Electron Microscopy* (R. A. Steinbrecht and K. Zierold, eds.), pp. 216–230. Springer-Verlag, Berlin and New York.

Bullock, G. R. (1984). The current status of fixation for electron microscopy: A review. *J. Microsc. (Oxford)* **133**, 1.

Calafat, J., Janssen, H., Demant, P., Hilgers, J., and Zavada, J. (1983). Specific selection of host cell glycoproteins during assembly of murine leukaemia virus and vesicular stomatitis virus: Presence of Thy-1 glycoprotein and absence of H-2, Pgp-1 and T-200 glycoproteins on the envelopes of these virus particles. *J. Gen. Virol.* **64**, 1241.

Carlemalm, E., Garavito, R. M., and Villiger, W. (1982). Resin development for electron microscopy and an analysis of embedding at low temperature. *J. Microsc. (Oxford)* **126**, 123.

Carrascosa, J. L., González, P., Carrascosa, A. L., Garciá-Barreno, B., Enjuanes, L., and Vinuela, E. (1986). Localization of structural proteins in African swine fever virus particles by immunoelectron microscopy. *J. Virol.* **58**, 377.

Causton, B. E. (1984). The choice of the resins for electron immunocytochemistry. In *Immunolabelling for Electron Microscopy* (J. M. Polak und I. M. Varndell, eds.), pp. 29–36. Elsevier, Amsterdam.

Cordell, J. L., Falini, B., Eiber, W. N., Ghosh, A. K., Abdulaziz, Z., McDonalds, S., Pulford, K. A. F., Stein, H., and Mason, D. Y. (1984). Immunoenzymatic labeling of monoclonal antibodies using immune complexes of alkaline phosphatase and monoclonal anti-alkaline phosphatase (APAAP complexes). *J. Histochem. Cytochem.* **32**, 219.

De Mey, J. (1986). The preparation and use of gold probes. In *Immunocytochemistry: Modern Methods and Applications* (J. Polak and S. Van Noorden, eds.), 2nd ed., pp. 115–145. Wright, Bristol.

Dourmashkin, R., Patterson, S., Shah, D., and Oxford, J. S. (1982). Evidence of diffusion artefacts in diaminobenzidine immuncytochemistry revealed during immune electron microscope studies of the early interactions between influenza virus and cells. *J. Virol. Methods* **5**, 27.

Evans, N. R. S., and Webb, H. E. (1984). Comparison of protein A–gold and ferritin immunoelectron microscopy of Semliki Forest virus in mouse brain using a rapid processing technique. *J. Histochem. Cytochem.* **32**, 372.

Fisher, K. A. (1975). "Half" membrane enrichment: Verification by electron microscopy. *Science* **190**, 983.

Gelderblom, H., Bauer, H., Ogura, H., Wigand, H., and Fischer, A. (1974). Detection of oncornavirus-like particles in HeLa cells. I. Fine structure and comparative morphological classification. *Int. J. Cancer* **13**, 246.

Gelderblom, H. (1975). The preparation and application of hybrid antibody for the detection of tumor-specific cell surface antigens. In *Applied Tumor Immunology* (H. Götz and E. S. Bücherl, eds.), pp. 131–153. de Gruyter, Berlin.

Gelderblom, H., and Schwarz, H. (1976). Relationship between the Mason-Pfizer monkey virus and HeLa virus: Immunoelectron microscopy. *J. Natl. Cancer Inst. (U.S.)* **56**, 635.

Gelderblom, H., Pauli, G., Reupke, H., Gregersen, J.-P., and Ludwig, H. (1982). Immuno electron microscopic (IEM) studies on herpesviruses and the detection of cross-reacting structural antigens. *Electron Microsc., Pap. Int. Congr., 10th, 1982* Vol. 3, pp. 129–130.

Gelderblom, H., Kocks, C., L'age-Stehr, J., and Reupke, H. (1985a). Comparative immunoelectron microscopy with monoclonal antibodies on yellow fever virus-infected cells: Preembedding labelling versus immunocryoultramicrotomy. *J. Virol. Methods* **10**, 225.

Gelderblom, H., Reupke, H., and Pauli, G. (1985b). Loss of envelope antigens of HTLV-III/LAV, a factor in AIDS pathogenesis? *Lancet* **2**, 1016.

Gelderblom, H. R., Hausmann, E. H. S., Özel, M., Pauli, G., and Koch, M. A. (1987). Fine structure of human immunodeficiency virus (HIV) and immunolocalization of structural proteins. *Virology* **156**, 171.

Gelderblom, H. R., Özel, M., Hausmann, E. H. S., Winkel, T., Pauli, G., and Koch, M. A. (1988). Fine structure of human immunodeficiency virus (HIV), immunolocalization of structural proteins and virus–cell relation. *Micron Microsc. Acta* **19**, 41.

Griffiths, G. W. and Jokusch, B. M. (1980). Antibody labeling of thin sections of skeletal muscle with specific antibodies: A comparison of bovine serum albumin (BSA) embedding and ultra-cryomicrotomy. *J. Histochem. Cytochem.* **28**, 969.

Griffiths, G. W., Simons, K., Warren, G., and Tokuyasu, K. T. (1983). Immuno electron microscopy using thin, frozen sections: Application to studies of the intracellular transport of Semliki Forest virus spike glycoproteins. In *Methods in Enzymology* 96, p. 466. Academic Press, New York.

Griffiths, G. W., McDowall, A., Back, R., and Dubochet, J. (1984). On the preparation of cryosections for immunocytochemistry. *J. Ultrastruct. Res.* **89**, 65.

Hämmerling, U., Aoki, T., De Harven, E., Boyse, E. A., and Old, L. J. (1968). Use of hybrid antibody with anti-G and antiferritin specificities in locating cell surface antigens by electron microscopy. *J. Exp. Med.* **128**, 1461.

Hassell, J., and Hand, A. R. (1974). Tissue fixation with diimidoesters as an alternative to aldehydes. Comparison of cross-linking and ultrastructure obtained with dimethylsuberimidate and glutaraldehyde. *Histochem. Soc.* **22**, 223.

Hausmann, E. H. S., Gelderblom, H. R., Clapham, P. R., Pauli, G., and Weiss, R. A. (1987). Detection of HIV envelope specific antibodies by immunoelectron microscopy and correlation with antibody titer and virus neutralizing activity. *J. Virol. Methods* **16**, 125.

Hayat, M. A. (1981). *Fixation for Electron Microscopy*. Academic Press, New York

Hayat, M. A. (1986). *Basic Techniques for Transmission Electron Microscopy*. Academic Press, Orlando, Florida.

Hayat, M. A. (1989). *Principles and Techniques of Electron Microscopy*, 3rd ed. Macmillan, London, and CRC Press, Boca Raton, Florida.

Horisberger, M. (1984). Electron opaque markers: A review. In *Immunolabelling for Electron Microscopy* (J. M. Polak and I. M. Varndell, eds.), pp. 17–26. Elsevier, Amsterdam.

Humbel, B. M., and Schwarz, H. (1989). Freeze-substitution for immunochemistry. In *Immunogold Labeling Methods* (A. J. Verkleij and J. L. M. Leunissen, eds.). CRC Press, Boca Raton, Florida (in press).

Karnovsky, M. J. (1965). A formaldehyde–glutaraldehyde fixative of high osmolality for use in electron microscopy. *J. Cell Biol.* **27**, 137A.

Keller, G.-A., Tokuyasu, K. T., Dutton, A. H., and Singer, S. J. (1984). An improved procedure for immunoelectron microscopy: Ultrathin plastic embedding of immunolabeled ultrathin frozen sections. *Proc. Natl. Acad. Sci. U.S.A.* **81**, 5744.

Lasfargues, E. Y., Kramarsky, B., Lasfargues, J. C., and Moore, D. H. (1974). Detection of mouse mammary tumor virus in cat kidney cells infected with purified B particles from RIII milk. *J. Natl. Cancer Inst. (U.S.)* **53**, 1831.

Leduc, E. H., and Bernhard W. (1967). Recent modifications of the glycol methacrylate embeddig procedure. *J. Ultrastruct. Res.* **19**, 196.

Mason, T. E., Pfifer, R. F., Spicer, S. S., Swallow, R. A., and Dreskin, R. B. (1976). An immunoglobulin–enzyme bridge method for localizing tissue antigens. *J. Histochem. Cytochem.* **17**, 563.

McLean, I. W., and Nakane, P. K. (1974). Periodate–lysine–paraformaldehyde fixative. A new fixative for immunoelectron microscopy. *J. Histochem. Cytochem.* **22**, 1077.

McLean, J. D., and Singer, S. J. (1970). A general method for the specific staining of intracellular antigens with ferritin–antibody conjugates. *Proc. Natl. Acad. Sci. U.S.A.* **65**, 122.

Nakane, P. K., and Pierce, G. D. (1966). Enzyme-labelled antibodies: Preparation and application for the localization of antigen. *J. Histochem. Cytochem.* **14**, 929.

Nakane, P. K., and Kawaoi, A. (1974). Peroxidase-labeled antibody. A new method of conjugation. *J. Histochem. Cytochem.* **22**, 1084.

Nermut, M. V., Hockley, D. J., and Gelderblom, H. G. (1987). Electron microscopy: Methods for study of virus/cell interactions. *Perspect. Med. Virol.* **3**, 21–36.

Newman, G. R., Jasani, B., and Williams, E. D. (1983). A simple postembedding system for the rapid detection of tissue antigens under the electron microscope. *Histochem. J.* **15**, 543.

Oshiro, L. S., Rose, H. M., Morgan, C., and Hsu, K. C. (1967). Electron microscopic study of the development of simian virus 40 by use of ferritin-labelled antibodies. *J. Virol.* **1**, 384.

Palmer, E., Sporborg, C., Harrison, A., Martin, M. L., and Feorino, P. (1985). Morphology and immunoelectron microscopy of AIDS virus. *Arch. Virol.* **85**, 189.

Richardson, S. C., Mercer, L. E., Sonza, S., and Holmes, I. H. (1986). Intracellular localization of rotavirus proteins. *Arch. Virol.* **88**, 251.

Robards, A. W., and Sleytr, U. B. (1985). Low-temperature methods in biological electron microscopy. In *Practical Methods in Electron Microscopy* (A. M. Glauert, ed.), Vol. 10. Elsevier Biomedical Press, Amsterdam.

Roberts, J. M. (1975). Tungsten coating—methods of improving glass microtome knives for cutting ultrathin sections. *J. Microsc. (Oxford)* **103**, 113.

Romano, E. L., and Romano, M. (1977). Staphylococcal protein A bound to colloidal goal: A useful reagent to label antigen–antibody sites in electron microscopy. *Immunochemistry* **14**, 711.

Roth, J. (1982). The preparation of protein A–gold complexes with 3 nm and 15 nm gold particles and their use in labelling multiple antigens on ultrathin sections. *Histochem. J.* **14**, 791.

Roth, J. (1984). The protein A–gold technique for antigen localization in tissue sections by light and electron microscopy. In *Immunolabelling for Electron Microscopy* (J. M. Polak and I. M. Varndell, eds.), pp. 113–121. Elsevier, Amsterdam.

Schenk, P., Pietschmann, S., Gelderblom, H., Pauli, G., and Ludwig, H. (1988). Monoclonal antibodies against herpes simplex virus type 1-infected nuclei defining and localizing the ICP8 protein, 65k DNA-binding protein and polypeptides of the ICP35 family. *J. Gen. Virol.* **69**, 99.

Schwarz, H., Hunsmann, G., Moenning, V., and Schäfer, W. (1976). Properties of mouse leukemia viruses. XI. Immunoelectron microscopic studies on viral structural antigens on the cell surface. *Virology* **69**, 169.

Shahrabadi, M. S., and Yamamoto, T. (1971). A method for staining intracellular antigens in thin sections with ferritin-labelled antibody. *J. Cell Biol.* **50**, 246.

Singer, S. J. (1959). The preparation of an electron dense antibody conjugate. *Nature (London)* **183**, 1523.

Singer, S. J., Tokuyasu, K. T., Keller, G. A., Takata, K., and Dutton, A. H. (1987). Immunoelectron microscopy and the molecular ultrastructure of cells. *J. Electron Microsc.* **36**, 63.

Sommer, J. R. (1977). To cationize glass. *J. Cell Biol.* **75**, 245a.

Sternberger, L. A. (1979). *Immunocytochemistry,* 2nd ed. Wiley, New York.

Stierhof, Y.-D., Schwarz, H., and Frank, H. (1986). Transverse sectioning of plastic-embedded immunolabeled cryosections: Morphology and permeability to protein A–colloidal gold complexes. *J. Ultrastruct. Res.* **97**, 187.

Stockem, W. (1970). Die Eignung von Pioloform F für die Herstellung elektronenmikroskopischer Trägerfilme. *Mikroskopie* **26**, 185.

Takamiya, H., Shimizu, F., and Vogt, A. (1975). A two-stage method for cross-linking antibody globulin to ferritin by glutaraldehyde. III. Size and antibody activity of the conjugates. *J. Immunol. Methods* **8**, 303.

Tao, H., Semao, X., Zinyi, C., Gan, S., and Yanagihara, R. (1987). Morphology and morphogenesis of viruses of hemorrhagic fever with renal syndrome. II. Inclusion bodies—ultrastructural markers of hantavirus-infected cells. *Intervirology* **27**, 45.

Tokuyasu, K. T. (1980). Immunochemistry on ultrathin frozen sections. *Histochem. J.* **12**, 381.

Tokuyasu, K. T. (1983). Present state of immunocryoultramicrotomy. *J. Histochem. Cytochem.* **31**, 164.

Van Bergen en Henegouwen, P. M. P., and Leunissen, J. L. M. (1986). Controlled growth of colloidal gold particles and implications for labelling efficiency. *Histochemistry* **85**, 81.

Venable, J. H., and Coggeshall, R. (1965). A simpified lead citrate stain for use in electron microscopy. *J. Cell Biol.* **25**, 407.

Willingham, M. C., Yamada, S. S., and Pastan, I. (1978). Ultrastructural antibody localization of α_2-macroglobulin in membrane-limited vesicles in cultured cells. *Proc. Natl. Acad. Sci. U.S.A.* **75**, 4359.

14

Study of Exocytosis with Colloidal Gold and Other Methods

PIETER BUMA

Department of Orthopaedy
St. Radboud Hospital
Nijmegen, The Netherlands

INTRODUCTION
TISSUE SPECIMENS
METHODS FOR DETECTING EXOCYTOSIS
    Rat Tissues
COMPARISON OF DIFFERENT METHODS
    Conventional Fixation
    TAGO Fixation
    The TARI Method
APPLICATIONS AND LIMITATIONS OF METHODS FOR
    DETECTING EXOCYTOSIS
IMMUNOCYTOCHEMICAL STAINING PROCEDURE
PROTEIN A–GOLD LABELING OF EXOCYTOTIC
    RELEASE SITES
COMPARISON OF PROTEIN A–GOLD METHOD WITH
    OTHER IMMUNOCYTOCHEMICAL METHODS
APPLICATIONS
REFERENCES

INTRODUCTION

Exocytosis is considered as the common cellular mechanism by which secretory products stored within membrane-bound granules are released (e.g., Douglas, 1974; Normann, 1976; Nagasawa, 1977). In many types of tissue, particularly if the chemical nature of the secretions is known, the secretory cells can be isolated, and when the secretory granules are large, the release activity can be studied by biochemical (e.g., pancreatic cells, Adelson and Miller, 1985; adrenal medullary cells, Baker and Knight, 1984; Perrin and Aunis, 1985; von Grafenstein *et al.*, 1986), electrophysiological (mast cells, Breckenridge and Almers, 1987), and light microscopic methods (adrenal medullary cells, Schmauder-Chock and Chock, 1987; mast cells, Breckenridge and Almers, 1987). In many tissues, particularly in those where different secretions are released from relatively small secretory granules (e.g., brain tissue), ultrastructural methods are needed to localize the cell's exocytotic release site(s) and to quantify its release activity (Nagasawa, 1977; Rademakers, 1977; Buma *et al.*, 1984). However, generally, routine fixation procedures are not suited to capture the fleeting process of exocytosis. Therefore other methods are needed for better visualization of the exocytotic event.

A few years ago it was shown that tissue fixation with tannic acid (TA) (tannic acid–glutaraldehyde–OsO_4, or TAGO, method, Roubos and van der Wal-Divendal, 1980) or tissue incubation in tannic acid in Ringer solution (tannic acid-Ringer incubation, or TARI, method; Buma *et al.*, 1984) improved the visualization of exocytosis. TA strongly binds to (proteinergic) secretions as well as to heavy metal (lead and uranyl) ions. As TA does not penetrate plasma membranes, only extracellular substances, including released contents of secretory granules, are stained. As a result, the detection of exocytosis is markedly facilitated (Roubos and van der Wal-Divendal, 1980; Buma *et al.*, 1984). Moreover, during TARI treatment of tissues exocytosis proceeds, but the exteriorized contents of the secretory granules are immediately captured by TA and do not diffuse into the extracellular space. Thus, the number of exocytosis phenomena at the ultrastructural level increases with increasing incubation time.

However, apart from the localization of the cell's release site(s) and the quantification of its release activity, it is very important to identify the chemical nature of the released products. Particularly in neural tissues with many different types of axon terminals with different transmitter contents, it is very important to use methods for the detection of exocytosis in combination with an identification method. Recently, a combination for the simultaneous detection of exocytosis and postembedding immunocytochemical identification of cell types has been described (Van Putten *et al.*, 1987). This chapter deals with the potency of different methods for the detection of exocytosis and with the simultaneous visualization of exocytotic release and the immunocytochemical identification of the chemical nature of the secretory products involved.

TISSUE SPECIMENS

The potency of the fixation methods for the detection of exocytosis was tested on the neuroendocrine caudodorsal cells (CDC) in the cerebral ganglia of the pond snail *Lymnaea stagnalis* and on the oxytocinergic fibers and terminals in the central nervous system (CNS) of the rat. With respect to *Lymnaea*, the CDC release an ovulation hormone from their neurohemal axon terminals in the periphery of the intercerebral commissure. Release takes place by exocytosis, particularly during a short (45-min) period of intense electrical activity (active state). However, most of the time (1–2 day intervals between ovipositions), the CDC are electrically inactive with concomitant very low exocytotic and hormone release activity (resting state; for references, see Roubos, 1984). Such resting and active CDC have been used to show the mode of action of the different methods for the detection of exocytosis.

Oxytocin is synthesized in the paraventricular and supraoptic nuclei and transported along axons to the neurohemal contact zone of the posterior pituitary, where the contents of the neurosecretory granules are released by exocytosis from the axon terminals. Since it is very difficult to show exocytosis in the posterior pituitary (Douglas *et al.*, 1971; Krisch *et al.*, 1972; Nagasawa, 1977), we tested the combination of tannic acid and immunocytochemical methods on this structure.

METHODS FOR DETECTING EXOCYTOSIS

Neuroendocrine CDC cells of adult snails, *L. stagnalis* (shell height ≈30 mm), were brought into the resting and active states as described elsewhere (Buma and Roubos, 1983). The two cerebral ganglia were dissected and treated in one of the following ways.

1. Conventional fixation: immersion fixation in a 0.05 M cacodylate-buffered (pH 7.2) solution of 1% glutaraldehyde (2 hr at 4°C); rinsing in the buffer (15 min at 4°C); postfixation in 1% OsO_4 (2 hr at 4°C) in the buffer.

2. Tannic acid–glutaraldehyde–OsO_4 method (TAGO fixation; Roubos and van der Wal-Divendal, 1980): as in conventional fixation, but with 1% TA added to the glutaraldehyde fixative.

3. Tannic acid–Ringer incubation method (TARI method; Buma et al., 1984): tissues were incubated for 1 hr in Ringer solution containing 1% TA and 30 mM NaCl, 4 mM CaCl$_2$, 2 mM MgCl$_2$, 1.5 mM KCl, 0.25 mM NaH$_2$PO$_4$, and 18 mM NaHCO$_3$; the pH of the Ringer solution was adjusted with concentrated NaCl or NaOH to 7.0. Incubations were carried out at 20°C. The tissues were conventionally fixed.

Rat Tissues

1. Conventional fixation: Male Wistar rats were anesthetized with sodium pentobarbital (60 mg/kg body weight) and perfused via the left ventricle with Ringer solution containing 154 mM NaCl, 5.6 mM KCl, 2.2 mM CaCl$_2$, 1 mM MgCl$_2$, 6 mM NaHCO$_3$, 2 mM Tris, and 10 mM glucose (pH 7.4) until the liver was bleached ($\approx$30 sec to 2 min). For routine electron microscopy, perfusion was continued with 500 ml of 0.1 M phosphate-buffered (pH 7.4) 1% paraformaldehyde and 1.25% glutaraldehyde at 20°C. Two rats were perfused in the same way with Ringer solution containing 50 mM KCl (substituted for NaCl) and 50 mM KCl added to the fixation fluid (high-K fixation, modified according to Gronblad, 1983). After dissection, the pituitaries and blocks containing the median eminence were again fixed in the same fixative for 16 hr at 4°C and then postfixed in 1% OsO$_4$ for 1 hr at 20°C.

2. TAGO fixation: as in conventional fixation, but with 1% TA added during the first 1–2 hr of fixation.

3. TARI-method: rats were perfused with the Ringer solution until the liver was bleached, after which the perfusion was continued for 15 min with the same solution containing 1% TA. Two rats were perfused in the same way, but with 50 mM KCl during the last 5 min of the TA perfusion (high-K stimulation). Subsequently, the perfusion was continued with normal Ringer solution without TA for 5 min. During the latter perfusion, unbound TA was removed, which is necessary for optimal fixation. All Ringer solutions were saturated with 95% O$_2$ and 5% CO$_2$, adjusted to pH 7.0 with concentrated NaOH or HCl, and heated prior to use to 38°C. To avoid precipitation of TA in the Ringer solution, solutions were stirred, kept on a hot plate, and not exposed to the open air. Tissues were conventionally fixed. After fixation, tissues were dehydrated in graded series of ethanol, embedded in Epon 812, and cut on a Reichert Ultracut E. For routine electron microscopy, sections were mounted on Formvar-coated one-hole grids and stained with lead citrate for 2–10 min. Uranyl staining was not applied as it diminishes the electron density between stationary and exocytosing secretory granules in TA-treated tissues.

COMPARISON OF DIFFERENT METHODS

Conventional Fixation

Irrespective of the mode and method of tissue preservation, the general ultrastructural preservation of the cerebral ganglia, intercerebral commissure, and pituitaries was good. After conventional fixation, exocytoses were clearly visible in active CDC (Fig. 14.1B). Only exocytoses with exocytosing contents facing the extracellular space (omega exocytosis) could be clearly discerned from stationary nonexocytosing secretory granules. Caps of exocytosis, with no opening to the extracellular space in the plane of the section, were not recognized because of the lack of difference in electron density between stationary and exocytosing secretory granules.

In the nonstimulated posterior pituitary only pinocytotic pits were found (Fig. 14.1H). After high-K stimulation, more pinocytotic pits were present (Fig. 14.1J). In a few pits exocytosing secretory granule contents could be visualized (Fig. 14.1J), located at some distance from the release site, suggesting their dissociation and outward diffusion.

TAGO Fixation

In TAGO-fixed tissues exocytosis phenomena were more clearly seen than in conventionally fixed tissues (Fig. 14.1C–F, K). This is due to the fact that exocytosing granule contents are stained very electron dense by TA. Also, other extracellular substances, namely collagen fibers, basal laminae, and the outer leaflet of plasma membranes, were stained electron dense (Fig. 14.1D, F). The omega exocytosis showed, in both the CDC and the posterior pituitary, the same inhomogeneity as in conventional fixed tissues (Fig. 14.1D, K). With the TAGO method caps of exocytosing granules were also visualized (Fig. 14.1E, F), which strongly facilitates the detection of the release process.

The TARI Method

In the TARI-treated tissues, exocytosing granule contents showed the same high electron density. However, exocytoses were far more abundant than in conventionally and TAGO-fixed tissues (Fig. 14.1G, L). Counts in resting CDC axon terminals that had been TARI treated for 1 hr showed that the number of exocytosis phenomena was ≈50 times that in conventionally fixed tissues (Buma *et al.*, 1984).

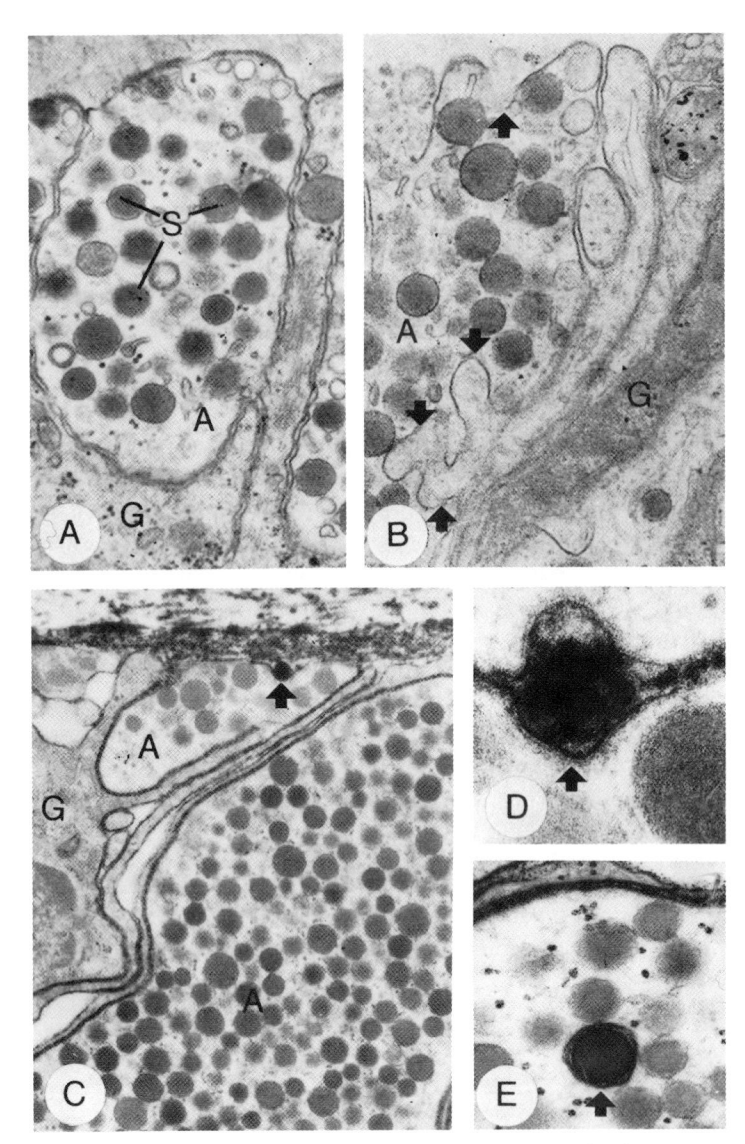

Fig. 14.1. Exocytoses (arrows) in axon terminals of the ovulation hormone-producing caudodorsal cells of the pond snail *Lymnaea stagnalis* (A–G), and in axon terminals of the posterior pituitary of the rat (H–L). Comparison of conventional fixation (A, B, H, J), TAGO fixation (C–F, K), and the TARI method (G, L). Two secretory states of the caudodorsal cells are shown: the resting state with low secretory activity (A, C–G), and the active state with very high secretory activity (B, F). **G,** Glial cell; **S,** secretory granules. A and B ×40,000; C ×25,000. (D) Note the irregular outline of granule contents, indicating their dissociation and outward diffusion; ×55,000. (E) Note cap of exocytosing secretory granule; ×35,000, F ×35,000, G ×35,000. (H, J) Note pinocytotic pits (arrowheads) and outward-diffusing and dissociating secretory granule contents (small arrows); ×60,000. K, L ×60,000. (A, C, D, G, L from Buma *et al.*, 1984; B from Buma and Roubos, 1983.)

14. Study of Exocytosis with Colloidal Gold 291

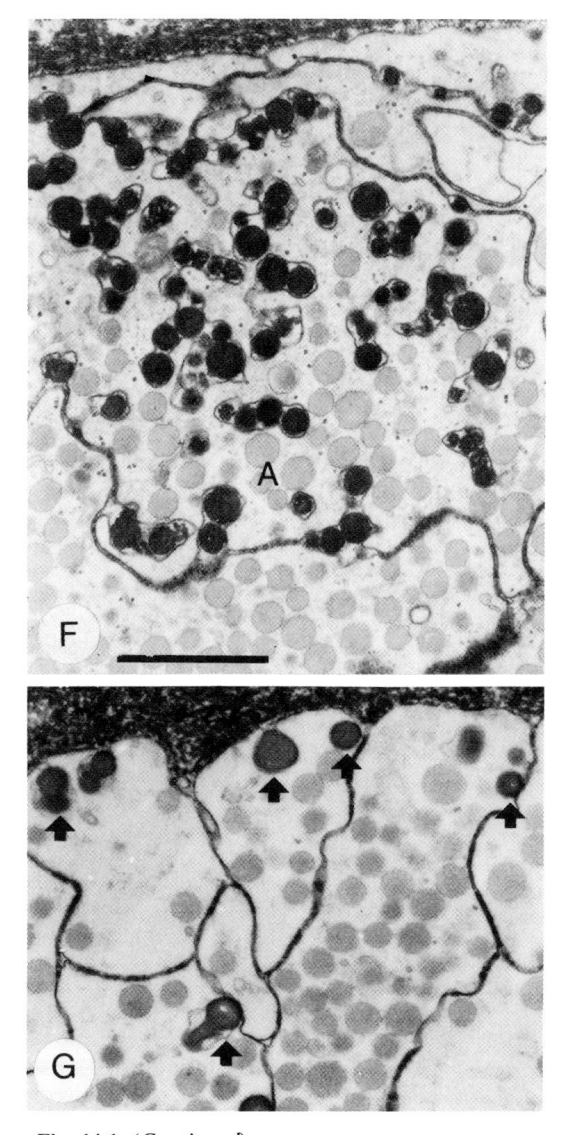

Fig. 14.1. *(Continued)*

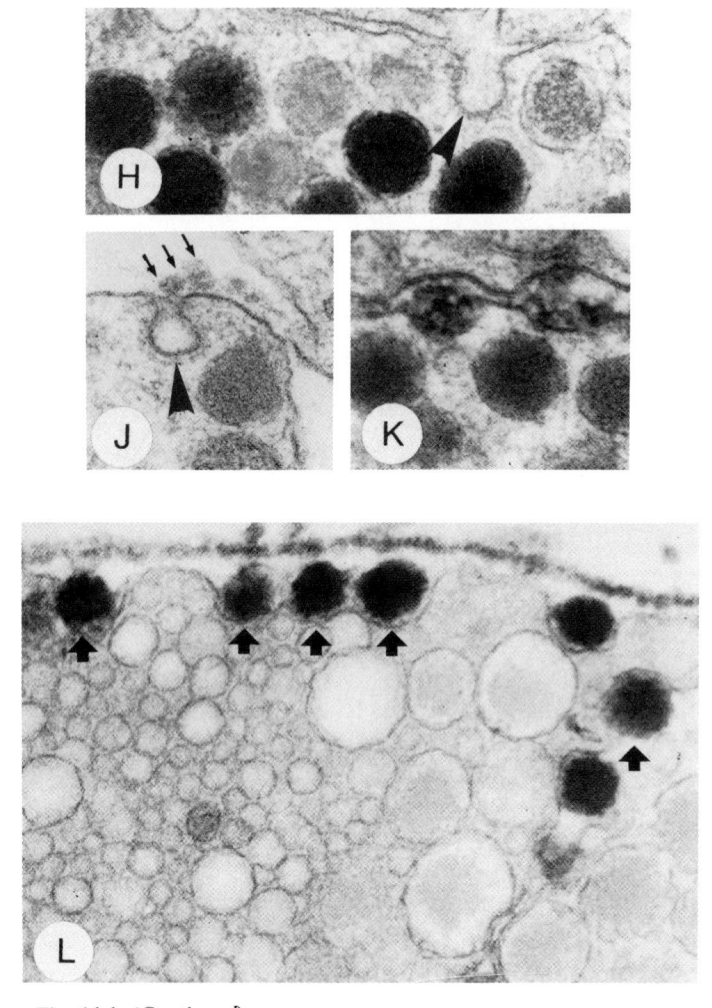

Fig. 14.1. *(Continued)*

APPLICATIONS AND LIMITATIONS OF METHODS FOR DETECTING EXOCYTOSIS

Some studies of exocytosis with conventional fixation methods have been successful. Among the cells displaying exocytosis in the CNS with conventional fixation procedures are different types of neuroendocrine cells in the snail *Lymnaea stagnalis* (Wendelaar Bonga, 1971; Buma and

Roubos, 1983; Roubos et al., 1987), the neuroendocrine AKH cells of the insect *Locusta migratoria* (Rademakers, 1977), and unspecialized nonsynaptic release sites in neurons in the CNS of earthworms and snails (Golding and Bayraktaroglu, 1984; Buma and Roubos, 1986). In the rat CNS, exocytoses have been described in the perivascular part of the palisade layer of the median eminence (Stoeckart et al., 1972; Zamora and Ramirez, 1983). Zhu et al. (1986) showed exocytoses at structurally nonspecialized sites of axon terminals within the trigeminal subnucleus caudalis. With respect to the present observations with the high-K stimulation (Gronblad, 1983), we found exocytoses in the neural lobe that were not found after normal fixation. However, even with the high-K method, exocytotic release phenomena were not found in this study in other parts of the rat CNS, e.g., the median eminence.

The TAGO method facilitates the detection of exocytosis in two ways: (1) quantitatively, caps of exocytosing secretory granules are visualized, and (2) qualitatively, all exocytotic events are stained electron dense (see also Roubos and van der Wal-Divendal, 1980). With the TAGO method exocytosis has been demonstrated in a variety of neurosecretory cells in the CNS of *Lymnaea stagnalis* and the insect *Locusta migratoria* (Roubos and van der Wal-Divendal, 1980). However, in many cell types the detection of exocytosis remains difficult or impossible; probably such cells release their secretions during very short periods of time and/or too quickly to see (Buma et al., 1984). The TARI method overcomes these problems in many cell types. With the TARI method exocytosis phenomena have been clearly demonstrated in neuroendocrine cells of snails (Buma et al., 1984), insects (Khan and Buma, 1985), and rats (Buma and Nieuwenhuys, 1987, 1988; Van Putten et al., 1987). Synaptic release sites have been found in snails and in the area postrema of the rat (Buma et al., 1984). Nonsynaptic release sites were found in the CNS of snails and insects and in the area postrema and median eminence of the rat (Buma et al., 1984; Buma and Nieuwenhuys, 1987, 1988). Furthermore, release from rat pinealocytes in the pineal gland has been shown (Noteborn et al., 1986; Masson-Pevet et al., 1987). However, even with the sensitive TARI method it is not possible to demonstrate release in brain tissues that are protected by the blood–brain barrier. Apparently the molecular weight of TA (1700) does not allow penetration through this barrier.

IMMUNOCYTOCHEMICAL STAINING PROCEDURE

Immunolabeling was performed on TAGO-fixed and TARI-treated posterior pituitaries with or without high-K stimulation. Pale gold sections of tissues with exocytosis were mounted on uncoated 300-mesh or Formvar-coated one-hole nickel grids. All incubation steps were performed on top of drops on Parafilm or wax in a moist chamber. Some sections were etched 1% H_2O_2 for 1–20 min or etched with saturated sodium metaperiodate (1–30 min). The grids were successively placed on (1) phosphate-buffered saline (PBS, pH 7.4) with 50 mM glycine for 10 min, (2) PBG (PBS containing 0.5% bovine serum albumin and 0.2% gelatin) for 15 min with three changes, (3) the primary antiserum diluted (1 : 20) in PBG for 24–72 hr at 4°C, (4) PBG for 30 min with six changes, (5) protein A–gold (17 nm; Jansen Pharmaceutica, Beerse, Belgium) diluted (1 : 25) in PBG for 1 hr, (6) PBG for 30 min with six changes and (7) PBS with six changes for 30 min, after which they were (8) rinsed with distilled water. The primary antiserum was a monoclonal anti-oxytocin–neurophysin (PS 36, a gift from Dr. M. H. Whitnall, Laboratory of Neurochemistry and Neuroimmunology, Bethesda, MD). The specificity of this antibody has been tested and described by several authors (Ben-Barak *et al.*, 1985; Castel *et al.*, 1985; Whitnall *et al.*, 1985b). All sections were stained with lead citrate for 1–10 min.

PROTEIN A–GOLD LABELING OF EXOCYTOTIC RELEASE SITES

After application of the immunocytochemical staining procedure with protein A–gold as the marker, fine-structural elements such as microtubuli, microvesicles, and the limiting membrane of secretory granules as well as the electron density of the exocytotic granules remained clearly visible in the axon terminals of the posterior pituitary (Fig. 14.2B–E). Only neurosecretory granules and their remnants in lysosomal structures appeared to be immunoreactive (Fig. 14.2A). No or negligible background staining was observed.

Etching appeared to be not essential for the immunostaining with this antibody. Mild etching (1% H_2O_2 for 1–20 min) increased the intensity of the immunolabeling. After intense etching (metaperiodate for 5–30 min, 10% H_2O_2 for 10 min), the electron density of the exocytosing secretory granule contents was reduced and did not allow the discrimination of exocytosis (see also Van Putten *et al.*, 1987).

COMPARISON OF PROTEIN A–GOLD METHOD WITH OTHER IMMUNOCYTOCHEMICAL METHODS

With respect to the choice of immunocytochemical staining procedure, it is evident that certain procedures are not suitable for combination with fixation methods for the detection of exocytosis, for the following reasons. Preembedding peroxidase–antiperoxidase (PAP) staining procedures with 3,3'-diaminobenzidine (DAB) as chromogen (Sternberger *et al.*, 1970) need detergents (saponin, Triton X-100, or dimethyl sulfoxide) to improve the penetration of antisera into the tissue sections, a procedure which negatively affects the ultrastructural preservation. Used without detergents, only the (damaged) surfaces of the sections are stained. Moreover, after the DAB reaction, immunostained structures are obscured by a highly electron-dense reaction product. In this way, the difference in electron density between stationary and exocytosing secretory granules in TA-treated tissues is diminished or completely absent. Furthermore, the DAB reaction product obscures the ultrastructural details. Postembedding staining with DAB as chromogen has the same disadvantages. Postembedding staining with protein A–gold and staining with other highly punctate markers (e.g., ferritin, gold coupled to goat anti-rabbit IgG) do not have these disadvantages and allow precise localization

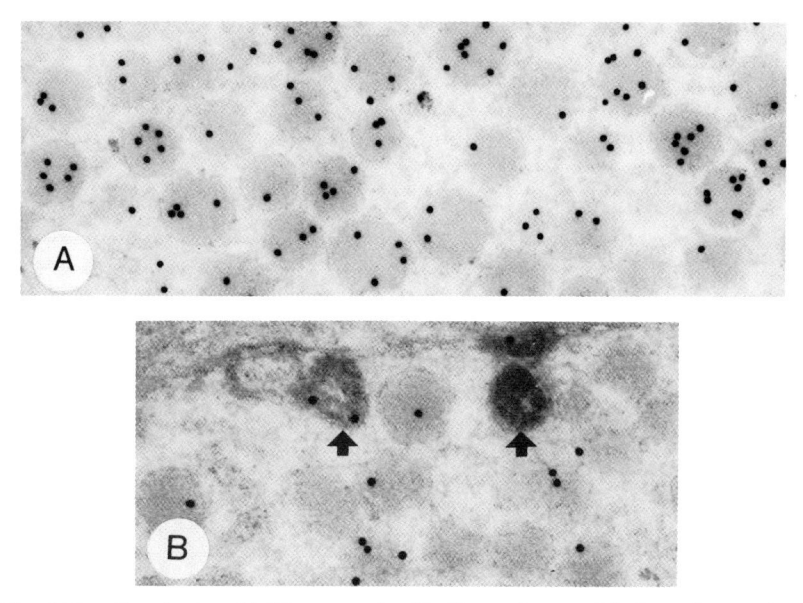

Fig. 14.2. Protein A–gold (diameter of gold, 17 nm) immunolabeling of oxytocinergic secretory granules in axon terminals in the posterior pituitary (A–E) and in varicose axons in the fiber layer of the median eminence (G, H). Exocytoses (arrows) were demonstrated with the TAGO method (B, C), TARI method (D–F), without (B, D) and with high potassium stimulation (C, E). (A) Note very specific immunostaining of oxytocinergic secretory granules; ×60,000. (B) Note irregular outline of granule contents indicating their dissociation and outward diffusion; ×55,000. (C) ×25,000. (D) ×55,000. (E) Note very large compound exocytoses (large arrows) after the TARI procedure with high potassium; ×25,000. (F) Non-immunoreactive (vasopressinergic) axon terminals of the posterior pituitary; ×25,000. (G) Varicose oxytocinergic axon in the fiber layer of the median eminence (from Buma and Nieuwenhuys, 1987); ×20,000. (H) Enlargement of (G); ×50,000.

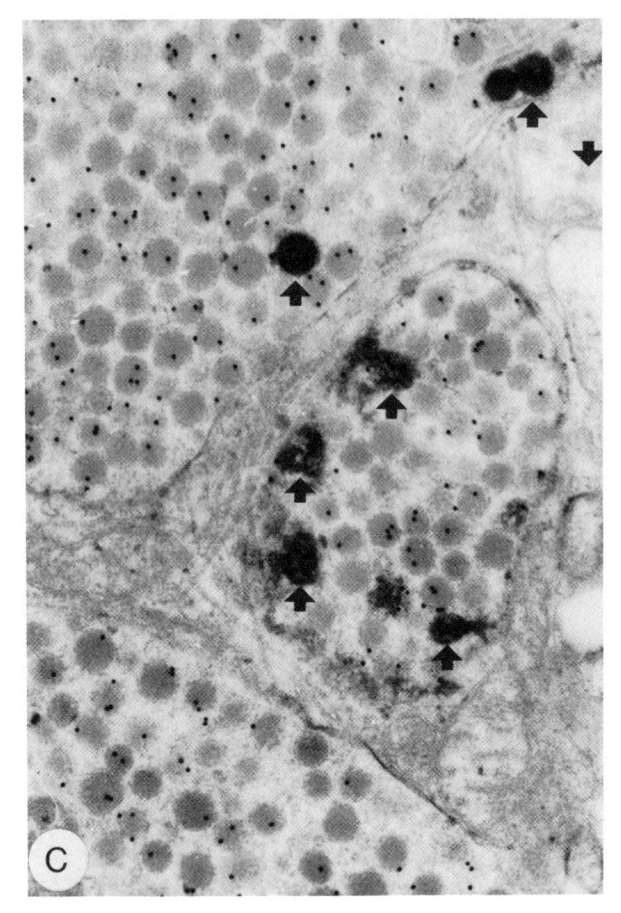

Fig. 14.2. (*Continued*)

of the antigenic site and simultaneous observations of the ultrastructural details, with maintenance of electron density between stationary and exocytosing secretory granules (see also Van Putten *et al.*, 1987). Moreover, these postembedding immunocytochemical methods have been found useful by many authors for the demonstration of transmitter substances in the CNS (Varndell *et al.*, 1982; Beauvillain *et al.*, 1984; Calas, 1985; Lamberts and Goldsmith, 1985; van den Pol, 1985; Whitnall *et al.*, 1985a; Batton, 1986; Castel *et al.*, 1985).

APPLICATIONS

With respect to the applications of the combination of methods, only a few studies have been carried out. Roubos *et al.* (1987) studied immuno-

14. Study of Exocytosis with Colloidal Gold 297

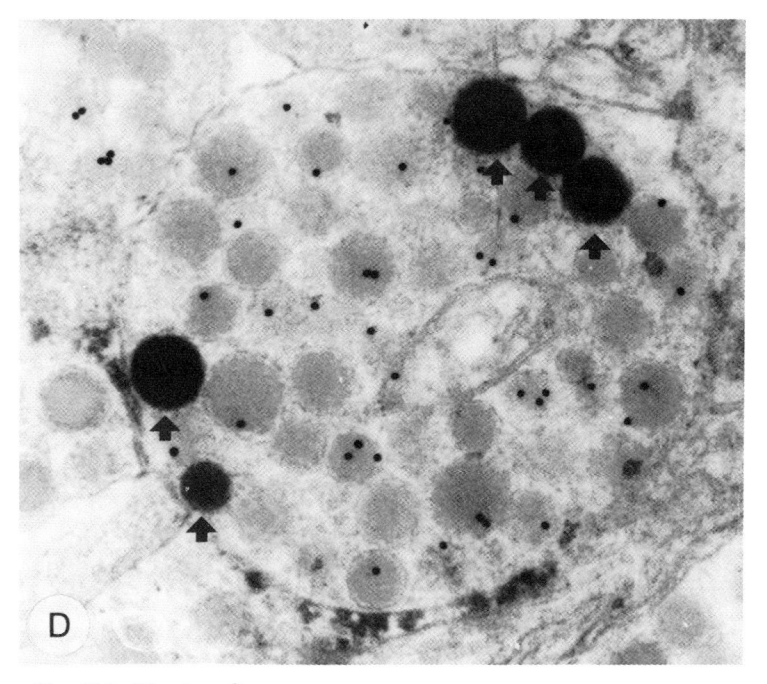

Fig. 14.2. (*Continued*)

cytochemically (goat anti-rabbit IgG–gold) the synthesis, maturation, and exocytotic release of the ovulation hormone (CDCH)-containing secretory granules of the CDC in the CNS of *Lymnaea*. Neurohemal exocytotic release of the contents of the secretory granules was clearly labeled with the antibody, indicating that CDCH is released by exocytosis. Since the CDC release at least nine different peptidergic secretions (for references, see Roubos, 1984) and show, in addition to the neurohemal release sites, somal and preterminal nonsynaptic release sites (Buma and Roubos, 1986), it would be very interesting to see whether the chemical nature of the released products differs between different release sites.

With respect to the rat CNS, it was shown that oxytocin and vasopressin fibers release not only secretory material from neurohemal axon terminals into the blood but also oxytocin and vasopressin from varicose fibers in the fiber layer of the median eminence (Buma and Nieuwenhuys, 1987, 1988) (Fig. 14.2G,H). The fibers occupied a subependymal position and were not located in the direct vicinity of the capillaries of the primary portal plexus. Therefore, it was suggested that vasopressin and oxytocin may travel through the large extracellular space in the median eminence and may exert their influence on distant targets with the appropriate receptors. Finally, the combination is also very suitable for nonnervous tis-

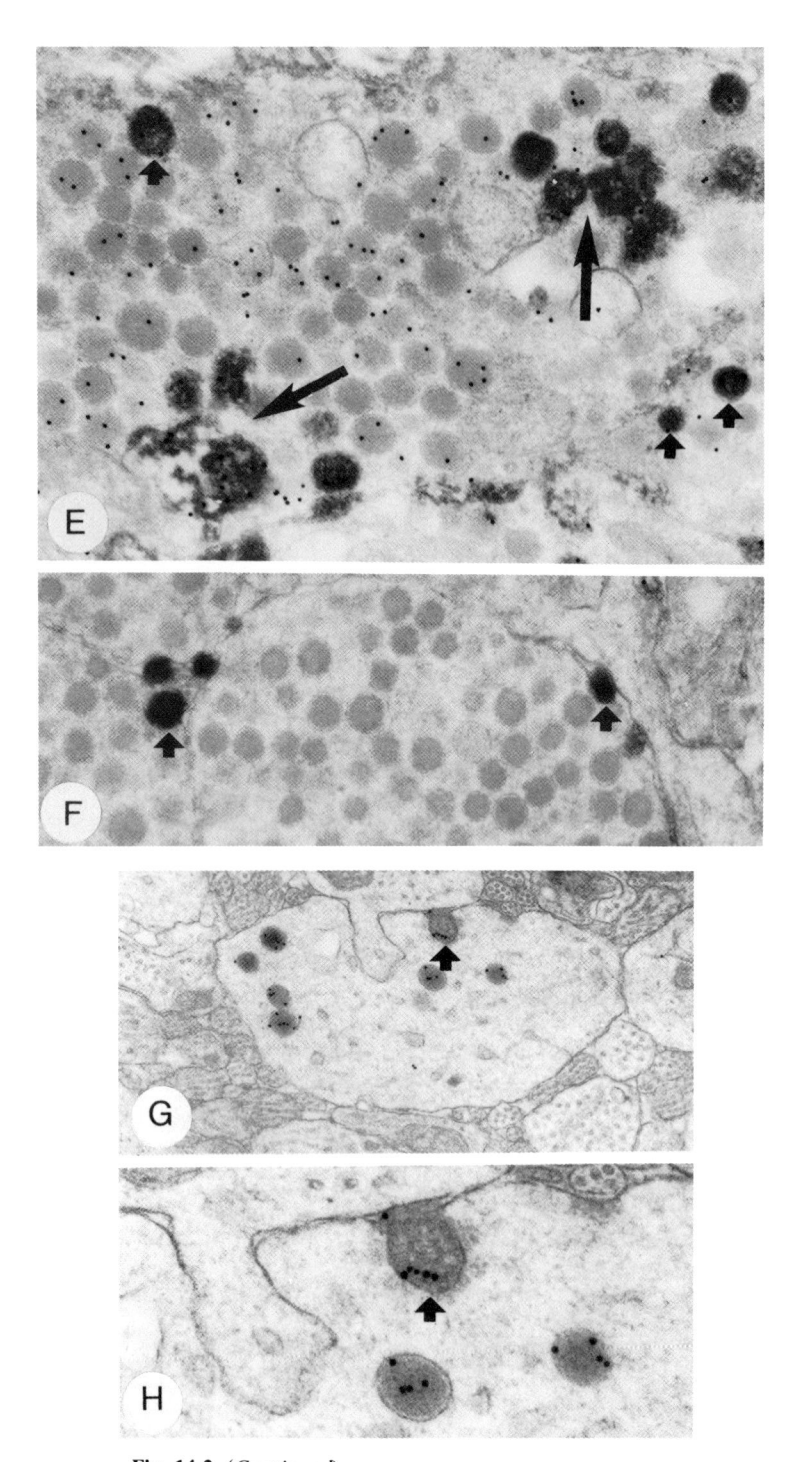

Fig. 14.2. (*Continued*)

sues, particularly when many different cell types are intermingled (e.g., the anterior pituitary; Van Putten *et al.,* 1987).

I wish to express my gratitude to Dr. M. H. Whitnall for his generous gift of the oxytocin–neurophysin antibody.

REFERENCES

Adelson, J. W., and Miller, P. E. (1985). Pancreatic secretion by nonparallel exocytosis: Potential resolution of a long controversy. *Science* **228,** 993.

Baker, P. F., and Knight, D. E. (1984). Calcium control of exocytosis in bovine adrenal medulla. *Trends Neuro Sci. (Pers. Ed.)* **4,** 120.

Batton, T. F. C. (1986). Ultrastructural characterization of neurosecretory fibres immunoreactive for vasotocin, isotocin, somatostatin, LHRH and CRF in the pituitary of a teleost fish *Poecillia latipinna. Cell Tissue Res.* **244,** 661.

Beauvillain, J.-C., Tramu, G., and Garaud, J.-C. (1984). Coexistance of substances related to enkephalin and somatostatin in granules of the guinea-pig median eminence: Demonstration by use of colloidal gold immunocytochemical methods. *Brain Res.* **301,** 389.

Ben-Barak, Y., Russell, J. T., Whitnall, M. H., Ozato, K., and Gainer, H. (1985). Neurophysin in the hypothalamo-neurohypophysial system. I. Production and characterization of monoclonal antibodies. *J. Neurosci.* **5,** 81.

Breckenridge, L. J., and Almers, W. (1987). Final steps in exocytosis observed in a cell with giant secretory granules. *Proc. Natl. Acad. Sci. U.S.A.* **84,** 1945.

Buma, P., and Roubos, E. W. (1983). Calcium dynamics, exocytosis, and membrane turnover in the ovulation hormone releasing caudo dorsal cells of *Lymnaea Stagnalis. Cell Tissue Res.* **233,** 143.

Buma, P., Roubos, E. W., and Buijs, R. M. (1984). Ultrastructural demonstration of exocytosis in neural, neuroendocrine and endocrine secretions with an *in vitro* tannic acid (TARI-) method. *Histochemistry* **80,** 247.

Buma, P., and Roubos, E. W. (1986). Ultrastructural demonstration of nonsynaptic release sites in the central nervous system of the snail *Lymnaea stagnalis,* the insect *Periplaneta americana,* and the rat. *Neuroscience* **17,** 867.

Buma, P., and Nieuwenhuys, R. (1987). Ultrastructural demonstration of oxytocin and vasopressin release sites in the neural lobe and median eminence of the rat by tannic acid and immunogold methods. *Neurosci. Lett.* **74,** 151.

Buma, P., and Nieuwenhuys, R. (1988). Ultrastructural characterization of exocytotic release sites in different layers of the median eminence of the rat. *Cell Tissue Res.* **252,** 107.

Calas, A. (1985). Morphological correlates of chemically specified neuronal interactions in the hypothalamo-hypophyseal area. *Neurochem. Int.* **7,** 927.

Castel, M., Morris, J., Ben-Barak, Y., Timberg, R., Sivan, N., and Gainer, H. (1985). Ultrastructural localization of immunoreactive neurophysins using monoclonal antibodies and protein A–gold. *J. Histochem. Cytochem.* **33,** 1015.

Douglas, W. W., Nagasawa, J., and Schulz, R. (1971). Electron microscopic studies on the mechanism of secretion of posterior pituitary hormones and significance of microvesicles ("synaptic vesicles"): Evidence of secretion by exocytosis and formation of microvesicles as byproduct of this process. *Mem. Soc. Endocrinol.* **19**, 353.

Douglas, W. W. (1974). Mechanism of release of neurohypophysial hormones: Stimulus–secretion coupling. In *Handbook of Physiology* (R. O. Greep and E. B. Astwood, eds.), Sect. 7, Vol. IV, Part 1, pp. 191–224. Williams & Wilkins, Baltimore, Maryland.

Golding, D. W., and Bayraktaroglu, E. (1984). Exocytosis of secretory granules—a probable mechanism for the release of neuromodulators in invertebrate neuropiles. *Experientia* **40**, 1277.

Gronblad, M. (1983). Improved demonstration of exocytotic profiles in glomus cells of rat carotid body after perfusion with glutaraldehyde fixative containing a high concentration of potassium. *Cell Tissue Res.* **229**, 627.

Khan, M. A., and Buma, P. (1985). Neural control of the corpus allatum in the Colorado potato beetle, *Leptinotarsa decemlineata:* An electron microscope study utilizing the in vitro tannic acid Ringer incubation method. *J. Insect Physiol.* **31**, 639.

Krisch, B., Becker, K., and Bargmann, W. (1972). Exocytosis im Hinterlappen der Hypophyse. *Z. Zellforsch Mikrosk. Anat.* **123**, 47.

Lamberts, R., and Goldsmith, P. C. (1985). Preembedding colloidal gold immunostaining of hypothalamic neurons: Light and electron microscopic localization of β-endorphin-immunoreactive pericaria. *J. Histochem. Cytochem.* **33**, 499.

Masson-Pevet, M., Pevet, P., and Noteborn, H. P. J. M. (1987). Ultrastructural demonstration of exocytosis in the pineal gland. *J. Pineal Res.* **4**, 61.

Nagasawa, J. (1977). Exocytosis: The common release mechanism of secretory granules in glandular cells, neurosecretory cells, neurons and paraneurons. *Arch. Histol. Jpn.* **40**, Supp., 31.

Normann, T. C. (1976). Neurosecretion by exocytosis. *Int. Rev. Cytol.* **46**, 1.

Noteborn, H. P. J. M., Roubos, E. W., Ebels, I., van der Ven, A. M. H., and Buma, P. (1986). Ultrastructural demonstration of secretion by exocytosis in rat pinealocytes with the use of the tannic acid method. *Cell Tissue Res.* **245**, 223.

Perrin, D., and Aunis, D. (1985). Reorganization of a-fodrin induced by stimulation in secretory cells. *Nature (London)* **315**, 589.

Rademakers, L. H. P. M. (1977). Effects of isolation and transplantation of the corpus cardiacum on hormone release from its glandular cells after flight in *Locusta migratoria. Cell Tissue Res.* **184**, 213.

Roubos, E. W., and van der Wal-Divendal, R. M. (1980). Ultrastructural analysis of peptide-hormone release by exocytosis. *Cell Tissue Res.* **207**, 267.

Roubos, E. W. (1984). Cytobiology of the ovulation-neurohormone producing caudo-dorsal cells of the snail *Lymnaea stagnalis. Int. Rev. Cytol.* **89**, 295.

Roubos, E. W., van der Ven A. M. H., and van Minnen, J. (1987). Immuno-electron microscopy of formation, degradation and exocytosis of the ovulation neurohormone in *Lymnaea stagnalis. Cell Tissue Res.* **250**, 441.

Schmauder-Chock, E. A., and Chock, S. P. (1987). Mechanism of secretory granule exocytosis: Can granule enlargement precede pore formation? *Histochem. J.* **19**, 413.

Sternberger, L. A., Hardy, P. H., Jr., Cuculis, J. J., and Meyer, H. G. (1970). The unlabeled antibody enzyme method of immunocytochemistry. Preparation and properties of soluble antigen–antibody complex (horseradish peroxidase– antihorseradish peroxidase) and its use in the identification of spirochetes. *J. Histochem. Cytochem.* **38**, 87.

Stoeckart, R., Janse, H. G., and Kreike, A. J. (1972). Ultrastructural evidence for exocytosis in the median eminence of the rat. *Z. Zellforch. Mikrosk. Anat.* **131**, 99.

van den Pol, A. N. (1985). Dual ultrastructural localization of two neurotransmitter-related antigens: Colloidal gold-labeled neurophysin-immunoreactive supraoptic neurons receive peroxidase-labeled glutamate decarboxylase- or gold-labeled GABA-immunoreactive synapses. *J. Neurosci.* **5**, 2940.

14. Study of Exocytosis with Colloidal Gold

Van Putten, L. J. A., Kiliaan, A. L., and Buma, P. (1987). Ultrastructural localization of exocytotic release sites in immunocytochemically characterized cell types. A combination of two methods. *Histochemistry* **86,** 375.

Van Putten, L. J. A., and Kiliaan, A. J. (1988). Immuno-electron microscopic study of the prolactin cells in the pituitary gland of the male Wistar rats during aging. *Cell Tissue Res.* **251,** 353.

Varndell, I. M., Tapia, F. J., Probert, L., Buchan, A. M. J., Gu, J., de Mey, J., Bloom, S. R., and Polak, J. M. (1982). Immunogold staining procedure for the localization of regulatory peptides. *Peptides (N.Y.)* **3,** 259.

von Grafenstein, H., Roberts, C. S., and Baker, P. F. (1986). Kinetic analysis of the triggered exocytosis/endocytosis secretion cycle in cultured bovine adrenal medullary cells. *J. Cell Biol.* **103,** 2343.

Wendelaar Bonga, S. E. (1971). Formation, storage, and release of neurosecretory material studied by quantitative electron microscopy in the snail *Lymnaea stagnalis* (L.). *Z. Zellforch. Mikrosk. Anat.* **113,** 490.

Whitnall, M. H., Castel, M., Key, S., and Gainer, H. (1985a). Immunocytochemical identification of dynorphin-containing vesicles in Brattleboro rats. *Peptides (N.Y.)* **6,** 241.

Whitnall, M. H., Key, S., Ben-Barak, Y., Ozato, K., and Gainer, H. (1985b). Neurophysin in the hypothalamo-neurohypophysial system. II. Immunocytochemical studies of the ontogeny of oxytocinergic and vasopressinergic neurons. *J. Neurosci.* **5,** 98.

Zamora, A. J., and Ramirez, V. D. (1983). Structural changes in nerve endings of rat median eminence superfused with media rich in potassium ions. *Neuroscience* **10,** 463.

Zhu, P. C., Thureson-Klein, A., and Klein, R. L. (1986). Exocytosis from large and dense cored vesicles outside the active synaptic zones of terminals within the trigeminal subnucleus caudalis: A possible mechanism for neuropeptide release. *Neuroscience* **19,** 43.

15

Colloidal Gold Labeling of Acrylic Resin-Embedded Plant Tissues

ELIOT MARK HERMAN

Plant Molecular Biology Laboratory
U.S. Department of Agriculture
Agricultural Research Service
Beltsville, Maryland

INTRODUCTION
FIXATION OF PLANT CELLS
PROCESSING AND EMBEDDING
   Lowicryl
      Lowicryl Embedding of Plant Tissues
      Ethanol Dehydration and Lowicryl Embedding at Subfreezing Temperature
      Dehydration with N,N'-Dimethylformamide at Room Temperature and Embedding at $-20°C$
      Future Prospects of Lowicryl in the Plant Sciences
   LR White
      Unosmicated Tissue
      Osmicated Tissue
      Chemical Catalyst Polymerization at Subzero Temperature
   LABELING SECTIONS WITH SPECIFIC ANTIBODIES
   Sources of Antibodies

Protocols for the Indirect Labeling of Acrylic Thin Sections
  Labeling Lowicryl Sections
  Labeling Unosmicated LR White Sections
  Removal of Osmium from LR White Sections and Immunogold Labeling
  Labeling Osmicated Tissue in LR White Sections without Removal of Osmium
CONTROLS; CHECKING FOR NONSPECIFIC OR PSEUDOSPECIFIC LABELING
  Control Solutions
  Control Tissues
OTHER ARTIFACTS COMMON TO PLANT TISSUES
  Cell Wall Labeling
  Pseudospecific Labeling
  Glycoprotein Cross-Reactivity
SUMMARY
REFERENCES

INTRODUCTION

Immunocytochemical observations offer the plant scientist the ability to investigate the compartmentation of physiologically important macromolecules. Recent technical innovations permit the routine immunocytochemical assay of plant tissues. The applications of immunocytochemistry to plant specimens have been reviewed, in general in Herman (1988) and in more specialized applications to the study of legume root nodules in Robertson *et al.* (1985), VandenBosch (1986), and Verma *et al.* (1986), and transgenic tobacco (Herman *et al.*, 1989).

The applications and methods of immunocytochemistry for plant specimens embedded in acrylic resins are reviewed here. The advent of acrylic resins has revolutionized immunocytochemical techniques, perhaps most notably in the plant sciences. Plant tissues have proved to be difficult subjects for preembedding and cryosectioning techniques. Resin embedding offers a simple and elegant solution to the problems of tissue preparation. The primary limitation is the availability of specific antibodies useful for immunocytochemical studies. While other methods may be equally sensitive in defining the compartmentation of macromolecules, indirect labeling of material embedded in acrylic resins offers the investigator a routine, sensitive, and reproducible assay method. Because the technique departs from conventional electron microscopic methods only in small details, immunocytochemical assays are easily accomplished by the trained microscopist.

FIXATION OF PLANT CELLS

Plant cells easily undergo plasmolysis during fixation, which leaves a large gap between the cell wall and the plasma membrane. The shrinkage of cell contents and possible disruption of intracellular organelles will result not only in the loss of the detailed architecture of the cell but also possibly in the redistribution of antigens. Therefore, it is critical that the osmotic pressure of the fixative be adjusted to permit the best possible fixation. The structure of chloroplasts and mitochondria is a relatively sensitive indicator of fixation quality and should be critically examined. The range of possible fixatives is common to all electron microscopy studies (for a detailed discussion, see Hayat, 1981). Primary formaldehyde fixation has been shown to preserve antigenicity (Craig and Goodchild, 1982) but will not result in good ultrastructural preservation. Glutaraldehyde alone or in combination with acrolein or formaldehyde is the only primary fixative which will simultaneously preserve antigenicity and structure. We have had success fixing with 4% formaldehyde and 2% glutaraldehyde in 0.1 M phosphate buffer at pH 7.4 (Herman and Shannon, 1984). Another effective combination is 4% formaldehyde, 0.75% acrolein, 0.75% glutaraldehyde in 500 mM sucrose, 25 mM phosphate buffer at pH 7.2 (Greenwood and Chrispeels, 1985). Other investigators use glutaraldehyde alone at concentrations between 2 and 3% in various buffers between 20 and 100 mM (for example, see Robertson *et al.*, 1984; Shaw and Henwood, 1985). We have found that the inclusion of the rapidly penetrating formaldehyde is useful in dense tissue such as seeds; however, leaf and root tissues are often well fixed without the inclusion of formaldehyde. Many investigators attempt to "lightly fix" in order to reduce the extent of loss of epitopes; however, this effect cannot easily be quantitated. In my laboratory we fix tissues to preserve structure first and epitopes second. In my opinion it does not matter how well a structure is labeled if it has lost fidelity to the living organism. Weakly fixed macromolecules may easily be redistributed during further processing with resulting artifactual antigen distribution.

PROCESSING AND EMBEDDING

Acrylic resins are now the standard for indirect postembedding immunocytochemical assays. Several problems not encountered in animal sciences are common in the preparation of plant tissues. The two major types of commercially available acrylic resins in wide use today are the

Lowicryls and LR White. The use of these resins and the embedding problems associated with their use are discussed in the subsequent sections.

Lowicryl

Lowicryl Embedding of Plant Tissues

Soon after Lowicryl became commercially available, several plant science investigators reported its use (Craig and Goodchild, 1982; Tomenius et al., 1983; Herman and Shannon, 1984; Robertson et al., 1984). Lowicryl is a hydrophilic acrylic resin which provides a medium for enhanced postembedding immunogold labeling (Roth et al., 1981; Carlemalm et al., 1982). Lowicryl will tolerate as much as 30% water in the dehydration solution. It is polymerized by illumination with long-wave UV light and may be processed at temperatures well below the freezing point of water. These two properties of Lowicryl presumably aid in the rentention of antigenicity. In almost all of the work to date on plant tissues the K4M formulation of Lowicryl has been used. The recently released K11M formula has been used in my laboratory, and we have found that it provides some improvement in structural preservation over the preceding formulation. Plant tissues may embed poorly in Lowicryl and, in much of the published work, appear to be highly extracted. The embedding problems that occur with Lowicryl may be derived from two unique constituents of plants. Most plant tissues contain various and abundant pigments. These pigments absorb light in the blue range (including near UV), and consequently plant tissues can be opaque to UV light in the frequency used for polymerization, resulting in poor polymerization of the resin within the tissue sample. Much of the pigment may be removed by extended extraction with organic solvents such as acetone, alcohols, and dimethylformamide. However, complete extraction requires a solvent concentration higher than the 70% (v/v) which might otherwise be used. The rate of extraction of pigments at high solvent concentration is dependent on the temperature; at −20°C a solvent wash of 12 hr is frequently necessary to remove the last vestiges of pigment. Second, the complex carbohydrates of plant cells are poorly embedded in Lowicryl. Much of the fine fibrillar detail of the cell wall is absent in published micrographs of Lowicryl-embedded tissues. The cell walls are frequently soft and fragile after embedding and may fragment, making sectioning and subsequent manipulations difficult. The large starch grains commonly found in the cytoplasm and chloroplasts are very difficult to embed. They often fall out during sectioning, leaving holes in the thin sections which enlarge during illumination with the electron beam. The experience of my laboratory is that root tips and seed cotyledons are very well embedded in Lowicryl. We have found that leaf tissue can be difficult, and woody tissues like bark have proved nearly impossible to embed. The lack of osmium postfixation and the extensive extraction of lipids have adverse consequences for the structural preservation of plant membranes. The vacuole, Golgi, and smooth endoplasmic reticulum membranes are in particular poorly preserved. The structural preservation of plant tissues embedded in Lowicryl K4M is illustrated in Figs. 15.1 and 15.2.

Ethanol Dehydration and Lowicryl Embedding at Subfreezing Temperature

This is an adaptation of the methods recommended by the developers of Lowicryl (Roth *et al.*, 1981; Carlemalm *et al.*, 1982). The aldehyde-fixed sample is simultaneously dehydrated and progressively chilled to subfreezing temperature, avoiding the formation of ice crystals. After the sample is dehydrated and equilibrated at -20 to $-35°C$, resin infiltration is begun. This is the most widely used method for Lowicryl embedding and has proved to be effective in the structural preservation of seed cotyledons, root nodules, and leaves (for examples, see Robertson *et al.*, 1984; Shaw and Harwood, 1985; Titus and Becker, 1985). The basic method is outlined below:

1. The tissue is cut into small pieces not exceeding 1–2 mm^3 and is fixed in a fixative determined to yield good structural preservation. The small size of tissue blocks is essential to ensure that the interior of the sample is throughly illuminated by the polymerization lamp. Osmium postfixation cannot be used as it would darken tissue, leading to unsuccessful polymerization.

2. The tissue is cooled to 0°C on ice and dehydrated stepwise through 50% solvent. Ethanol is the most commonly used solvent but methanol, acetone, and *N,N'*-dimethylformamide (DMF) appear to be equally effective. The timing of the solvent steps does not appear to be critical, although in practice we use 15 to 30 min as a minimum time.

3. The tissue is then transferred to 70% solvent at $-20°C$. Various mixtures of ice, NaCl, and CaCl$_2$ can be used to provide subzero temperature baths. However, we find it much easier to equilibrate solutions in a freezer prior to use. The solvent concentration must be above 50% before the tissue is cooled below freezing, otherwise damaging ice crystals may form. The samples are then brought to 90% solvent at -20 to $-35°C$ until most of the pigment is removed. This will require from 30 min to 12 hr, depending on the tissue. Lowicryl infiltration is then begun with Lowicryl : solvent ratios of 1 : 1 and 2 : 1, for 1 hr each, then 100% Lowicryl

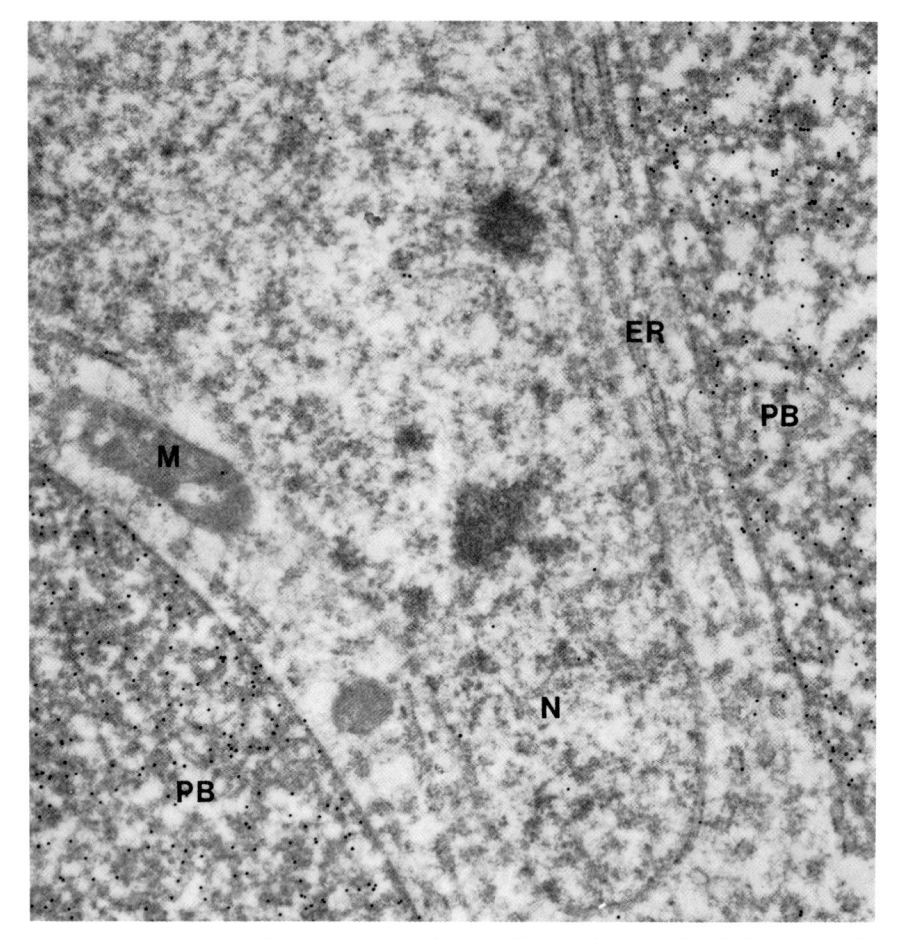

Fig. 15.1. A portion of a storage parenchyma cell of a midmaturation jack bean cotyledon embedded in Lowicryl K4M is shown. The organelles shown in this micrograph include the protein bodies or protein storage vacuoles (PB), nucleus (N), mitochondrion (M), and endoplasmic reticulum (ER). This section is labeled with antisera directed against the acid hydrolase and vacuole marker enzyme α-mannosidase and indirectly localized with protein A–colloidal gold. The protein body is specifically labeled with the gold particles; ×22,500.

for at least 12 hr. The resin should be changed two or three times prior to embedding in BEEM capsules.

4. The tissue is embedded and the blocks polymerized, still at -20 to $-35°C$. The samples are well polymerized in 6 hr by 25 μW/cm² (~10 cm from a 25-W fluorescent long-wave UV light) and then further hardened by illumination at 20°C for at least 12 hr with 25 μW/cm². The resulting

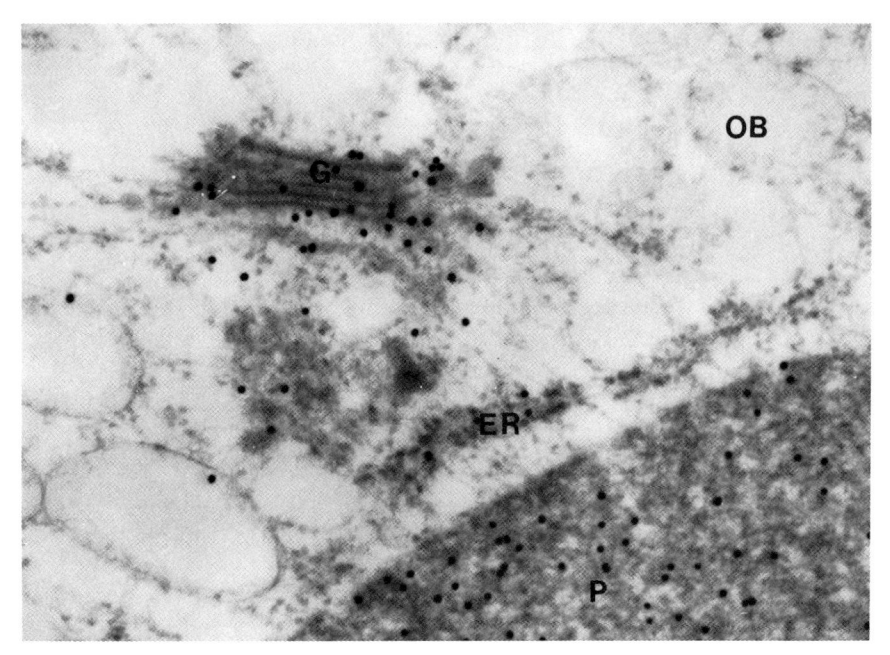

15. Labeling of Acrylic Resin-Embedded Plant Tissues

Fig. 15.2. The Golgi apparatus (G), endoplasmic reticulum (ER), oil bodies (OB), and protein body (P) of a Lowicryl K4M-embedded maturing soybean cotyledon is shown. The sections were labeled with anti-α-galactosidase serum and colloidal gold–protein A. Note although the Golgi apparatus is readily identified, the structural preservation of the cisterna membranes is poor. The gold particles are localized on the ER, Golgi apparatus, and protein bodies, illustrating the transport of α-galactosidase through the endomembrane system for deposition in the protein bodies. Observations of this type demonstrate the role of the Golgi apparatus in mediating the transport of specific macromolecules; ×48,000. (Reprinted with the permission of the American Society of Plant Physiologists.)

blocks are easily cut with glass or diamond knives, but the cutting speed and angle of attack must be higher for Lowicryl blocks than for epoxy blocks. It should be noted that bleached plant tissues embedded in Lowicryl are very translucent and difficult to see during block trimming. Many microtomes are now equipped with transmitted illumination systems behind the block which aid in visualization of embedded tissue. The use of such illumination is recommended for all unosmicated blocks.

Dehydration with N,N'-Dimethylformamide at Room Temperature and Embedding at −20°C

This procedure was developed by Herman and Shannon (1984) to facilitate the removal of plant cells pigments from fixed tissues. The resulting bleached tissue is infiltrated with Lowicryl and is easily polymerized by long-wave UV light. The structural preservation of jack bean and soybean cotyledon tissue embedded in Lowicryl K4M is illustrated in Figs. 15.1 and 15.2.

1. Small pieces of tissue (~ 1 mm^3) are fixed with 4% formaldehyde (v/v), 2% glutaraldehyde in 0.1 M phosphate buffer for 1 hr at room temperature and then overnight at 7°C. This ensures that the tissue is well fixed, preserving both structure and the location of intracellular antigens.

2. The tissue is then dehydrated stepwise in DMF through 20, 50, 75, and into 90% (v/v) at room temperature with 1 hr for each step. Pigment extraction is completed during the 90% wash. The tissue must remain in the 90% solution until it is completely bleached (1–12 hr), which is aided by at least one change of 90% DMF.

3. The Lowicryl (K4M) is then added, mixed 1 : 1 with 90% DMF for 1 hr before transferring tissue to 100% Lowicryl. The liquid Lowicryl and tissue samples are kept in a dark container except during solution changes. Lowicryl penetrates into plant tissue very slowly and long incubations are recommended as well as several changes of pure resin. For example, seed tissues are well infiltrated with three changes of resin over a 24-hr period.

4. The samples are placed into freshly made Lowicryl and loaded into BEEM capsules. The capsules are cooled to $-20°C$ prior to initiating illumination with long-wave light. This procedure has been modified by Altman *et al.* (1984), who rapidly dehydrate the fixed tissue in DMF, infiltrate with Lowicryl, and polymerize at 7°C. The entire process is accomplished within a single working day. This procedure does not appear to allow sufficent infiltration time for the plant tissues that we have tried, but it may be appropriate for some samples, in particular membrane fractions or organelles.

Future Prospects of Lowicryl in the Plant Sciences

Lowicryl will continue to be a widely used embedding resin for plant tissues. Its high sensitivity, low background, and excellent antigen preservation make Lowicryl an excellent medium for immunocytochemical studies. The primary limitation of Lowicryl is the extracted appearance of the tissue, which may indicate that the intracellular localization of some macromolecules is not retained. Lowicryl would appear to have good prospects for use as an embedding medium after cryofixation and freeze substitution as well, since it can be polymerized at low temperatures.

LR White

LR White is a commercially available aromatic acrylic resin (Newman *et al.*, 1983) which has found wide application in immunocytochemistry

15. Labeling of Acrylic Resin-Embedded Plant Tissues

(Newman, Ch. 4, this volume; Craig and Miller, 1984; Greenwood and Chrispeels, 1985; McCurdy and Pratt, 1986; VandenBosch and Newcomb, 1986). Like Lowicryl, it tolerates partial dehydration of the tissue and results in excellent antigen preservation and labeling density. LR White is usually thermally polymerized, although a catalyst is also available. It has been reported to be less extractive of plant tissues than comparable Lowicryl embedding (for comparison see Herman, 1987). LR White can be used in the embedding of both osmicated and unosmicated tissue, allowing useful comparisons to be made. We have found that LR White will often successfully embed tissues which have proved difficult with Lowicryl, for example, bark and some seeds such as tobacco. Another asset of LR White is that tissue color is not a relevant factor in polymerization.

Unosmicated Tissue

The embedding of unosmicated tissue in LR White is accomplished by simply dehydrating the fixed plant tissue in a graded ethanol series. We usually dehydrate at 7°C using temperature-equilibrated ethanol in 30-min steps. While it is not necessary to dehydrate tissue completely in order to embed it in LR White, we have found that for certain types of plant tissues a complete dehydration results in a more successful embedding. Lipid droplets appear to react poorly with the resin, resulting in semipolymerized tissue samples. This is easily corrected by completely extracting the lipid components with 100% ethanol prior to resin infiltration. Infiltration times for plant specimens are longer than for animal tissue specimens. We have found that 1 hr each in 1 : 1 and 2 : 1 LR White : solvent followed by a minimum of 12 hr in pure resin generally results in excellent infiltration. To ensure complete infiltration or for convenience, tissue samples can remain in pure LR White for several days. We always change the resin at least two or three times prior to embedding.

Polymerization of LR White at 50°C is reported to result in limited cross-linking of the resin and consequent formation of long-chain polymers, while polymerization at 65°C results in extensive cross-linking (Newman and Hobot, 1987). We have found that at 50°C polymerization frequently results in excessively soft tissue in the blocks, which cannot be thin-sectioned. Our present routine is to incubate the blocks at 65°C for 2 days, which yields a moderately hard block that is easily sectioned to silver-gray thickness with glass, diamond, or sapphire knives.

The structural preservation of cellular membranes after embedding in LR White is variable. In unosmicated plant cells embedded in LR White, Golgi apparatus is poorly preserved, for reasons that are still unclear. The vacuole membrane and chloroplast envelope are not well visualized in thin sections of unosmicated tissue. The matrix components, however, appear to be better preserved than the boundary membranes in unosmicated tissue samples embedded in LR White. Figure 15.3 shows a portion of a transgenic tobacco embryo, illustrating the structural preservation of unosmicated specimens embedded in LR White.

Osmicated Tissue

The ability to utilize thermally polymerized LR White on osmicated tissue has proved to be an important asset in several immunocytochemical studies. Many ultrastructural components are poorly preserved in the absence of osmium postfixation. However, it is important to note that several organelles of plant tissues have distinctly different appearances when embedded in LR White than those embedded in epoxy resins such as Spurr's or Epon. Osmium derivatives of storage lipids appear to react poorly with LR White embedding. The reserve oil bodies of seeds appear misshapen and granular (E. M. Herman, unpublished observations). The

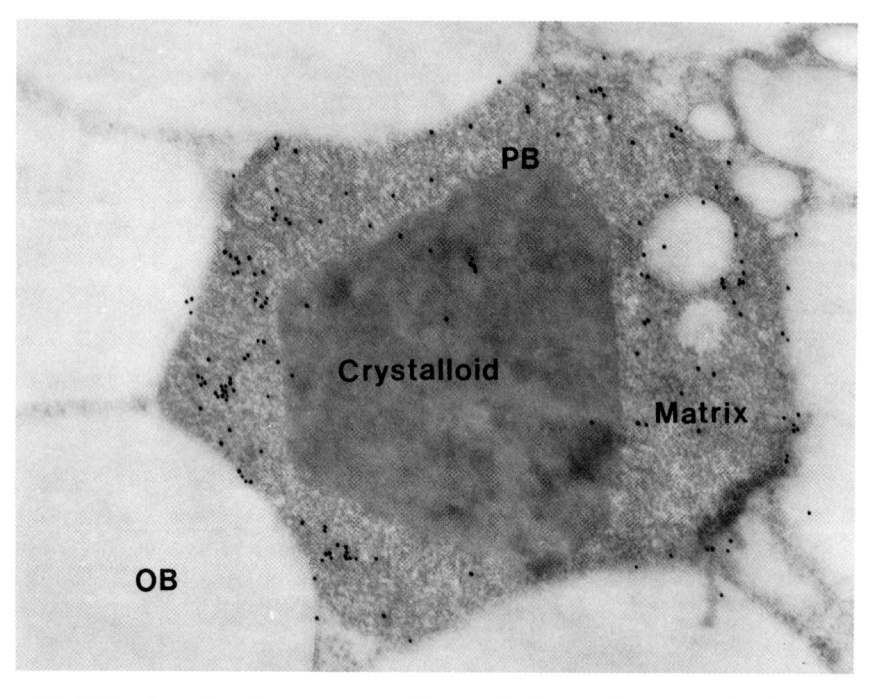

Fig. 15.3. A portion of a storage parenchyma cell of a maturing transgenic tobacco embryo embedded in LR White is shown. Note that the contents of the oil bodies (OB) are completely extracted in the unosmicated tissue. The bean lectin phytohemagglutinin expressed in the transgenic tobacco is shown to be localized in the matrix portion of the protein body (PB) by indirect colloidal gold–goat-anti-rabbit IgG. This observation illustrates the utility of immunocytochemical protocols in the examination of expression and targeting of heterologously expressed proteins; ×47,400.

Golgi apparatus cisternae are not well perserved even after osmium postfixation. Other organelles such as the chloroplast membranes may appear as negative images against an electron-dense matrix. We have noted that it is often necessary to dehydrate osmicated plant tissue completely for embedding in LR White. Partially hydrated osmicated plant tissue often does not polymerize well even though the surrounding block does.

Osmium tetroxide irreversibly destroys many antigenic sites, which limits the usefulness of postfixed material for immunocytochemical studies. However, some antigens are not destroyed, and in these limited situations immunocytochemical studies may be conducted on tissue samples with structural preservation comparable to that of conventionally prepared epoxy resin samples. We have observed that some vacuolar proteins such as storage proteins, lectins, and acid hydrolases may be labeled in postfixed tissue samples. The immunogold labeling of sections of osmicated tissue embedded in LR White may require the removal of osmium with sodium metaperiodate and HCl (Bendayan and Zollinger, 1983; Craig and Goodchild, 1982) to unmask the antigenic epitopes.

Chemical Catalyst Polymerization at Subzero Temperature

LR White may be polymerized at subzero temperature through the use of a chemical catalyst. This has the advantage that the processing of tissue by progressive temperature reduction simultaneously with dehydration, which is normally used for Lowicryl, may be adapted for the less extractive LR White. Harris and Croy (1985) have embedded pea seeds in this way. This procedure may be used only with unosmicated specimens. The samples may be prepared and processed by the progressive temperature lowering procedure outlined for Lowicryl while substituting LR White. In order to preclude premature polymerization, the chemical catalyst is added at 1 drop per 10 ml of resin only when the samples are embedded. The capsules containing the samples and the catalyzed resin are incubated at $-20°C$. It is important to note that this is a vigorous exothermic reaction and as a consequence the capsules must be placed in a heat sink to preclude excessive temperature rise during polymerization.

LABELING SECTIONS WITH SPECIFIC ANTIBODIES

Sources of Antibodies

Recent technical innovations in immunological techniques have presented the investigator with a wide variety of possible sources of antibodies for cytochemical studies. These choices fall into three major categories: (1) polyclonal sera, (2) affinity-purified antibodies from polyclonal sera, and (3) monoclonal antibodies. Each type may be successfully used for immunocytochemical studies. Crude antiserum has the advantage of ease of preparation and low cost, while affinity-purified and monoclonal antibodies can be expensive and time consuming to prepare but offer highly specific selectivity.

Crude polyclonal antiserum is frequently used for labeling plant proteins. Antiserum against a wide variety of proteins has been produced as a consequence of other studies and is therefore available for new cytochemical studies. Proteins of moderate abundance (0.05–1% of the total) can usually be labeled by antiserum dilutions of 1 : 50–1 : 1000 of primary antiserum into a buffered saline/blocking solution. The exact concentration of antiserum must be experimentally determined by serial dilutions. The labeling times may be varied; we have used as short as 5 min and as long as an hour. In general, we have observed that longer labeling results in the gradual increase in nonspecific labeling.

Affinity purification of antibodies is quite useful since the specific antibody may be used at high dilution in a blocking buffer to help minimize spurious cross-reactivity. The limitation of this technique is the requirement for larger volumes of serum and quantities of antigen to use in affinity purification.

Monoclonal antibodies have been used in many studies. Their advantage is that each antibody is a single species of immunoglobulin having specificity for a single epitope on the antigen. The production of monoclonals is a complex and expensive process and is not often justified if the sole objective is cytochemical localization. The antibodies may be produced in milligram quantities per milliliter by ascites fluid or at 10 µg/ml in culture supernatants. Immunocytochemical studies use such small quantities of antibodies that the crude culture supernatant may often be used. We have used the culture supernatants at 1 : 50 dilutions to obtain highly specific labeling. In our experience monoclonals frequently yield excellent specific-to-nonspecific label ratios and often have little of the nonspecific binding problems often observed in our use of polyclonal antibodies.

Protocols for the Indirect Labeling of Acrylic Thin Sections

The following list illustrates some of the labeling protocols used by various investigators to obtain specific labeling of plant proteins. These protocols contain examples of possible dilution solutions chosen on the basis of the quality of the labeling presented, the organelle targeted, and the diversity of methods as well as apparent absence of nonspecific labeling. Other protocols may yield equally effective or superior labeling, so the examples presented here are suggested as starting points.

Labeling Lowicryl Sections

1. Titus and Becker (1985) have slightly modified the procedure of Geuze *et al.* (1983) to obtain highly specific labeling of the matrix proteins of glyoxysomes and peroxisomes of cucumber cotyledons embedded in Lowicryl K4M. The grids are initially blocked with 10 mM Tris-HCl, pH 7.4, 0.5 M NaCl, 0.3% Tween 20 (TBST) containing 1% bovine serum albumin (BSA). The grids are then labeled in 1 : 100–1 : 2000 dilution of the primary antiserum in TBST/BSA for 1 hr, washed with TBST/BSA, and indirectly labeled with 10-nm protein A–colloidal gold in TBST for 10 min. Other investigators have found that second antibody gold probes are equally as effective as indirect labels in this procedure. The protocol of Geuze *et al.* (1983) has been used by many other investigators with little modification.

2. Shaw and Harwood (1985) obtained excellent labeling of chloroplast proteins in Lowicryl K4M sections using a high-Tris buffer without NaCl. The lack of salt is unusual for immunological buffers but the results obtained demonstrate that a salt-free buffer may be effective. Thin sections of Lowicryl-embedded tissue were incubated in 1 : 10 to 1 : 1000 dilution of a primary polyclonal serum in 0.5 M Tris-HCl, pH 7.4, 0.1% gelatin, 1% Tween-20, 1% ovalbumin for 1 hr. The sections were then washed with distilled water and indirectly labeled for 1 hr with Janssen goat anti-rabbit IgG–colloidal gold.

Labeling Unosmicated LR White Sections

1. McCurdy and Pratt (1986) have obtained excellent results with monoclonal antibody labeling of thin LR White sections to localize the cytoplasmic photoreceptor protein phytochrome. Grids were blocked for 30 min in 10 mM Tris-HCl, 0.5 M NaCl, 0.3% Tween 20, 1% BSA (TBS-BSA) and 5% lamb serum and labeled with 5 µg/ml monoclonal antibodies in TBS-BSA plus 1% lamb serum for 1 hr. After washing three times in TBS-BSA, grids were indirectly labeled for 1 hr with goat anti-rabbit IgG–gold diluted 1 : 1 with TBS-BSA plus 1% lamb serum.

2. Strum *et al.* (1988) found high background labeling of the cell wall and nucleus in unosmicated LR White-embedded tobacco embryos labeled with crude antisera. The object of this study was to localize the bean seed lectin phytohemagglutinin (PHA) in transgenic tobacco. In order to correct the nonspecific labeling, grids were blocked with 10% normal goat serum in Tris-buffered saline with Tween 20 (50 mM Tris-HCl,

pH 7.4, 0.15 M NaCl, 0.2% Tween 20; TBST) for 10 min. The grids were then incubated in 0.1–0.25 μg/μl affinity-purified anti-PHA antibodies in TBST–goat serum for 1 hr. The bound antibodies were indirectly localized with goat anti-rabbit IgG–colloidal gold diluted 1:3 in TBST–goat serum for 15 min. The grids were then stained for 15 min with 5% (w/v) uranyl acetate. This procedure was ineffective with crude antisera, but using very dilute solutions of purified antibodies could block nonspecific cell wall binding. Figure 15.3 shows the localization of PHA in the vacuolar protein bodies of transgenic tobacco seeds as an example of labeling of unosmicated tissue embedded in LR White.

3. Hoffman et al. (1987) used an alternative procedure for eliminating cell wall artifacts using crude antiserum as the primary labeling reagent. The maize storage protein zein was localized in transgenic tobacco seeds using primary zein antiserum as the labeling reagent. The blocker BLOTTO (Johnson et al., 1984) was slightly modified to consist of 5% (w/v) nonfat dry milk in Tris-buffered saline with Tween 20 (TBST; 50 mM Tris-HCl, pH 7.4, 0.15 M NaCl, 0.1% Tween 20) and used to dilute the primary antiserum. The grids were incubated in a 1:50 dilution of antiserum in BLOTTO for 1 hr, washed with TBST, and the bound antibodies indirectly localized with undiluted commercial goat anti-rabbit IgG–colloidal gold. After washing with TBST and distilled water, the grids were stained with 5% uranyl acetate for 30 min. We have observed that BLOTTO dilution buffer is very stringent and will reduce the label density with most antibody preparations. However, we have also observed that it is one of the most effective solutions to the problem of nonspecific or pseudospecific label.

Removal of Osmium from LR White Sections and Immunogold Labeling

The procedure of Bendayan and Zollinger (1983) for the removal of osmium from epoxy resin sections has been adapted by Craig and Miller (1984) for the immunogold labeling of osmicated seed tissue embedded in LR White. Thin sections on grids are treated with saturated aqueous NaIO$_4$ for 10 to 30 min, washed in distilled water, put in 0.1 N HCl for 10 min, and washed again with distilled water. This removes the osmium from the sections by oxidation. The sections are then labeled with affinity-purified antibody (10–50 μg/ml) diluted in 10 mM Na phosphate buffer (pH 7.1), 500 mM NaCl, 0.1% (v/v) Tween 20, 1 mg/ml bovine serum albumin for 10 min. The bound antibodies were then indirectly labeled with protein A–gold or goat anti-rabbit IgG–gold diluted in the same buffer for 10 min. The sections were then stained with 2% (w/v) uranyl acetate for 5–7 min.

This procedure will not work on all antigens; however, when it is effective it has been shown to result in the enhancement of label density from 4- to 15-fold (for example, see Craig and Goodchild, 1982; VandenBosch, 1986). Other antigens are irreversibly destroyed by the osmium fixation, and the subsequent removal of osmium will not restore immunoreactivity. The periodate treatment used to remove the osmium will itself destroy glycan antigens. We have observed that glycan antigens often are not affected by the osmium postfixation and may be labeled with glycan-specific monoclonal antibodies if the sections are not pretreated with periodate (E. M. Herman, unpublished observations). Parallel controls in which the osmium is not removed should always accompany this protocol.

Labeling Osmicated Tissue in LR White Sections without Removal of Osmium

Herman et al. (1988) observed that it was not necessary to remove the osmium from postfixed *Sophora japonica* leaf and bark tissues in order to localize members of a tissue-specific family of lectins in LR White-embedded samples. Ultrathin sections mounted on nickel grids were labeled for 30 minutes with 1 : 20 to 1 : 500 dilutions of the primary seed lectin antisera diluted in TBST (50 mM Tris-HCl, pH 7.4, 0.15 M NaCl, 0.5% v/v Tween 20). The grids were then indirectly labeled with protein A–gold in TBST for 5 min. After washing with TBST and distilled water the grids were stained with 5% uranyl acetate for 20 min and alkaline lead citrate for 5 min. The retention of osmium in the tissue results in the enhanced electron density of subcellular organelles. An example of *Sophora* leaf tissue in which the lectin is localized in vacuolar protein deposits is shown in Fig. 15.4. This procedure is limited in its application to the labeling of antigens which are not destroyed or masked by osmium postfixation.

CONTROLS; CHECKING FOR NONSPECIFIC OR PSEUDOSPECIFIC LABELING

We always do simultaneous controls with every experimental assay. Without a control it is impossible to understand how an assay may have resulted in a false positive. If an antigen is simultaneously localized in several unrelated subcellular compartments, this may indicate that the assay is not accurate and must be carefully examined and reconsidered. The most common sites of false positives are the cell wall, nucleus, and vacuolar inclusions. Simultaneous labeling in these compartments is almost certainly an indication of a nonspecific binding problem. Controls

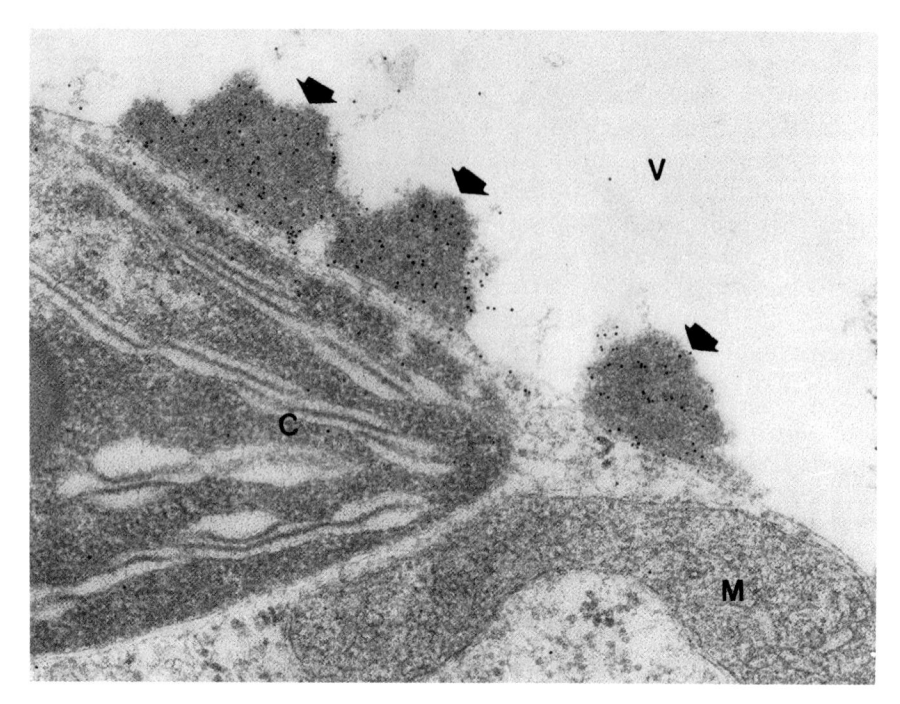

Fig. 15.4. A portion of a *Sophora japonica* leaf embedded in LR White resin after postfixation with OsO$_4$ is shown. This section is labeled with seed lectin antiserum and indirectly labeled with protein A–colloidal gold. The immunogold labeling of a partially filled leaf vacuole (V) is shown. Note that the immunogold labeling is restricted to the aggregated protein (arrows) precipitated against the vacuolar membrane, while the vacuolar sap is unlabeled. Also shown is a portion of chloroplast (C) and mitochondrion (M); ×50,000. (Reprinted with the permission of the American Society of Plant Physiologists.)

come in two broad varieties. The first consists of control solutions which do not contain the specific antibody but which control for all other nonspecific binding of the components of the assay solutions. In the second type, the labeling reaction contains the specific antibody but the tissue either lacks the specific antigen or has been modified to remove the antigenic epitope prior to labeling.

Control Solutions

Suitable controls are preimmune or nonimmune sera or IgG at identical concentrations to the specifically labeled sample. This is the most commonly used control for antisera, affinity-purified antibodies, and monoclonal antibody labeling reactions. This single variety of control is widely accepted as a demonstration of specificity, particularly if the antibody has been previously characterized by biochemical criteria. In experiments using antisera as the primary labeling reagent it is possible to immunoabsorb the specific antibodies from the antisera, leaving all of the nonspecific antibodies. When properly done this is an excellent and appropriate control.

Control Tissues

Mutant plant lines which do not express the targeted protein are excellent controls. In experiments involving transgenic plants and the localization of a heterologously expressed protein, an untransformed plant will provide control tissue (for example, see Greenwood and Chrispeels, 1985). The advantage of a control plant as opposed to a control solution is that the entire specific labeling reaction is run. Selected chemical removal or modification of an antigenic epitope is another possible control. Antibodies directed at carbohydrate epitopes common to plant proteins may be controlled by pretreating parallel sections with periodate and HCl, which oxidizes the carbohydrate, often abolishing specific labeling. Alternatively, carbohydrate residues can be selectively removed with commercially available glycosidases. This has the added advantage of partially characterizing the antigenic epitope in addition to serving as a control.

OTHER ARTIFACTS COMMON TO PLANT TISSUES

Cell Wall Labeling

The single most common problem in the immunocytochemistry of plant tissues is the nonspecific labeling of the cell wall. Varying the blocking solutions will often reduce or eliminate much of this problem; solutions containing nonfat milk, normal serum, or Tween 20 have been found to be useful. The most consistent method we have found is to use very dilute solutions of the specific antibody (affinity purified or monoclonals work better) in a relatively concentrated blocking solution. One method which seems to be effective in my laboratory is to use 10% normal goat serum in TBST containing no more than 0.5 $\mu g/\mu l$ specific antibody (0.5% w/v) with a goat antibody coupled to colloidal gold as an indirect label. We have also found that BLOTTO (Johnson *et al.*, 1984) (5% nonfat dry milk in TBS/PBS) is very effective at blocking nonspecific antibodies or nonspecific binding of specific antibodies. This has worked particularly well with monoclonal antibodies. However, we have also noted that the stringency of the blocker is such that it may also reduce specific labeling.

Pseudospecific labeling

Protein A–colloidal gold may lead to pseudospecific artifacts. We have noted on several occasions that PA–gold is specifically localized in the vacuole of seed tissues without prior incubation in IgG. This pseudospecific reaction has proved impossible to block effectively and is avoided only by the use of anti-IgG–gold indirect labels.

Glycoprotein Cross-Reactivity

Many proteins of interest in plants contain glycan chains. Glycans may be highly immunogenic and a single type of glycan side chain may be common to several distinct and unrelated proteins. This may result in artifactual labeling of many different organelles and proteins. Before using antiserum raised against glycoproteins, it must be examined for specificity on Western blots of extracts of the total organismal protein. There are, however, distinct advantages in using glycan epitopes of plant proteins. Osmium postfixation does not destroy glycans. We have had excellent results using polyclonal antisera against vacuole glycoproteins such as leaf lectins on aldehyde-fixed and osmium-postfixed LR White-embedded material (shown as an example in Fig. 15.4).

SUMMARY

The current methods of indirect acrylic plastic thin-section immunocytochemistry will permit the localization of most moderately abundant proteins. For the future, immunocytochemical techniques in plant science research appear to offer a rich diversity of potential avenues of exploration. For example, immunocytochemical observations will no doubt prove to be an essential component of future studies on the architecture of the cell wall, the targeting and accumulation of heterologously expressed proteins in transgenic plants, and the physiological response to physical and biological stress. Immunocytochemical observations provide a bridge between molecular biology and the morphological manifestations of plant growth and development.

Some of the work used as examples in this chapter is the product of collaborations with Drs. Maarten Chrispeels (University of California, San Diego), Leland Shannon (University of California, Riverside), and Les Hoffman (Agrigenetics). The comments of Drs. Karen Herman (NIH) and Diane Melroy (USDA) are gratefully acknowledged. Research in my laboratory is supported by in-house USDA funds and the USDA Office of Competitive Grants.

REFERENCES

Altman, L. G., Schneider, B. G., and Papermaster, D. S. (1984). Rapid embedding of tissues in Lowicryl K4M for immunoelectron microscopy. *J. Histochem. Cytochem.* **32,** 1217.

Bendayan, M., and Zollinger, M. (1983). Ultrastructural localization of antigenic sites on osmium-fixed tissues applying the protein A–gold technique. *J. Histochem. Cytochem.* **31,** 101

Carlemalm, E., Garavito, R. M., and Villiger, W. (1982). Resin development for electron microscopy and an analysis of embedding at low temperature. *J. Microsc. (Oxford)* **126,** 132.

Carlemalm, E., Villiger, W., Hobot, J. A., Acetarin, J. D., and Kellenberger, E. (1982). Low temperature embedding with Lowicryl resins: Two new formulations and some applications. *J. Microsc. (Oxford)* **140,** 55.

Craig, S., and Goodchild, D. J. (1982). Post-embedding immunolabeling. Some effects of tissue preparation on the antigenicity of plant proteins. *Eur. J. Cell Biol.* **28,** 251.

Craig, S., and Miller, C. (1984). LR White resin and improved on-grid immunogold detection of vicilin, a pea seed storage protein. *Cell Biol. Int. Rep.* **8,** 879.

Geuze, H., Slot, J. W., Strous, G. J. A. M., Lodish, H. F., and Schwartz, A. L. (1983). Intracellular site of asialoglycoprotein receptor-ligand uncoupling: Double-label immunoelectron microscopy during receptor-mediated endocytosis. *Cell (Cambridge, Mass.)* **32,** 277.

Greenwood, J. S., and Chrispeels, M. J. (1985). Correct targeting of the bean storage protein phaseolin in the seeds of transformed tobacco. *Plant Physiol.* **79,** 65.

Harris, N., and Croy, R. R. D. (1985). The major albumin protein from pea (*Pisum sativum* L.). Localisation by immunocytochemistry. *Planta* **165,** 522.

Hayat, M. A. (1981). *Fixation for Electron Microscopy.* Academic Press, New York.

Herman, E. M., and Shannon, L. M. (1984). Immunocytochemical evidence for the involvement of Golgi apparatus in the deposition of seed lectin of *Bauhinia purpurea* (Leguminosae). *Protoplasma* **121,** 163.

Herman, E. M. (1987). The immunogold localization and synthesis of an oil body membrane protein in developing soybean seeds. *Planta* **172,** 336.

Herman, E. M. (1988). Immunocytochemical localization of macromolecules with the electron microscope. *Annu. Rev. Plant Physiol. Plant Mol. Biol.* **39,** 139.

Herman, E. M., Hankins, C. N., and Shannon, L. M. (1988). The leaf and bark lectins of *Sophora japonica* are sequestered in protein storage vacuoles. *Plant Physiol.* **86,** 1027.

Herman, E. M., Chrispeels, M. J., Hoffman, L. M. (1989). Vacuole accumulation of storage protein and lectin expressed in transgenic tobacco seeds. *Cell Biol. Int. Reports* **13,** 37.

Hoffman, L. M., Donaldson, D. D., Bookland, R., Rashka, K., and Herman, E. M. (1987). Synthesis and protein body deposition of maize 15-kd zein in transgenic tobacco seeds. *EMBO J.* **6,** 3213.

Johnson, D. A., Gautsch, J. W., Sportsman, J. R., and Elder, J. H. (1984). Improved technique utilizing nonfat dry milk for analysis of proteins and nucleic acids transferred to nitrocellulose. *Gene Anal. Technol.* **1,** 3.

McCurdy, D. W., and Pratt, L. H. (1986). Immunogold electron microscopy of phytochrome in *Avena:* Identification of intracellular sites responsible for phytochrome sequestering and enhanced pelletability. *J. Cell Biol.* **103,** 2541.

Newman, G. R., Jasani, B., and Williams, E. D. (1983). A simple post-embedding system for rapid demonstration of tissue antigens under the electron microscope. *Histochem. J.* **15,** 543.

Newman, G. R., Jasani, B., and Williams, E. D. (1983). A simple post-embedding system for rapid demonstration of tissue antigens under the electron microscope. *Histochem. J.* **15,** 543.

Newman, G. R., and Hobot, J. A. (1987). Modern acrylics for post-embedding immunostaining techniques. *J. Histochem. Cytochem.* **35,** 971.

Robertson, J. G., Wells, B., Bissseling, T., Farnden, K. J. F., and Johnston, A. W. B. (1984). Immuno-gold localization of leghaemoglobin in nitrogen-fixing root nodules of pea. *Nature (London)* **311,** 254.

Robertson, J. G., Wells, B., Brewin, N. J., and Williams, M. A. (1985). Immuno-gold localization of cellular constituents in the legume—*Rhizobium* symbiosis. *Oxford Surv. Mol. Biol.* **2,** 69.

Roth, J., Bendayan, M., Carlemalm, E., Villiger, W., and Garavito, M. (1981). Enhancement of structural preservation and immunocytochemical staining in low temperature embedded pancreatic tissue. *J. Histochem. Cytochem.* **29,** 663.

Shaw, P. J., and Harwood, J. A. (1985). Immuno-gold localization of cytochrome f, light-harvesting complex, ATP synthase, and ribulose 1,5-bisphosphate carboxylase/oxygenase. *Planta* **165,** 333

Strum, A., Volker, T., Herman, E. M., and Chrispeels, M. J. (1988). Correct targeting and glycosylation of the bean vascular protein phytohemagglutinin in transgenic tobacco. *Planta* **175,** 170.

Titus, D. E., and Becker, W. M. (1983). Investigation of the glyoxysome-peroxisome transition in germinating cucumber cotyledons using double-label immunoelectron microscopy. *J. Cell. Biol.* **101,** 1288.

Tomenius, K., Claphan, D., and Oxelfelt, P. (1983). Localization by immunogold cytochemistry of viral antigen in sections of plant cells infected with red clover mottle virus. *J. Gen. Virol.* **64,** 2669.

VandenBosch, K. A. (1986). Light and electron microscopic visualization of uricase by immunogold labeling of sections of resin-embedded soybean nodules. *J. Microsc. (Oxford)* **143,** 187.

VandenBosch, K. A., and Newcomb, E. H. (1986). Immunogold localization of nobule-specific uricase in developing soybean root nodules. *Planta* **167,** 425.

Verma, D. P. S., Fortin, M. G., Stanley, J., Mauro, V. P., Purohit, S., and Morison, N. (1986). Nodulins and nodulin genes of *Glycine max. Plant Mol. Biol.* **7,** 51.

16

Preembedding Immunogold Staining of Cell Surface-Associated Antigens Performed on Suspended Cells and Tissue Sections

CORRADO FERRARI

Department of Histology and General Embryology
University of Parma
Parma, Italy

GIUSEPPE DE PANFILIS

and

GIAN CARLO MANARA

Department of Dermatology
University of Parma
Parma, Italy

INTRODUCTION
IMMUNOGOLD LABELING OF SUSPENDED CELLS
  Blood Cell Preparation
  Epidermal Cell Suspension Preparation
  Prefixation
  Immunogold Labeling of a Single Antigen
  Immunogold Labeling of Two Distinct Antigens on the Same Cell Surface
  Immunogold Labeling of Three Distinct Antigens on the Same Cell Surface
  Immunogold–Silver Staining
IMMUNOGOLD LABELING OF TISSUE SECTIONS
  Rat Kidney Processing
    Localization of MoAb8 in Rat Kidney
  Human Skin Processing
    Cell Surface Labeling
CONCLUDING REMARKS
REFERENCES

INTRODUCTION

In recent years, numerous studies have been performed in order to visualize cytoplasmic, extracellular, or cell surface antigens by means of immunogold approaches in the electron microscope. To detect cytoplasmic and extracellular antigens, postembedding techniques have been favored. This kind of immunogold labeling has been performed both on thin frozen sections (Slot and Geuze, 1981) and on thin sections of resin-embedded tissues (Roth et al., 1978; Probert et al., 1981; Bendayan, 1984a,b; Usellini et al., 1984; Lemanski et al., 1985; Warhol et al., 1985a). On the other hand, to detect cell surface-associated antigens, preembedding immunogold methods have been employed on suspended cells (Horisberger and Rosset, 1977; Romano and Romano, 1977; Schmitt et al., 1984; Romani et al., 1985; Manara et al., 1986a; De Panfilis et al., 1988a; Horisberger, 1989). Moreover, the possibility of varying the size of the colloi-dal gold particles renders this marker particularly suitable for detecting two (Bendayan, 1982; Roth, 1982; Dezutter-Dambuyant et al., 1985; De Panfilis et al., 1988b) or more (Doerr-Schott and Lichte, 1986; De Panfilis et al., 1988c; Doerr-Schott, 1989) distinct antigens simultaneously using multiple-labeling techniques. In this chapter we describe preembedding immunogold approaches for detecting cell surface-associated antigens both on suspended cell samples and on tissue sections.

IMMUNOGOLD LABELING OF SUSPENDED CELLS

Labeling techniques for surface antigens performed on both peripheral blood lymphocytes and epidermal cell suspensions are described below.

Blood Cell Preparation

Human peripheral heparinized blood was obtained from randomly selected normal donors. Mononuclear cells were isolated by Ficoll–Hypaque gradient centrifugation (Böyum, 1968) and depleted of adherent cells by incubation on plastic petri dishes for 2 hr at 37°C. Cells were then resuspended in RPMI 1640 containing 50 µg/ml gentamycin, at a concentration of 5×10^6 cells/ml. (See Manara, 1988.)

Epidermal Cell Suspension Preparation

Fresh human skin slices were obtained during abdominal surgery by means of a keratotome set at 0.4 mm. They were incubated for 1 hr at 37°C in 10 mM phosphate buffered-saline (PBS) solution (pH 7.4) containing 0.25% trypsin. The epidermis was then separated from the dermis with fine forceps. The epidermal sheets were placed in RPMI 1640 medium supplemented with 10% fetal calf serum, which avoids further trypsinization, and then pipetted and shaken vigorously for a couple of minutes. After filtration through sterile gauze, the cells were washed in RPMI 1640 and enriched for Langerhans cells by Ficoll–Hypaque. Cells were finally washed in RPMI 1640 and resuspended in the same medium, containing 50 µg/ml gentamycin, at a concentration of 5×10^6 cells/ml. Epidermal cell viability generally ranged between 80 and 90%, as determined by trypan blue exclusion. (See De Panfilis *et al.*, 1988b.)

Prefixation

Cells were prefixed in 0.1% glutaraldehyde in 0.1 M cacodylate buffer (pH 7.4) or in 3% formaldehyde in PBS (pH 7.4); the choice of the prefixative solution depends on the antigen sensitivity to the fixative. The prefixation procedure is generally performed at 4°C for 15 min. After prefixation, cells were washed three times in the same buffer used in the prefixative solution, containing 50 mM glycine to quench aldehyde groups.

Immunogold Labeling of a Single Antigen

Cells were incubated for 30 min at 4°C in PBS containing 1% bovine serum albumin (BSA), 0.2% sodium azide, and 10% decomplemented human AB serum to saturate Fc receptors. Cells were washed twice in PBS–0.1% BSA and incubated with the specific monoclonal antibody (MoAb) diluted in PBS–1% BSA for 30 min at 4°C. Subsequently, cells were washed in PBS–0.1% BSA and then in 0.02 M Tris-HCl buffer (pH 8.2) containing 0.02 M sodium azide and 0.1% BSA (0.1% BSA buffer). Cells were incubated for 1 hr at 4°C in a goat antimouse antibody coupled

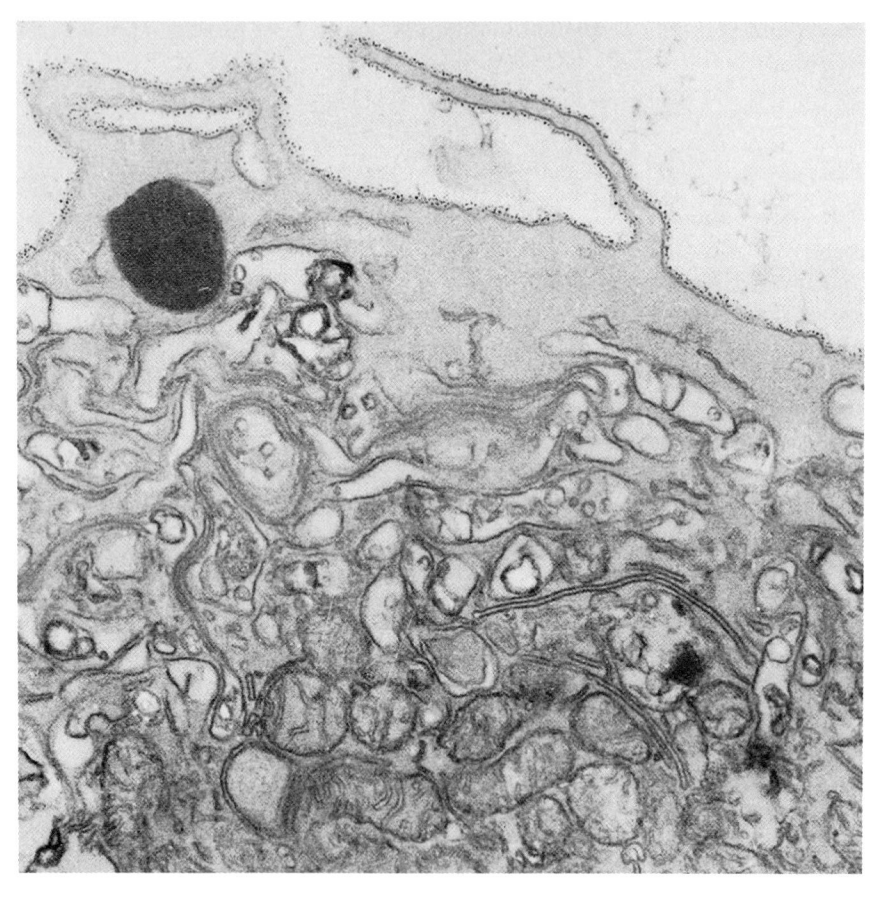

Fig. 16.1. Immunogold labeling of a single antigen. Epidermal cell suspension. HLA-DR-positive dendritic cell. The cell membrane is decorated by 5-nm colloidal gold particles. Section was not poststained; ×42,000.

16. Immunogold Staining of Cell Surface-Associated Antigens 327

to colloidal gold particles (commercially available 5- to 40-nm particles) diluted 1 : 10 in 0.02 M Tris-HCl buffer (pH 8.2) containing 0.02 M sodium azide and 1% BSA (1% BSA buffer) and finally washed three times in 0.1% BSA buffer. The specificity of the gold labeling was checked by (1) omitting incubation with the MoAb, (2) substituting the MoAb with mouse purified immunoglobulins or with a nonspecific MoAb of the same isotype, and (3) using uncoupled gold particles of the same size instead of the gold-coupled immunoreagent. (See Figs. 16.1 and 16.2; Manara *et al.*, 1986a; De Panfilis *et al.*, 1988a.)

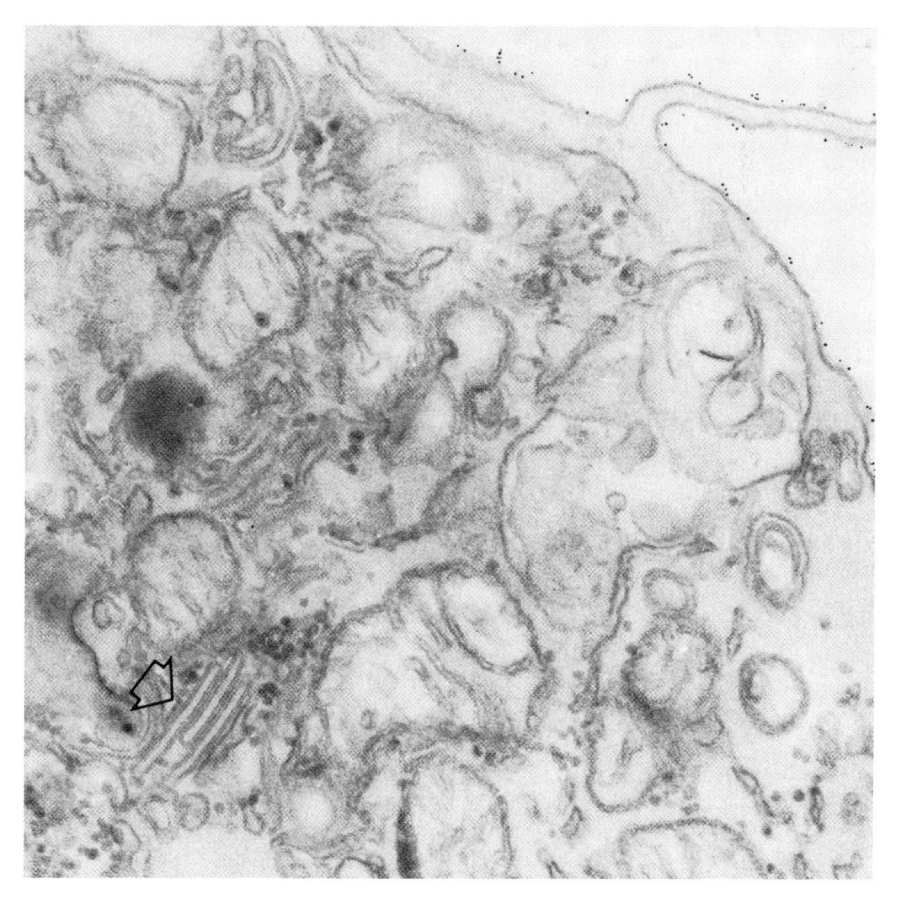

Fig. 16.2. Immunogold labeling of a single antigen. Epidermal cell suspension. A CD11c-positive Langerhans cell showing 5-nm colloidal gold particles bound to its surface. Arrow indicates Birbeck granules. Section was not poststained; ×45,000.

Immunogold Labeling of Two Distinct Antigens on the Same Cell Surface

Cells were incubated for 30 min at 4°C in PBS containing 10% decomplemented human AB serum, 1% BSA, and 0.2% sodium azide, and then washed three times in PBS–0.1% BSA. Cells were then incubated with the first MoAb diluted in PBS–1% BSA for 30 min at 4°C and were washed in PBS–0.1% BSA and then in 0.1% BSA buffer. The first MoAb was detected by incubating the cells for 1 hr at 4°C, with a goat antimouse antibody coupled to colloidal gold particles 20 nm in diameter diluted 1 : 10 in 1% BSA buffer. This incubation was followed by washes in 0.1%

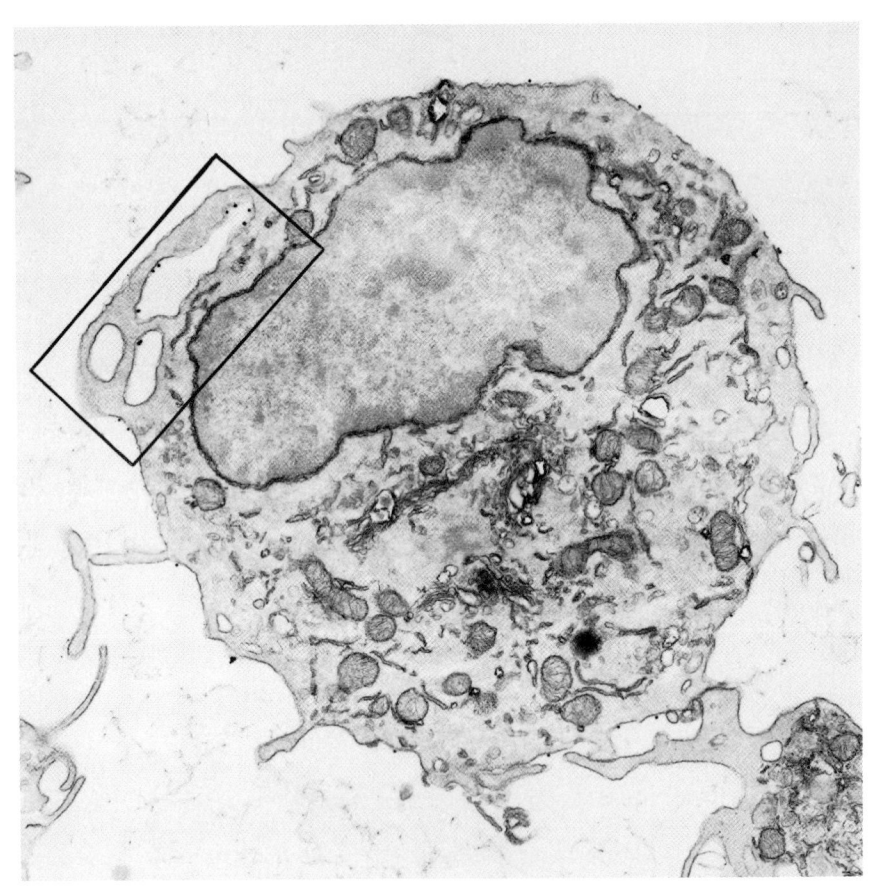

Fig. 16.3. Immunogold labeling of two distinct antigens. Peripheral blood lymphocytes. A lymphocyte displaying colloidal gold particles on its surface. At this magnification only 20-nm gold particles are visible. Section was not poststained; ×18,000.

BSA buffer and then in PBS–0.1% BSA. To saturate the still available antigen binding sites of the goat antimouse antibody, cells were incubated in mouse purified immunoglobulins diluted in PBS–1% BSA for 30 min at 4°C. Cells were washed three times in PBS–0.1% BSA, incubated with the second gold-conjugated (5-nm) MoAb diluted in PBS–1% BSA for 1 hr at 4°C, and then washed three times in PBS–0.1% BSA. The labeling specificity can be checked by (1) avoiding incubation with the first MoAb, (2) substituting the first MoAb with mouse purified immunoglobulins or with a nonspecific MoAb of the same isotype, (3) preincubating the cells with unlabeled second MoAb before incubating with the gold-coupled MoAb, and (4) using uncoupled gold particles of the same size instead of the gold-coupled immunoreagents. (See Figs. 16.3 and 16.4; De Panfilis et al., 1988b.)

Immunogold Labeling of Three Distinct Antigens on the Same Cell Surface

Cells were incubated for 30 min at 4°C in PBS containing 10% decomplemented human AB serum, 1% BSA, and 0.2% sodium azide and then washed three times in PBS–0.1% BSA. The cells were incubated with the first MoAb diluted in PBS–1% BSA for 30 min at 4°C and were washed in PBS–0.1% BSA and then in 0.1% BSA buffer. They were incubated with a goat antimouse antibody coupled to colloidal gold particles 40 nm in diameter, diluted 1:10 in 1% BSA buffer for 1 hr at 4°C, and washed in 0.1% BSA buffer and then in PBS–0.1% BSA. Subsequently, cells were incubated in mouse purified immunoglobulins diluted in PBS–1% BSA for 30 min at 4°C, washed three times in PBS–0.1% BSA, and incubated with the second biotinated MoAb diluted in PBS–1% BSA for 30 min at 4°C. They were washed in PBS–0.1% BSA and then in 0.1% BSA buffer. The biotinated MoAb was detected by incubation in streptavidin–colloidal gold particles (20 nm in size) diluted 1:10 in PBS–1% BSA for 1 hr at 4°C. After three washes in PBS–0.1% BSA, the third antigen was directly detected by incubating the cells with the third MoAb conjugated with colloidal gold particles (5 nm) diluted 1:10 in PBS–1% BSA for 1 hr at 4°C. Finally, cells were washed three times in PBS–0.1% BSA. The specificity of the labeling can be checked by (1) substituting the three MoAbs with buffer solutions, (2) substituting the three MoAbs with mouse purified immunoglobulins or with nonspecific MoAbs of the same isotype, and (3) using uncoupled gold particles of the same size instead of the gold-coupled immunoreagents.

After labeling, cells were fixed in 1% glutaraldehyde in 0.1 M cacodylate buffer (pH 7.4) for 3 hr at room temperature, washed in 0.15 M caco-

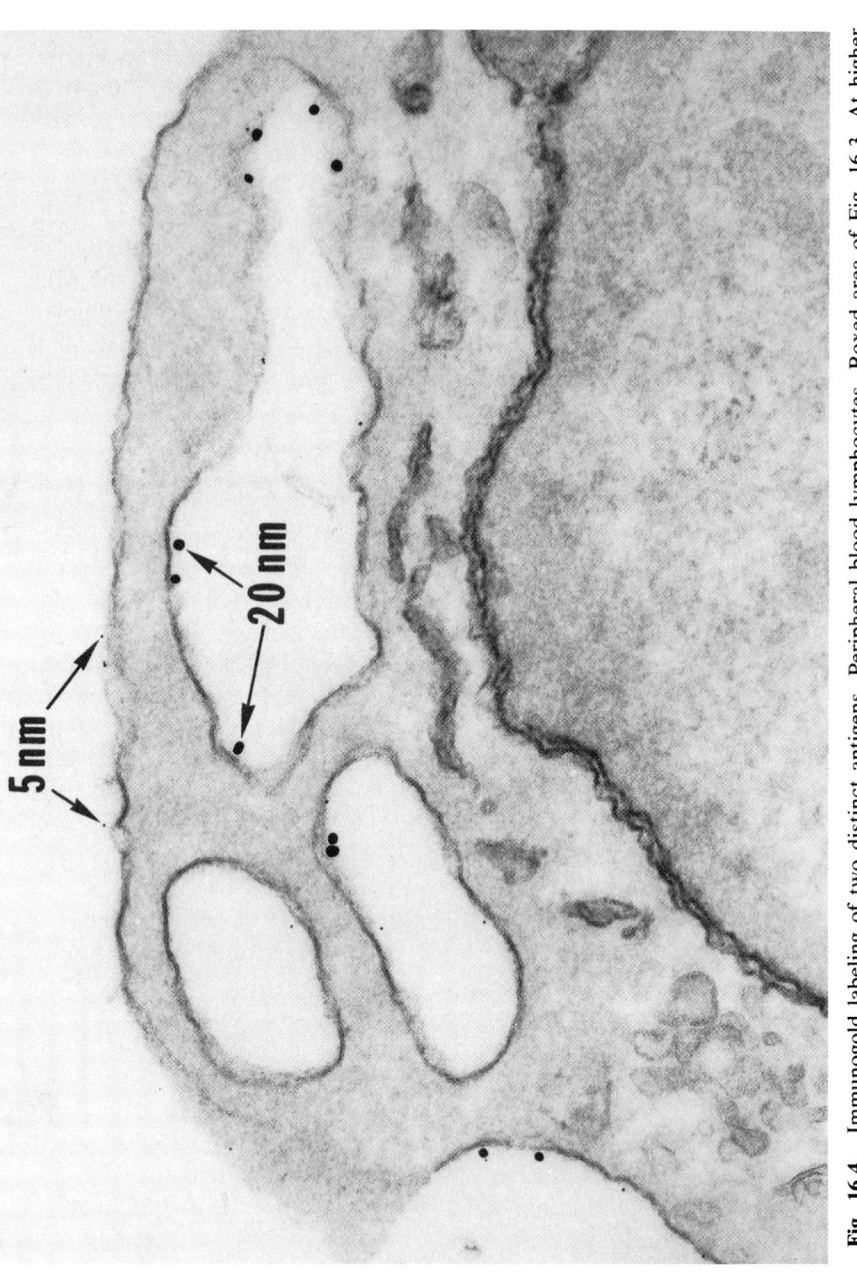

Fig. 16.4. Immunogold labeling of two distinct antigens. Peripheral blood lymphocytes. Boxed area of Fig. 16.3. At higher magnification both 20-nm (HNK-1 antigen) and 5-nm (CD8 antigen) colloidal gold particles are detectable; ×51,000.

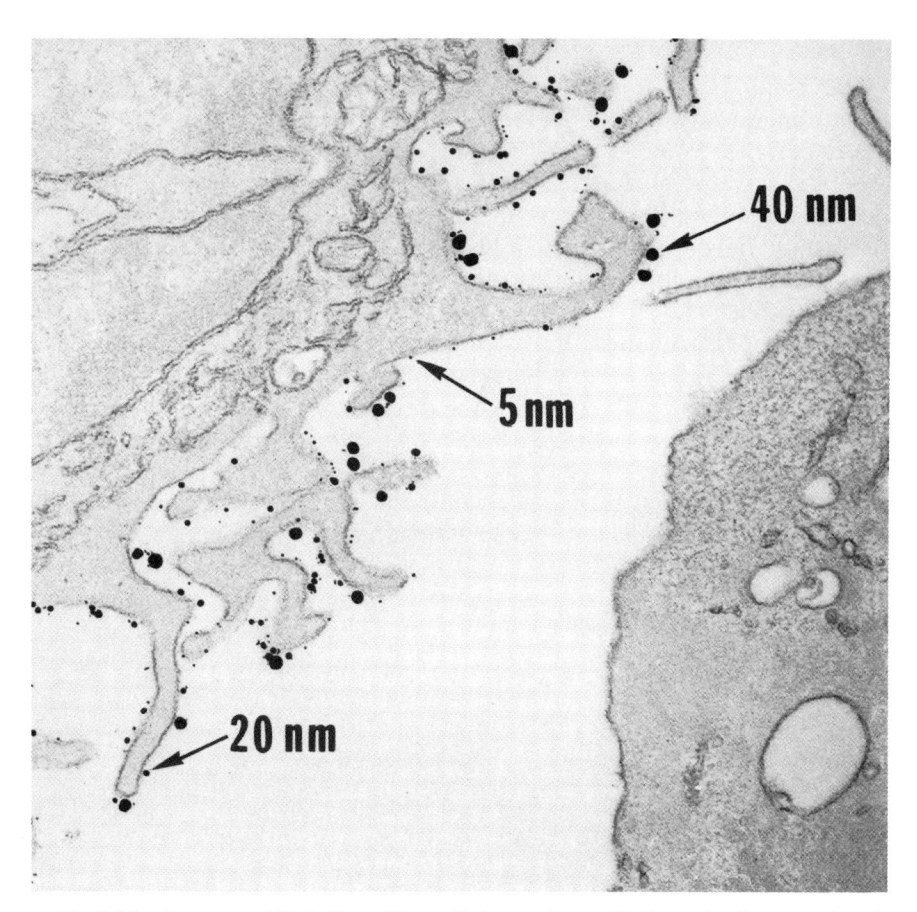

16. Immunogold Staining of Cell Surface-Associated Antigens

Fig. 16.5. Immunogold labeling of three distinct antigens. Epidermal cell suspension. A dendritic epidermal cell and a keratinocyte are visible. On the dendritic cell membrane 40-nm (CD1a antigen), 20-nm (HLA-DR antigen), and 5-nm (CD4 molecule) gold particles are detectable. No gold particles are present on the keratinocyte membrane. Section was not poststained; ×45,000.

dylate buffer (pH 7.4), and packaged in 2% Bacto-agar at 45°C. Samples were postfixed in potassium ferricyanide-reduced osmium tetroxide (Karnovsky, 1971), dehydrated in graded acetones, and embedded in Durcupan ACM. Thin sections were examined without poststaining or counterstained with uranyl acetate and lead citrate. (See Fig. 16.5; De Panfilis *et al.*, 1988c.)

Immunogold–Silver Staining

The immunogold–silver staining (IGSS) method has been largely used in light microscopy immunocytochemistry (Holgate *et al.*, 1983; Hacker *et al.*, 1985; De Waele *et al.*, 1986; Morhenn *et al.*, 1987; Hacker, 1989). Here we describe an IGSS method for detecting plasmalemmal-associated antigens of suspended cells in the transmission electron microscope.

After the immunogold labeling performed by using 5-nm colloidal gold particles and glutaraldehyde fixation, cells were thoroughly washed in distilled water. The silver enhancement reagent was prepared by mixing equal volumes of enhancer and initiator solutions of the "IntenSE II Kit" (Janssen). When the silver enhancement reagent is added to the cells, metallic silver precipitates on heavy metals and, as a result, enlarges small colloidal gold particles. The reaction is time dependent; we suggest that 6 min is the time required to reach a suitable enhancement. Within this time we obtain a reaction specific for colloidal gold particles. A longer enhancement time could give rise to a disturbing background due to silver precipitates formed spontaneously by self-nucleation. The reaction is also temperature dependent. We suggest performing the silver enhancement procedure at room temperature (22°C). A higher temperature would require a shorter enhancement time. Finally, the cells would be free of heavy metal contamination. The silver enhancement procedure was followed by thorough washes in distilled water. (See Fig. 16.6; Manara *et al.*, 1988b.)

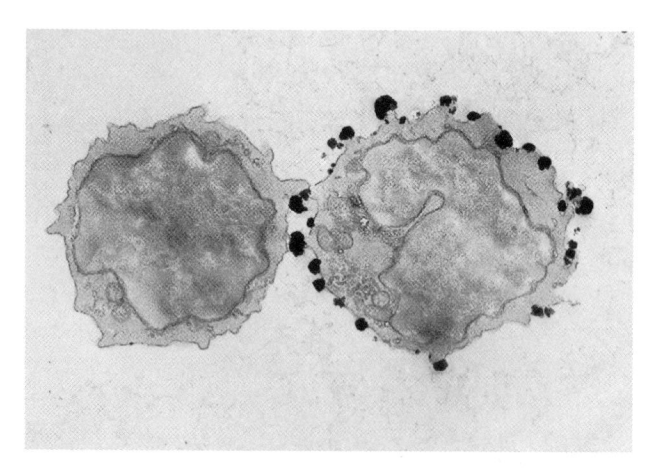

Fig. 16.6. Immunogold–silver staining of a cell surface antigen. Peripheral blood lymphocytes. Two facing lymphocytes, one of which shows on its surface large electron-dense deposits due to metallic silver precipitation on 5-nm colloidal gold particles. The CD8 molecule has been labeled. Section was not poststained; ×8100.

IMMUNOGOLD LABELING OF TISSUE SECTIONS

We describe here the immunogold labeling of cell surface-associated antigens in tissue sections of rat kidney and human skin fixed by vascular perfusion and immersion, respectively.

Rat Kidney Processing

Sprague-Dawley rats were intravenously injected with murine ascites containing IgG MoAb and killed 24 to 72 hr later. This antibody was isolated and characterized by Drs. P. Ronco and P. Verroust, Inserm 64, Paris, and referred as MoAb8. It reacts with a protein present in the brush border of the renal proximal tubule (Ronco *et al.*, 1984a,b). The aorta distal to the left renal artery was cannulated, the aorta was ligated between the two kidney arteries, and the renal vein was cut. The perfusion pressure was kept at 120 mm Hg; as a result the flow rate was 10 ml/min. Kidneys were flushed with Dulbecco's minimal essential medium for 5 min at 20°C, then perfusion fixed with 50 ml of a mixture of 4% formaldehyde and 0.1% glutaraldehyde in 0.1 M cacodylate buffer (pH 7.4) for 5 min at 20°C. After fixation, kidneys were perfused with PBS containing 50 mM glycine to quench residual aldehyde groups. Tissue slices were prepared, transferred to a cryoprotectant solution of 10% dimethyl sulfoxide (DMSO) in PBS for 1 hr at 22°C, snap-frozen in isopentane, and stored in liquid nitrogen until used. (See Ferrari *et al.*, 1988.)

Localization of MoAb8 in Rat Kidney

Cryostat sections (10–15 μm thick) were incubated in suspension with 5-nm colloidal gold particles–goat antimouse IgG antibody diluted in 10 volumes of 1% BSA buffer for 5 hr at 20°C, and then thoroughly washed in PBS–0.1% BSA. The specificity of the immunolabeling can be checked by (1) avoiding the intravenous injection of the MoAb8, (2) substituting the MoAb8 with an equivalent amount of a murine monoclonal immunoglobulin having the same immunochemical properties but no antibody activity, and (3) substituting the 5-nm colloidal gold-coupled immunoreagent with 5-nm uncoupled colloidal gold particles. (See Fig. 16.7; Ferrari *et al.*, 1988.)

Human Skin Processing

Human skin biopsies were taken from normal human skin of healthy volunteers or from skin lesions of patients suffering from inflammatory or neoplastic skin disorders. Skin slices were prepared, washed in Hanks'

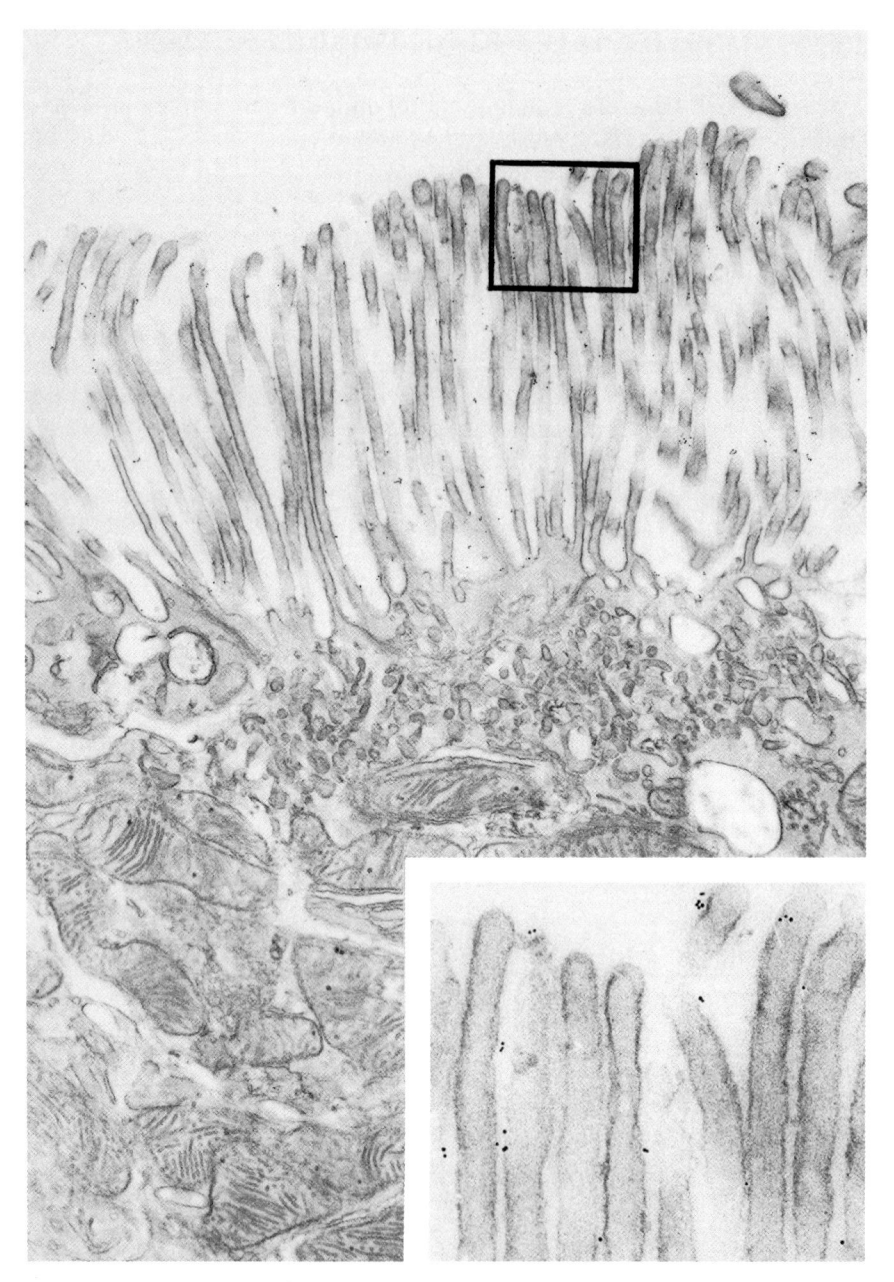

Fig. 16.7. Immunogold labeling of tissue sections (rat kidney). Numerous 5-nm colloidal gold particles are scattered along the microvilli of a proximal tubule brush border. Section was not poststained; ×23,400. Inset: A higher magnification of the checked area shows 5-nm gold granules linked to the microvilli surface; ×63,400.

balanced salt solution, prefixed by immersion in 4% formaldehyde in 0.1 M cacodylate buffer (pH 7.4) for 15 min at 20°C, washed in PBS (50 mM with respect to glycine), cryoprotected in PBS (10% with respect to DMSO), snap-frozen in isopentane, and stored in liquid nitrogen. (See De Panfilis *et al.*, 1986a,b; Ferrari *et al.*, 1988.)

Cell Surface Labeling

Free-floating cryostat sections (10–15 μm thick) were incubated in the specific MoAb diluted in PBS–1% BSA for 12 hr at 4°C. After washing for 24 hr in PBS–0.1% BSA, sections were further incubated in 5-nm colloidal gold particles–goat antimouse IgG antibody diluted in 10 volumes of 1% BSA buffer (pH 8.2) for 12 hr at 4°C, and finally washed in 0.1% BSA buffer (pH 8.2) for 24 hr. The specificity of the labeling was checked in control experiments by (1) omitting the incubation with the MoAb, (2) substituting the MoAb with mouse purified immunoglobulins of the same isotype, and (3) using 5-nm uncoupled colloidal gold particles instead of the 5-nm coupled immunoreagent.

After the immunogold labeling, tissue sections were extensively washed in PBS–0.1% BSA, fixed in 1% glutaraldehyde in 0.1 M cacodylate buffer (pH 7.4) for 3 hr at room temperature, postfixed in osmium tetroxide reduced with potassium ferricyanide (Karnovsky, 1971), dehydrated in a graded acetone, and embedded in Durcupan ACM. Silver sections were examined without poststaining. (See Figs. 16.8 and 16.9; De Panfilis *et al.*, 1986a,b; Ferrari *et al.*, 1988.)

CONCLUDING REMARKS

Each of the currently available immunoelectron microscopy (IEM) procedures has several advantages as well as certain drawbacks which must be considered before a method is chosen. The first problem in IEM is to adjust the delicate balance between adequate preservation of ultrastructural details and retention of antigenicity. In our experience (i.e., preembedding methods), when it is possible, a prefixation of the samples should be done with 0.1% glutaraldehyde or 3–4% formaldehyde (freshly prepared from paraformaldehyde); the selection of the prefixative solution is based on the antigen characteristics. Prefixation with glutaraldehyde results in good preservation of ultrastructure, but it sometimes destroys the antigenicity of proteins. On the other hand, formaldehyde preserves antigenicity much better, while stabilizing cells less satisfactorily.

The most commonly used IEM technique has been, for a long period, the immunoperoxidase technique (De Panfilis *et al.*, 1983; Manara *et al.*,

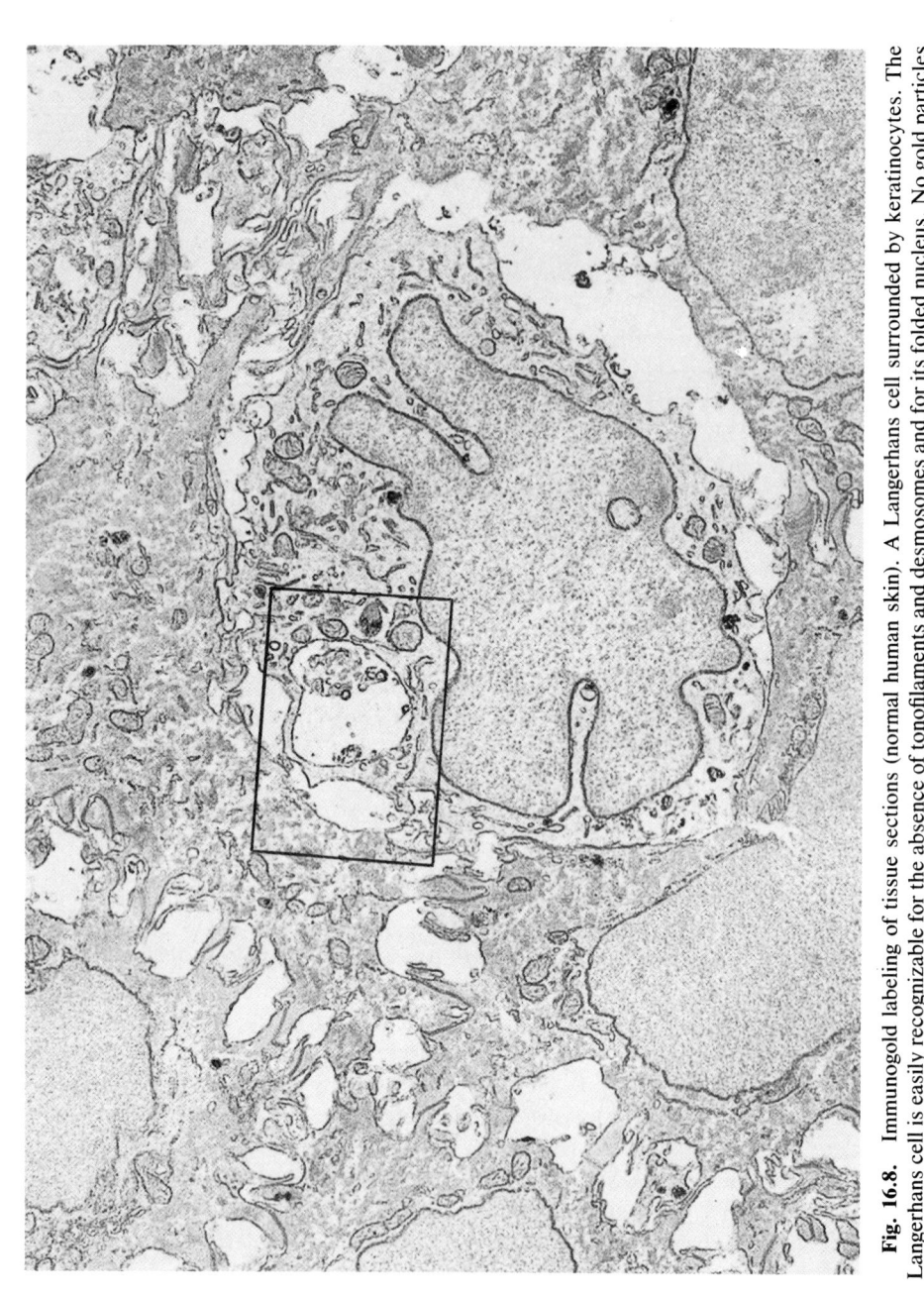

Fig. 16.8. Immunogold labeling of tissue sections (normal human skin). A Langerhans cell surrounded by keratinocytes. The Langerhans cell is easily recognizable for the absence of tonofilaments and desmosomes and for its folded nucleus. No gold particles are detectable at this magnification. Section was not poststained; ×12,000.

16. Immunogold Staining of Cell Surface-Associated Antigens 337

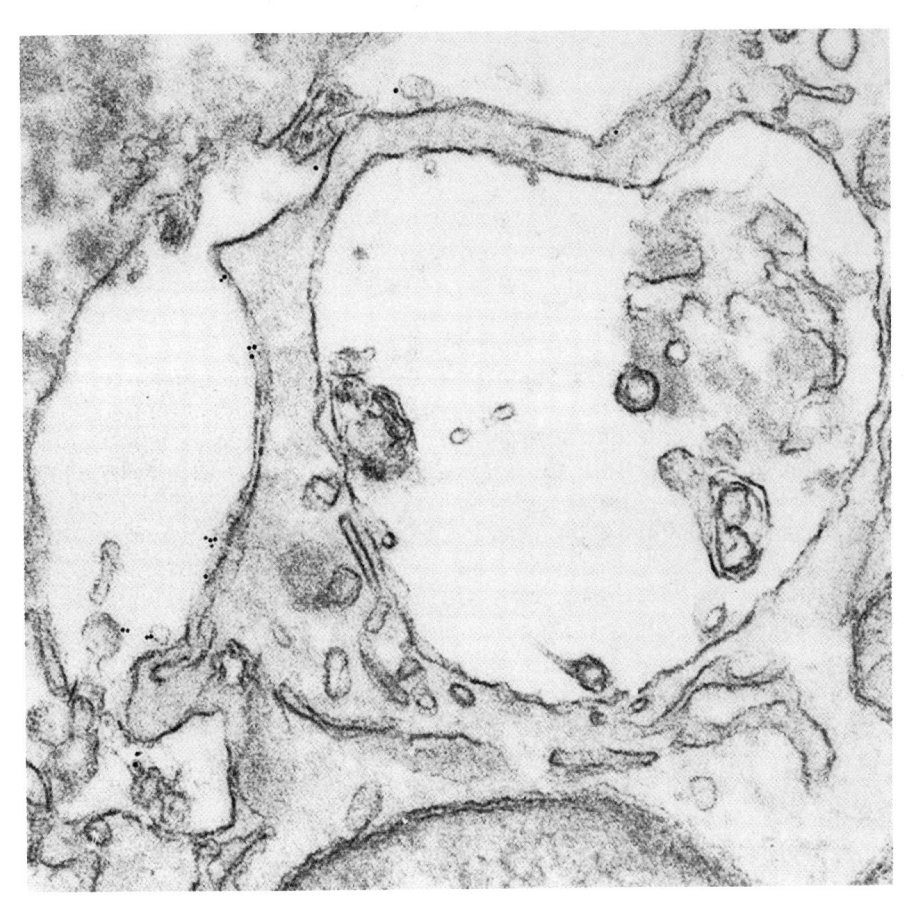

Fig. 16.9. Immunogold labeling of tissue sections (normal human skin). A higher magnification of the boxed area of Fig. 16.8, allows the detection of 5-nm colloidal gold particles bound to the Langerhans cell membrane (CD11b-positive); ×58,500.

1983, 1984a; De Vos *et al.*, 1985; Sansoni *et al.*, 1987). It is relatively simple to perform, does not require any special equipment, and possesses a high sensitivity. On the other hand, since a histochemical enzyme reaction is used for the visualization of the antigen, artifacts can rise, due to diffusion of the peroxidase–diaminobenzidine (DAB) reaction products from their sites of generation and reabsorption on adjacent cell structures (Novikoff, 1980; Courtoy *et al.*, 1983). Colloidal gold particles were introduced as electron-dense markers in IEM by Faulk and Taylor (1971), since they possess a characteristic shape and a well-defined size and are not subject to diffusion and reabsorption artifacts. Preembedding immunogold staining of suspended cells is detectable by simple techniques not requiring any special equipment and possessing a high sensitivity as well as a high specificity. Colloidal gold particles are in diameters ranging from 5 to 40 nm.

Cells labeled with 40-nm gold particles are easily detectable in the transmission electron microscope (TEM) (Manara et al., 1986a). On the other hand, because of steric hindrance, the accessibility of the gold-conjugated antibody to cell surface-associated antigenic sites depends on the size of the gold particles (Horisberger and Tacchini-Vonlanthen, 1983). As a result, the use of 40-nm gold particles largely minimizes the real number of antigenic sites. Therefore, when semiquantitative analyses of antigenic sites are to be performed, 5-nm gold particles should be favored (Manara et al., 1988a). Nevertheless, in order to detect 5-nm colloidal gold particle-labeled cells, examination of the specimens at high magnification is required. This represents a limitation, especially when a low percentage of cells in the sample are labeled. Therefore, the procedure could take advantage of the silver enhancement. In particular, after the 5-nm gold labeling and fixation, a useful procedure consists in subjecting one half of the cell suspension to a silver enhancement procedure. The large size of the electron-dense marker obtained by the IGSS allows an easy and quick qualitative examination of the samples, even at low magnification (Manara et al., 1988b). When the percentage of labeled cells has been established, one still has the option to perform semiquantitative estimation of the antigen by counting the number of gold particles on cells from the other half of the sample (not subjected to the silver enhancement procedure). When semiquantitative analyses are not required, 15-nm (De Panfilis et al., 1988a) to 20-nm (Dezutter-Dambuyant et al., 1984) colloidal gold particles are usually employed. Gold particles 40 nm in diameter should be chosen when a peroxidase–colloidal gold double labeling is performed (Manara et al., 1984b, 1985, 1986b, 1987; De Panfilis et al., 1988d). In fact, colloidal gold particles of smaller size could be hidden by the peroxidase–DAB reaction products.

The availability of colloidal gold particles in different well-defined sizes allows the simultaneous visualization of two (Bendayan, 1982; Roth, 1982; Dezutter-Dambuyant et al., 1985; De Panfilis et al., 1988b) or more (Doerr-Schott and Lichte, 1986; De Panfilis et al., 1988c; Doerr-Schott, 1989) different surface antigens on the same cell. The multiple-labeling protocols we present here avoid interferences among the colloidal gold particles of different diameters by using unreactive mouse immunoglobulins which bind to the still available antigen binding site of the secondary gold-conjugated antibody, linked to the first MoAb. Thus, the second and, when employed, the third MoAbs cannot be trapped unspecifically. Moreover, the gold-coupled streptavidin binds only to the biotinated

16. Immunogold Staining of Cell Surface-Associated Antigens

MoAb. On the other hand, such multiple-labeling approaches are not suited to establish a relationship between the quantity of surface antigens and the number of gold particles. In fact, the intensity of the gold labeling is dependent on the size of the gold particles (Horisberger and Tacchini-Vonlanthen, 1983).

Preembedding immunogold labeling in tissue sections is ideally suited for detecting cell surface-associated antigens (De Panfilis et al., 1986a,b; Ferrari et al., 1988). Most papers dealing with immunogold staining of tissue antigens *in situ* at the TEM level refer to postembedding techniques employed in the detection of cytoplasmic and extracellular antigens in hydrophobic resin-embedded samples. Recently, new hydrophilic acrylic resins, such as Lowicryl and London resins have proved to be very useful for the detection of several antigens. The embedding process leads to less extensive antigen denaturation when hydrophilic acrylic resins are used instead of hydrophobic resins. A limitation of postembedding techniques, dealing with the detection of cell surface antigens, concerns their relatively low sensitivity, because only antigenic sites exposed on the surface of the section are available for the immunogold labeling. In fact, some antigens, particularly membrane proteins present in relatively small amounts, are not detectable even in hydrophilic acrylic resin-embedded specimens, while they are demonstrable by preembedding approaches, in which large amounts of antigens are available for antibody binding (Kerjaschki et al., 1986; Bendayan et al., 1987). Nevertheless, immunogold labeling of cell surface-associated antigens on enterocytes (Roth et al., 1985), keratinocytes (Warhol et al., 1985b), and leukocytes and histiocytes (Warhol and Longtine, 1986) has been performed on thin sections of low-temperature Lowicryl K4M-embedded specimens. On the other hand, a limitation of the postembedding staining of hydrophilic resin-embedded specimens is the lack of osmium tetroxide postfixation. As a result, membranes appear indistinct, giving rise to a loss of ultrastructural details. The preembedding immunogold staining in tissue sections described in the present chapter allows postfixation with osmium tetroxide and therefore permits good ultrastructural resolution of membranes.

The present preembedding immunogold method is best used for the detection of cell surface antigens, since gold conjugates are usually too large to penetrate within intracellular compartments through intact plasmalemma. Moreover, in order to achieve penetration of gold conjugates within 10–15-μm-thick tissue sections, 5-nm colloidal gold particles need to be used as electron-dense markers and, as a consequence, specimens must be examined at high magnification. Finally, in our experience, both ultrastructural details and antigenic preservation are assured by performing a mild fixation with 4% formaldehyde (added, when possible, with

0.1% glutaraldehyde) and by incubating the specimens in the presence of cryoprotectants such as 10% DMSO before sectioning on a cryostat at −30°C. In conclusion, we maintain that the immunogold preembedding and postembedding approaches complement each other to a certain extent. In particular, the preembedding method described by us is ideally suited for the detection of cell surface-associated antigens.

This work was supported in part by Grant 87.00096.04 from CNR, Rome. We thank Landino Allegri for helpful discussion and Claudio Torresani for valuable collaboration. Finally, the excellent technical assistance of Giorgio Medici is gratefully acknowledged.

REFERENCES

Bendayan, M. (1982). Double immunocytochemical labeling applying the protein A–gold technique. *J. Histochem. Cytochem.* **30**, 81.

Bendayan, M. (1984a). Protein A–gold electron microscopic immunocytochemistry: Methods, applications, and limitations. *J. Electron. Microsc. Tech.* **1**, 243.

Bendayan, M. (1984b). Enzyme–gold electron microscopic cytochemistry: A new affinity approach for the ultrastructural localization of macromolecules. *J. Electron. Microsc. Tech.* **1**, 349.

Bendayan, M., Nanci, A., and Kan, F. W. K. (1987). Effect of tissue processing on colloidal gold cytochemistry. *J. Histochem. Cytochem.* **35**, 983.

Böyum, A. (1968). Isolation of mononuclear cells and granulocytes from human blood. *Scand. J. Clin. Lab. Invest., Suppl.* **97**, 77.

Courtoy, P. J., Picton, D. H., and Farquhar, M. G. (1983). Resolution and limitations of the immunoperoxidase procedure in the localization of extracellular matrix antigens. *J. Histochem. Cytochem.* **31**, 945.

De Panfilis, G., Manara, G. C., Ferrari, C., Manfredi, G., and Allegra, F. (1983). Imbalance in phenotypic expression of T cell subpopulations during different evolutional stages of lichen planus lesions. *Acta Derm.-Venereol.* **63**, 369.

De Panfilis, G., Manara, G. C., and Ferrari, C. (1986a). Immunogold labelling of epidermal Langerhans cells on tissue sections of normal human skin. *Br. J. Dermatol.* **115**, 351.

De Panfilis, G., Ferrari, C., and Manara, G. C. (1986b). An *in situ* immunogold method applied to the identification of plasma membrane-associated antigens of skin-infiltrating cells. *J. Invest. Dermatol.* **87**, 510.

De Panfilis, G., Manara, G. C., Ferrari, C., Torresani, C., and Sansoni, P. (1988a). Hairy cell leukemia cells express CD1a antigen. *Cancer (Philadelphia)* **61**, 52.

De Panfilis, G., Manara, G. C., Ferrari, C., Torresani, C., Donelli, S., and Caleffi, E. (1988b). Reactivity of Langerhans cells with anti-CD11C monoclonal antibody. In *Second Workshop on Langerhans Cells* (J. Thivolet and D. Schmitt, eds.), pp. 149–157. John Libbey Eurotext Ltd.

De Panfilis, G., Manara, G. C., Ferrari, C., and Torresani, C. (1988c). Simultaneous colloidal gold immunoelectronmicroscopy labeling of CD1a, HLA-DR and CD4 surface antigens of human epidermal Langerhans cells. *J. Invest. Dermatol.* **91**, 547.

De Panfilis, G., Rowden, G., Manara, G. C., Ferrari C., Torresani, C., and Sansoni, P. (1988d). The S-100β protein in normal human peripheral blood is uniquely present within a discrete suppressor-T-cell compartment. *Cell. Immunol.* **114,** 398.

De Vos, R., De Wolf-Peeters, C., van den Oord, J. J., and Desmet, V. (1985). A recommended procedure for ultrastructural immunohistochemistry of small human tissue samples. *J. Histochem. Cytochem.* **33,** 959.

De Waele, M., De Mey, J., Renmans, W., Labeur, C., Jochmans, K., and van Camp, B. (1986). Potential of immunogold–silver staining for the study of leukocyte subpopulations as defined by monoclonal antibodies. *J. Histochem. Cytochem.* **34,** 1257.

Dezutter-Dambuyant, C., Cordier, G., Schmitt, D., Faure, M., Laquoi, C., and Thivolet, J. (1984). Quantitative evaluation of two distinct cell populations expressing HLA-DR antigens in normal human epidermis. *Br. J. Dermatol.* **111,** 1.

Dezutter-Dambuyant, C., Schmitt, D., Faure, M., Horisberger, M., and Thivolet, J. (1985). Immunogold technique applied to simultaneous identification of T6 and HLA-DR antigens on Langerhans cells by electron microscopy. *J. Invest. Dermatol.* **84,** 465.

Doerr-Schott, J., and Lichte, C. M. (1986). A triple ultrastructural immunogold staining method. Application to the simultaneous demonstration of three hypophyseal hormones. *J. Histochem. Cytochem.* **34,** 1101.

Doerr-Schott, J. (1989). Colloidal gold for multiple-labeling method. In *Colloidal Gold: Principles, Methods, and Applications, Vol. 1* (M. A. Hayat, ed.), pp. 145–190. Academic Press, San Diego, California.

Faulk, W. P., and Taylor, G. M. (1971). An immunocolloid method for the electron microscope. *Immunochemistry* **8,** 1081.

Ferrari, C., De Panfilis, G., Allegri, L., and Manara, G. C. (1988). Detection of cell surface antigens in tissue sections by means of pre-embedding immunogold staining. *Stain Technol.* **63,** 39.

Hacker, G., Springall, D., van Noorden, S., Bishop, A., Grimelius, L., and Polak, J. (1985). The immunogold–silver staining method—a powerful tool in histopathology. *Virchows Arch. A: Pathol. Anat.* **406,** 449.

Hacker, G. (1989). Silver-enhanced colloidal gold for light microscopy. In *Colloidal Gold: Principles, Methods, and Applications, Vol. 1* (M. A. Hayat, ed.), pp. 297–321. Academic Press, San Diego, California.

Holgate, C., Jackson, P., Cowen, P., and Bird, C. (1983). Immunogold–silver staining: New method of immunostaining with enhanced sensitivity. *J. Histochem. Cytochem.* **31,** 938.

Horisberger, M., and Rosset, J. (1977). Colloidal gold, a useful marker for transmission and scanning electron microscopy. *J. Histochem. Cytochem.* **25,** 295.

Horisberger, M., and Tacchini-Vonlanthen, M. (1983). Stability and steric hindrance of lectin-labeled gold markers in transmission and scanning electron microscopy. In *Lectins* (T. C. Bog-Hansen and G. A. Spengler, eds.), Vol. 3, p. 189. de Gruyler, Berlin.

Horisberger, M. (1989). Colloidal gold for scanning electron microscopy. In *Colloidal Gold: Principles, Methods, and Applications, Vol. 1* (M. A. Hayat, ed.), pp. 217–227. Academic Press, San Diego, California.

Karnovsky, M. J. (1971). Use of ferricyanide reduced osmium tetroxide in electron microscopy. *J. Cell Biol.* **51,** 284a.

Kerjaschki, D., Sawada, H., and Farquhar, M. G. (1986). Immunolectron microscopy in kidney research: Some contributions and limitations. *Kidney Int.* **30,** 229.

Lemanski, L. F., Paulson, D. J., Hill, C. S., Davis, L. A., Riles, L. C., and Lim, S. S. (1985). Immunoelectron microscopic localization of α-actinin on Lowicryl-embedded thin-sectioned tissues. *J. Histochem. Cytochem.* **33,** 315.

Manara, G. C., Scandroglio, R., De Panfilis, G., and Ferrari, C. (1983). Cell subpopulations in the paracortical area of the normal human lymph node defined by monoclonal antibodies. *Basic Appl. Histochem.* **27**, 267.

Manara, G. C., De Panfilis, G., Ferrari, C., Bonati, A., and Scandroglio, R. (1984a). The fine structure of HNK-1 (Leu-7) positive cells. A study using an immunoperoxidase technique. *Histochemistry* **81**, 153.

Manara, G. C., De Panfilis, G., Ferrari, C., and Scandroglio, R. (1984b). Immunoperoxidase–immunogold double labelling in immunoelectronmicroscopy of large granular lymphocytes. *J. Immunol. Methods* **75**, 189.

Manara, G. C., De Panfilis, G., and Ferrari, C. (1985). Ultrastructural characterization of human large granular lymphocyte subsets defined by the expression of HNK-1 (Leu-7), Leu-11, or both HNK-1 and Leu-11 antigens. *J. Histochem. Cytochem.* **33**, 1129.

Manara, G. C., Ferrari, C., Scandroglio, R., Rocchi, G., Pagani, L., and De Panfilis, G. (1986a). Characterization of two morphologically distinct Leu-7$^+$ cell subsets with respect to Leu-15 antigen. Evaluation of Leu-15 determinant distribution on both E rosetting and non-adherent non-E rosetting cell populations *Scand. J. Immunol.* **23**, 225.

Manara, G. C., Sansoni, P., Ferrari, C., and De Panfilis, G. (1986b). Natural killer cells expressing the Leu-11 antigen display phagocytic activity for 2-aminoethylisothiouronium bromide hydrobromide-treated sheep red blood cells. *Lab. Invest.* **55**, 412.

Manara, G. C., Arancia, G., Fiorentini, C., Ferrari, C., and De Panfilis, G. (1987). Ultrastructural differences between Leu-7$^+$,11$^-$ NK cells and Leu-11$^+$ NK cells. *J. Immunol. Methods* **98**, 155.

Manara, G. C. (1988). Ultrastructural features of NK cell subsets. *EOS—Riv Immunol. Imunofarmacol.* **8**, 138.

Manara, G. C., Ferrari, C., Pagani, L., Torresani, C., Bologna, G., Donelli, S., and De Panfilis, G. (1988a). Evaluation of HLA-DR expression on Langerhans cells and indeterminate cells with regard to ultrastructural features. *J. Invest. Dermatol.* **91**, 391.

Manara, G. C., Ferrari, C., Torresani, C., and De Panfilis, G. (1988b). Immunogold silver staining of suspended cells in transmission electron microscopy. *Clin. Exp. Dermatol.* (in press).

Morhenn, V. B., Roth, S., and Roth, R. (1987). Use of a monoclonal antibody (VM-2) plus the immunogold–silver technique to stain basal cell carcinoma cells. *J. Am. Acad. Dermatol.* **17**, 765.

Novikoff, A. B. (1980). DAB cytochemistry: Artifact problems in its current uses. *J. Histochem. Cytochem.* **28**, 1036.

Probert, L., De Mey, J., and Polak, J. M. (1981). Distinct subpopulations of enteric p-type neurones contain substance P and vasoactive intestinal polypeptide. *Nature (London)* **294**, 470.

Romani, N., Stingl, G., Tschachler, E., Witmer, M. D., Steinman, R. M., Shevach, E. M., and Schuler, G. (1985). The Thy-1-bearing cell of murine epidermis. A distinctive leukocyte perhaps related to natural killer cells. *J. Exp. Med.* **161**, 1368.

Romano, E. L., and Romano, M. (1977). Staphylococcal protein A bound to colloidal gold: A useful reagent to label antigen–antibody sites in electron microscopy. *Immunocytochemistry* **14**, 711.

Ronco, P., Melcion, C., Geniteau, M., Ronco, E., Reininger, L., Galceran, M., and Verroust, P. (1984a). Production and characterization of monoclonal antibodies against rat brush border antigens of the proximal convoluted tubule. *Immunology* **53**, 87.

Ronco, P., Allegri, L., Melcion, C., Pirotsky, E., Appay, M. D., Bariety, J., Pontillon, F., and Verroust, P. (1984b). A monoclonal antibody to brush border and passive Heymann nephritis. *Clin. Exp. Immunol.* **55**, 319.

Roth, J., Bendayan, M., and Orci, L. (1978). Ultrastructural localization of intracellular antigens by the use of protein–A gold complex. *J. Histochem. Cytochem.* **26,** 1074.

Roth, J. (1982). The preparation of protein A–gold complexes with 3 nm and 15 nm gold particles and their use in labelling multiple antigens on ultra-thin sections. *Histochem. J.* **14,** 791.

Roth, J., Lentze, M. J., and Berger, E. G. (1985). Immunocytochemical demonstration of ectogalactosyltransferase in absorptive intestinal cells. *J. Cell Biol.* **100,** 118.

Sansoni, P., Rowden, G., Manara, G. C., Ferrari, C., and De Panfilis, G. (1987). S-100-positive T cells are largely restricted to a CD8-positive, 9.3-negative subset. *Virchows Arch. B* **53,** 301.

Schmitt, D., Faure, M., Dambuyant-Dezutter, C., and Thivolet, J. (1984). The semi-quantitative distribution of T4 and T6 surface antigens on human Langerhans cells. *Br. J. Dermatol.* **111,** 655.

Slot, J. W., and Geuze, H. J. (1981). Sizing of protein A–colloidal gold probes for immunoelectron microscopy. *J. Cell Biol.* **90,** 533.

Usellini, L., Buchan, A. M. J., Polak, J. M., Capella, C., Cornaggia, M., and Solcia, E. (1984). Ultrastructural localization of motilin in endocrine cells of human and dog intestine by the immunogold technique. *Histochemistry* **81,** 363.

Warhol, M. J., Lucocq, J. M., Carlemalm, E., and Roth, J. (1985a). Ultrastructural localization of keratin proteins in human skin using low-temperature embedding and protein A–gold technique. *J. Invest. Dermatol.* **84,** 69.

Warhol, M. J., Roth, J., Lucocq, J. M., Pinkus, G. S., and Rice, R. H. (1985b). Immunoultrastructural localization of involucrin in squamous epithelium and cultured keratinocytes. *J. Histochem. Cytochem.* **33,** 141.

Warhol, M. J., and Longtine, J. A. (1986). A low-temperature embedding colloidal gold technique for immunoelectron microscopy. *Diagn. Immunol.* **4,** 47.

17

Colloidal Gold in High-Voltage Electron Microscopy—Ruthenium Red Method and Whole-Cell Mount

KUNIAKI TAKATA
and
HIROSHI HIRANO

Department of Anatomy
Kyorin University School of Medicine
Shinkawa, Mitaka
Tokyo, Japan

INTRODUCTION
COLLOIDAL GOLD–LECTIN–RUTHENIUM RED
    STAINING
  Labeling Procedure
  Sectioning and Microscopic Observation
LABELING OF WHOLE-CELL MOUNTS
  Labeling Procedure
COMPARISON WITH SCANNING IMAGES
REFERENCES

INTRODUCTION

Electron microscopes operating at accelerating voltages of ~1000 kV have recently been available for the examination of biological specimens. The high penetrating power of electron beams in the high-voltage electron microscope (HVEM) allows the examination of relatively thick specimens. This microscope has been used to observe a variety of biological specimens (for a rewiew, see Glauert, 1974). For example, small organisms were observed in the living state in a specially designed specimen holder called an environmental cell (Parsons *et al.*, 1972). Stereo-high-voltage observation of critical-point dried cells clearly showed the three-dimensional organization of the fibrous elements in the cytoplasm (Wolosewick and Porter, 1976, 1979). Observation of thick resin sections revealed the three-dimensional organization of specific structures such as Golgi apparatus (Mayahara *et al.*, 1978; Yamada and Ishikawa, 1981), T tubules in muscle cells (Yamada and Ishikawa, 1981; Ishikawa and Tsukita, 1983), and extension of neuronal processes (Kosaka and Hama, 1979), which otherwise can be shown only by the time-consuming reconstruction of serial sections. Recently, reversible embedment cytochemistry in thick sections was used for viewing the localization of colloidal gold-labeled antibodies with an HVEM (Gorbsky and Borisy, 1986).

Colloidal gold conjugated to specific antibodies, lectins, etc. is especially suitable as an immunocytochemical marker for high-voltage electron microscopy. It is extremely electron dense and easy to discern. It clearly stands out against the relatively high background density of the thick specimens used in high-voltage electron microscopy. On the other hand, ferritin particles, which are less electron dense, are difficult to distinguish except for large aggregates (Kawakami and Hirano, 1986) in thick sections. Since contrast of the image decreases as the accelerating voltage increases, colloidal gold labeling is also very convenient for the observation of low-contrast images obtained by high-voltage electron microscopy.

Two- and three-dimensional distribution of specific molecules in cells has been determined by various methods. Scanning electron microscopy of cells labeled with ferritin (Marchalonis *et al.*, 1978; Hirano, 1980), hemocyanin (Weller, 1974), latex particles (Molday *et al.*, 1975), or colloidal gold (Horisberger *et al.*, 1975; Horisberger, 1989) has provided such information (for a review, see Molday, 1983; Hodges *et al.*, 1984). In red blood cells, cell membrane specimens prepared by rupturing cells at an air–water interface were labeled with ferritin-conjugated lectins for examining the distribution of sugar residues on the cell surface (Nicolson and

17. Colloidal Gold in HVEM

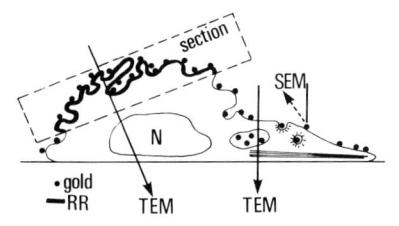

Fig. 17.1. A scheme showing the colloidal gold–lectin–ruthenium red staining (left), the whole-cell mount (center), and SEM (right). RR, Ruthenium red; N, nucleus of a cell. For detail see the text.

Singer, 1971; Beppu et al., 1983). Surface replicas of colloidal gold-labeled cells have also been used (Hopkins and Trowbridge, 1983). The colloidal gold particles remain on the replica, so that their distribution on the cell surface can be examined with the transmission electron microscope (TEM).

We describe two methods of treatment of colloidal gold-labeled cells (Fig. 17.1): (1) the colloidal gold–ruthenium red method for thick sections (Takata et al., 1984a,b) and (2) the whole-cell mount method (Hopkins and Trowbridge, 1983; Kasamatsu et al., 1983; Takata and Hirano, 1984; Peters and Mosher, 1987). Both are best suited for examination with a 1000-kV electron microscope. In most cases, however, examination with a 200-kV electron microscope also gives satisfactory results. Observation even at an accelerating voltage of 100–125 kV is worth trying.

COLLOIDAL GOLD–LECTIN–RUTHENIUM RED STAINING

Because of the overlapping of structures in thick resin sections and relatively low contrast in high-voltage electron microscopy, it is difficult to distinguish the cell surface. The purpose of ruthenium red treatment is to specifically enhance the contrast of the cell surface, so that the distribution of colloidal gold particles as well as the topography of the cell surface can be clearly observed (Takata et al., 1984a,b).

Labeling Procedure

Cells are cultured in standard plastic tissue culture dishes. Nonadherent cells such as erythrocytes and lymphocytes can be attached to the dishes by coating the substratum with poly-L-lysine. Alternatively, such cells can be treated as a suspension in the procedure described below. In that case, washing is carried out by centrifugation instead of simply decanting the solution.

Cells are labeled for the specific binding of lectins or antibodies using colloidal gold as a marker. We usually use colloidal gold particles 10–20 nm in diameter, which give strong enough contrast for the particles to be recognized against the ruthenium red-stained cell surface. We have not tested smaller gold particles (such as 3 or 5 nm), but perhaps they could be used if the intensity of ruthenium red stain and section thickness were reduced.

Since it is not the purpose of this contribution to describe the procedure of colloidal gold labeling in detail, we simply show here the two-step method of concanavalin A (Con A)–colloidal gold labeling (Ackerman and Freeman, 1979). For detailed reviews of the preparation of the colloidal gold conjugates and the labeling, see Roth (1983), De Mey (1986), and this volume. In our published work on Con A–horseradish peroxidase–colloidal gold staining (Takata *et al.,* 1984a,b), cells were incubated with Con A (50 μg/ml; Miles-Yeda, Israel) in phosphate-buffered saline (PBS) for 1 hr. They were then washed with PBS and incubated with horseradish peroxidase–colloidal gold conjugate. After washing with PBS to remove the unattached colloidal gold conjugate, specimens were ready for ruthenium red stain. Any specimens labeled with the colloidal gold method for cell surface antigens, receptors, and lectin binding sites can be used.

Cells are treated with ruthenium red in order to specifically enhance the contrast of the cell surface (Luft, 1971a,b; Hayat, 1986, 1989). Ruthenium red is a cation and is thus considered to react with polyanions. Since the cell surface is, in general, negatively charged (Weiss, 1969), ruthenium red serves as an effective dye to stain the cell surface. Ruthenium red is relatively a large molecule and hence does not penetrate the cell membrane. Only the contrast of the cell surface increases. Surface-connected tubular structures and cellular invaginations are stained with ruthenium red as well (Luft, 1971b). The contrast of the cytoplasm, on the other hand, does not change at all. The procedure described below is essentially based on the method by Luft (1971b) as modified by Yokoyama *et al.* (1980).

1. Fix cells in dishes for 30–60 min at 4°C (277 K) in a mixture of ruthenium red (0.5 mg/ml), 1.2% glutaraldehyde, and 0.067 M sodium piperazine-N,N'-bis(2-ethanesulfonate) (PIPES) buffer (pH 7.3). This solution can be prepared by mixing equal volumes of ruthenium red (1.5 mg/ml distilled water), 3.6% aqueous glutaraldehyde, and 0.2 M PIPES buffer (pH 7.3). The mixture should be prepared just prior to use. If pure ruthe-

nium red is not available, purify it according to the procedure described by Luft (1971b).
  2. Wash with 0.15 M PIPES buffer (pH 7.3).
  3. Fix for 30 min at room temperature in ruthenium red (0.5 mg/ml)–0.67% OsO_4–0.067 M PIPES buffer (pH 7.3). This solution can be prepared by mixing equal volumes of ruthenium red (1.5 mg/ml distilled water), 2% aqueous OsO_4, and 0.2 M PIPES buffer (pH 7.3). The mixture is unstable and should be prepared just before use.
  4. Wash thoroughly with 0.15 M PIPES buffer (pH 7.3).
  5. Dehydrate through a graded series of ethanol.
  6. Add propylene oxide to detach cells from plastic dishes. Transfer the detached cells to glass tubes.
  7. Wash in propylene oxide twice, and embed in epoxy resin as usual.

Note: Since the mixture of ruthenium red, OsO_4, and PIPES buffer is relatively unstable, cacodylate buffer as used by Luft (1971b) can be used. We found, however, that PIPES buffer consistently gives adequate contrast on the cell surface.

Sectioning and Microscopic Observation

Cut 0.1–3-μm thick sections with glass knives, pool them on the water in the trough, and transfer them on uncoated copper grids. Examine the grids with the TEM without further staining.

Using sections of 0.3–0.5-μm thickness, observation at an accelerating voltage of 200 kV or higher is recommended (Takata *et al.*, 1984a,b). If such a microscope is not available, sections can be examined at an accelerating voltage of 100–125 kV (Rieder *et al.*, 1985). The contrast of the image increases as the accelerating voltage decreases. Care should be taken not to get excessive contrast during the process of film development and printing, especially in micrographs taken at 100–125 kV. The detail will be lost in high-contrast images. For thicker sections of 1–3 μm, good results can be obtained only when examination is done at an accelerating voltage of 1000 kV (Takata *et al.*, 1984b). Obtain a pair of photomicrographs at tilt angles of ±6–15° (Fig. 17.2). Record the axis of tilting for stereomicroscopic observation of negatives and prints.

Because sections are thick, it is easy to include a wide expanse of the cell surface in a single section, especially when sections are cut parallel or tangential to the cell surface (Fig. 17.1). In these cases the images obtained are, in some respects, similar to those obtained by high-resolution scanning electron microscopy of the cell surface (Takata *et al.*, 1984b). However, the colloidal gold–ruthenium red method has several advantages over scanning electron microscopy. Micrographs of superior quality

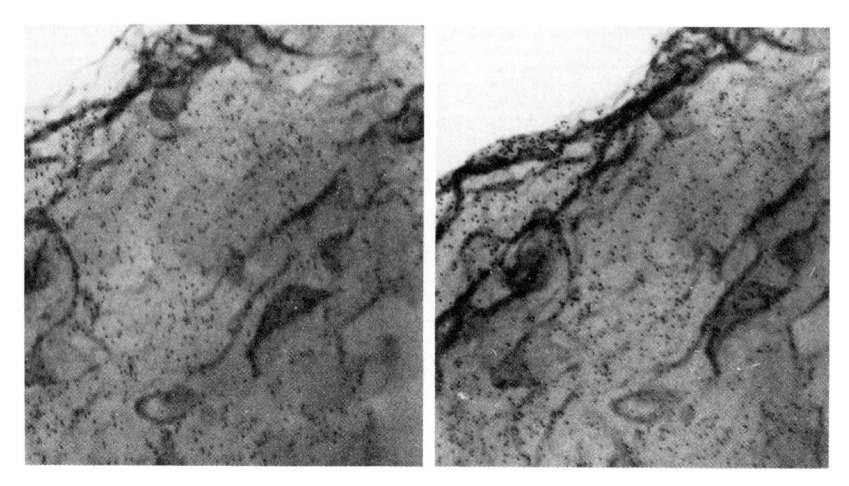

Fig. 17.2. A 3-μm-thick section of a macrophage stained by the Con A–colloidal gold–ruthenium red method. A stereo pair of electron micrographs taken at an accelerating voltage of 1000 kV and at tilt angles of ±8°; ×26,000.

can be obtained very easily because of the inherent high resolving power of the TEM. In fact, individual colloidal gold particles can easily be discerned. This allows the use of smaller colloidal gold particles as a marker to increase the sensitivity of the label. This method is also suitable for the examination of the colloidal gold distribution in cells with irregular cell surface topography. Colloidal gold particles in the invaginations, pits, and cytoplasmic vesicles, as well as behind microvilli and blebs, can be seen in a single stereo pair of transmission electron micrographs.

In comparison with the cell surface replica method (Hopkins and Trow-

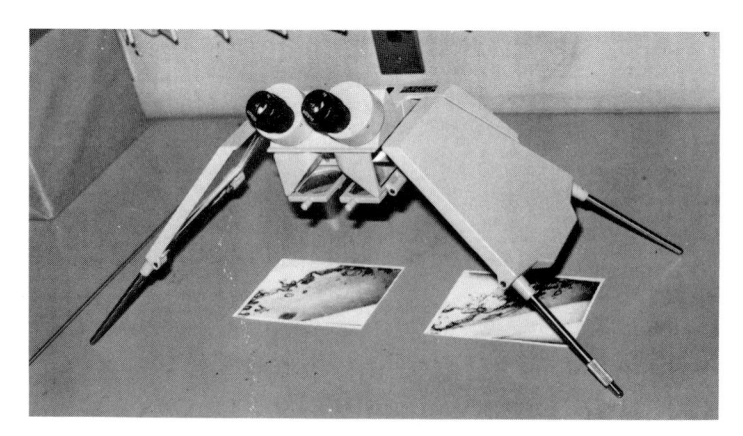

Fig. 17.3. A stereo viewer (stereoscope).

bridge, 1983), the image of the cell surface is similar, but the colloidal gold particles in the endocytic vesicles are more easily recognized in the colloidal gold–ruthenium red method. If required, a section adjacent to the thick one may be cut and stained with uranyl acetate and lead citrate to check the ultrastructure of the cell (Takata et al., 1984a).

For stereomicroscopic observation, one can practice so that three-dimensional images can be visualized without using special equipment. It is more convenient, however, to use a stereo viewer (stereoscope) (Fig. 17.3). Recently, an automatic analysis system has been developed to extract three-dimensional information from high-voltage electron microscopic stereo images of biological specimens (Arii and Hama, 1987).

LABELING OF WHOLE-CELL MOUNTS

The high penetrating power of the electron beam of the HVEM permits the visualization of a single whole cell. With the advent of the 1000-kV model, dried cells were examined as a whole and an intracellular lattice structure, termed the microtrabecular lattice, was reported (Wolosewick and Porter, 1976, 1979). Whole-cell mounts have also been used to compare the ultrastructure of cells with their immunofluorescence images. By comparing the immunofluorescence micrographs and the electron micrographs of the whole-cell mount of the same cell, identification of the components in electron micrographs has been made possible (Osborn et al., 1978).

The immunoperoxidase method was also used in a whole-cell mount for comparing the light microscopic images with the electron microscopic ones (Henderson and Weber, 1979). For the electron microscopic markers, heavy meromyosin or S1 fragment was used for decoration of actin cables in whole-mount cytoskeleton specimens of mouse macrophages (Painter et al., 1981) and cultured chick embryo neurons (Letourneau and Ressler, 1983). Because a whole-cell mount, even if it is the detergent-extracted cytoskeleton, is electron dense compared with the thin resin section, ferritin is not adequate as a marker except in special cases (Webster et al., 1978).

Colloidal gold was used to show the localization of actin and viral structural polypeptide Vpl by the double-labeling method in SV40-infected TC7 cells (Kasamatsu et al., 1983). Hopkins and Trowbridge (1983) showed the distribution of transferrin and transferrin receptors in A431 cells in whole-cell mounts using the colloidal gold labeling method. Peters and Mosher (1987) demonstrated the localization of the cell surface binding sites of fibronectin in cultured human fibroblasts using colloidal gold-labeled fibronectin with a 1000-kV electron microscope. Binding sites for

Con A and wheat germ agglutinin (WGA) in bovine spermatozoa during epididymal maturation was studied in whole-cell mounts at 100 kV (Friess and Sinowatz, 1984). We used colloidal gold to label the cell surface Con A binding sites in mouse macrophages (Takata and Hirano, 1984). Colloidal gold is extremely electron dense and uniform in shape compared with other electron microscope markers. It is therefore very easily distinguished over a relatively thick macrophage cytoplasm. Endocytosed colloidal gold-labeled Con A in cytoplasmic vesicles and vacuoles can also be traced easily (Fig. 17.4). The size of the colloidal gold particles for labeling may be determined by considering the thickness of the specimens, intensity of the label desired, and the accelerating voltage of the TEM.

Labeling Procedure

1. Prepare Formvar-coated gold or platinum grids attached on a glass coverslip. This can be sterilized by UV irradiation.
2. Place the coverslips with grids as mentioned above on a standard plastic tissue culture dish. Cells are cultured as usual. In the case of nonadherent cells, one can coat the Formvar-coated grid with poly-L-lysine to facilitate their attachment.
3. Label the cells with antibodies or lectins using colloidal gold as a marker.
4. Fix the sample in 1% OsO_4 for 10 min. Wash with distilled water and stain with 0.5% aqueous uranyl acetate for 10 min (Wolosewick and Porter, 1976).
5. Dehydrate in a graded series of ethanol or acetone.
6. Dry the specimen by the critical-point drying method using liquid CO_2.
7. Coat the specimen with carbon.
8. Examine cells on the Formvar-coated grids at an accelerating voltage of 200 kV. Take a pair of photomicrographs at tilt angles of ±6–15°. Record the axis of tilting for stereomicroscopic observation. A voltage of 200 kV is enough for the observation of the flattened regions of cultured cells. For the examination of the perinuclear and nuclear regions, however, the use of a 1000-kV microscope is required.

Note: Instead of culturing cells on expensive gold or platinum grids, cells can be cultured on Formvar film. Specimens are labeled by the colloidal gold method, osmicated, dehydrated, and critical-point dried as above. Then place copper grids on the Formvar film with labeled cells for electron microscopic observation. This procedure is advantageous when one is interested in a particular type of cell in culture. One can mark such cells under the light microscope and place grids on them for electron microscopy.

17. Colloidal Gold in HVEM   353

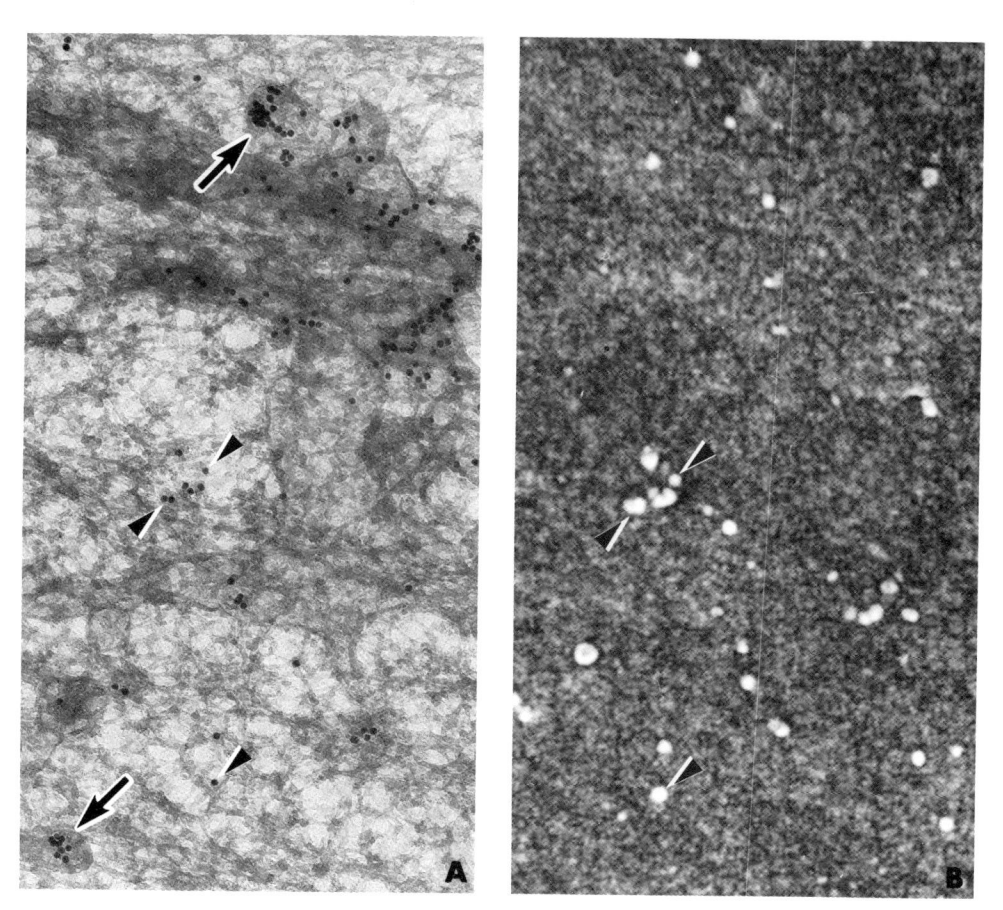

Fig. 17.4. Transmission (A) and scanning (B) electron micrographs of the same field of a whole-cell mount labeled with Con A–colloidal gold. Macrophages were labeled at 4°C (277 K) and incubated in phosphate-buffered saline for 40 min at 36°C (309 K). Endocytosis of colloidal gold particles (arrows) is evident. Arrowheads show examples of corresponding gold particles in transmission and scanning images on the cell surface; ×55,000.

COMPARISON WITH SCANNING IMAGES

After the observation with the TEM, the same specimen can be examined with a scanning electron microscope (Fig. 17.1). The specimen is coated with gold or platinum. Secondary electron images of the same field as seen by the TEM can then be recorded (Takata and Hirano, 1984). The images obtained by scanning electron microscopy are basically similar to those from transmission electron microscopy, although colloidal gold markers are seen as very bright particles instead of electron-dense ones

(Fig. 17.4). If specimens are coated heavily with gold or platinum, the contrast of the colloidal gold particles decreases, making it difficult to discern the label. The development of scanning electron microscopy using the backscattered electron imaging mode allows colloidal gold particles to be seen very clearly (de Harven et al., 1984; de Harven and Soligo, 1989; Takata et al., 1988).

Due to the much greater resolving power of the TEM, colloidal gold particles can be more easily recognized with it than with the SEM. At the edges of blebs and microvilli, colloidal gold particles are rather difficult to see because of the "edge effect" in scanning electron microscopy. The colloidal gold particles in deep invaginations or in the endocytic vesicles and vacuoles cannot be seen by scanning electron microscopy (Figs. 17.1. and 17.4). Thus, transmission electron microscopic observation of the whole-cell mount is especially useful in such cases. Another advantage of the TEM is that the cytoplasmic organization can be seen as well as the colloidal gold label. The disadvantage is that only a relatively small region of the cell can be observed with the TEM. This can partly be overcome by using larger colloidal gold particles and observing at higher accelerating voltages.

REFERENCES

Ackerman, G. A., and Freeman, W. H. (1979). Membrane differentiation of developing hemic cells of the bone marrow demonstrated by changes in concanavalin A surface labeling. *J. Histochem. Cytochem.* **27**, 1413.

Arii, T., and Hama, K. (1987). Method of extracting three-dimensional information from HVTEM stereo images of biological materials. *J. Electron Microsc.* **36**, 177.

Beppu, H., Nakajima, M., Nishiyama, F., Uono, M., and Hirano, H. (1983). Concanavalin A binding sites on the erythrocytes of normal and genetically dystrophic chickens. *J. Neurol. Sci.* **59**, 401.

de Harven, E., Leung, R., and Christensen, H. (1984). A novel approach for scannning electron microscopy of colloidal gold-labeled cell surfaces. *J. Cell Biol.* **99**, 53.

de Harven, E., and Soligo, D. (1989). Backscattered electron imaging of the colloidal gold marker on cell surfaces. In *Colloidal Gold: Principles, Methods, and Applications, Vol. 1* (M. A. Hayat, ed.), pp. 229–249. Academic Press, San Diego, California.

De Mey, J. (1986). The preparation and use of gold probes. In *Immunocytochemistry: Modern Methods and Applications* (J. M. Polak and S. Van Noorden eds.), 2nd ed., p. 115. Wright, Bristol.

Friess, A. E., and Sinowatz, F. (1984). Con A- and WGA-binding sites on bovine epididymal spermatozoa: TEM of specimens *in toto*. *Biol. Cell.* **50**, 279.

Glauert, A. M. (1974). The high voltage electron microscope in biology. *J. Cell Biol.* **63**, 717.

Gorbsky, G., and Borisy, G. G. (1986). Reversible embedment cytochemistry (REC): A versatile method for the ultrastructural analysis and affinity labeling of tissue sections. *J. Histochem. Cytochem.* **34**, 177.

Hayat, M. A. (1986). *Basic Techniques for Transmission Electron Microscopy.* Academic Press, Orlando, Florida.
Hayat, M. A. (1989). *Principles and Techniques of Electron Microscopy,* 3rd. ed. Macmillan, London, and CRC Press, Boca Raton, Florida.
Hendeson, D., and Weber, K. (1979). Three-dimensional organization of microfilaments and microtubules in the cytoskeleton. Immunoperoxidase labelling and stereo-electron microscopy of detergent-extracted cells. *Exp. Cell Res.* **124**, 301.
Hirano, H. (1980). Ultracytochemical studies on the macromolecular architecture and transmembrane control of the cell membrane. *Acta Histochem. Cytochem.* **13**, 127.
Hodges, G. M., Smolira, M. A., and Livingston, D. C. (1984). Scanning electron microscope immunocytochemistry in practice. In *Immunolabeling for Electron Microscopy* (J. M. Polak and I. M. Varndell, eds.), p. 189. Elsevier, Amsterdam.
Hopkins, C. R., and Trowbridge, I. S. (1983). Internalization and processing of transferrin and the transferrin receptor in human carcinoma A431 cells. *J. Cell Biol.* **97**, 508.
Horisberger, M., Rosset, J., and Bauer, H. (1975). Colloidal gold granules as markers for cell surface receptors in the scanning electron microscope. *Experientia* **31**, 1147.
Horisberger, M. (1989). Colloidal gold for scanning electron microscopy. In *Colloidal Gold: Principles, Methods, and Applications, Vol. 1* (M. A. Hayat, ed.), pp. 217–227. Academic Press, San Diego, California.
Ishikawa, H., and Tsukita, S. (1983). High voltage electron microscopy of the T-system in the mouse diaphragm. In *Muscular Dystrophy: Biomedical Aspects* (S. Ebashi and E. Ozawa, eds.), p. 167. Springer-Verlag, Berlin and New York.
Kasamatsu, H., Lin, W., Edens, J., and Revel, J. P. (1983). Visualization of antigens attached to cytoskeletal framework in animal cells: Colocalization of simian virus 40, Vpl polypeptide, and actin in TC7 cells. *Proc. Natl. Acad. Sci. U.S.A.* **80**, 4339.
Kawakami, H., and Hirano, H. (1986). Rearrangement of the open-canalicular system of the human blood platelet after incorporation of surface-bound ligands. A high-voltage electron-microscopic study. *Cell Tissue Res.* **245**, 465.
Kosaka, T., and Hama, K. (1979). Ruffed cell: A new type of neuron with a distinctive initial unmyelinated portion of the axon in the olfactory bulb of goldfish *(Carassius auratus).* I. Golgi impregnation and serial thin sectioning studies. *J. Comp. Neurol.* **186**, 301.
Letourneau, P. C., and Ressler, A. H. (1983). Differences in the organization of actin in the growth cones compared with the neurites of cultured neurons from chick embryos. *J. Cell Biol.* **97**, 963.
Luft, J. H. (1971a). Ruthenium red and violet. I. Chemistry, purification, methods of use for electron microscopy and mechanism of action. *Anat. Rec.* **171**, 347.
Luft, J. H. (1971b). Ruthenium red and violet. II. Fine structural localization in animal tissues. *Anat. Rec.* **171**, 369.
Marchalonis, J. J., Bucana, C., Hoyer, L., Warr, G. W., and Hanna, M. G. (1978). Visualization of a guinea pig T lymphocyte surface component cross-reactive with immunoglobulin. *Science* **199**, 433.
Mayahara, H., Ishikawa, T., Ogawa, K., and Chang, J. P. (1978). The three-dimensional structure of the Golgi complex in cultured fibroblasts. An ultracytochemical study with thin and thick sections. *Acta Histochem. Cytochem.* **11**, 239.
Molday, R. S., Dreyer, W. J., Rembaum, A., and Yen, S. P. S. (1975). New immunolatex spheres: Visual markers of antigens on lymphocytes for scanning electron microscopy. *J. Cell Biol.* **64**, 75.
Molday, R. S. (1983). Labeling of cell surface antigens for SEM. In *Techniques in Immunocytochemistry* (G. R. Bullock and P. Petrusz, eds.), Vol. 2, pp. 117–154. Academic Press, London.

Nicolson, G. L., and Singer, S. J. (1971). Ferritin-conjugated plant agglutinins as specific saccharide stains for electron microscopy: Application to saccharides bound to cell membranes. *Proc. Natl. Acad. Sci. U.S.A.* **68**, 942.

Osborn, M., Webster, R. E., and Weber, K. (1978). Individual microtubules viewed by immunofluorescence and electron microscopy in the same PtK2 cell. *J. Cell Biol.* **77**, R27.

Painter, R. G., Whisenand, J., and McIntosh, A. T. (1981). Effects of cytochalasin B on actin and myosin association with particle binding sites in mouse macrophages: Implications with regard to the mechanism of action of the cytochalasin. *J. Cell Biol.* **91**, 373.

Parsons, D. F., Matricardi, V. R., Subjeck, J., Uydess, I., and Wray, G. (1972). High-voltage electron microscopy of wet whole cancer and normal cells. Visualization of cytoplasmic structures and surface projections. *Biochim. Biophys. Acta* **290**, 110.

Peters, D. M. P., and Mosher, D. F. (1987). Localization of cell surface sites involved in fibronectin fibrillogenesis. *J. Cell Biol.* **104**, 121.

Rieder, C. L., Rupp, G., and Bowser, S. S. (1985). Electron microscopy of semithick sections: Advantages for biomedical research. *J. Electron Microsc. Tech.* **2**, 11.

Roth, J. (1983). The colloidal gold marker system for light and electron microscopic cytochemistry. In *Techniques in Immunocytochemistry* (G. R. Bullock and P. Petrusz, eds.), Vol. 2, pp. 217–284. Academic Press, London.

Takata, K., and Hirano, H. (1984). Whole-cell-mount cytochemistry by the colloidal gold labeling method. Combined transmission and scanning electron microscopic study of Con A binding sites in mouse macrophages. *Histochemistry* **81**, 435.

Takata, K., Arii, T., Yamagishi, S., and Hirano, H. (1984a). Use of colloidal gold and ruthenium red in stereo high-voltage electron microscopic study of Con A-binding sites in mouse macrophages. *Histochemistry* **81**, 441.

Takata, K., Arii, T., Yamagishi, S., and Hirano, H. (1984b). Concanavalin A (Con A) receptors in mouse macrophages demonstrated in thick sections with a 1,000kV transmission electron microscope. *Proc. Jpn. Acad. Ser. B* **60**, 273.

Takata, K., Akimoto, Y., Ogura, K., Yamagishi., S., and Hirano, H. (1988). Colloidal gold label observed with a high resolution backscattered electron imaging in mouse in lymphocytes. *J. Electron Microsc.* **37**, 346.

Webster, R. E., Henderson, D., Osborn, M., and Weber, K. (1978). Three-dimensional electron microscopical visualization of the cytoskeleton of animal cells: Immunoferritin identification of actin- and tubulin-containing structures. *Proc. Natl. Acad. Sci. U.S.A.* **75**, 5511.

Weiss, L. (1969). The cell periphery. *Int. Rev. Cytol.* **26**, 63.

Weller, N. K. (1974). Visualization of concanavalin A-binding sites with scanning electron microscopy. *J. Cell Biol.* **63**, 699.

Wolosewick, J. J., and Porter K. R. (1976) Stereo high-voltage electron microscopy of whole cells of the human diploid line, WI-38. *Am. J. Anat.* **147**, 303.

Wolosewick, J. J., and Porter K. R. (1979). Microtrabecular lattice of the cytoplasmic ground substance. Artifact or reality. *J. Cell Biol.* **82**, 114.

Yamada, E., and Ishikawa, H. (1981). Dense tissue and special stains. *Methods Cell Biol.* **22**, 123.

Yokoyama, M., Chang, J. P., and Moller, P. C. (1980). Cytochemical study of concanavalin A binding sites and their mobility in normal, cystic fibrosis, and SV40 transformed human fibroblasts in vitro. *J. Histochem. Cytochem.* **28**, 543.

18

Correlative Light and Electron Microscopic Immunocytochemistry on Reembedded Resin Sections with Colloidal Gold

HENDERSON MAR

Department of Pathology
School of Medicine
University of Washington
Seattle, Washington

and

THOMAS N. WIGHT

Department of Pathology
School of Medicine
University of Washington
Seattle, Washington

INTRODUCTION
TECHNICAL REQUIREMENTS
  Fixation
  Embedding

    Sectioning
    Removal of Resin
  IMMUNOSTAINING FOR LIGHT MICROSCOPY
    Antibodies
    Peroxidase Screening
    Colloidal Gold Immunostaining
    Microscopic Techniques
  ELECTRON MICROSCOPY OF IMMUNOSTAINED THICK
      SECTIONS
    Reembedding Techniques for Electron Microscopy
    Electron Microscopy of Reembedded 2-μm Sections
  ELECTRON MICROSCOPY OF DEPLASTICIZED ADJACENT
      THIN SECTIONS
    Preparation and Handling of Grids
    Deplasticization of Thin Sections
    Colloidal Gold Immunocytochemical Staining
    Reembedding of Thin Sections
  GENERAL CHARACTERISTICS OF DEPLASTICIZED SECTIONS
    Section Thickness
    Rate of Resin Removal
    Preservation of Antigenicity
    Preservation of Ultrastructure
  LIMITATIONS
  REFERENCES

INTRODUCTION

The goal of immunocytochemistry is to localize specific antigens with antibodies of known specificity. To obtain optimal results, consideration should be given to the following: (1) using the same type of fixative for light and electron microscopy so that a comparative examination can be made on the same tissue samples, (2) optimizing the preservation of morphology for simultaneous light and electron microscopic examination, (3) incorporating procedures which eliminate excessive background immunostaining, (4) optimizing the penetration of antibodies into the tissue and cells, and (5) using an immunostaining protocol which is applicable at both light and electron microscopic levels on the same tissue. Although not all the conditions stated above need be satisfied in most immunocytochemical studies, all are essential when a correlative study at the light and electron microscopic levels is undertaken. Localizing specific antigens at both the light and electron microscopic levels using the same processing and staining procedures eliminates inconsistencies in the interpretation of immunostaining results from several dissimilar methods.

18. Correlative Immunocytochemistry on Reembedded Sections

Most correlative electron microscopic immunocytochemical studies have been done on fixed, cultured, or isolated cells prior to embedding using colloidal gold (De Brabander *et al.*, 1977; Geoghegan *et al.*, 1978; De Mey *et al.*, 1981; Horobin, 1982; Lamberts and Goldsmith, 1985) or a combination of gold and fluorescent marker (Horisberger, 1981; Alexander *et al.*, 1985). However, the preembedding immunostaining technique has certain disadvantages. Permeabilizing treatments are often necessary to obtain labeling of intracellular antigens in cultured cells and/or tissue sections. The use of penetration agents or detergents often leads to partial destruction of labile cytoplasmic structures, such as membranes and filaments.

Postembedding immunocytochemical staining, on the other hand, does not require penetrating agents since membrane barriers of most cells and organelles are disrupted during sectioning, allowing access of immunoreagents. The advantages of postembedding immunocytochemical staining using either peroxidase (Baskin *et al.*, 1979a, 1982) or colloidal gold (Horisberger and Rosset, 1977; Horisberger, 1979; De Mey *et al.*, 1981; Lackie *et al.*, 1985) include high-resolution localization and excellent morphological preservation.

We have found that although 1–2-μm-thick resin sections exhibit intense postembedding immunostaining at the light microscopic level after the resin has been removed, thin sections on nickel grids may immunostain weakly or not at all. In both cases, the resin matrix in and around the cells presumably prevents antibody binding to a sufficient number of antigenic sites to give satisfactory localization (Mar *et al.*, 1987; Mar and Wight, 1988). This problem has made correlative light and electron microscopic immunocytochemistry on the same tissue difficult.

This chapter describes two procedures that avoid not only the penetration problem but also the problems associated with resin interference for electron microscopic immunocytochemistry. The first procedure involves removal of Epon from 2-μm-thick sections (Lane and Europa 1965; Holec and Ciampor, 1978) followed by colloidal gold immunostaining and reembedment of the sections for subsequent thin sectioning (Mar *et al.*, 1987). The procedure enables intracellular antigens to be visualized by light and electron microscopy on the same section. This procedure is useful in situations where antigenic sites and the number of stainable cells are extremely limited, thus requiring that the same immunostained section be examined at the ultrastructural level.

The second procedure involves removal of Epon or Araldite from ultrathin sections while on the grid followed by colloidal gold immunostaining and reembedment of the sections in dilute Epon (Keller *et al.*, 1984; Mar and Wight, 1988). Both procedures permit localization of tissue antigens which were previously masked by the embedding resin. These procedures yield staining of high specificity, with minimal background staining and well-preserved ultrastructure. The results demonstrate that tissue antigens can be visualized at both the light and electron microscopic levels either on the same section or on adjacent thin sections cut from the same block. Both procedures are useful in situations where problems with postembedding electron microscopic immunostaining exist and where correlative light and electron microscopic immunostaining is essential.

TECHNICAL REQUIREMENTS

Fixation

All tissues are fixed at room temperature in a mixture of 3% formaldehyde and 0.25% glutaraldehyde in 0.05 M phosphate buffer (pH 7.4). Human superficial femoral artery is taken by biopsy and immersion fixed for 2 hr. Rat kidney and pituitary are fixed by vascular perfusion, removed, cut into small pieces, and fixed for the balance of 2 hr. Tissues are rinsed three or four times in buffer over 2 hr and stored overnight at 4°C in fresh buffer.

Other fixatives tested for antigen preservation in subsequent immunostaining protocols include 3% formaldehyde, a mixture of 3% formaldehyde and 2.5% glutaraldehyde, and 3% glutaraldehyde; all were prepared in 0.05 M phosphate buffer (ph 7.4). Stock 10% formaldehyde is made fresh from powdered reagent, heated to boiling, cleared with several drops of 10 N NaOH, and filtered, and the pH is adjusted to 7.4 with 1 N NaOH. The solution is adjusted to the original volume with distilled water before diluting with glutaraldehyde and 0.1 M buffer. Glutaraldehyde is diluted from 8% stock solution.

Embedding

Tissues are dehydrated in ethanol/propylene oxide and embedded in either Epon 812 or Araldite 502. Other embedding resins (Spurr, Polybed 812, Lowicryl K4M, Medcast, and LR White) were tried but proved not as satisfactory in the subsequent steps of resin removal (see below). Tissues should not be stained en bloc with uranyl acetate due to possible inhibition of immunoreactivity. Tissues were polymerized at 60°C for 48 hr.

Sectioning

Two methods of section collection and preparation are used for the immunostaining protocols we describe. The first method requires that thick sections are to be picked up and transferred to a small drop of distilled water on a gelatin-coated glass slide and dried at 45°C on a hot plate for 4–6 hr. A small circle is scribed on the bottom of the slide with a carbide pencil to help locate sections. The entire thickness of these reembedded sections (2 µm) is thin sectioned at 70–200 nm (silver–gold). Every thin section is collected on copper grids and counterstained for ultrastructural examination. The second method involves serial thin sectioning of embedded tissue at 70–90-nm thickness to be used for thin-section deplasticization techniques. Here, thin sections are picked up individually on nickel grids prepared as described later.

Removal of Resin

Removal of resin from 2-µm-thick sections on slides to be used for immunostaining and subsequent reembedding is discussed in this section; a discussion of resin removal from thin sections on grids will be presented in a later section. However, the preparation of solutions and the techniques of resin removal and processing are common to both. Saturated NaOH solution is made by adding 13 g of NaOH pellets to 150 ml of 100% ethanol (Lane and Europa, 1965; Holec and Ciampor, 1978). The solution is made in a 150-ml polyethylene bottle, stirred for 1 hr, and allowed to remain at 22°C. It is capped for 3–5 days before decanting off the supernatant for use. The Na-ethoxide solution is ready for use when it shows a light golden color. Tightly capped bottles can be stored in the refrigerator when not used immediately. Storing in the refrigerator will also retard the further maturation of the ethoxide solution.

Epon sections (2 µm thick) on slides are assembled into plastic racks, placed in polyethylene staining dishes (Lab Tek, Miles Scientific, Naperville, IL) containing ethoxide solution, and diluted 1 : 1 with fresh 100% ethanol for 15–20 min (Araldite sections for 35–45 min). One slide is taken out after 15 min, rinsed several times in 100% ethanol, and examined while still wet under a dissecting microscope to check the progress of resin removal. Adequate removal of sufficient quantities of resin is characterized by indistinct section edges. (It is important to keep the slide and section wet at all times. If the section begins to fog while examining the slide out of alcohol solutions, return the slide immediately to alcohol staining dishes to prevent drying damage and/or NaOH residue formation.) Inadequate removal after the suggested times above may indicate that the ethoxide solution is not sufficiently aged and thus requires longer removal times. Conversely, short removal times or complete removal of the section from the slide can indicate that the ethoxide solution is too strong and has aged longer than acceptable for the controlled removal of resin. Rather than further diluting the stock solution, it should be discarded.

After adequate resin removal has been determined, slides are quickly transferred to four consecutive rinses (5 min each) in 100% ethanol (slides are totally submerged). Slides are agitated in the slide racks to facilitate complete washing. Slides are hydrated to water through a graded alcohol series (95%, 70%, 50%, 30%, distilled water, 1–2 min each), placed in 0.3% H_2O_2 for 15 min (for peroxidase screening, see below), and rinsed in running distilled water for 10 min. Slides are individually removed from the racks, and water is blotted from the bottom and on top around the section itself. Tris-buffer saline (TBS) is immediately placed onto the section, and the slide is placed into a flat-bottomed moist chamber with a lid. Sections are incubated at 22°C for 30–60 min.

During this incubation period and all subsequent incubations, it is critical that sections do not dry. Moist chambers without a perfectly flat bottom can cause solutions to run off to one end of the slide. Insufficient volumes of antisera and prolonged periods of having the cover off the incubation chamber can not only cause drying but also alter the concentration of solutions. Ignoring these technical aspects can result in lack of or decrease in specific immunostaining and produce unwanted background staining.

IMMUNOSTAINING FOR LIGHT MICROSCOPY

Antibodies

Antibody against rat growth hormone (a gift from Dr. John A. Parsons, Department of Anatomy, University of Minnesota, Minneapolis) was produced in guinea pigs and characterized as previously described (Baskin *et al.*, 1979a). A polyclonal antibody against tissue matrix dermatan sulfate proteoglycan (DSPG) from human skin fibroblasts (a generous gift from Dr. Hans Kresse, Institute for Physiological Chemistry, University of Munster, FRG) was generated in rabbits according to procedures outlined elsewhere (Glossl *et al.*, 1984). Polyclonal antibody against lamb kidney Na^+,K^+-ATPase (donated by Dr. William L. Stahl, Veterans Administration Medical Center, Seattle, WA) was generated according to the procedure of McCans *et al.* (1975) and the specificity of the antiserum was determined as described by Baskin and Stahl (1982). A monoclonal antibody designated HHF 35, which is specific for muscle actin isotypes (supplied by Dr. Allen M. Gown, Department of Pathology, University of Washington, Seattle), was generated according to procedures of Gown and Vogel (1982), using Sodium dodecyl sulfate (SDS)-extracted protein fractions from human myocardium. The characterization of HHF 35 and its specificity has been described elsewhere (Tsukada *et al.*, 1987). Immunogold reagents are from Janssen Life Science Products (Beerse, Belgium) and unconjugated secondary antibodies from Cooper Biomedical (Malvern, PA).

Peroxidase Screening

Colloidal gold solutions are used to accomplish the final staining for light microscopy. However the peroxidase–diaminobenzidine (DAB) technique (Sternberger, 1986) is used to screen tissues for the presence and location of immunoreactivity (Mar *et al.*, 1987). This process also allows for the determination of optimal antibody dilutions. In addition, when screening large numbers of tissue samples, the large volume of colloidal gold solutions which would be required makes the DAB procedure a more practical and cost-effective technique. The slides are immunostained as described previously (Mar *et al.*, 1987). Briefly, after resin removal and hydration of the sections to water, slides are placed into 0.3% H_2O_2 for 10 min to block endogenous peroxidase activity and rinsed again for 10 min in distilled water.

To reduce nonspecific staining, one or two drops of 10% normal goat serum + 2% bovine serum albumin (Sigma) in TBS are immediately placed on the sections, and the slides are returned to the covered moist chamber. After 30 min, the serum is drained from the slides, carefully wiped from around the section, and a drop of primary antiserum is applied to the section. Sections are incubated with primary antiserum overnight at 4°C. Secondary antibody (goat anti-rabbit IgG, goat anti-mouse IgG or IgM) is then applied for 1 hr after a brief rinse with TBS. This is followed by another rinse and incubation with rabbit or mouse peroxidase–antiperoxidase for 1 hr. Slides are rinsed thoroughly and the sites of immunoreactivity are visualized by reaction with DAB. Slides are counterstained with hematoxylin, dehydrated, and mounted with Pro-Texx (Scientific Products, McGaw Park, IL). Tissues fixed with concentrations of glutaraldehyde > 2.5% show greatly reduced immunocytochemical staining. However, low concentrations of glutaraldehyde (<1%) aided in the retention of ultrastructure without significantly reducing antigenicity.

Colloidal Gold Immunostaining

Slides are kept in a covered moist chamber between incubations and washes. Solution volumes of at least 100 µl are applied to each slide. Procedures are the same as those outlined above for peroxidase staining through the overnight incubation of primary antiserum at 4°C. Prior to application of the gold secondary antibody, slides are rinsed thoroughly with TBS, two or three times over 15 min. A goat anti-mouse or anti-rabbit IgG (or IgM) conjugated to 10-, 15-, 20-, and 40-nm gold particles

(Janssen, Beerse, Belgium) was chosen for application to 2-μm serial sections of the same block. The smaller gold particles (10 and 15 nm) give greater resolution for subsequent electron microscopy, whereas larger gold particles (20 and 40 nm) give better visual localization with light microscopy (see Figs. 18.3, 18.7A, and 18.9A,B).

Colloidal gold conjugates are diluted 1 : 10 with TBS immediately prior to use and centrifuged (all except 40-nm gold) in a microfuge (45 sec), and supernatant is applied to sections rinsed with TBS. Slides are incubated in a covered moist chamber at 22°C for 30–60 min. Slides are rinsed gently with TBS from a squirt bottle and put into a slide rack within a staining dish containing fresh phosphate-buffered saline (PBS). The immunostained sections are then postfixed in a solution of 3% glutaraldehyde in PBS either applied dropwise to the slide or immersed in a staining dish for 10 min. Another rinse of PBS from a squirt bottle is followed by soaking in PBS prior to light microscopy and/or processing for electron microscopy.

Microscopic Techniques

Light microscopy is carried out on a Zeiss Photomicroscope III. Tissue sections on slides stained by the peroxidase method (see above) are viewed under bright-field illumination with Zeiss plan-apo objectives. Micrographs are taken with Kodak 2415 Tech Pan film at 100 ASA.

Sections immunolabeled with immunoglobulin-conjugated colloidal gold are observed without additional chemical enhancement or counterstaining. Slides are examined after mounting under a coverslip (Corning No. 1) with a drop of glycerol. Sections are viewed and photographed under phase-contrast or Nomarski interference optics. After micrographs have been taken, slides are soaked in several changes of PBS to remove coverslips. Sections not used for further electron microscopy are dehydrated and mounted under a coverslip with Pro-Texx mounting medium. Sections used for electron microscopy are postfixed in 1% OsO_4 in PBS applied dropwise to the slide for 5 min, followed by rinsing in fresh PBS, and processed for electron microscopy.

ELECTRON MICROSCOPY OF IMMUNOSTAINED THICK SECTIONS

Reembedding Techniques for Electron Microscopy

Slides with sections are kept submerged and transferred quickly during dehydration in an alcohol series of 30, 50, 70, 95, (two times in each), and 100% (three times for 2 min in each). Slides are quickly transferred to two changes of propylene oxide, for 5 min each, followed by a 1 : 1 mixture of propylene oxide and Epon for 1 hr in a staining dish or Coplin jar. They are then individually removed from the mixture and several drops of this mixture are immediately applied to the sections. Excess resin is wiped completely from around the sections and from the bottom of the slide. A drop of 100% Epon is applied to the sections and allowed to infiltrate for 1 hr. Another drop of Epon is placed on the section and allowed to infiltrate for an additional hour. Excess resin is again wiped completely from around the sections.

BEEM embedding capsules or gelatin capsules filled to near capacity with 100% Epon are quickly inverted over the sections. Slides should be relatively clean with little or no resin seeping out from under the lip of the capsules. If seepage continues to occur from any capsule despite care in inverting the capsule over the section, the lip of the capsule may not be level with the surface of the slide, and the capsule should be replaced. Slides with mounted capsules are heated at 60°C for 24–48 hr to polymerize the resin. After cooling, capsules are removed along with sections from the slides by warming quickly on a hot plate at 80–90°C for 10–12 sec and bending the block to one side while simultaneously pushing down on the slide. Because of the volatility of ethanol and propylene oxide, staining racks should be used in experiments with multiple slides to allow for the rapid transfer of slides to successive changes of solutions during reembedment.

Electron Microscopy of Reembedded 2-μm Sections

Polymerized blocks are removed from the BEEM capsules and trimmed, and serial thin (90 nm, i.e., gold) and/or semithin (200 nm) sections are made. All sections are picked up on Formvar-coated 150–200-mesh copper grids. Grids are floated on drops of saturated aqueous uranyl acetate for 30 min, rinsed in double-distilled water, and floated on drops of fresh lead citrate for 3 min. They are rinsed in a stream of double-distilled water and dried. All grids are examined for the presence of labeling, because the depth of gold penetration is dependent on gold particle size, the method of tissue preparation, and tissue types.

ELECTRON MICROSCOPY OF DEPLASTICIZED ADJACENT THIN SECTIONS

Preparation and Handling of Grids

Grids for thin sections are prepared by stripping 0.3% Formvar from precleaned glass slides onto distilled water and placing 300-mesh nickel grids, shiny surface down, on top of the Formvar. Formvar should be of an appropriate thickness (silver diffraction color) to provide adequate support for thin sections. Grids with Formvar are picked up with a sheet of Parafilm (American Can Co., Greenwich, CT) and allowed to dry in a clean petri dish. Grids are then carefully removed from the Parafilm by scribing around them with the tip of a sharp forceps. Grids are placed on a clean glass slide, Formvar side (shiny side of grid) down. A medium coat of carbon is evaporated over the grids (sections will be picked up on the Formvar side of the grid). It is worth noting that only grids without holes or tears in the Formvar should be used in subsequent immunostaining procedures. This is important since it will be difficult to dry the back of grids with filter paper and allow the other side (section side) to remain wet without having solutions wick through the holes; consequently, care should be taken to avoid damage to the Formvar. Sections 70–90 nm thick (silver–gold) when picked up on the carbon-stabilized Formvar grid, will show a red to purple diffraction color.

During subsequent procedures grids are always picked up by the edge and handled with anticapillary forceps to avoid mechanical damage to the film and section and to minimize carryover of solutions between drops. Transfer of floating grids between drops of aqueous solutions is done carefully so as not to wet the back side of the grid (section side is always kept wet). Grids are transferred between alcoholic solutions quickly and are always kept submerged to prevent volatilization and drying of the solvents from the grid surface. Forceps with grids are held in a forcep rack (EM Indus, Redmond, WA), and the rack is moved quickly between beakers of alcohols (Fig. 18.1).

Deplasticization of Thin Sections

The alcoholic NaOH solution (see above for preparation of ethoxide solution) was diluted 1 : 3 with ethanol just before use in small polyethylene beakers. The beaker should be kept covered with a small petri dish between uses, because the solution is very volatile and NaOH crystals will form on the surface. Grids held by anticapillary forceps in racks are totally immersed in this solution for the removal of resin. Resin removal times are determined by the use of test grids, which are immersed in the sodium ethoxide solutions for various durations followed by several rinses in 100% ethanol.

Test grids are dried and examined unstained in the electron microscope to determine the amount of resin removed. Removal times can vary between 2 and 10 min, depending on block hardness and types of resin used. Typical times for resin removal are 2–5 min for Epon and 5–10 min for

18. Correlative Immunocytochemistry on Reembedded Sections 367

Fig. 18.1. Grids are transferred between beakers of solvents during resin removal (10–15 min) and reembedding (10 min) with the aid of a forcep rack, which not only eliminated fatigue in holding the grids but also allowed both hands to be free for other manipulations.

Araldite. Adequate removal of sufficient quantities of resin is characterized by indistinct edges of the trimmed resin when viewed in the transmission electron microscope (TEM). After the resin removal times have been determined, experimental grids are deplasticized and rinsed in three consecutive washes of 100% ethanol (1 min each). Grids are immediately hydrated to water through a graded alcohol series (95%—two times, 70%, 50%, 30%, double-distilled water, 1 min each). The back of the grid is blotted quickly and immediately floated (section side down) onto a droplet of 0.05 M TBS on a clean piece of Parafilm while the section side of the grid is still wet. Grids are allowed to remain in a covered moist chamber until all grids are ready for subsequent immunostaining.

Colloidal Gold Immunocytochemical Staining

Grids are kept covered in moist chambers (petri dishes with moistened filter paper on the bottom) between incubations and rinses. Serum and antibody incubations as well as rinses are performed on clean pieces of Parafilm or dental wax. Droplets of buffer for rinses are administered by syringes with disposable 2-μm pore size Millipore filters (Millipore, Bedford, MA). The immunostaining protocol is as follows: Grids are transferred from TBS to droplets of 10% normal goat serum Millipore filtered in TBS for 10 min, followed by rinsing in TBS four times for a total period of 10 min. Primary antiserum is diluted in filtered TBS just before use (i.e., guinea pig anti-rat growth hormone, 1 : 500; rabbit anti-lamb kidney Na^+,K^+-ATPase 1 : 100; rabbit anti-human myocardium actin, HHF 35, 1 : 1000; and rabbit anti-human skin fibroblast DSPG, 1 : 20).

Grids are incubated overnight on droplets of antisera at 4°C. Each grid is rinsed on 6 droplets of TBS for a total period of 15 min prior to incubation on colloidal gold solutions. Goat anti-rabbit IgG or goat anti-mouse IgG or IgM conjugated to 10-, 15-, 20-, or 40-nm colloidal gold particles is diluted 1 : 10 with filtered TBS immediately before use. Grids are incubated on gold solutions for 45 min before rinsing in 7–10 drops of fresh TBS. This step is followed by rinsing in filtered PBS four times for 10 min and fixation in 2% glutaraldehyde in PBS for an additional 10 min. Grids are rinsed four times in PBS and floated on droplets of 1% OsO_4 in PBS. After fixing for 10 min, grids are rinsed three times in distilled water and floated on drops of 2% neutral uranyl acetate for 30 min. This solution is prepared by mixing equal volumes of 4% aqueous uranyl acetate and 0.3 M oxalic acid and the pH is adjusted to 7.2 with 10% NaOH (Tokuyasu, 1978).

Reembedding of Thin Sections

Immediately after staining on drops of neutral uranyl acetate, grids are held in anticapillary forceps (inserted into forcep racks) and dehydrated in an alcohol series of 30, 50, 70, 90, and 100% (three changes for 1 min each). Grids are immediately immersed in 2% Epon or Araldite in 100% ethanol (2 min) and blotted immediately between a folded piece of No. 50 Whatman filter paper, being careful not to abrade the surface of the grids (Keller *et al.*, 1984). The grids are set up by their edges on a piece of silicon rubber (back side of an embedding mold from Pelco, Inc., Redding, CA, with small shallow slits cut into it; Fig. 18.2). Polymerization is carried out overnight in a covered petri dish at 60°C before removal from slits and counterstained with 4% aqueous uranyl acetate and lead citrate.

GENERAL CHARACTERISTICS OF DEPLASTICIZED SECTIONS

Section Thickness

About 1.5–2.0-μm-thick sections are cut with a glass knife for correlative light and electron microscopy on the same section. Sections of this thickness not only give good resolution and sharpness for light micros-

18. Correlative Immunocytochemistry on Reembedded Sections

Fig. 18.2. After immersion in dilute Epon, grids are set up into slots cut into a piece of silicon rubber and polymerized at 60°C for 24–48 hr.

copy after staining but also are of sufficient thickness to allow resectioning of the same section for electron microscopy after reembedment. Thin sections cut from blocks to be used directly for colloidal gold immunostaining are 60–80 nm thick (silver–light gold). Sections thicker than 90 nm failed to give adequate transmission at 80 kV after reembedding. The removal of resin does not appear to affect the morphology preserved in the section of tissue. In fact, resin reembedded thin sections show topographical variations as resin infiltrates and adsorbs to the components of the deplasticized section (Figs. 18.5 and 18.6), contributing contrast and definition of certain components in the tissue.

Rate of Resin Removal

The removal of resin from sections and the rate of removal are dependent on three factors: (1) type of resin used, (2) thickness of the section, and (3) concentration of the ethoxide solution. Tissues embedded in Araldite 502 or Epon 812 are preferred because these resins dissolve readily in the ethoxide solution. Medcast (Pelco, Redding, CA), Polybed 812 (Polyscience, Warrington, PA), and Spurr (Pelco) require much longer removal times. Removal of Lowicryl K4M (Polysciences) or LR White (E. F. Fullam, Schenectady, NY) from tissue sections has not been successful.

Section thickness inevitably influences the time required for adequate resin removal. We routinely cut 1.5–2.0-μm-thick sections for correlative immunocytochemical staining. Resin removal times are ~20 min for Epon and 30–40 min for Araldite, when the ethoxide solution is diluted 1:1 with 100% absolute ethanol. Deplasticization of 70–90-nm Epon thin sec-

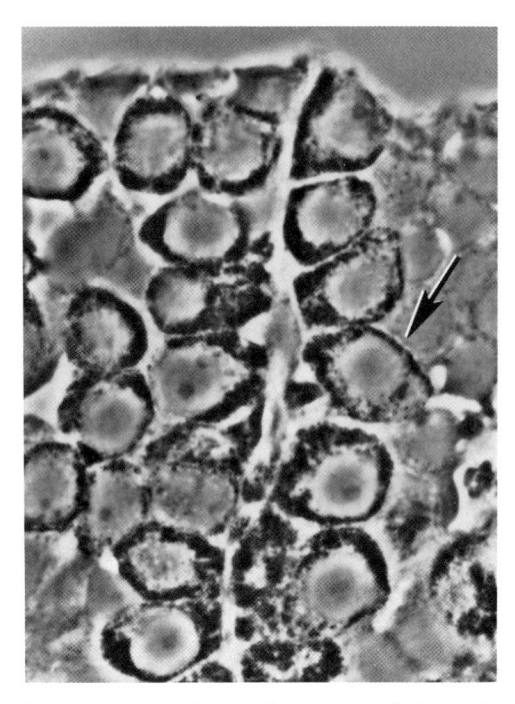

Fig. 18.3. Rat pituitary section stained with immunoglobulin-conjugated 20-nm colloidal gold particles for growth hormone. Phase-contrast micrograph of a deplasticized 2-μm-thick section showing dense cytoplasmic immunolabeling surrounding the nucleus of growth hormone cells. Note that the cell with the arrow is the cell in the electron micrograph of Fig. 18.4; ×1700. (From Mar *et al.*, 1987; reprinted by permission of The Histochemical Society, Inc.)

tions on grids requires 3–3.5 min immersion in a 1 : 3 mixture of alcoholic-NaOH and 100% ethanol, and up to 10 min for Araldite sections. It is necessary to establish the required time for removal by using test slides and grids (as discussed earlier) for each experiment since the strength of the ethoxide solution changes on standing. Optimal strength is reached when the solution is of a light to medium orange color (∼3–5 days after preparation). Lighter-colored solutions require longer removal times, and darker solutions generally result in rapid tissue damage.

Preservation of Antigenicity

Resin removal does not appear to decrease the antigenicity of the tissue or alter immunoreactive sites. In fact, resin removal from thick sections actually increases the intensity of immunostaining (Erlandsen *et al.* 1979; Rodning *et al.*, 1980; Horobin and Proctor, 1982). Furthermore, modifi-

18. Correlative Immunocytochemistry on Reembedded Sections 371

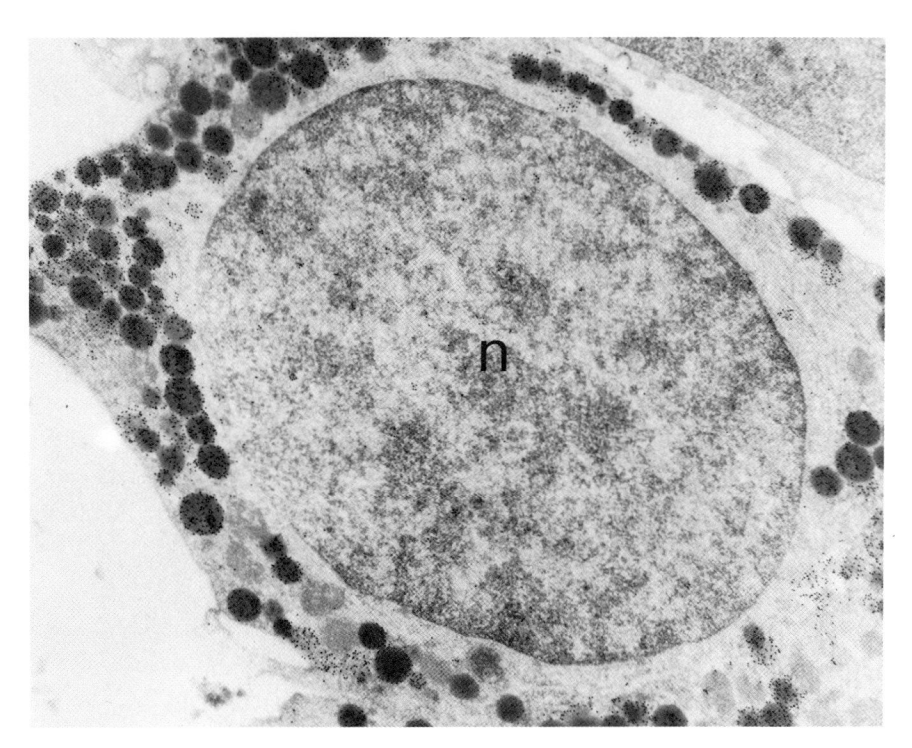

Fig. 18.4. Rat pituitary growth hormone cell labeled with 20-nm colloidal gold particles. Section in Fig. 18.3 was reembedded in Epon, and semithin sections were collected on grids and stained with uranyl acetate and lead citrate. n, Nucleus; ×15,000. (From Mar *et al.*, 1987; reprinted by permission of The Histochemical Society, Inc.)

cation of the removal procedures for use on resin thin sections does not seem to destroy the sites of immunoreactivity, nor does it appear to extract intracellular or extracellular components. For example, tissues such as pituitary which have been immunostained by traditional postembedding methods at the electron microscopic level with anti-growth hormone (Baskin *et al.*, 1979b, 1980) yielded identical staining patterns when they were deplasticized and immunostained on grids with the same antibodies (not shown).

Removal of resin from thin sections has been particularly useful where postembedding immunostaining was successful for light microscopy but electron microscopic immunostaining of adjacent thin sections from the same block was not possible through traditional methods. For example, when resin was removed from 1.5-μm kidney sections localization of Na^+,K^+-ATPase was possible, but localization of the antigenic sites at the electron microscopic level on Epon thin sections yielded consistent

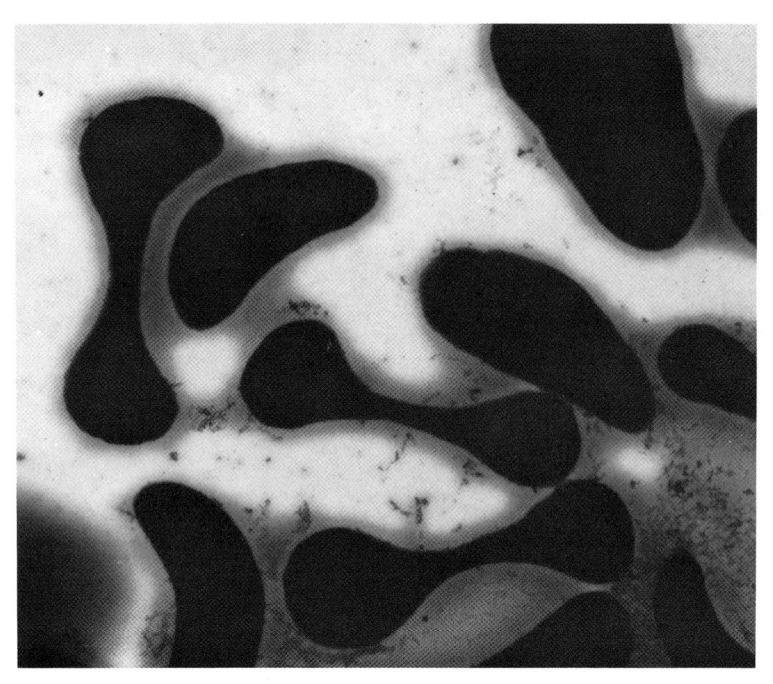

Fig. 18.5. Red blood cells. Thin section was deplasticized while on the grid, and the tissue was reembedded by immersion of the grid in dilute Epon. Note residual resin adsorbing to the red blood cells during reembedding, demonstrating that the original resin embedment of the thin section was removed; ×6500.

background staining on all areas of the section, including those outside the tissue where only resin was present. Both the anti-smooth muscle cell actin and anti-dermatan sulfate antibodies gave intense staining on deplasticized 1-μm-thick sections of blood vessels but failed to give any localization (including background) on thin sections. Removal of the embedding resin from around these thin sections resulted in specific localization of antigenic sites with low background staining (Figs. 18.7–18.9). We conclude that the removal of resin from either 1-μm-thick or thin sections with sodium ethoxide does not significantly reduce antigenicity originally preserved in the tissue and that the procedures outlined above can effectively increase the specificity and intensity of immunostaining.

Preservation of Ultrastructure

The procedure as described does not appear to destroy tissue ultrastructure. However, the lack of osmium in the initial fixation prior to primary embedment often results in a decrease in membrane contrast. Well-preserved ultrastructure is observed in 1-μm sections which have been

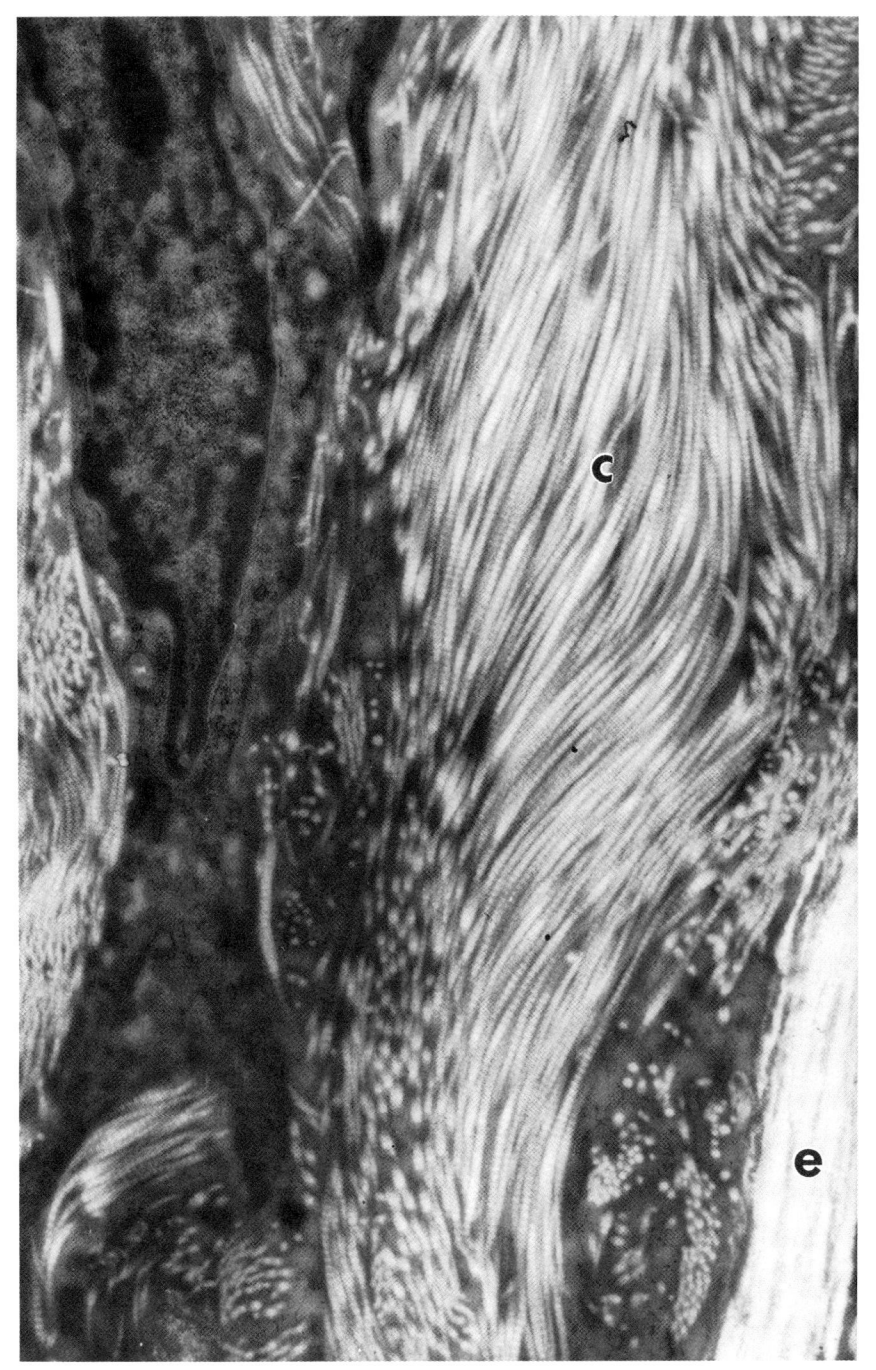

Fig. 18.6. Human superficial femoral artery. Deplasticized and reembedded thin section. Note preservation of ultrastructure and definition of collagen fibrils (c) and elastin (e); ×16,400.

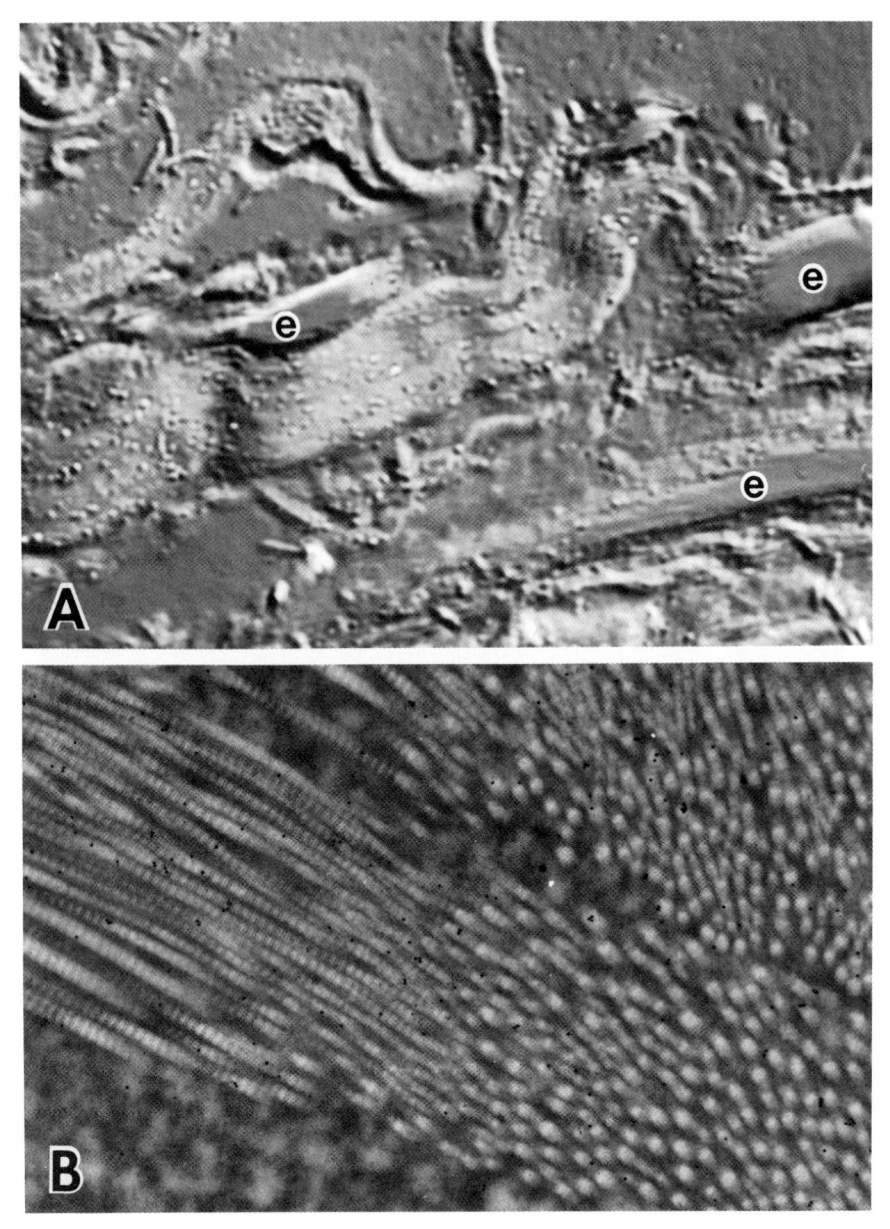

Fig. 18.7. Human superficial femoral artery. Sections immunostained for dermatan sulfate proteoglycan (DSPG) with immunoglobulin-conjugated colloidal gold. (A) 1.5-μm-thick section photographed under Nomarski interference optics after resin removal and immunolabeling with 40-nm gold particles. Note labeling of large collagen bundle and lack of staining on elastin (e); ×2100. (B) Electron micrograph of a thin section which had been deplasticized, immunostained, and reembedded in dilute Epon. Note labeling of 10-nm gold particles on collagen fibrils; ×19,000.

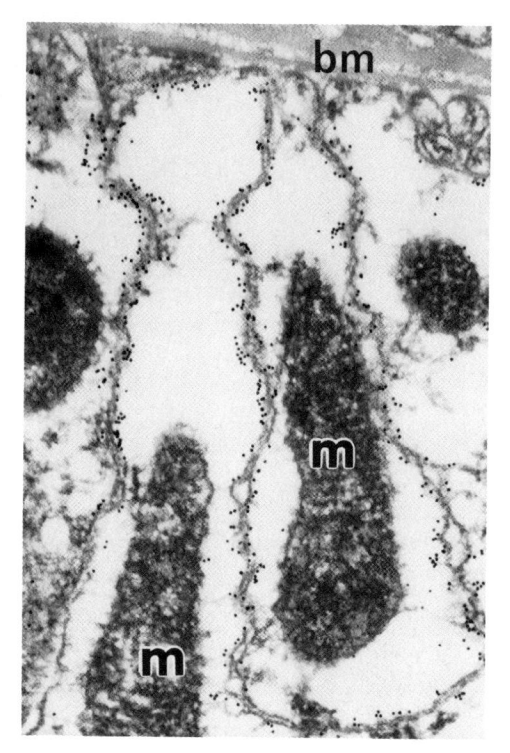

18. Correlative Immunocytochemistry on Reembedded Sections 375

Fig. 18.8. Rat kidney stained with immunoglobulin conjugated to 10-nm colloidal gold particles for Na^+,K^+-ATPase. Thin section was deplasticized and reembedded after immunolabeling. Micrograph shows localization of gold particles to the lateral membrane processes of the distal convoluted tubules. m, Mitochondria; bm, basement membrane; ×38,100.

deplasticized, immunostained, reembedded in Epon, and resectioned for electron microscopy (see Fig. 18.4). Additional support for the conclusion that resin removal and reembedment does not result in significant loss of ultrastructure is obtained when thin (70–90-nm) sections are deplasticized and reembedded in dilute Epon while on the grid (Figs. 18.6–18.9). Cytoplasmic granules, endoplasmic reticulum, mitochondria, and actin filaments are distinct and well preserved. Extracellular matrix structures such as collagen and elastin appear intact and sharply delineated.

LIMITATIONS

The procedures described can yield specific localization of antigenic sites at the electron microscopic level on the same resin-embedded tis-

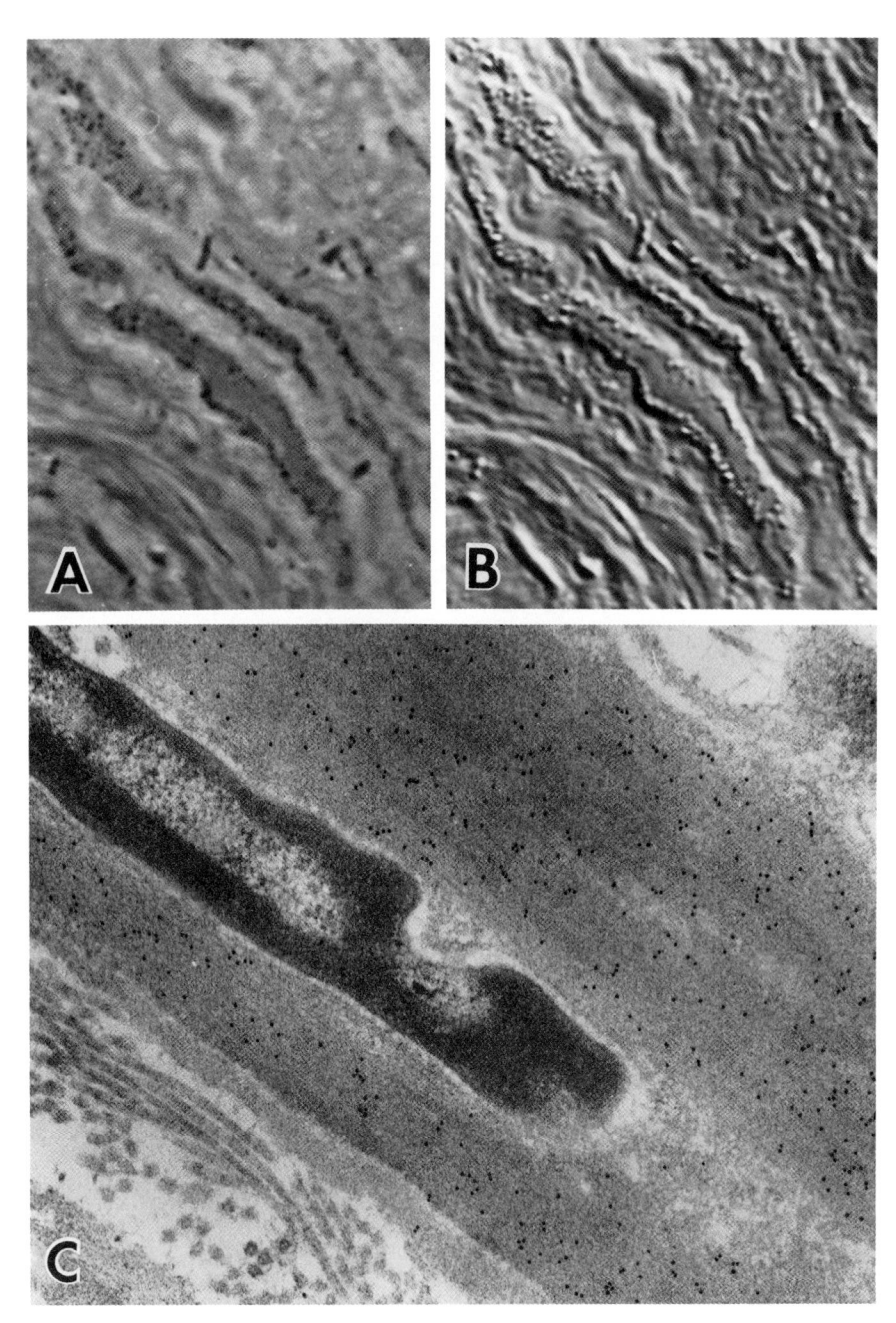

Fig. 18.9. Human superficial femoral artery. Deplasticized sections stained with HHF 35 antibody against smooth muscle cell actin and IgG-conjugated colloidal gold. A smooth muscle cell in a 1.5-μm-thick deplasticized section immunostained with 40-nm colloidal gold particles and photographed under phase-contrast optics (A) and Nomarski interference optics (B). Note that gold particles are localized at the peripheral margins of the cell; ×2100.

sues which immunostain at the light microscopic level. In many cases, we have been successful in localizing antigenic sites at the electron microscopic level where we have failed through traditional postembedding methods. However, there are limitations to these procedures: (1) possible extraction or destruction of antigenic sites with alcoholic sodium hydroxide, (2) technical difficulties in handling large numbers of grids within time schedules, and (3) difficulties with removal of some resins which are only weakly soluble in the alcoholic sodium hydroxide solution. Although any of these factors can lead to lack of precise immunolocalization, we find that tissues which stain at the light microscopic level when resin is removed will also stain at the electron microscopic level when treated by similar procedures.

REFERENCES

Alexander, R. B., Issaacs, W. B., and Barrack, R. R. (1985). Immunogold probes for electron microscopy: Evaluation of staining by fluorescence microscopy. *J. Histochem. Cytochem.* **33**, 995.

Baskin, D. G., Erlandsen, S. L., and Parsons, J. A. (1979a). Immunocytochemistry with osmium fixed tissue. I. Light microscopic localization of growth hormone and prolactin with the unlabeled antibody–enzyme method. *J. Histochem. Cytochem.* **27**, 867.

Baskin, D. G., Erlandsen, S. L., and Parsons, J. A. (1979b). Influence of hydrogen peroxide or alcoholic sodium hydroxide on the immunocytochemical detection of growth hormone and prolactin after osmium fixation. *J. Histochem. Cytochem.* **27**, 1290.

Baskin, D. G., Erlandsen, S. L., and Parsons, J. A. (1980). Functional classification of cell types in the growth hormone and prolactin secreting rat MtTW$_{15}$ mammosomatotropic tumor with ultrastructural immunocytochemistry. *Am. J. Anat.* **158**, 455.

Baskin, D. G., and Stahl, W. L. (1982). Immunocytochemical localization of Na$^+$,K$^+$-ATPase in the rat kidney. *Histochemistry* **73**, 535.

Baskin, D. G., Mar, H., Gorray, K. C., and Fujimoto, W. Y. (1982). Electron microscopic immunoperoxidase staining of insulin using 4-chloro-1-naphthol after osmium fixation. *J. Histochem. Cytochem.* **30**, 710.

De Brabander, M., De Mey, J., Joniau, M., and Geuens, G. (1977). Immunocytochemical visualization of microtubules and tubulin at the light and electron microscopic level. *J. Cell. Sci.* **28**, 283.

De Mey, J., Moeremans, M., Geuens, G., Nuydens, R., and De Brabander, M. (1981). High resolution light and electron microscopic localization of tubulin with the IGS (immunogold staining) method. *Cell. Biol. Int. Rep.* **5**, 889.

Erlandsen, S. L., Parsons, J. A., and Rodning, C. B. (1979). Technical parameters of immunostaining of osmicated tissue in epoxy sections. *J. Histochem. Cytochem.* **27**, 1286.

(C) Electron micrograph of a thin section which had been deplasticized, immunostained with HHF 35 and 10-nm colloidal gold particles, and reembedded in dilute Epon. Gold is clearly localized to areas in smooth muscle cells corresponding to those seen at the light microscopic level (A and B); ×47,600.

Geoghegan, W. D., Scillian, J. J., and Ackerman, G. A. (1978). The detection of human B lymphocytes by both light and electron microscopy utilizing colloidal gold labeled anti-immunoglobulin. *Immunol. Commun.* **7**, 1.

Glossl, J., Beck, M., and Kresse, H. (1984). Biosynthesis of proteodermatan sulfate in human fibroblasts. *J. Biol. Chem.* **259**, 14144.

Gown, A. M., and Vogel, A. M. (1982). Monoclonal antibodies to intermediate filament proteins of human cells: Unique and cross-reacting antibodies. *J. Cell Biol.* **95**, 414.

Holec, B., and Ciampor, F. (1978). Methods of re-embedding tissues and cell suspensions for electron microscopy. *Folia Morphol. (Prague)* **26**, 154.

Horisberger, M., and Rosset, J. (1977). Colloidal gold, a useful marker for transmission and scanning electron microscopy. *J. Histochem. Cytochem.* **25**, 195.

Horisberger, M., (1979). Evaluation of colloidal gold as a cytochemical marker for transmission and scanning microscopy. *Biol. Cell.* **36**, 253.

Horisberger, M. (1981). Colloidal gold: A cytochemical marker for light and fluorescent microscopy and for transmission and scanning electron microscopy. *Scanning Electron Microsc.* **11**, 9.

Horobin, R. W., and Proctor, J. (1982). Estimating the effect of etching agents on plastic sections. *J. Microsc. (Oxford)* **126**, 169.

Keller, G. A., Tokuyasu, K. T., Dutton, A. H., and Singer, S. J. (1984). An improved procedure for immunoelectron microscopy: Ultrathin plastic embedding of immunolabeled ultrathin frozen sections. *Proc. Natl. Acad. Sci. U.S.A.* **81**, 5744.

Lackie, P. M., Hennessy, R. F., Hacker G. W., and Polak, J. M. (1985). Investigation of immuno-gold–silver staining by electron microscopy. *Histochemistry* **83**, 545.

Lamberts, R., and Goldsmith, P. C. (1985). Pre-embedding colloidal gold immunostaining of hypothalamic neurons: Light and EM localization of β endorphin-immunoreactive perikarya. *J. Histochem. Cytochem.* **33**, 499.

Lane, B. P., and Europa, D. L. (1965). Differential staining of ultrathin sections of Epon embedded tissues for light microscopy. *J. Histochem. Cytochem.* **13**, 579.

Mar, H., Tsukada, T., Gown, A. M., Wight, T. N., and Baskin, D. G. (1987). Correlative light and electron microscopic immunocytochemistry on the same section with colloidal gold. *J. Histochem. Cytochem.* **35**, 419.

Mar, H., and Wight, T. N. (1988). Colloidal gold immunostaining on deplasticized ultrathin sections. *J. Histochem. Cytochem.* **36**, 1387.

McCans, J. L., Lindenmayer, G. E., Pitts, B. J. R., Ray M. V., Raynor, B. D., Butler, V. P., Jr., and Schwartz, A. (1975). Antigenic differences in (Na^+,K^+)-ATPase preparations isolated from various organs and species. *J. Biol Chem.* **25**, 7251.

Rodning, C. B., Erlandsen, S. L., Coulter, H. D., and Wilson, I. D. (1980). Immunohistochemical localization of IgA antigens in sections embedded in epoxy resin. *J. Histochem. Cytochem.* **28**, 197.

Sternberger, L. A. (1986). *Immunocytochemistry,* 3rd ed. Wiley, New York.

Tokuyasu, K. T. (1978). A study of positive staining of ultrathin frozen sections. *J. Ultrastruct. Res.* **63**, 287.

Tsukada, T., Tippens, D., Gordon, D., Ross, R., and Gown, A. M. (1987) HHF 35, A muscle-actin-specific monoclonal antibody. I. Immunocytochemical and biochemical characterization. *Am. J. Pathol.* **126**, 51.

19

Streptavidin–Gold Labeling for Ultrastructural *in Situ* Nucleic Acid Hybridization

ROBERT A. WOLBER

Department of Anatomic Pathology
Vancouver General Hospital
University of British Columbia
Vancouver, British Columbia, Canada

and

THEODORE F. BEALS

Department of Pathology
Ann Arbor Veterans Administration Center
The University of Michigan School of Medicine
Ann Arbor, Michigan

INTRODUCTION
PRINCIPLES OF HYBRIDIZATION TECHNIQUE
   Fixation
   Probes
PREEMBEDDING ULTRASTRUCTURAL
   HYBRIDIZATION
   Cell Pretreatments
   Hybridization

Quantitation
Improvements
POSTEMBEDDING HYBRIDIZATION ON THIN SECTION
APPLICATIONS
REFERENCES

INTRODUCTION

Many publications have appeared on the technique of *in situ* nucleic acid hybridization and its application to various biological specimens. Refinements of the methodology initially described by Gall and Pardue (1969) have led to a multiplicity of protocols for autoradiographic and nonisotopic detection of large single-copy nucleic acid sequences and highly reiterated sequences within tissue sections, cytological preparations, cultured cells, and chromosome spreads (Brahic and Haase, 1978; Langer-Safer *et al.*, 1982; Manuelidis *et al.* 1982; Brigati *et al.*, 1983; Gee and Roberts, 1983; Lawrence and Singer, 1985; Haase, 1986; Singer *et al.*, 1986; Unger *et al.*, 1986; Lawrence *et al.*, 1988). While applications of *in situ* hybridization have generally been restricted to the light microscope (LM) level, early attempts were made to extend *in situ* hybridization to the electron microscope (EM) level in tissue sections and whole cells, using radioactively labeled nucleic acid probes (Jacob *et al.*, 1971, 1974; Croissant *et al.*, 1972; Gueskens and May, 1974; Steinert *et al.*, 1976). The subsequent development of nonisotopic nucleic acid probes (Manning *et al.*, 1975; Langer *et al.*, 1981; Singer and Ward, 1982; Leary *et al.*, 1983) has been applied to the ultrastructural detection of DNA sequences in chromosomes and centromeres (Wu and Davidson, 1981; Hutchison *et al.*, 1982; Hutchison, 1984; Manuelidis and Ward, 1984; Cremers *et al.*, 1987) and, more recently, to the detection of RNA sequences in thin sections of resin-embedded tissue (Binder *et al.*, 1986; Binder, 1987; Webster *et al.*, 1987) and in fixed tissue slices (Harris and Croy, 1986; Webster *et al.*, 1987).

This review is intended to summarize our experience with preembedding ultrastructural localization of viral DNA and RNA (Wolber *et al.*, 1988, 1989) using biotinylated viral DNA probes and streptavidin–gold as an electron-dense marker. The focus of the discussion is on methodologic aspects of the preembedding technique. Because the literature on *in situ* hybridization technique is extensive, and no single methodology is universally used, the literature is not exhaustively referenced. Instead, articles which provide access to other references are preferentially cited.

PRINCIPLES OF HYBRIDIZATION TECHNIQUE

In situ hybridization allows detection of specific nucleic acid sequences within a tissue, cell, or chromosome, while providing the exact spatial location of the identified sequences. Specificity of binding of nucleic acid probes to target molecules within the specimen is dependent upon coupling of complementary strands with appropriate nucleotide base sequences. The base-pairing reaction is characterized by a specific thermal denaturation temperature (T_m), below which hybridization takes place. The T_m is about 90°C under usual conditions, and maximal hybridization occurs at ~25°C below the T_m of duplexes formed *in situ*. The T_m is reduced under conditions of alkaline pH and low ionic strength and by increasing amounts of polar anionic molecules (e.g., formamide) in the solution (Marmur and Doty, 1962; Shildkraut and Lifson, 1965; McConaughy *et al.*, 1969). T_m is also reduced as the percentage of adenine and thymidine bases and as the percentage of base pair mismatches between complementary sequences increases. Hence, the hybridization reaction can proceed under conditions of high stringency, favoring single strandedness, or low stringency, favoring double strandedness. High-stringency hybridization implies high specificity of binding (specificity is also determined by probe length). Stringency conditions are controlled during the hybridization reaction and during posthybridization washes. No single set of conditions is critical to the success of the hybridization technique, yet most protocols utilize a hybridization cocktail containing sodium citrate–sodium chloride buffer (SSC), formamide, dextran sulfate, and sheared nonspecific DNA and tRNA (to competitively inhibit nonspecific nucleic acid binding).

In situ hybridization can be used to localize both DNA and RNA. If DNA is to be detected, the specimen must first be heated to a temperature greater than the T_m in order to separate double-stranded DNA segments. Extensive heating is unnecessary for RNA localization, since RNA is single stranded in its native form.

Fixation

Prior to hybridization, tissues or cells must be fixed to preserve morphology and to retain the target nucleic acids through hybridization conditions and posthybridization washing steps. Optimal fixation provides for nucleic acid retention, while allowing accessibility of target nucleic acids to probe molecules. The efficacy of various fixatives for RNA hybridization has been extensively reviewed by Singer *et al.* (1986). They found that freshly prepared solutions of formaldehyde (4%) gave optimal RNA retention at the LM level, provided prehybridization tissue digestion steps were not used. Glutaraldehyde was also found to be an effective fixative when used in conjunction with protease digestion (~1 µg/ml protease for every 1% of glutaraldehyde used), since glutaraldehyde produces extensive cross-linking of cellular proteins, which appear to inhibit probe penetration into the cell. Carnoy's fixative and ethanol–acetic acid were found to be ineffective for RNA retention, although they have been advocated along with methanol–acetic acid for DNA hybridization (McAllister and Rock, 1985; Lawrence et al., 1988). Buffered formalin provides excellent fixation for DNA retention in paraffin-embedded tissue sections (Unger et al., 1986).

Probes

Nucleic acid probes can be composed of double- or single-stranded DNA or single-stranded RNA. Double-stranded probes are generated by nick translation and, with large probes, overlapping sequences from unhybridized portions can be used to magnify signals by probe networking (Lawrence and Singer, 1985). Single-stranded probes provide increased efficiency of hybridization under saturating conditions (Cox et al., 1984) in the absence of competitive reannealing of complementary probe sequences. Single-stranded RNA probes offer the advantage that RNA–RNA hybrids are the most stable nucleic acid duplexes, hence hybridization and wash conditions can be very stringent. In addition, unutilized probe can be eliminated by posthybridization treatment with RNase, minimizing background labeling. Small oligonucleotide probes are very efficient due to increased tissue penetration and are rapidly becoming widely available.

The hybridized probe sequence can be detected autoradiographically using ^{35}S-, ^{32}P-, or tritium-labeled probes, or by using nonisotopic, biotinylated probes (Manning et al., 1975; Langer et al., 1981; Singer and Ward, 1982). Biotinylated probes are chemically stable and retain reactivity for prolonged periods when stored at −20°C. They can be detected using antibiotin antibodies, avidin, or streptavidin coupled to enzymatic detection systems (Brigati et al., 1983), fluorescent molecules (Langer-Safer et al., 1982), or gold particles (Hutchison et al., 1982) with silver enhancement for light microscopic work.

PREEMBEDDING ULTRASTRUCTURAL HYBRIDIZATION

The utility of light microscopic *in situ* hybridization in localizing specific nucleic acid sequences when applied to problems in diagnostic pathology, virology, cytogenetics, and developmental cell biology has been amply demonstrated. Clearly, extension of the LM technique to the EM level in whole-cell preparations would represent a significant advance. It would allow subcellular localization of sites of viral DNA replication, cellular or viral RNA transcription, and, if combined with ultrastructural immunocytochemistry, RNA translation into specific proteins. Possible approaches to this problem include hybridization of thin cryosections or of fixed tissue sections (Harris and Croy, 1986), postembedding labeling of resin thin sections (Binder *et al.*, 1986), or preembedding labeling of fixed cells permeabilized by treatment with a detergent and protease (Wolber *et al.*, 1988, 1989).

We approached ultrastructural hybridization using a preembedding technique because it allowed the least degree of deviation from established hybridization protocols at the LM level and because of concern that the embedding resin matrix would render cellular nucleic acids inaccessible to probe molecules. Furthermore, we were interested in localizing highly reiterated viral sequences in cultured cells, and preembedding hybridization of cells in aqueous solution is well adapted to that experimental system.

Possible disadvantages of this approach are that some ultrastructural membrane detail is sacrificed, and probe molecules and gold markers may not completely penetrate into fixed whole cells. Although autoradiographic (Gueskens and May, 1974) and immunoperoxidase (Harris and Croy, 1986) procedures have previously demonstrated labeling of fixed whole cells and tissues by nucleic acid probes at the EM level, we were specifically interested in using fully quantifiable and spatially precise gold particles for localizing nucleic acid sequences, which have not been generally accepted for use in preembedding techniques (Sternberger, 1986). We found that precise use of a detergent and protease pretreatments allowed penetration of cell cytoplasm and nuclei by nucleic acid probes and gold particles with good preservation of ultrastructural detail.

Cell Pretreatments

Our initial experiments involved ultrastructural localization of human cytomegalovirus (CMV) DNA in neonatal human foreskin fibroblast cultures. Later, we worked with human herpes simplex virus type II (HSV) cultures due to the shorter incubation period necessary for development of cytopathic effect (CPE).

Cultures of human foreskin fibroblasts were established by digestion of fresh neonatal foreskins in 0.5% trypsin (Difco, Detroit MI) in phosphate-buffered saline (PBS) (pH 7.2) at 37°C for 1 hr. Cells were grown to confluent monolayers in 75-cm^2 sterile culture flasks in Eagle's minimal essential medium (EMEM) (Gibco, Grand Island, NY) with 10% fetal calf serum (FCS) and 0.1% gentamicin. Cytomegalovirus or HSV viral culture supernatants in EMEM with 2% FCS were layered onto cultured cells at an approximate multiplicity of infection of 1–2 plaque-forming units per cell and adsorbed for 2 hr at 37°C. Inoculating fluid was then replaced with EMEM–2% FCS and cultures were incubated at 37°C. Early CPE appeared at 4 days in CMV cultures and was widespread at 18 days. Well-developed CPE was present in HSV cultures at 18–20 hr (overnight).

Cultures were harvested by gentle scraping with a rubber spatula and fixed in cold, buffered 5% formaldehyde or in a mixture of cold 3% formaldehyde and 3% glutaraldehyde overnight (18–24 hr). Trials of shorter fixation times generally resulted in excessive loss of cells during subsequent posthybridization wash steps. All fibroblast processing was carried out in clean glassware rinsed with 0.4% diethyl pyrocarbonate (DEPC) (Sigma, St. Louis, MO) and autoclaved prior to use in order to inhibit contaminating RNase activity. In addition, latex gloves were worn during handling of all specimens and glassware.

Fixed fibroblasts (10^6 cells/sample) were washed twice for 10 min in Triton X-100 (Sigma) in PBS. Preliminary experiments at multiple concentrations of Triton X-100 revealed that a solution of 0.02–0.03% would selectively extract cell lipid membranes without severely disrupting ultrastructural morphology (Fig. 19.1) (Wolber *et al.*, 1988). Higher concentrations of Triton X-100, when used in conjunction with prehybridization protease digestion, resulted in excessive stripping of cell cytoplasm from nuclei. Omission of detergent pretreatments resulted in marked reduction of gold labeling.

Cells were washed an additional time in PBS and then incubated in a solution of proteinase K (Sigma) in PBS for 10 min. The concentration and temperature of incubation were determined by preliminary studies at the LM level. For DNA hybridization, a concentration of 500 μg/ml was used at room temperature. Incubation at this concentration at 37°C resulted in excessive cell digestion. For RNA hybridization, a 10 μg/ml concentration at 37°C was used. Higher concentrations of proteinase K appear to be required for detection of formaldehyde-fixed viral DNA (Unger *et al.*, 1986), while much lower concentrations allow detection of RNA if glutaraldehyde is included as a fixative (Lawrence and Singer, 1985). The studies of myoblast actin RNA expression by these scientists also indicate that omission of protease digestion provides optimal RNA detection in specimens fixed in formaldehyde alone. In our experience, omission of protease treatment eliminated hybridization signal in formaldehyde- or formaldehyde plus glutaraldehyde-fixed fibroblasts at the EM level. Following protease treatment, cells were again washed twice in PBS.

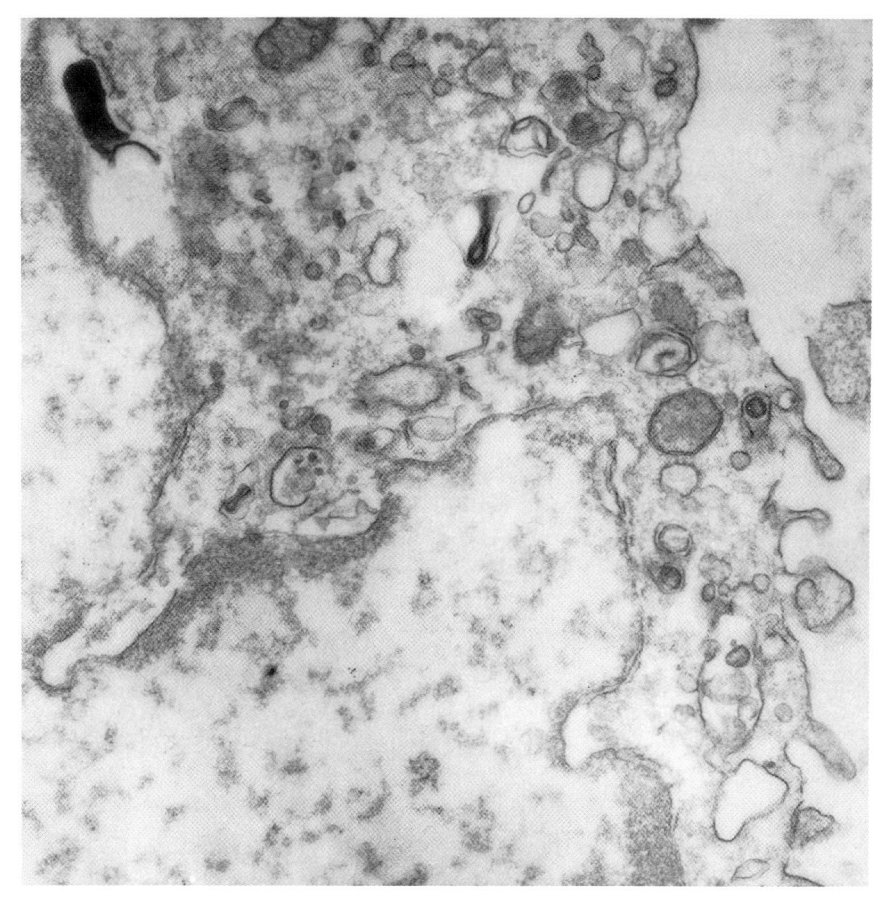

Fig. 19.1. Effect of pretreatment with Triton X-100. There are some breaks in the nuclear envelope and plasma membrane, but the cytoplasmic organelles are easily recognized. Fibroblast culture cell infected with herpes simplex virus processed for RNA *in situ* hybridization with 5-nm gold-labeled probe; ×31,000.

Hybridization

Control specimens are very important in the evaluation of hybridization labeling. Negative controls for these experiments consisted of uninfected fibroblasts, infected fibroblasts hybridized with biotinylated nonhomologous viral probe or plasmid DNA, and infected cells incubated with pancreatic DNase (Amersham International, Amersham, UK) (10 μg/ml) or RNase (Sigma) (100 μg/ml) in PBS for 1 hr, prior to hybridization. DNase solution was freshly prepared for each use. RNase solution was boiled for 15 min to eliminate contaminating DNase activity and then stored at 4°C for up to 1 month. An additional set of controls which has been recommended is specimens saturated with varying ratios of unlabeled and labeled probe. The decrease in hybridization signal is then compared with specimens hybridized with labeled probe alone.

Cells were subsequently washed twice in 2 × SSC (8.82 g sodium citrate, 17.53 g sodium chloride in 1 liter of RNase-free water), pelleted, and resuspended in hybridization diluent. Initially, separate hybridization diluents were used for DNA and RNA hybridization. However, equivalent DNA hybridization was obtained with the RNA medium, which was subsequently used for all experiments. The diluent was prepared in lots of 100 ml:

>   20 ml of 50% dextran sulfate (Sigma) in DEPC-treated water
>   15 ml of 2 × SSC
>   2 ml of 50 × Denhardt's solution (Sigma)
>   10 mg salmon sperm DNA (Sigma)
>   100 mg inorganic sodium pyrophosphate
>   18 ml DEPC-treated water
>   50 ml deionized formamide

The diluent was stored at −30°C between uses.

The nucleic acid probes were obtained from ENZO Biochemicals, New York, NY. They consisted of double-stranded biotinylated DNA cleaved from plasmid BR 322. The CMV probe was a mixture of two sequences, 25.2 and 17.2 kb in length, constituting 17% of the CMV genome. The HSV probe contained an 8-kb sequence from HSV-I and a 16-kb sequence from HSV-II. Individual probe fragments ranged in size from 200 to 1000 base pairs. Light microscopic characterization has indicated no cross-reactivity between these probes or with uninfected human cells. Initial studies utilized several concentrations of probe in probe diluent. Optimal labeling was obtained at a final CMV probe concentration of 20 μg/ml and HSV probe concentration of 6 μg/ml.

For DNA hybridization, cells were suspended in glass centrifuge tubes in 100 μl of probe mixture, covered with aluminum foil, and heated to 100°F for 20 min in a convection oven. They were then incubated overnight in a humidified chamber at 37°C. For RNA hybridization, cells were suspended in 100 μl of probe diluent alone and incubated at 60°C for 30 min. 100 μl of diluent containing probe at twice the final concentration was boiled for 3 min and added to each specimen. Specimens were then placed in a humidified chamber at 37°C and incubated overnight.

Following hybridization, cells were washed once in PBS–0.03% Triton X-100 for 2 min, once in 2 × SSC at 37°C for 10 min, twice in PBS–0.03%

Triton X-100 for 5 min each, and once in PBS–1% bovine serum albumin (BSA). Cells were then pelleted and resuspended in 200 µl of 5-nm streptavidin–gold (Janssen, Piscataway, NJ) at 1 : 10 dilution in PBS–1% BSA or in 15–20-nm streptavidin–gold (Polyscience, Warrington, PA) at 1 : 2 dilution in PBS–1% BSA. Diluted streptavidin–gold was centrifuged at 2000 rpm for 10 min immediately prior to use in order to eliminate gold aggregates. Specimens were then incubated overnight at room temperature with gentle agitation. Cells were then washed twice (30 min each) in PBS–1% BSA and postfixed in 2% glutaraldehyde in Sörensen's buffer for at least 1 hr.

For electron microscopy, cells were pelleted in 28% albumin, postfixed in OsO_4, dehydrated in graded alcohols, and embedded in a resin. Toluidin blue-stained sections 1 µm thick were used to select cell clumps for thin sectioning. Thin sections (60 nm) were cut, mounted on copper grids, and stained with salts of uranium and lead.

Quantitation

Initial experiments using 20-nm gold particles to label CMV nucleic acids succeeded in localizing viral DNA only (Wolber et al., 1988). Gold particles were identified within cell nuclei only, predominantly in areas surrounding electron-dense nuclear inclusions (Fig. 19.2). Cytomegalovirus cultures harvested at times of early CPE development (4 days) showed, within some cells, close association of gold particles with spherical electron-dense bodies ~70 nm in diameter, suggesting a condensed form of unencapsidated viral DNA. The 20-nm gold particles could be counted at a low magnification, in contrast to 5-nm gold particles (Fig. 19.2). An average of thirty 20-nm particles were identified per cell profile, resulting in an estimate of 2000 gold particles per cell nucleus. Gold particles were present as singlets and in tight clusters of up to six. Only rare gold particles were identified in DNase-treated and nonhomologous probe negative controls. RNase-treated specimens showed labeling equivalent to that of untreated positive specimens.

More recent experiments used 5-nm gold particles in HSV-infected fibroblasts harvested after 18–20 hr incubation (Wolber et al., 1989). In these experiments, HSV RNA was localized within cell nuclei and cytoplasm (Fig. 19.3). The 5-nm gold particles were present as singlets and in clusters of up to 38 particles. The 5-nm gold particles were too small for reliable identification at a low magnification; hence gold singlets and clusters were quantitated by linear scanning of cell profiles at 44,000×. Nuclear gold particles were found both in proximity to viral nucleocapsids and in areas surrounding amorphous aggregates of electron-dense mate-

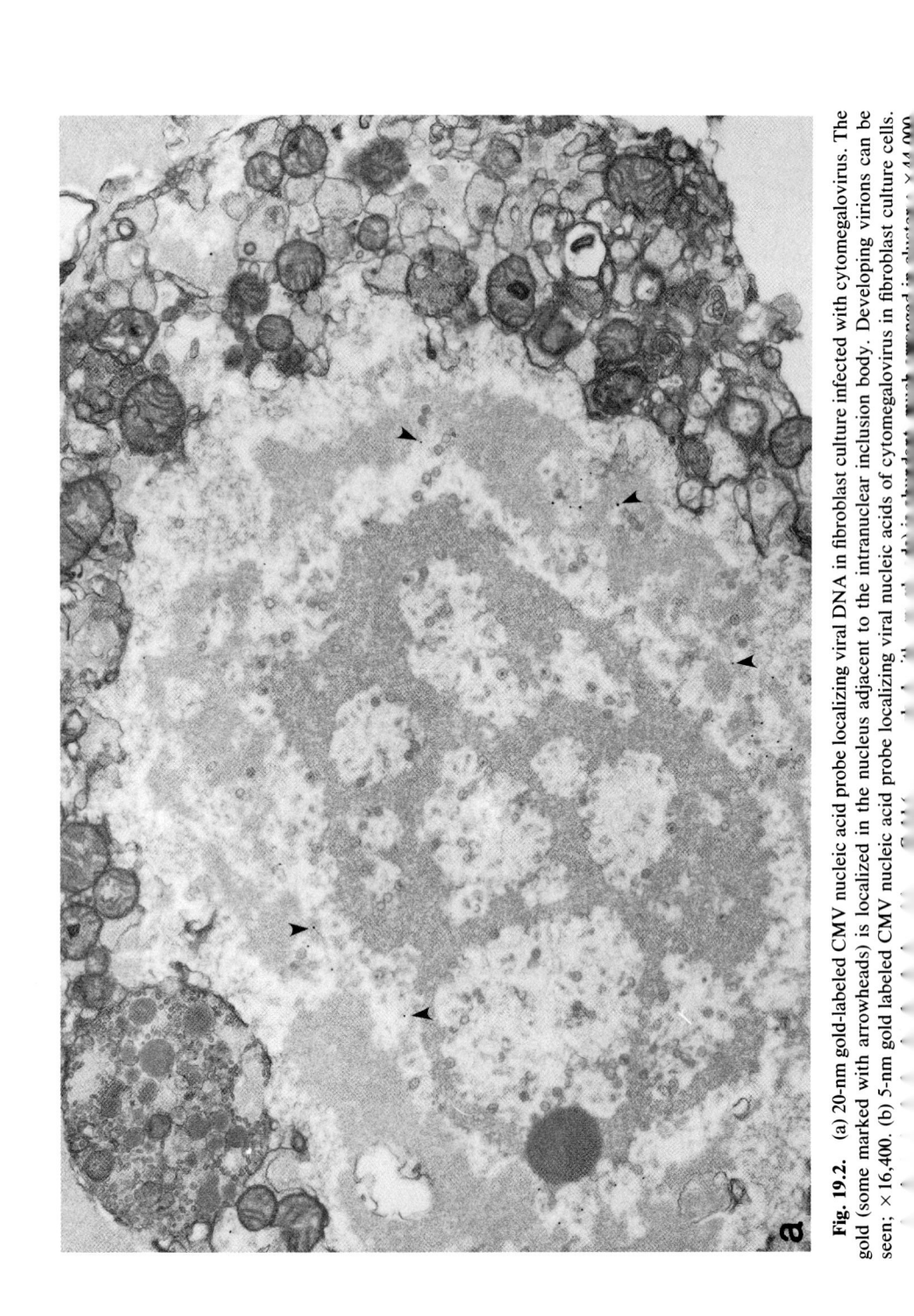

Fig. 19.2. (a) 20-nm gold-labeled CMV nucleic acid probe localizing viral DNA in fibroblast culture infected with cytomegalovirus. The gold (some marked with arrowheads) is localized in the nucleus adjacent to the intranuclear inclusion body. Developing virions can be seen; ×16,400. (b) 5-nm gold labeled CMV nucleic acid probe localizing viral nucleic acids of cytomegalovirus in fibroblast culture cells.

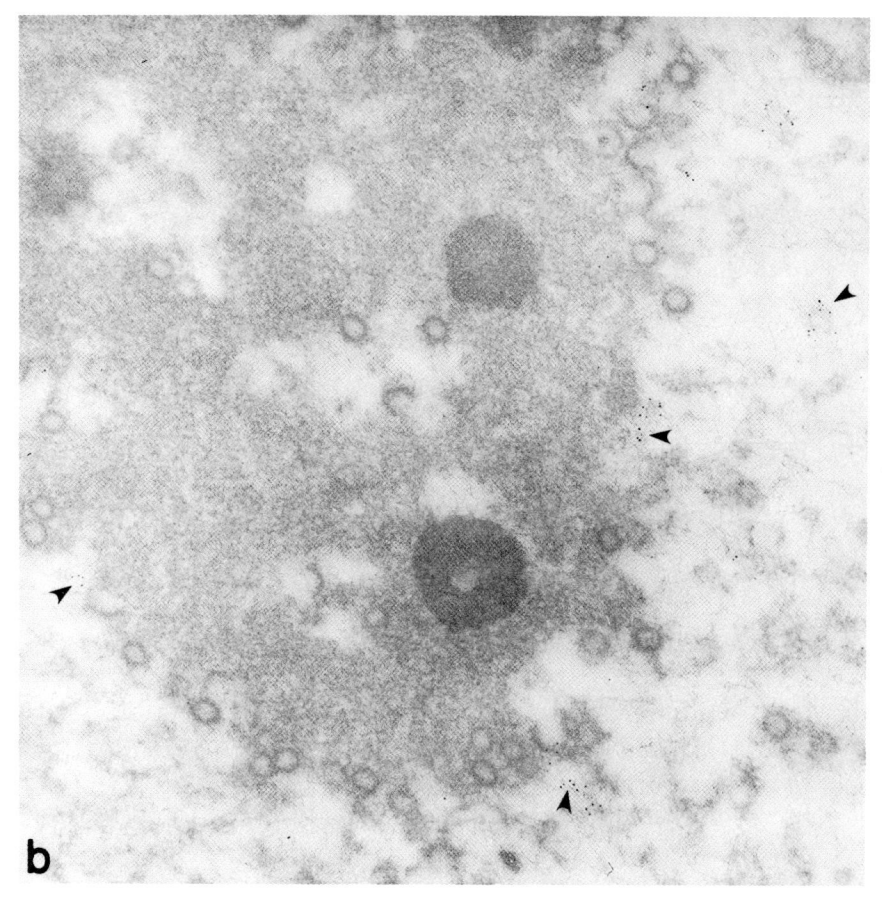

19. Ultrastructural *in Situ* Nucleic Acid Hybridization 389

Fig. 19.2. (*Continued*)

rial, often near the nuclear envelope. Cytoplasmic particles were primarily found free in the cytosol and were never present within mitochondria or endoplasmic reticulum. Comparisons of gold labeling within positive and negative control specimens demonstrated means of 17.5 and 7.2 gold clusters per nuclear and cytoplasmic cell profile, respectively, when formaldehyde plus glutaraldehyde was used as a fixative. Means of 13.2 and 4.6 gold clusters per nuclear and cytoplasmic profile were present in specimens fixed in formaldehyde alone. The increase in clustered gold particles in formaldehyde plus glutaraldehyde-fixed specimens was accompanied by a significant decrease in gold singlets and improved preservation of cytoplasmic intermediate filaments. Nonhomologous probe and

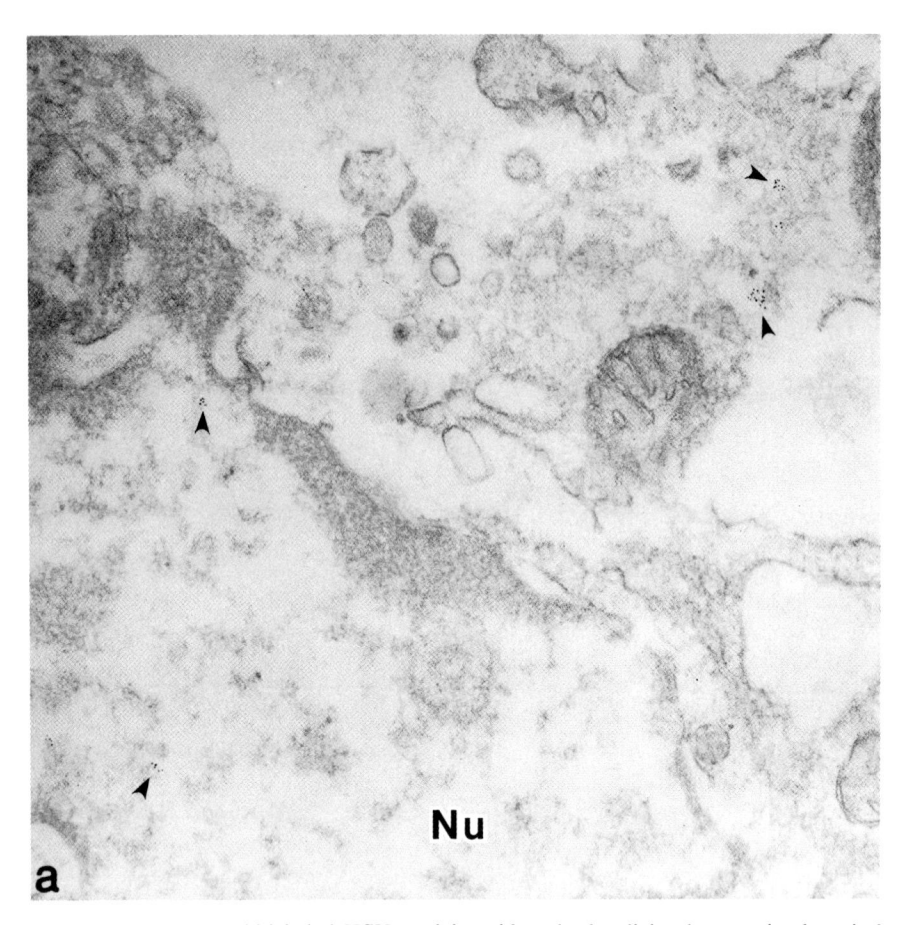

Fig. 19.3. 5-nm gold-labeled HSV nucleic acid probe localizing herpes simplex viral RNA in infected fibroblast culture cells. (a) Clusters of gold (some marked with arrowheads) are present in the nucleus (Nu) as well as in the cytoplasm; ×39,200. (b) Two virions within a nucleus. Clusters and individual gold label are abundant, frequently associated with moderately electron-dense material; ×65,700.

RNase-treated controls indicated that at least 36% of gold singlets represented background label, while virtually no gold clusters were observed in these specimens. DNase treatment did not affect nuclear or cytoplasmic gold labeling. We concluded that clustered 5-nm gold particles represented unequivocal nucleic acid labeling and that formaldehyde plus glutaraldehyde was the superior fixative for RNA detection by this preembedding technique.

We have applied 5-nm gold particles to CMV-infected fibroblasts and have succeeded in localizing viral DNA only (Fig. 19.2). The reason for

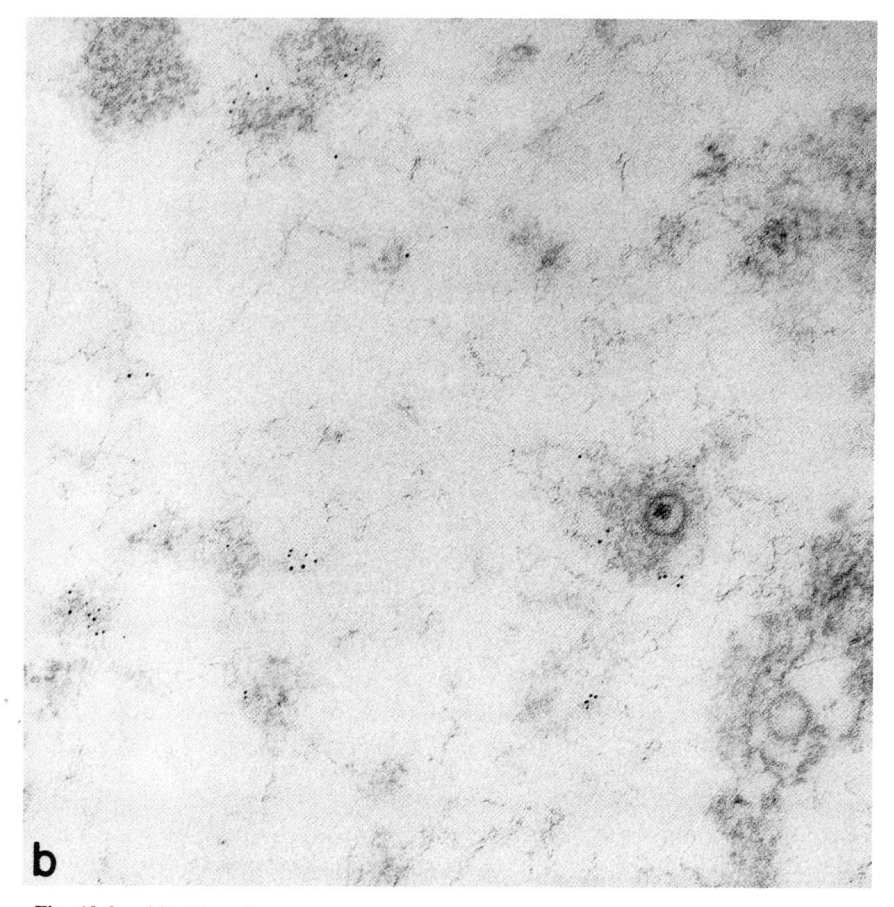

19. Ultrastructural *in Situ* Nucleic Acid Hybridization 391

Fig. 19.3. (*Continued*)

this is unknown, but it may be related to low levels of RNA transcription of sequences detected by the probe in CMV-infected cells (Wathen and Stinski, 1982). In addition, we have been as yet unable to localize viral DNA and RNA simultaneously by this technique.

We have not made direct measurements of the sensitivity of this method. One reason for this is that the multiplicity of binding of gold particles and probe molecules to target nucleic acid sequences is unknown. Theoretically, the size differential between the sequences detected by the probe molecules, the individual probe lengths, and the streptavidin-conjugated gold particles would allow multiplicity of binding, and evidence for multiplicity of binding has been demonstrated (Wolber *et al.*, 1989).

Improvements

The method described above is well suited to cells in suspension. It is not, at present, directly applicable to whole tissues or tissue sections, although preembedding ultrastructural hybridization of tissue sections has been described (Harris and Croy, 1986; Webster et al., 1987). Harris and Croy (1986) localized mRNA for pea legumin on thin sections of pea seeds using avidin–peroxidase and avidin–ferritin, which do not provide the spatial preciseness or quantifiability of gold markers. Webster et al. (1987) localized glycoprotein RNA in 25-μm vibrotome-cut fixed tissue sections using 20-nm streptavidin–gold particles without detergent extraction or protease digestion, although quantification was not provided and ultrastructural morphology was apparently not superior to that of our technique. The postembedding EM procedure of Binder et al. (1986) may be better adapted to tissue sections. This technique was also utilized in the studies of Webster et al. (1987), although comparisons of the sensitivity of pre- and postembedding methods were not provided.

The combination of Triton X-100 and protease treatments resulted in loss of the majority of identifiable ribosomes from the cell cytoplasm, which is undesirable if sites of RNA translation are to be precisely identified. Variations of the prehybridization permeabilization steps, possibly used in conjunction with smaller oligonucleotide probes, may overcome this problem. Adaptations of permeabilization steps may be required for EM localization of RNA sequences in cells other than fibroblasts, due to variations in cell structural integrity. Finally, light microscopic techniques for combined *in situ* hybridization and antigen immunolabeling (Blum et al., 1984; Brahic et al., 1984; Wolber and Lloyd, 1988) can doubtless be adapted to the EM level by utilizing gold particles of two different sizes linked to streptavidin and antibody molecules.

POSTEMBEDDING HYBRIDIZATION ON THIN SECTIONS

Binder and colleagues have published reports of ultrastructural *in situ* hybridization of mitochondrial rRNA and U-I nuclear RNA in thin sections of *Drosophila* ovary tissue embedded in Lowicryl K4M (Binder et al., 1986; Binder, 1987). Webster and associates (1987) also used a similar technique in their studies of myelin-forming Schwann cells. These investigators localized hybridized RNA sequences in a two-step procedure using antibiotin antibodies followed by a secondary antibody or protein A conjugated to gold particles. Using an additional anti-protein A–protein A amplification procedure, Binder (1987) observed clustering of gold particles due to the amplification step. Clustering in this postembedding technique was interpreted as disadvantageous due to diminished spatial resolution. It was not interpreted as a marker of specific nucleic acid labeling. This is in contrast to our studies, which showed clustering of gold particles due to binding of multiple probe molecules to RNA strands or of multiple gold particles to biotinylated probe without subsequent amplification steps. Binder (1987) also utilized acridine orange staining of the thin sections to demonstrate accessibility of RNA when embedded within the water-soluble resin medium. In the procedure used by Binder (1987), a combination of formaldehyde and glutaraldehyde (0.1%) was found to provide better RNA retention than formaldehyde alone, similar to our findings. However, he also found the use of pronase digestion to be detrimental to RNA detection, which may be due to the very low concentrations of glutaraldehyde used.

APPLICATIONS

The ability to localize viral and cellular DNA and RNA sequences precisely at the ultrastructural level is likely to be important in the investigation of viral replicative metabolism and cellular gene expression. With it, investigators can morphologically identify sites of gene replication, transcription, and translation and the interaction of RNA sequences with specific cell structures. Of particular interest is the possible role that the intermediate filament network may play in viral and cellular RNA transport and translation (Fulton *et al.*, 1980; Cervera *et al.*, 1981; Chatterjee *et al.*, 1984; Lawrence and Singer, 1986). Detergent-extracted cells in general provide good preservation of cytoskeletal elements, both for transmission electron microscopy and high-voltage embedment-free electron microscopy (Penman, 1985). A preembedding hybridization technique should be well suited to these applications.

REFERENCES

Binder, M., Tourmente, S., Roth, J., Renaud, M., and Gehring, W. J. (1986). *In situ* hybridization at the electron microscopic level: Localization of transcripts on ultrathin sections of Lowicryl K4M-embedded tissue using biotinylated probes and protein-A–gold complexes. *J. Cell Biol.* **102**, 1646.

Binder, M. (1987). *In situ* hybridization at the electron microscope level. *Scanning Microsc.* **1**, 331.

Blum, H., Haase, A. T., and Vyas, G. (1984). Molecular pathogenesis of hepatitis B virus infection: Simultaneous detection of viral DNA and antigens in paraffin embedded liver sections. *Lancet* **3**, 771.

Brahic, M., and Haase, A. T. (1978). Detection of viral sequences of low reiteration frequency by *in situ* hybridization. *Proc. Natl. Acad. Sci. U.S.A.* **75**, 6125.

Brahic, M., Haase, A. T., and Case, E. (1984). Simultaneous *in situ* detection of viral RNA and antigens. *Proc. Natl. Acad. Sci. U.S.A.* **81**, 5445.

Brigati, D. J., Myerson, D., Leary, J. J., Spanholz, B., Travis, S. Z., Fong, C. K. Y., Hsiung, G. D., and Ward, D. C., (1983). Detection of viral genomes in cultured cells and paraffin embedded tissue sections using biotin labeled hybridization probes. *Virology* **126**, 32.

Cervera, M., Dreyfuss, G., and Penman, S. (1981). Messenger RNA is translated when associated with the cytoskeletal framework in normal and VSV infected HeLa cells. *Cell (Cambridge, Mass.)* **23**, 113.

Chatterjee, P., Cervera, M., and Penman, S. (1984). Formation of vesicular stomatitis virus nucleocapsid cytoskeleton framework bound N protein: Possible model for structure assembly. *Mol. Cell. Biol.* **14**, 2231.

Cox, K. H., DeLeon, D. V., Angerer, L. M., and Angerer, R. C. (1984). Detection of mRNAs in sea urchin embryos by *in situ* hybridization using asymmetric RNA probes. *Dev. Biol.* **101**, 485.

Cremers, A. F. M., Jansen indeWal, N., Wiegart, J., Dirks, R. W., Weisbeek, P., van der Ploeg, M., and Landegent, J. E. (1987). Nonradioactive *in situ* hybridization: A comparison of several immunocytochemical detection systems using reflection contrast electron microscopy. *Histochemistry* **86**, 609.

Croissant, O., Dauguet, C., Jeanteur, P., and Orth, G. (1972). Application de la technique d'hybridization moléculaire *in situ* à la mise en évidence au microscope électronique de la réplication végétative de l-Adn viral dans les papillomes par le virus de Shope chez de lapin cottontail. *C. R. Hebd. Seances Acad. Sci., Ser. D* **274**, 614.

Fulton, A. B., Wan, K. M., and Penman, S. (1980). The spatial distribution of polyribosomes in 3T3 cells and the associated assembly of proteins into the skeletal framework. *Cell (Cambridge, Mass.)* **20**, 849.

Gall, J. G., and Pardue, M. L. (1969). Formation and detection of RNA-DNA hybrid molecules in cytological preparations. *Proc. Natl. Acad. Sci. U.S.A.* **63**, 378.

Gee, C., and Roberts, J. (1983). *In situ* hybridization histochemistry: A technique for the study of gene expression in single cells. *DNA* **2**, 157.

Gueskens, M., and May, E. (1974). Ultrastructural localization of SV-40 viral DNA in cells, during lytic infection, by *in situ* molecular hybridization. *Exp. Cell Res.* **87**, 175.

Haase, A. T. (1986). Analysis of viral infections by *in situ* hybridization. *J. Histochem. Cytochem.* **1**, 27.

Harris, N., and Croy, R. R. D. (1986). Localization of mRNA for pea legumin: *In situ* hybridization using biotinylated cDNA probe. *Protoplasma* **130**, 57.

Hutchison, N., Langer-Safer, P., Ward, D., and Hamkalo, B. (1982). *In situ* hybridization at the electron microscope level: Hybrid detection by autoradiography and colloidal gold. *J. Cell Biol.* **95**, 609.

Hutchison, N. (1984). Hybridization histochemistry: *In situ* hybridization at the electron microscope level. In *Immunolabelling for Electron Microscopy* (J. M. Polak and I. M. Varndell, eds.), p. 341. Elsevier, Amsterdam.

Jacob, J., Todd, K., Birnstiel, M. L., and Bird, A. (1971). Molecular hybridization of ^{3}H-labeled ribosomal RNA with DNA in ultrathin sections prepared for electron microscopy. *Biochim. Biophys. Acta* **228** 761.

Jacob, J., Gillies, K., MacLeod, D., and Jones, K. (1974). Molecular hybridization of mouse satellite DNA complementary RNA in ultrathin sections prepared for electron microscopy. *J. Cell Sci.* **14**, 253.

Langer, P., Waldrop, A., and Ward, D. (1981). Enzymatic synthesis of biotin labeled polynucleotides: Novel nucleic acid affinity probes. *Proc. Natl. Acad. Sci. U.S.A.* **78,** 6633.

Langer-Safer, P., Levine, M., and Ward, D. (1982). An immunological method for mapping genes on *Drosophila* polytene chromosomes. *Proc. Natl. Acad. Sci. U.S.A.* **79,** 4381.

Lawrence, J. B., and Singer, R. H. (1985). Quantitative analysis of *in situ* hybridization methods for the detection of actin gene expression. *Nucleic Acids Res.* **13,** 1777.

Lawrence, J. B., and Singer, R. H. (1986). Intracellular localization of messenger RNAs for cytoskeletal proteins. *Cell (Cambridge, Mass.)* **45,** 407.

Lawrence, J. B., Villnave, C. A., and Singer, R. H. (1988). Sensitive, high resolution chromatin and chromosome mapping *in situ:* Presence and orientation of two closely integrated copies of EBV in a lymphoma line. *Cell (Cambridge, Mass.)* **52,** 51.

Leary, J. J., Brigati, D. J., and Ward, D. (1983). Rapid and sensitive calorimetric method for visualizing biotin-labeled DNA probes hybridized to DNA or RNA immobilized on nitrocellulose: Bio-blots. *Proc. Natl. Acad. Sci. U.S.A.* **80,** 4045.

Manning, J., Hershey, N., Broker, T., Pellegrini, M., Mitchell, H., and Davidson, N. (1975). A new method of *in situ* hybridization. *Chromosoma* **53,** 107.

Manuelidis, L., Langer-Safer, P., and Ward, D. (1982). High resolution mapping of satellite DNA using biotin labeled DNA probes. *J. Cell Biol.* **95,** 619.

Manuelidis, L., and Ward, D. (1984). Chromosomal and nuclear distribution of the *Hind*III 1.9 kb human DNA repeat segment. *Chromosoma* **91,** 28.

Marmur, J., and Doty, P. (1962). Determination of the base pair composition of DNA from its thermal denaturation temperature. *J. Mol. Biol.* **5,** 109.

McAllister, H. A., and Rock, D. L. (1985). Comparative usefulness of tissue fixatives for *in situ* viral nucleic acid hybridization. *J. Histochem. Cytochem.* **33,** 1026.

McConaughy, J. L., Laird, C. D., and McCarthy, M. J. (1969). Nucleic acid reassociation in formamide. *Biochemistry* **8,** 3284.

Penman, S. (1985). Virus metabolism and cellular architecture. In *Virology* (B. N. Fields *et al.,* eds.), p. 169. Raven Press, New York.

Schildkraut, C. L., and Lifson, S. (1965). Dependence of the melting temperature of DNA on salt concentration. *Biopolymers* **3,** 195.

Singer, R. H., and Ward, D. (1982). Actin gene expression visualized in chicken muscle tissue culture by using *in situ* hybridization with a biotinated nucleotide analog. *Proc. Natl. Acad. Sci. U.S.A.* **79,** 7331.

Singer, R. H., Lawrence, J. B., and Villnave, C. (1986). Optimization of *in situ* hybridization using isotopic and nonisotopic detection methods. *BioTechniques* **4,** 230.

Steinert, G., Thomas, C., and Brachet, J. (1976). Localization by *in situ* hybridization of amplified ribosomal DNA during *Xenopus laevis* oocyte maturation (a light and electron microscopy study). *Proc. Natl. Acad. Sci. U.S.A.* **73,** 833.

Sternberger, L. A. (1986). *Immunocytochemistry,* pp. 82, 126. Wiley, New York.

Unger, E. R., Budgeon, L. R., Myerson, D., and Brigati, D. J. (1986). Viral diagnosis by *in situ* hybridization: Description of a rapid simplified colorimetric method. *Am. J. Surg. Patho.* **10,** 1.

Wathen, M. W., and Stinski, M. F. (1982). Temporal patterns of human cytomegalovirus transcription: Mapping the viral RNAs synthesized at immediate early, early and late times after infection. *J. Virol.* **41,** 462.

Webster, H. de F., Lampeath, L., Favilla, J. T., Lemke, G., Tesin, D., and Manuelidis, L. (1987). Use of a biotinylated probe and *in situ* hybridization for light and electron microscopic localization of P$_o$mRNA in myelin-forming Schwann cells. *Histochemistry* **86,** 441.

Wolber, R. A., and Lloyd, R. V. (1988). Cytomegalovirus detection by nonisotopic *in situ* DNA hybridization and viral antigen immunostaining using a two-color technique. *Hum. Pathol.* **19,** 736.

Wolber, R. A., Beals, T. F., Lloyd, R. V., and Maassab, H. F. (1988). Ultrastructural localization of viral nucleic acid by *in situ* hybridization. *Lab. Invest.* **59,** 144.

Wolber, R. A., Beals, T. F., and Maassab, H. F. (1989). Ultrastructural localization of herpes simplex virus RNA by *in situ* hybridization. *J. Histochem. Cytochem.* **37,** 97.

Wu, M., and Davidson, N. (1981). Transmission electron microscopic method for gene mapping on polytene chromosomes by *in situ* hybridization. *Proc. Natl. Acad. Sci. U.S.A.* **78,** 7059.

20

Detection of Proteins with Colloidal Gold*

ROLAND ROHRINGER

Research Station
Agriculture Canada
Winnipeg, Manitoba, Canada

INTRODUCTION
METHODOLOGY
   General, Nonspecific Detection of Proteins on Blots
      Preparation of Gold Colloids Using White Phosphorus
      Preparation of Gold Colloids Using Citric Acid
      Treatment of Blots before Staining With Gold Sols
      pH and Colloidal Gold Particle Size
      Staining Protocol
   Detection of Specific Proteins on Blots
   General and Specific Protein Staining Performed
      Sequentially on the Same Blot
   Negative Gold Stain
   Quantitative Determination of Protein
APPLICATIONS
   Selected Examples of the Use of Gold Colloids on Blots
REFERENCES

*Listed as publication number 1311, Agriculture Canada, Research Station, Winnipeg.

INTRODUCTION

Although protein–gold conjugates have been used for some time as target-specific electron-dense markers in cytochemistry (Faulk and Taylor, 1971), it was only recently that the visual color signal inherent in gold colloids and their adsorption to macromolecules have been exploited for detection of proteins in biochemistry. Undoubtedly, the use of gold markers in ultrastructural research stimulated this development, providing one of the rare examples of technology transfer from microscopy to analytical biochemistry rather than the reverse. Thus, it is not surprising that the first use of colloidal gold for noncytochemical detection of protein involved the same types of gold-conjugated probes that were first used in microscopy, i.e., gold-labeled antibodies and protein A–gold conjugates. These probes stained antigens that had been immobilized on nitrocellulose membranes by dot-spotting or by blotting from polyacrylamide gels (Brada and Roth, 1984; Hsu, 1984; Moeremans et al., 1984).

In a short abstract, Gounon and Moreau (1984) also mentioned this technique as well as a second more generally useful method with which proteins blotted onto nitrocellulose membranes can be detected unspecifically using "naked," nonconjugated gold colloids. This publication does not appear to have had an impact on the further development of this technique, and the blotting assay for protein with nonconjugated gold colloids was rediscovered independently by two separate groups (Moeremans et al., 1985; Rohringer and Holden, 1985). In principle, the method is similar to that of protein detection with india ink (Hancock and Tsang, 1983), whereby negatively charged colloids bind to the positive charge on immobilized proteins; charge interaction, however, may not be the only mechanism involved in the binding. Detection of nitrocellulose-blotted proteins with nonconjugated colloidal gold offers several advantages: it is quick, inexpensive, easy, reproducible, and it results in a permanent record. Its sensitivity is about the same as that of the silver stain for protein in gels. Furthermore, depending on the protein and the size of gold colloids, the stain may be polychromatic. Thus, differently colored "landmark" proteins are often apparent to facilitate interblot comparison of complex protein mixtures. An additional extremely sensitive colloidal gold stain was reported by Righetti et al. (1986) for proteins separated in cellulose acetate membranes. Protein stains with nonconjugated gold colloids and with india ink have one common disadvantage: because of the negatively charged colloids they cannot be used for proteins immobilized on positively charged membranes such as "Zeta Probe" or positively charged nylon.

Specific stains with gold-conjugated or enzyme-labeled probes can be used sequentially on the same blot before or after a nonspecific reagent

20. Detection of Proteins with Colloidal Gold

consisting of "naked" gold colloids. Thus, specific antigens on a blot can be localized in a "background" of nonimmunogenic proteins (Daneels et al., 1986; Moeremans et al., 1987b). Since staining with nonconjugated gold colloids preserves immunoreactivity of blotted proteins (Egger and Bienz, 1987), the general protein stain can be used first, allowing the separation to be evaluated before the second, specific probe is used (Egger and Bienz, 1987; Schapira and Keir, 1988). The only known alternatives for double staining of this type are the use of india ink followed by immunoprobing with antibodies and ^{125}I-labeled protein A (Glenney, 1986), or fluorescent staining followed by immunoprobing with enzyme-linked antibodies (Szewczyk and Summers, 1987). Obviously, these double stains are not limited to immunochemical protocols. Any protein for which a specific probe is available may be localized in a mixture of other proteins on a blot, e.g., glycoproteins with gold-conjugated lectins or lectins with gold-conjugated neoglycoproteins.

At about the same time as gold stains were developed for proteins on blots, another method became available that uses gold colloids to detect proteins in gels (Casero et al., 1985; Budowle and Gamble, 1986) with a sensitivity comparable to that of the more complex silver stains. A possible disadvantage of this gold stain is the fact that it takes much longer to develop. However, since "nonspecific" protein stains generally do differ in their sensitivity for different proteins, it is useful to have another stain available to optimize detection of the protein(s) of interest.

In addition to the qualitative methods outlined above, two methods were described that allow quantitation of proteins with colloidal gold, one for dot-spotted protein (Hunter and Hunter, 1987), the other for protein in solution (Stoscheck, 1987; Ciesiolka and Gabius, 1988). Both techniques are extremely simple to perform. They can be used for protein assays in the low nanogram range and are thus intermediate in sensitivity between the dye-binding assays (Bradford, 1976; Nakamura et al., 1985) and the bioluminescence-enhanced detection system of Hauber and Geiger (1987).

In the following, recommended methods will be described in detail. A concluding section contains selected examples of the use of gold stains in blotting methodology. Because of the novelty of this technology, further technical improvements and new areas of application may be expected in the near future.

METHODOLOGY

General, Nonspecific Detection of Proteins on Blots

This method is limited to proteins immobilized on nitrocellulose or polyvinylidene diflouride membranes. Solutions of colloidal gold are commercially available ("AuroDye," Janssen Life Sciences Products, B-2340 Beerse, Belgium) or can be made easily in the laboratory at low cost and stored for several months at 2°C (275 K). In the following, two methods are presented for producing colloidal gold, one for 5-nm particles and the other for particles of larger size. This is followed by pretreatment of the blots, pH adjustment of gold sols, and the staining protocol.

Preparation of Gold Colloids Using White Phosphorus

The 5-nm particles produced with this procedure (Horisberger, 1979; Van Bergen en Henegouwen and Leunissen, 1986) impart an orange-red color to the colloidal suspension and to the majority of proteins stained with it. In a scrupulously clean Erlenmeyer flask, add 6 ml of a 0.5% gold chloride solution (e.g., Fisher Scientific SO-G-80) to 240 ml of distilled water and neutralize with 0.2 N K_2CO_3 (~5.4 ml). Stir for at least 3 min until the yellowish color has disappeared. An ethereal solution of white phosphorus is then added. The ether solution consists of 4 parts diethyl ether and 1 part phosphorus-saturated ether. The mixture is shaken and allowed to stand for 30 min at 25°C (298 K), during which time it develops a rust brown color. It is subsequently heated with a Bunsen burner under reflux until an orange-red color has developed. This takes from 5 to 30 min, depending on the efficiency of heating.

To avoid flocculation of the gold, careful consideration should be given to the condition of the gold chloride used: no problems have been reported with commercially available 0.5% gold chloride solution, but the use of crystalline chloroauric acid ($HAuCl_4$) often has resulted in flocculation if it became hydrated and brittle before it was dissolved in water (Morris and Saelinger, 1986).

Preparation of Gold Colloids Using Citric Acid

Particle sizes between ~12 and 150 nm can be produced with this method (Frens, 1973), although only those up to a diameter of ~30 nm are useful for staining protein on blots. Colors of these colloidal suspensions and subsequent protein stains range from wine red via purple to violet with increasing particle size and depending on the protein. To produce particles in a range between 12 and 30 nm, add 3 to 1.5 ml of a 1% Na_3-citrate solution (w/v) to 240 ml of a boiling 0.02% solution of gold chloride. Note that particle size increases with decreasing amount of citric acid added. The boiling solution turns faintly blue at first and suddenly changes to give stable colors in the range indicated above. The reduction is practically complete after 5 min and further heating does not affect the solution substantially. This method has been discussed in detail by Handley in Volume 1 of this series.

Treatment of Blots before Staining with Gold Sols

Commercially available gold sol (AuroDye) contains Tween 20 as a stabilizer. Before staining with this gold sol, blots should be washed with a solution of this detergent (0.1 to 0.3%) in phosphate-buffered saline (PBS) (Moeremans et al., 1987a). Sensitivity of protein detection can be increased 5- to 100-fold, depending on the protein, when blots are treated for 5 min with 1% KOH or NaOH, then washed three times (15 min each) with PBS containing 0.3% Tween 20 (v/v) followed by distilled water, before staining with AuroDye (Sutherland and Skerritt, 1986). Pretreatment with base will also enhance protein detection with other gold sols (R. Rohringer, unpublished results).

pH and Colloidal Gold Particle Size

Gold sols prepared in the laboratory have a very low ionic strength and can be adjusted to a desired pH for selective protein detection and low background staining (Moeremans et al., 1985; Rohringer and Holden, 1985). All proteins, including acidic proteins such as α_1-acid glycoprotein (pI 2.7), can be detected at pH 3.5, whereas only basic proteins adsorb gold colloids at pH $\geq$ 8. For most proteins, the optimum signal-to-background ratio is obtained when the gold sol has been adjusted to pH 5. pH adjustment is critical when sols consisting of large particles are used: those of ~30 nm diameter are very useful for staining certain proteins with very high sensitivity at pH 4.5, but incubation at pH values <4.5 lead to unacceptably high background, while pH values >5.0 diminish sensitivity.

A note of caution regarding pH adjustment of gold sols: to avoid adsorption of gold particles to the measuring electrode, a drop of polyethylene glycol 20,000 (Baker) is added to a small aliquot of the sol, which is then adjusted to the desired pH with dilute K_2CO_3 or H_3PO_4; a proportional volume of K_2CO_3 or H_3PO_4 is then added to the amount of sol required for staining, and the pH of this solution is confirmed.

Staining Protocol

The procedure is extremely simple. After a wash in distilled water to remove blotting buffers or reagents used for possible pretreatment, the blots are incubated in the gold colloid suspension in a clean glass dish on a rotary or rocking shaker at room temperature. Staining is usually complete in 30 min.

Detection of Specific Proteins on Blots

After immobilization on nitrocellulose or other blotting matrices, specific proteins can be detected with gold conjugates if suitable probes are available that have affinities to these proteins. This technique is very versatile. Known examples include detection of glycoproteins with gold-conjugated lectins (Seitz, 1987) and of various antigens with appropriate gold-labeled antibodies (Hsu, 1984; Brada and Roth, 1984; Moeremans *et al.,* 1987a; Seitz, 1987). Immunoprobing can take various forms well known from immunocytochemistry. For example, rather than localizing the blotted antigen directly with gold-labeled antibody, unlabeled antibody (e.g., raised in rabbit) may be used first, followed by incubation with a second gold-labeled antibody (e.g., goat antirabbit) or with gold-labeled protein-A. Alternatively, after incubation with the unlabeled first antibody, the blot may be treated with biotinylated second antibody followed by avidin–gold conjugate. The last three methods are preferable because they often are more sensitive and because they do not require the operator to carry out the conjugation since gold-conjugated secondary and tertiary probes are commercially available (Janssen Life Sciences Products, Beerse, Belgium; Sigma Chemical Co, St. Louis, MO).

Before attempting specific probing with gold conjugates, checks should be carried out to determine whether the blotted protein binds gold probes nonspecifically. This is recommended based on the conclusions of Behnke et al. (1986) that even fully stabilized gold conjugates may contain, constantly or intermittently, "naked areas" on gold particles surfaces. If these are available for interactions with components that have a high electrostatic affinity for the charged gold surface, nonspecific binding of the probe may be the result. The authors presented a protocol for immunocytochemistry that can be adapted for analyzing blots with specific conjugated gold probes. In the following immunostaining procedure, essentially after Moeremans *et al.* (1987a), these tests are identified with an asterisk (*) and need be done only once in preliminary experiments for each combination of probe and blotted proteins. The staining procedure can be adapted for other purposes, for example, to detect glycoproteins with gold-labeled lectins or with unlabeled lectins followed by gold-labeled antilectins.

1. Apply protein to nitrocellulose or other membrane, either by dot-spotting or after transfer from gels (Beisiegel, 1986).

2. Block remaining protein binding sites by washing blot for 30 min in PBS, pH 7.2, containing 5% bovine serum albumin (BSA), and then 3 × 5 min in PBS containing 0.1% BSA. All steps are at room temperature.

*3. Using gold colloids stabilized with polyethylene glycol (Horisberger and Rosset, 1977) or methylcellulose at an optical density of 0. 5 at 520 nm, and at the pH and ionic strength used in the planned experiment, determine whether the blotted protein binds one of these nonspecific con-

20. Detection of Proteins with Colloidal Gold 403

jugates. If no such binding occurs, proceed with the intended specific probe.

*4. If the blotted protein binds one of the gold conjugates in step 3, a number of additional tests are necessary to find an "inert" competing protein with a higher affinity for gold than the blotted protein. This is then added to the bathing solution containing the gold-labeled probe. In tests by Behnke *et al.* (1986), 0.25% gelatin of a Bloom number between 60 and 100 (Merck) has been effective.

5. Incubate for 2 hr with first antibody (e.g., raised in rabbit) in PBS containing 0.1% BSA (or the "competing" protein chosen in step 4). The concentration of the first antibody depends on its titer but is typically 1 µg/ml.

6. Wash blot 3 × 10 min in PBS containing 0.1% BSA.

7. Incubate for 2 hr with gold-labeled probe (e.g., anti-rabbit serum raised in goat and linked to gold colloids: Auro Probe BL GAR, Janssen Life Sciences Products; or rabbit IgG–gold conjugate G3766, Sigma Chemical; or protein A–gold conjugate P9660, Sigma Chemical). Before incubation, the immunogold probe is diluted with PBS/0.1% BSA, 0.4% gelatin to an optical density of 0.02 at 520 nm.

8. Wash 2 × 5 min with PBS/0.1% BSA nd 2 × 5 min with distilled water.

*9. In a control experiment, confirm binding specificity of the gold probe. For instance, if protein A–gold is used as the probe to localize bound antibody, binding should be inhibitable with unlabeled protein A.

The signal obtained with gold staining can be intensified (Moeremans *et al.*, 1984) by a factor of ~10 with silver enhancement, making the assay more sensitive than enzyme-coupled assays. Thus, the method is a useful alternative for protein assays that involve the use of radioactively labeled ligands.

10. Wash blot for 2 min in 0.2 M citrate buffer (pH 3.85).

11. Incubate for 5 to 15 min in developer while protecting blots from diffuse daylight with aluminum foil. The developer is made by mixing 60 ml of distilled water, 10 ml of 2 M citrate buffer (pH 4.85), 15 ml of distilled water containing 0.85 g of hydroquinone, and 15 ml of distilled water containing 0.11 g of silver lactate. The hydroquinone and silver lactate solutions are prepared immediately before use in vessels protected from light.

12. Treat blots for 2 min in a photographic fixer (such as "Agefix" by Agfa Gevaert) which has been diluted 1 : 10 with distilled water.

13. Wash blots with excess distilled water and air-dry.

General and Specific Protein Staining Performed Sequentially on the Same Blot

Localizing immunoreactive proteins in a background of other proteins on the same blot is very helpful for determining immunospecificity. Three methods with gold sols are available for this purpose.

In the first (Daneels *et al.*, 1986), the blot is blocked with Tween 20 and incubated with a first antibody (ABI, e.g., raised in rabbit), followed by a gold-conjugated second antibody (ABII, e.g., goat antirabbit); the gold signal is intensified by silver stain enhancement; nonimmunoreactive proteins are then revealed with nonconjugated ("naked") gold:

Tween 20 blocking → ABI → ABII–gold → silver enhancement → gold sol

Using nonconjugated gold colloids in the last step is possible with Tween 20 as the blocking agent. Proteinaceous blocking agents, such as BSA or gelatin, would lead to a heavily stained background. The silver enhancement results in good contrast of immunoreactive proteins from those that are not immunoreactive.

The second procedure (Egger and Bienz, 1987) has several important advantages and is described in detail here. In this procedure, the staining sequence is reversed because it had been shown that staining with colloidal gold preserves immunoreactivity of blotted proteins:

Gold sol → protein blocking → ABI → ABII–peroxidase

Color differentiation between the red signal of the general protein stain and the dark purple signal of the peroxidase stain is excellent. While Tween 20 is the only blocking reagent known to work in the procedure described by Daneels and colleagues, the "gold-sol-first" procedure of Egger and Bienz can use exogenous protein as blocking reagent, thereby further suppressing the background and allowing optimal antigen detection. Also, quality of the blots can be checked after the general protein stain before possibly expensive immunoreactions are performed. To further economize, only the relevant regions of a blot need be excised prior to antigen detection. The individual steps involved in this double stain are as follows:

1. Wash blot briefly in 25 mM Tris-HCl (pH 7.4) and process immediately or air-dry and store at 4°C (277 K).
2. Treat blot with 0.1 M Tris-HCl (pH 7.4).
3. Stain proteins with gold sol (e.g., page 401) by incubating blot for 30 min to 4 hr at room temperature on a shaker.
4. Rinse blot with 0.1 M Tris-HCl (pH 7.4).

5. Block for 1 or 2 hr in 0.1 M Tris-HCl (pH 7.4) containing 0.25% gelatin and 3% ovalbumin, or in a blocking agent found optimal for the specific probe used.

6. Incubate blot with shaking overnight at room temperature in a suitably diluted solution of the first antibody (e.g., 1 : 10 diluted mouse hybridoma supernatant).

7. Wash blot in 50 mM Tris-HCl (pH 7.4) containing 150 mM NaCl, 5 mM EDTA, 0.25% gelatin, and 0.4% NP40.

8. Incubate blot for 2.5 hr at room temperature and with constant agitation in a suitably diluted solution of the second antibody linked to an enzyme marker (e.g., 1 : 500 diluted rabbit antimouse–peroxidase conjugate).

9. Wash blot in 50 mM Tris-HCl (pH 7.4) supplemented with 450 mM NaCl, 5 mM EDTA, and 0.4% Sarkosyl.

10. Develop blot by incubation in 0.018% 4-chloro-1-naphthol and 0.006% H_2O_2.

The third method (Schapira and Keir, 1988) for detecting immunoreactive proteins in a background of other proteins uses a gold sol originally developed for protein staining on cellulose acetate (Righetti et al., 1986) and a biotin–avidin system for antibody detection:

Gold sol → protein blocking → ABI → biotinyl ABII → streptavidin–peroxidase

It has been reported that with this system the nitrocellulose blot can be reprobed several times to identify individual polypeptides separated two-dimensionally and blotted from sodium dodecyl sulfate–polyacrylamide gels. The general gold stain is at least as sensitive as silver staining and is prepared as follows, according to Schapira and Keir (1988):

1. Prepare a 1% (w/v) solution of stannous chloride by dissolving 1 g of tin chloride in 5 ml of 1 M HCl and adding distilled water to 100 ml.

2. To 750 ml of distilled water, add 10 ml of 1% (w/v) aqueous gold chloride followed by 4 ml of the stannous chloride solution prepared as described in step 1.

3. To the mixture, add 100 ml of 20% (v/v) aqueous Tween 20 and 250 ml of a solution containing 9.66 g anhydrous citric acid per liter of distilled water. All additions are made dropwise with constant stirring.

4. Mix solution for 1 hr at room temperature and store in the dark at 4°C (277 K).

Before staining, wash nitrocellulose blot in distilled water; then proceed as follows:

5. Incubate in two 150-ml changes of the colloidal gold stain with constant agitation. Proteins begin to stain within 15 min but take 3 to 4 hr to complete. Overnight staining is possible without any increase in background.

6. Photograph blot to record the two-dimensional protein pattern.

7. Block blot for 1 hr at 37°C (310 K) in a 1% (w/v) solution of BSA in PBS (pH 7.4).

8. Wash blots during 30 min in three changes of PBS containing 0.1% Tween 20.

9. Incubate blots with the first antibody (ABI, e.g., raised in rabbit), suitably diluted with PBS, depending on the titer.

10. Wash as in step 8.

11. Incubate blot for 1 hr in a biotinylated second antibody (ABII, e.g., biotinyl–anti-rabbit IgG raised in goat, Sigma Chemical) appropriately diluted with PBS, again depending on the titer.

12. Wash as in step 8.

13. Incubate blot for 30 min in PBS containing streptavidin (or avidin) covalently linked to peroxidase (Sigma Chemical).

14. Wash as in step 8.

15. To develop color, incubate blot in a solution prepared by dissolving 12 mg of 4-chloro-1-naphthol in 4 ml of methanol and made to 24 ml with PBS, plus 12 µl H_2O_2.

16. Wash with distilled water and photograph with red filter to optimize color difference between the purple-black reaction product and the red gold stain.

Negative Gold Stain

Unlike the preceding stains that are used to reveal proteins after their transfer to blots, this stain detects proteins in gels as clear, unstained "bands" in a stained gel background. The mechanism responsible for this negative staining effect is not known. Sensitivity of this stain approaches that of silver stains and it reveals certain proteins that are not detectable in gels with either silver or Coomassie brilliant blue. It is extremely simple to perform. The following is an account of the procedure of Casero *et al.* (1985) as modified by Budowle and Gambel (1986).

1. Fix gel for 25 min with continuous shaking in a solution containing 5% trichloroacetic acid, 5% copper sulfate, and 0.35% sulfosalicylic acid.

2. Wash for 30 min with distilled water.

3. Prepare gold stain as follows: to a solution that is 0.1% (v/v) with respect to Tween 20 and 0.19% (w/v) with respect to citric acid, add chloroauric acid ($HAuCl_4$) immediately before use to a final concentration of 0.01% (w/v).

4. Place gel in a plastic bag containing the staining solution and shake overnight in the dark at room temperature.

5. Place gel in water and wipe gel surface gently with a wet cotton swab or strong tissue paper (e.g., "Kim Wipe") to remove precipitated gold.

6. Wash gel in 1% (v/v) glycerol for 15 min and air-dry, if storage is desired.

Quantitative Determination of Protein

Of the two assays that are described here, the assay by Hunter and Hunter (1987) quantitates dot-spotted protein in the range between 1.25 and 20 ng with colloidal gold. It is 20 times more sensitive than the dye-binding, dot-spot assay of Nakamura *et al.* (1985) and about as sensitive as the speedier but more involved silver stain described by Merril and Pratt (1986). As in all of these dot assays, sensitivity is a function of the amount of immobilized protein per unit area of the blotting matrix. Therefore, it is important to control the volume of the sample and the salt and detergent concentrations because these affect spreading of the sample. Under optimal conditions, standard errors between 3 and 5% can be expected near the high and low ends of the range, respectively. DNA, RNA, and polysaccharides reportedly do not interfere with protein quantitation using this gold stain. The only pieces of equipment required are a microsyringe, a sample applicator (e.g., "Super-Z," Helena Laboratories UK Ltd.), and a scanning densitometer.

1. Take up 1 µl of solution (containing 1.25 to 20 ng protein) in a microsyringe, transfer to slot applicator, and apply to nitrocellulose membrane.

2. After the sample has dried, wash membrane for 15 min at room temperature in PBS (containing 0.3% Tween 20) and rinse with distilled water.

3. Transfer membrane to gold sol and shake overnight at room temperature. The inventors of this procedure used AuroDye (Janssen Life Sciences Products); presumably, gold colloid preparations made in the laboratory and adjusted to pH 3.5 will give similar results.

4. Briefly wash membrane with water and scan with a densitometer, preferably at the reflectance maximum of the colloids used.

The second type of quantitative assay uses gold colloids for measuring protein in solution by making use of the fact that adsorption of dissolved proteins to gold colloids is accompanied by slight color change of the solution. It measures the difference of spectra of colloidal gold preparations with and without protein. Different proteins are detected with different sensitivities but few substances interfere with these extremely simple and quick assays. The costs per assay are substantially lower than those for dye-binding assays. Sensitivity of the assay pioneered by Stoschek (1987) is in the range between 20 and 200 ng protein; thus it is 25 to 50 times more sensitive than the assays by Bradford (1976) and Lowry *et al.* (1951), respectively, and requires glass or plastic cuvettes and a conventional laboratory spectrophotometer. The modification reported by Ciesiolka and Gabius (1988) described in the following achieves a further 8- to 10-fold increase in sensitivity with a higher concentration of colloidal gold and by using sols that have been stabilized with polyethylene glycol rather than Tween 20 contained in AuroDye. It was adapted for use with extremely small sample sizes and employs a microtiter plate reader for automation.

1. Pipette protein solution (containing 1 to 200 ng protein diluted with distilled water) into wells of microtiter plates and add distilled water to a volume of 10 µl.

2. Add 190 µl of gold sol prepared by reduction of gold chloride with sodium citrate according to Frens (1973) (see this chapter, page 400). Gold granule particle size should be between 17 and 30 nm. The sol, adjusted to pH 3.8 with citric acid and stabilized with microfiltered polyethylene glycol (Carbowax 20M) to a final concentration of 0.01%, is stable for extended periods if stored in the refrigerator.

3. Mix samples gently and determine absorbance at 590 nm. Color development is complete after 1–2 min and shows no further change during 24 hr.

APPLICATIONS

Selected Examples of the Use of Gold Colloids on Blots

Since no protein stain is equally sensitive for all proteins, gold stains have been compared with other stains to determine which is best for a particular project. Colloidal gold has been found useful for revealing blotted protein molecular weight standards (Bjerrum and Hinnerfeldt, 1987) and it was superior to Coomassie brilliant blue for the detection of proteinaceous contaminants in purified rabbit muscle ATPase (Gould *et al.*, 1987). These applications involved blotting from one-dimensional sodium dodecyl sulfate–polyacrylamide gels. The same technique was used by Lin *et al.* (1987) to identify viral structural proteins isolated from potato yellow dwarf virus-infected potato leaves. That gold staining is also useful for evaluating the results of crossed immunoelectrophoresis was shown in a study of human platelet proteins: nitrocellulose membrane replicas

20. Detection of Proteins with Colloidal Gold

stained with colloidal gold revealed all proteins that were seen after immunostaining of similar blots using a phosphatase-coupled second antibody (Bjerrum et al., 1987).

Gold staining of protein on nitrocellulose blots prepared by diffusion from isoelectric focusing (IEF) gels has not been satisfactory because of the high background stain due to immobilized ampholites (R. Rohringer, unpublished results). However, three groups of researchers have reported successful gold staining of blots from two-dimensional gels, even though IEF had been used in the first dimension (Segers and Rabaey, 1985; Daneels et al., 1986; Schapira and Keir, 1988). In the first two instances, the authors used a second blotting matrix, mounted cathodal to the gel, to serve as a background-reducing screen during electrophoretic transfer, while Schapira and Keir used a different gold stain that may be more generally useful for this purpose.

Although not within the terms of reference of this chapter, detection of blotted nucleic acids with gold stains (Saman, 1986) is of interest here. It may reduce the need for radiolabeled nucleic acids as markers in molecular hybridization and thus become a tool in molecular genetics. On nylon-based membranes, 0.1 µg DNA and 0.5 µg ribosomal RNA was detected with a negative stain using gold colloids. This stain, however, was less sensitive than a positive stain using colloidal iron.

REFERENCES

Behnke, O., Ammitzboll, T., Jessen, H., Klokker, M., Nilausen, K., Tranum-Jensen, J., and Olsson, L. (1986). Non-specific binding of protein-stabilized gold sols as a source of error in immunocytochemistry. *Eur. J. Cell Biol.* **41**, 326.

Beisiegel, U. (1986). Protein blotting. *Electrophoresis (Weinheim, Fed. Repub. Ger.)* **7**, 1.

Bjerrum, O. J., and Hinnerfeldt, F. R. (1987). Visualization of molecular-weight standards after electroblotting—detection by means of corresponding antibodies. *Electrophoresis (Weinheim, Fed. Repub. Ger.)* **8**, 439.

Bjerrum, O. J., Selmer, J. C., and Lihme, A. (1987). Native immunoblotting—transfer of membrane-proteins in the presence of non-ionic detergent. *Electrophoresis (Weinheim, Fed. Repub. Ger.)* **8**, 388.

Brada, D., and Roth, J. (1984). "Golden blot"—detection of polyclonal and monoclonal antibodies bound to antigens on nitrocellulose by protein A–gold complexes. *Anal. Biochem.* **142**, 79.

Bradford, M. M. (1976). A rapid and sensitive method for the quantitation of microgram quantities of protein utilizing the principle of protein–dye binding. *Anal. Biochem.* **72**, 248.

Budowle, B., and Gambel, A. M. (1986). Negative gold staining for electrophoretic protein profile interpretations. *Acta Histochem. Cytochem.* **19**, 647.

Casero, P., del Campo, G. B., and Righetti, P. G. (1985). Negative aurodye for polyacrylamide gels: The impossible stain. *Electrophoresis (Weinheim, Fed. Repub. Ger.)* **6**, 367.

Ciesiolka, T., and Gabius, H.-J. (1988). An 8- to 10-fold enhancement in sensitivity for quantitation of proteins by modified application of colloidal gold. *Anal. Biochem.* **168,** 280.

Daneels, G., Moeremans, M., De Raemaeker, M., and De Mey, J. (1986). Sequential immunostaining (gold/silver) and complete protein staining (AuroDye) on Western blots. *J. Immunol. Methods* **89,** 89.

Egger, D., and Bienz, K. (1987). Colloidal gold staining and immunoprobing of proteins on the same nitrocellulose blot. *Anal Biochem.* **166,** 413.

Faulk, W. P., and Taylor, G. M. (1971). An immunocolloid method for the electron microscope. *Immunochemistry* **8,** 1081.

Frens, G. (1973). Controlled nucleation for the regulation of the particle size in monodisperse gold suspensions. *Nature (London), Phys. Sci.* **241,** 20.

Glenney, J. (1986). Antibody probing of Western blots which have been stained with india ink. *Anal. Biochem.* **156,** 315.

Gould, G. W., Colyer, J., East, J. M., and Lee, A. G. (1987). Silver ions trigger Ca^{2+} release by interaction with the (Ca^{2+}–Mg^{2+})-ATPase in reconstituted systems. *J. Biol. Chem.* **262,** 7676.

Gounon, P., and Moreau, N. (1984). Une nouvelle technique simple et très sensible de coloration et d'immuno-détection des protéines transférées sur film de nitrocellulose. *Biol. Cell.* **52,** 102A.

Hancock, K., and Tsang, V. C. W. (1983). India ink staining of proteins on nitrocellulose paper. *Anal. Biochem.* **133,** 157.

Hauber, R., and Geiger, R. (1987). A new, very sensitive, bioluminescence-enhanced detection system for protein blotting. *J. Clin. Chem. Clin. Biochem.* **25,** 511.

Horisberger, M., and Rosset, J. (1977). Colloidal gold, a useful marker for transmission and scanning electron microscopy. *J. Histochem. Cytochem.* **25,** 295.

Horisberger, M. (1979). Evaluation of colloidal gold as a cytochemical marker for transmission and scanning electron microscopy. *Biol. Cell.* **36,** 253.

Hsu, Y.-H. (1984). Immunogold for detection of antigen on nitrocellulose paper. *Anal. Biochem.* **142,** 221.

Hunter, J. B., and Hunter, S. M. (1987). Quantification of proteins in the low nanogram range by staining with the colloidal gold stain AuroDye. *Anal. Biochem.* **164,** 430.

Lin, N. S., Hsu, Y. H., and Chin, R. J. (1987). Identification of viral structural proteins in the nucleoplasm of potato yellow dwarf virus-infected cells. *J. Gen. Virol.* **68,** 2723.

Lowry, O. H., Rosebrough, N. J., Farr, A. L., and Randall, R. J. (1951). Protein measurement with the Folin phenol reagent. *J. Biol. Chem.* **193,** 265.

Merril, C. R., and Pratt, M. E. (1986). A silver stain for the rapid quantitative detection of proteins or nucleic acids on membranes or thin layer plates. *Anal. Biochem.* **156,** 96.

Moeremans, M., Daneels, G., Van Dijck, A., Langanger, G., and De Mey, J. (1984). Sensitive visualization of antigen–antibody reactions in dot and blot immune overlay assays with immunogold and immunogold/silver staining. *J. Immunol. Methods* **74,** 353.

Moeremans, M., Daneels, G., and De Mey, J. (1985). Sensitive colloidal metal (gold or silver) staining of protein blots on nitrocellulose membranes. *Anal. Biochem.* **145,** 315.

Moeremans, M., Daneels, G., De Raemaeker, M., De Wever, B., and De Mey, J. (1987a). The use of colloidal metal particles in protein blotting. *Electrophoresis (Weinheim, Fed. Repub. Ger.)* **8,** 403.

Moeremans, M., Daneels, G., De Raemaeker, M., De Wever, B., and De Mey, J. (1987b). Overall staining and immunodetection of protein blots with colloidal metal particles. *Hoppe-Seyler's Z. Biol. Chem.* **368,** 441.

Morris, R. E., and Saelinger, C. B. (1986). Problems in the production and use of 5 nm avidin–gold colloids. *J. Microsc. (Oxford)* **143,** 171.

20. Detection of Proteins with Colloidal Gold

Nakamura, K., Tanaka, T., Kuwahara, A., and Takeo, K. (1985). Microassay for proteins on nitrocellulose filter using protein dye-staining procedure. *Anal. Biochem.* **148**, 311.

Righetti, P. G., Casero, P., and Del Campo, G. B. (1986). Gold staining in cellulose acetate membranes. *Clin. Chim. Acta* **157**, 167.

Rohringer, R., and Holden, D. W. (1985). Protein blotting: Detection of proteins with colloidal gold, and of glycoproteins and lectins with biotin-conjugated and enzyme probes. *Anal. Biochem.* **144**, 118.

Saman, E. (1986). A simple and sensitive method for detection of nucleic acids fixed on nylon-based filters. *Gene Anal. Technol.* **3**, 1.

Schapira, A. H. V., and Keir, G. (1988). Two-dimensional protein mapping by gold stain and immunoblotting. *Anal. Biochem.* **169**, 167.

Segers, J., and Rabaey, M. (1985). Sensitive protein stain on nitrocellulose blots. *Protides Biol. Fluids* 589.

Seitz, J. (1987). Karenzin—a hormone-dependent secretary protein of the rat seminal vesicle with actin-blinding properties. *Hoppe-Seyler's Z. Biol. Chem.* **368**, 442.

Stoscheck, C. M. (1987). Protein assay sensitive at nanogram levels. *Anal. Biochem.* **160**, 301.

Sutherland, M. W., and Skerritt, J. H. (1986). Alkali enhancement of protein staining on nitrocellulose. *Electrophoresis (Weinheim, Fed. Repub. Ger.)* **7**, 401.

Szewczyk, B., and Summers, D. F. (1987). Fluorescent staining of proteins transferred to nitrocellulose allowing for subsequent probing with antisera. *Anal. Biochem.* **164**, 303.

Van Bergen en Henegouwen, P. M. P., and Leunissen, J. L. M. (1986). Controlled growth of colloidal gold particles and implications for labelling efficiency. *Histochemistry* **85**, 81.

$dn = 11 = 0.8\,nm$

$n \propto M \propto d^3$

$\left(\dfrac{1}{0.8}\right)^3 = \dfrac{1}{.5} = 2 \times 11$

$\left(\dfrac{2}{0.8}\right)^3 = 8 \times 2 \times 11$
$= 27 \times 22$

d_{nm}	0.8	1.1	2	3	4	5
n	11	22	176	594	1342	2750

$MW \text{ of } 0.8\,nm\ (11\ atoms) = 11x$

" $1\,nm\ (22\ atoms) = 22x$

" $2\,nm\ (176\ atoms) = 176x$

21

Undecagold–Antibody Method

JAMES F. HAINFELD

Biology Department
Brookhaven National Laboratory
Upton, New York

INTRODUCTION
HISTORY OF THE UNDECAGOLD CLUSTER
METHODOLOGY
   Preparation of Undecagold
   Purification of the Undecagold Cluster
   Activation of Gold Cluster
   Preparation of Fab' Fragments
   Reaction of Fab' with Au_{11} and Purification
   Quantitation of Gold Cluster Labeling
ELECTRON MICROSCOPY
APPLICATIONS
REFERENCES

INTRODUCTION

A new antibody label has been synthesized (Hainfeld, 1987, 1988) that is 5 to 10 times smaller than other antibody markers used for electron microscopy. It is formed by covalently linking Fab' antibody fragment to an undecagold cluster, which has a core of 11 gold atoms 0.8 nm in diameter (Fig. 21.1). The distance of this electron-dense core to the antigenic

A

~ 2 nm

~ 0.82 nm

● = Au
X = CN

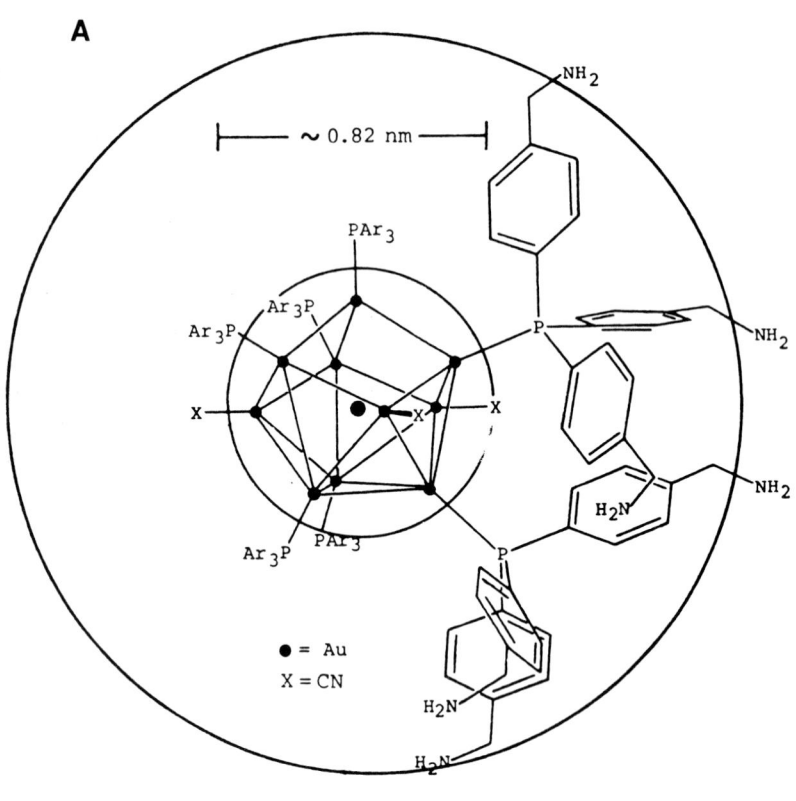

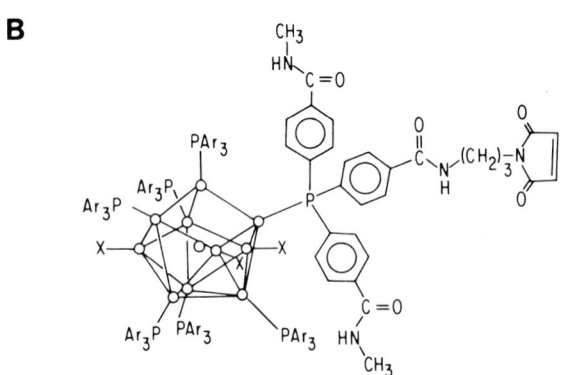

B

Ar = ⌬—CONHCH₃

X = CN

○ = Au

21. Undecagold–Antibody Method 415

site is 4.5–5.0 nm, which is therefore the limiting resolution of this label. The advantages of the label include its smaller size (Fig. 21.2), which improves resolution and permits the label to reach sites that are inaccessible to larger antibody labels. Also, the gold is covalently attached (see Fig. 21.3) and is stable under harsher conditions, such as during affinity purification, compared with the noncovalent attachment of colloidal gold (Bendayan, 1989; Handley, 1989). Another advantage is that the coupling of undecagold is at a specific sulfhydryl residue in the hinge region of the antibody, which is known from x-ray crystallography to be at the opposite end of the Fab' fragment from the antigen combining region. Thus, attachment of undecagold does not alter antibody specificity or destroy antibodies during the coupling process, as do some other tagging procedures. The disadvantages of this new label include the fact that it is so small that it is difficult to visualize in the conventional transmission electron microscope (CTEM). It is easily seen in the dark-field high-resolution scanning transmission electron microscope (STEM) however, where efficiency of collection in dark field is ~80% as opposed to ~5% in the tilted-beam dark field CTEM. Another limitation to its use is that it is not currently commercially available.

Because the undecagold is a new antibody label with distinct advantages over the widely used and successful colloidal gold-conjugated antibodies, this chapter will give the technical details concerning the preparation of undecagold and its use in labeling antibodies.

HISTORY OF THE UNDECAGOLD CLUSTER

The first synthesis of an undecagold cluster was reported in 1969 by McPartlin *et al.*, but this compound was insoluble in water and was used for organometallic chemical studies. Singer later siezed upon this compound as a potential label for electron microscopy, and Bartlett *et al.* (1978) modified its synthesis to make it water soluble (Fig. 21.1a). However, the cluster was so small that it was nearly invisible in the transmission electron microscope (TEM). It was, however, easily visualized in

---

Fig. 21.1. Undecagold clusters. (A) The derivative synthesized by Bartlett *et al.* (1978), which has an 11-gold-atom core (0.82 nm in diameter) and a shell of 7 triphenylphosphines (2 nm in diameter). This cluster has 21 amino groups at its periphery (3 per phosphine) that confer water solubility and may be used for attachment to other molecules. (B) The monomaleimidopropyl gold cluster synthesized by Safer *et al.* (1986), which is similar, but the phosphines are further modified so that 20 of 21 terminal groups are nonreactive methyls and the remaining one contains a maleimide, reactive with sulfhydryl groups. This cluster is therefore monofunctional.

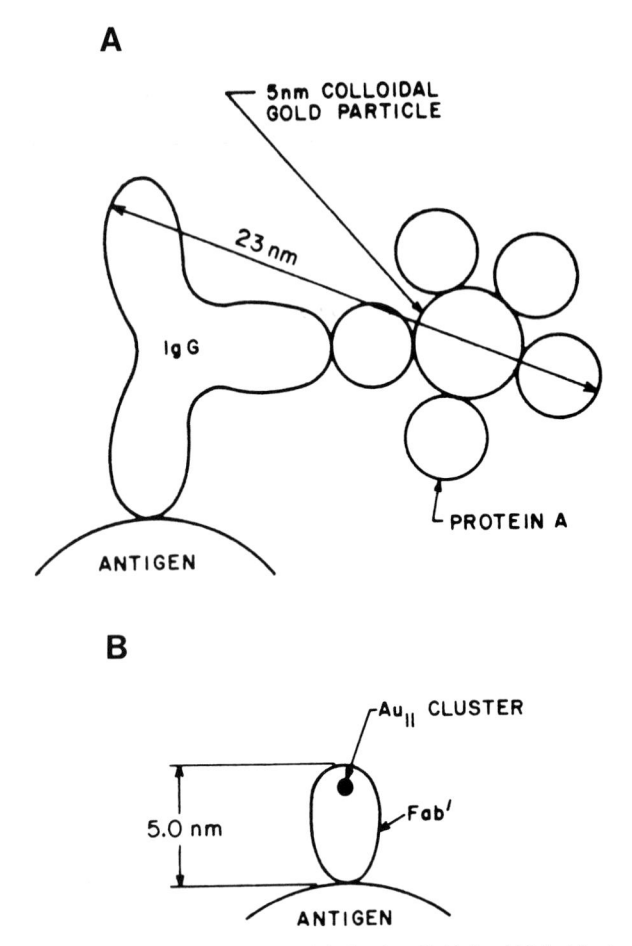

Fig. 21.2. Size comparison of antibody labels. A colloidal gold label is shown in (A) that relies on staphylococcal protein A binding first to the colloidal gold and then to the Fc region of an IgG; this gives an overall dimension of 23 nm (or larger if a larger colloidal gold particle is used). The label formed from an undecagold cluster covalently bound to an Fab' fragment shown in (B) gives an overall dimension of 5.0 nm.

the 0.25-nm-resolution STEM, and its scattering properties and behavior in the electron beam were studied (Wall *et al.*, 1982). In the form synthesized by Bartlett *et al.* (1978; Fig. 21.1a), the cluster has 21 primary amine groups at its periphery attached to 7 triphenylphosphines, which surround a core of 11 gold atoms. This form was cross-linked to various biomolecules through these amino groups. A modified biotin was attached and

21. Undecagold-Antibody Method

used to label the biotin binding sites on avidin to ~1 nm resolution (Safer et al., 1982). This was the first time that such a high resolution was achieved with labels on isolated biomolecules with the electron microscope. The cluster was also used to label glutamine synthetase (Lipka et al., 1984) and the carbohydrate moieties on the glycoprotein haptoglobin (Lipka et al., 1983).

The 21 amines, however, often led to excessive cross-linking, so monofunctional derivatives were prepared independently by Reardon and Frey (1984) and Safer et al. (1986). The one synthesized by Safer is shown in Fig.21.1b, and has 20 of the peripheral organic moieties blocked with a methyl group, whereas the remaining moiety is a single maleimide reactive with sulfhydryls. This compound has been used to label free sulfhydryls on actin (Safer et al., 1985), fibrinogen (Mosesson et al., 1986), myosin, the pyruvate dehydrogenase complex (Yang et al., manuscript in preparation), and other molecules (Hainfeld and Wall, 1986).

Because of the small size of the undecagold, a small antibody label incorporating it was sought. In order to maintain the advantage of small size, antibody fragments were used, specifically the Fab' fragment, which has a free sulfhydryl group. The successful preparation and testing of this label have been described in two prior publications (Hainfeld, 1987,

METHODOLOGY

Preparation of Undecagold

The preparation of undecagold has already been described in the literature in fair detail. The 21-amine compound synthesis is given by Bartlett et al. (1978) and two monofunctional derivatives are described by Reardon and Frey (1984) and Safer et al. (1986). Two covalently linked undecagold clusters that have a single arm for linkage to proteins or other molecules has been made (Yang and Frey, 1984) as well as several other modifications to the organic shell surrounding the core of 11 gold atoms (Yang et al., 1984). However, no undecagold clusters are available commercially at this writing, and neither are the appropriately derivatized phosphines, which are important reagents for cluster synthesis. A brief description of undecagold cluster synthesis will be given here, but it is recommended that this is best left to one trained in organic syntheses since traces of water can easily ruin the production at some steps and successful synthesis is not a trivial matter.

The method of synthesis described here most closely resembles that described by Safer *et al.* (1986), although the previously mentioned papers are useful references for more details.

First, a solution of tolyl magnesium bromide (I) is cooled to 0°C and 45 g of phosphorus trichloride is added.

$$\text{MgBr-C}_6\text{H}_4\text{-CH}_3 \quad \text{(I)}$$

The mixture is refluxed for 7 hr at 90°C, cooled, and then poured into 1.4 kg of ice and 90 ml of concentrated HCl. The mixture is extracted with benzene, dried, and recrystallized in 95% ethanol. Tris-*p*-tolylphosphine (II) is produced.

$$(\text{CH}_3\text{-C}_6\text{H}_4\text{-})_3\text{P} \quad \text{(II)}$$

42 g of compound II is then suspended in 150 ml of pyridine and heated to 40°C, and 66 g of Na_2CO_3 in 500 ml of water is added. The phosphine is oxidized by adding 10 g of $KMnO_4$. After heating to 125°C, 200 g of $KMnO_4$ is added. The pyridine is removed by distillation and the product filtered through Celite. The filtrate is cooled to 0°C and adjusted to pH 1 with concentrated HCl, and the precipitate is collected by filtration and washed with cold water. Tris-*p*-carboxyphenylphosphine oxide (III) is produced.

$$(\text{HOOC-C}_6\text{H}_4\text{-})_3\text{P=O} \quad \text{(III)}$$

37 g of compound III is mixed with 15 ml of acetyl chloride and 500 ml of methanol and heated to 50°C overnight. The methanol is evaporated and the product purified by adding 400 ml of boiling benzene, cooling, and then adding 800 ml of petroleum ether. The product, tris-*p*-carbomethoxyphenlphosphine oxide (IV), is collected by filtration.

$$(\text{CH}_3\text{OC(=O)-C}_6\text{H}_4\text{-})_3\text{P=O} \quad \text{(IV)}$$

32 g of compound IV is placed in 300 ml of benzene and 28 g of trichlorosilane and heated to 90°C for 5 hr. The solvent is removed and the residue dissolved in 300 ml of benzene, and then 150 ml of 28% NH_4OH is

21. Undecagold–Antibody Method

added. The solid is filter collected and dried and is tris-*p*-carbomethoxyphenylphosphine (V).

$$\left(CH_3OC(=O)-\underset{}{\bigcirc}- \right)_3 P \qquad \text{(V)}$$

25 g of compound V is added to 32 g of KOH in 200 ml of water and 200 ml of methanol and refluxed for 5 hr. The cooled solution is adjusted to pH 1 with 25% HCl and the precipitate collected by filtering and washed with water. Tris-*p*-carboxyphenylphosphine (VI) is produced.

$$\left(HOOC-\underset{}{\bigcirc}- \right)_3 P \qquad \text{(VI)}$$

0.8 g of compound VI is dissolved in 20 ml of dimethylformamide (DMF) and was mixed with 2 g of 1,1-carbonyl diimidazole. After 1 hr 60 mmol of methylamine in 13 ml of DMF and 220 mg of 1,3-diaminopropane are added. After 90 min the solvent is removed and 30 ml of methanol and cold 0.5% Na_2CO_3 in 300 ml of water are added. The precipitate is filtered, washed with water, and dried. It is mostly a mixture of 6 parts of tris-*p*-*N*-methylcarboxamidophenylphosphine (VII) and 1 part of *p*-*N*-aminopropylcarboxamidophenyl-di-*p*-*N*-methylcarboxamidophenylphosphine (VIII).

$$\left(CH_3-NH-C(=O)-\underset{}{\bigcirc}- \right)_3 P \qquad \text{(VII)}$$

$$\left(CH_3-NH-C(=O)-\underset{}{\bigcirc}- \right)_2 P-\underset{}{\bigcirc}-C(=O)-NH-(CH_2)-NH_2 \qquad \text{(VIII)}$$

This mixture (VII and VIII) is the one used to prepare the gold cluster since the cluster has 7 phosphines surrounding the 11-gold-atom core, and statistically the predominant product should therefore have 6 phosphines of VII and 1 of VIII, giving only one terminal amine and making it monofunctional. This amine is later converted into a maleimide group, which reacts with sulfhydryl groups. Some percentage of clusters will have other ratios of VII and VIII, but these can be removed during the purification, described below.

0.14 mol of the phosphine mixture in 3 ml of methanol is mixed with an equimolar amount of AuCN in 5 ml of 100% ethanol and stirred for 30 min. An equimolar amount of a 0.1% ethanolic solution of NaBH$_4$ is slowly added over 2 hr. The undecagold cluster is formed and has a reddish-brown color. Eight drops of acetone are added to quench any excess borohydride. To improve its solubility in aqueous buffers, the cyanide is exchanged with the chloride counterion by mixing with 8 ml of water and 3 g of Dowex (chloride form) overnight and then filtering off the ion exchange resin. The yield of cluster is typically 30%, which for this case produces 3.9 μmol.

Purification of the Undecagold Cluster

Once successful cluster formation has been achieved, the reaction mixture contains the cluster, excess phosphines, and gold phosphines in a solvent of methanol, ethanol, and water. The undecagold cluster must be separated out. Also, several undecagold species are produced, some that are nonfunctional (no amines), some that are monofunctional (one amine), as well as small amounts of difunctional, trifunctional, etc. The monofunctional form should be isolated.

The first step is to purify the clusters from the general reaction mixture. This may be done by size exclusion chromatography since the undecagold cluster has a molecular weight of ~5000 and the phosphines have molecular weights of ~250. Safer *et al.* (1986) described the use of a 2.5 × 100 cm column of BioRad P-10 running in 0.6 M triethylammonium bicarbonate and 5% methanol. The triethylammonium bicarbonate buffer is prepared by bubbling CO$_2$ into a mixture of 1 part triethylamine and 5 parts water for about a day to make 1.2 M triethylammonium bicarbonate. This is diluted and methanol added to form the above buffer. The advantage of this buffer is that it is volatile and can be removed fairly easily by rotary evaporation. However, 0.5 M NaCl is another suitable buffer choice (buffered by 0.1 M Tris-HCl, pH 7) for this column. P-10 has a fractionation range to molecular weight 15,000, so that colloids or larger aggregates may be removed. The disadvantage of this column is its 6–8-hr run time. For most runs, little material larger than undecagold is formed, and therefore another material with faster flow properties can be used such as the Amicon desalting gel GH25, which has a molecular weight cutoff of 3000. The cluster comes through in the void volume but runs are shortened to 30 min to 1 hr because of its higher flow rate under pressure.

The gold cluster peak from the column is pooled. For some columns the cluster peak may somewhat overlap with the phosphine–gold phosphine peak. To determine which fractions to pool, a UV–visible spectrum is taken (from 260 to 600 nm) and the 280 nm/420 nm absorption ratio may

21. Undecagold–Antibody Method

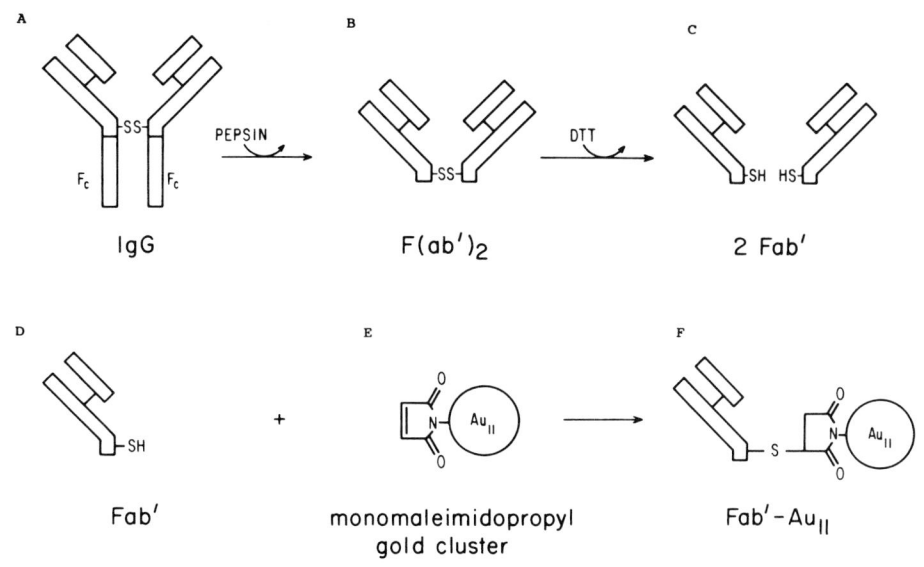

Fig. 21.3. Reaction scheme for labeling Fab′ with undecagold. IgG is digested with pepsin to yield F(ab′)$_2$ fragments joined by a disulfide (A, B). The constant fragment (Fc) region fragments are removed on a high-pressure liquid chromatography (HPLC) gel exclusion column. Next, 20 mM DTT is used to reduce the disulfide link. DTT is removed with an HPLC gel exclusion column. Undecagold is activated immediately prior to use by reaction of monoaminopropyl-Au$_{11}$ and N-methoxycarboxyl maleimide to form monomaleimidopropyl-Au$_{11}$ (E). The product is purified on a carboxylic CM-Sepharose column. The Fab′ and activated Au$_{11}$ are incubated overnight at 4°C (D–F). Unreacted Au$_{11}$ is removed with an HPLC sizing column.

be used as a guide to cluster purity. The undecagold has a ratio of 3.6 (ϵ_{420} = 4.71 × 10^4), whereas the phosphine components absorb only in the UV region (280 nm) but not at 420 nm. A 280/420 ratio > 3.6 indicates an impure cluster.

The pooled pure peak is next dried by rotary evaporation. Care must be used because of the large flakes formed by the triethylamine. The residue is rinsed with 5 ml of methanol and reevaporated four times at 45°C to remove most of the buffer.

Due to the mixture of phosphines used to form the cluster, a statistical mixture of clusters is produced, the monofunctional form being the one desired. Ion exchange chromatography is used to isolate this form. The column used is a 1 × 50 cm CM-Sepharose Fast Flow (Pharmacia) run in a gradient with A = water and B = 0.25 M NaCl, 10 mM triethylammonium-HCl (pH 7.0), and 5% methanol. The gold cluster sample is dissolved in a 5% methanol–water solution for injection and the gradient is

422   James F. Hainfeld

run from 0 to 100% B. Bands of yellow-colored gold cluster will separate; the first band to come off is the nonfunctional cluster and the next band is the monofunctional cluster desired, with one amine.

Next, the purified cluster must be desalted. This may be accomplished by dialysis using dialysis tubing with a cutoff at molecular weight 3000 (Spectrapor) or by binding the cluster in the salt buffer from the previous column to a reverse-phase biphenyl silica column (Vydac 219), rinsing with distilled water to remove the salt, and eluting with 0.3 M triethylammonium bicarbonate buffer in 70% methanol. The buffer is removed by rotary evaporation, as described above. The cluster is brought up in methanol at a concentration of 5 nmol/μl and stored in the refrigerator. A STEM image of undecagold clusters is shown in Fig. 21.4.

Activation of Gold Cluster

The reactive maleimide group is attached to the single amine of the cluster immediately prior to use. The reason for this attachment is that maleimides have a half-life of only a few hours in aqueous solutions. The recovery of gold through this step is 50–80% and the gold cluster is mixed with the antibody in 5–10-fold molar excess, so an appropriate amount is pipetted out of the methanol stock solution. For example, 200 μg of Fab′

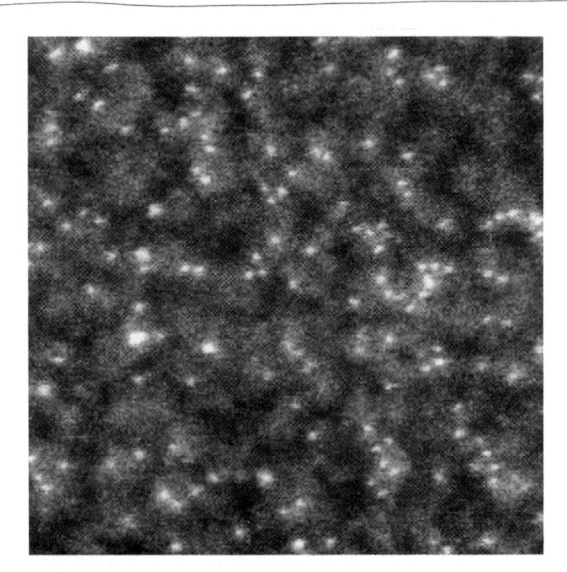

Fig. 21.4. STEM dark-field micrograph of undecagold clusters. A 10^{-5} M solution of the cluster in Fig. 21.1B was applied to a thin (2.5-nm) carbon film and, after a 1-min adsorption time, rinsed with deionized water and air dried. Full width, 62 nm.

21. Undecagold–Antibody Method

(mol. wt. 50,000) is 4 nmol, so 50 nmol of Au_{11} (for a 10 : 1 ratio) should be used.

50 nmol of Au_{11} in 10 μl of methanol is mixed with 1 mg of N-methoxycarboxyl maleimide (NMCM) and 10 μl of a 0.5 M $NaHCO_3$ solution on ice for 20 min. The reaction is diluted with 1 ml of distilled water and applied to a 3-ml gravity-fed ion exchange column, CM-Sepharose C50, which has been cycled with distilled water, 0.1 M phosphate buffer (pH 7), and distilled water. The undecagold adheres to the top of this column and the excess NMCM is removed by washing with 9 ml of distilled water. The Au_{11} is eluted, using a 0.1 M Na-phosphate (pH 7.0) buffer. The gold fractions are combined and quantitated using a spectrophotometer set at 420 nm and the extinction coefficient $\epsilon_{420} = 4.71 \times 10^4$. The compound produced is the monomaleimide undecagold cluster (Fig. 21.1B).

Preparation of Fab' Fragments

Starting with IgG, $F(ab')_2$ fragments must be prepared. This is done by appropriate digestion with the enzyme pepsin. Rabbit polyclonal IgG is converted to $F(ab')_2$ in good yield by using 2% by weight pepsin to the amount of IgG in a 0.1 M sodium citrate buffer (pH 3.5) at 37°C overnight. Mouse monoclonals require the reaction to be slowed somewhat to prevent overdigestion by increasing the pH (e.g., to 4.2) or decreasing the digestion time. A study of the optimal conditions was made by Parham (1983), and his basic conclusion is that $F(ab')_2$ fragments can be prepared from murine monoclonal subtypes IgG1 and IgG2a, with poorer yields from IgG2b. Also, each monoclonal antibody should be specifically optimized if maximal yields are sought by varying the pH and digestion time.

The $F(ab')_2$ produced must be purified. Frequently, *Staphylococcus* protein A columns are used to bind and remove the Fc fragment. However, pepsin digests the Fc into smaller pieces and since these are easily separated from the 100-kDa $F(ab')_2$ fragment on a gel filtration column, the staphylococcal protein A column need not be used. Although many gel filtration columns are acceptable, one that works well is the DuPont Zorbax GF-250. The running buffer is 0.2 M Na phosphate (pH 7.0).

The next step is to produce Fab' fragments from the $F(ab')_2$. The $F(ab')_2$ is adjusted to 0.5–1 mg/ml in 0.1 M Na-phosphate buffer (pH 7.0) and a freshly dissolved (or frozen stored) solution of dithiothreitol (DTT) is added to make the final concentration 20 mM in DTT. The solution is kept under N_2 at room temperature for 2 hr. The mixture is then concentrated by spinning in a Centricon 30 device (Amicon) and applied to a gel filtration column (DuPont GF-250) with a running buffer of 0.2 M Na-phosphate (pH 7.0). The Fab' peak is pooled and quantitated by UV absorption at 280 nm using $E^{1\%} = 15.4$ and the formula

$$\text{Concentration (mg/ml)} = \frac{OD_{280}}{1.54}$$

Reaction of Fab' with Au$_{11}$ and Purification

The Fab' is mixed with the activated undecagold (with the malemide arm added) such that there is a 5–10-fold molar excess of undecagold. The gold activation and Fab' preparation are coordinated so that each is freshly completed, since both are labile. The mixture is kept at 4°C overnight.

The next step is to remove any unbound Au$_{11}$ and to quantitate gold binding. The protein is concentrated using a Centricon 30 centrifuge filter and the 50–100 µl produced is injected onto a DuPont Zorbax GF-250 column running in 0.2 M Na-phosphate buffer (pH 7.0).

Quantitation of Gold Cluster Labeling

It would be convenient to assess quickly the amount of gold cluster reaction at this point and quantitatively determine how many gold clusters are bound to each Fab'. This is possible by using a UV–visible spectrophotometer. Since the spectra of the two components are known, the concentration of each may be determined in a mixture by measuring the absorption at two wavelengths and solving two simultaneous equations:

$$A_1 = \epsilon_{x1}C_x + \epsilon_{y1}C_y \quad \text{at } \lambda_1$$
$$A_2 = \epsilon_{x2}C_x + \epsilon_{y2}C_y \quad \text{at } \lambda_2$$

where A_1 is the absorption measured at λ_1, ϵ_{x1} is the extinction coefficient of component x at λ_1, and C_x is the concentration of component x. These equations assume that the mixture is a linear combination of the pure spectra, which implies that the component spectra are not altered as a result of being in the mixture. Measurements may be done at more than two wavelengths and a least-squares fit done to improve the accuracy. The gold cluster is a favorable case since it has an absorption peak at 420 nm, giving it the yellow color, whereas the Fab' absorbs at 280 nm but has no absorption at 420 nm. The gold cluster absorbs at 280 nm, but by taking readings at 280 and 420 nm and solving the above equations, good accuracy can be achieved. Using the extinction coefficients for Fab' ($\epsilon_{420} = 0$, $\epsilon_{280} = 7.5 \times 10^4$) and for undecagold ($\epsilon_{420} = 4.71 \times 10^4$, $\epsilon_{280} = 16.8 \times 10^4$), the labeling ratio of gold cluster to Fab' reduces to

$$\frac{Au_{11}}{Fab'} = \frac{OD_{420} \times 7.5}{OD_{280} \times 4.71 - OD_{420} \times 16.8}$$

For a good experiment, values of 1–2 are obtained (1–2 gold clusters per Fab'); poorer labeling may yield ratios as low as 0.05 (5% of the Fab' has an Au_{11}) or zero if there is a problem with reagents or protocol. Fractions from the final purification are scanned in the UV–visible, either during the column run by using an in-line spectrophotometer such as the Hewlett Packard 1040A diode array detector or after the run using a spectrophotometer.

ELECTRON MICROSCOPY

A STEM image of undecagold-labeled Fab' fragments is shown in Fig. 21.5. The labeling is virtually quantitative with nearly all antibody fragments having a gold cluster. Occasionally more than one cluster is attached to the antibody; up to four have been observed. These multiple clusters are always adjacent to each other at one end and never spaced randomly around the Fab'. Although the exact reason for multiple attachment is not clear, it is probably due to one of the other sulfhydryl groups

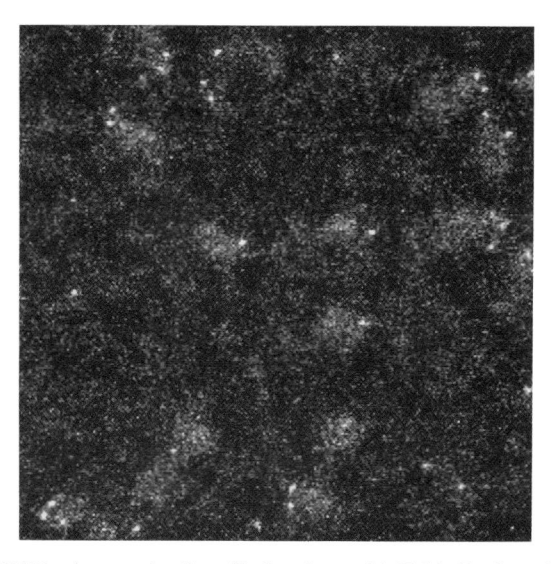

Fig. 21.5. STEM micrograph of purified undecagold–Fab'. Each antibody Fab' fragment is 5 × 3 × 2 nm and most have one or more gold clusters (white dots) attached at one end. Three clear Fab' molecules appear in the center of this field, each with one undecagold cluster bound. Full width, 62 nm.

in the hinge region (e.g., rabbit has one and mouse has three) or a more reactive amine on the Fab' surface reacting with the maleimide.

Unstained and unshadowed samples for the STEM are prepared as previously described (Wall and Hainfeld, 1986). Briefly, a thin carbon film (2.5 nm) over a holey film is used to give better contrast (lower background). The sample (in 20 mM ammonium acetate, pH 7.0) is applied to the grid at a concentration of 5–30 μg/ml, allowed to adsorb for 1 min, and then wicked to a thin film of fluid and washed eight times with the above buffer. It is wicked to a thin drop and quickly frozen in liquid nitrogen slush produced by pumping on a Dewar flask of liquid nitrogen in a desiccator jar with a mechanical pump. The grid is then freeze-dried overnight. The grid is viewed in a high-resolution (0.25-nm) STEM constructed at Brookhaven National Laboratory (Bittner and Wall, 1975). This instrument is capable of visualizing single heavy atoms (Wall et al., 1978), so the undecagold is easily seen. The good visibility is due to the high resolution and good collection efficiency in the dark field. The images shown (Figs. 21.4–21.6) are taken in the dark-field mode with no staining or shadowing. In fact, such treatments tend to obscure the gold clusters.

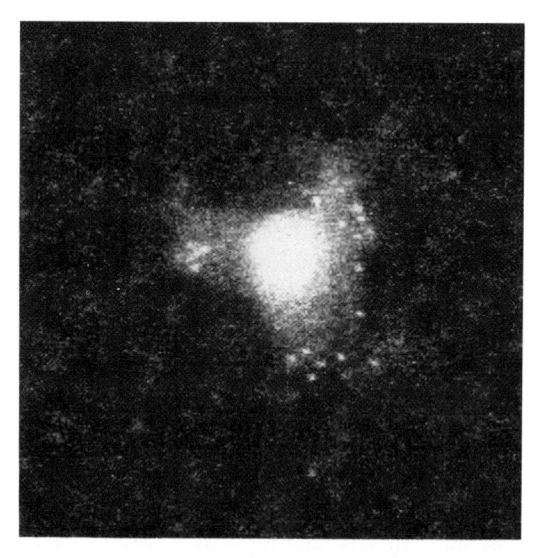

Fig. 21.6. Micrograph of a ferritin molecule with Fab'-Au$_{11}$ molecules bound to its periphery. The antibody used was polyclonal rabbit antiferritin and the Fab'-Au$_{11}$ is shown in Fig. 21.5. This Fab'-Au$_{11}$ was incubated with horse ferritin for 1 hr at 37°C and column purified to remove excess Fab'. The ferritin iron oxide core is 8.0 nm with a protein shell of 24 similar subunits 12.5 nm in diameter. The gold cluster core (0.8 nm) is one-tenth the size of this common electron microscope label. Full width, 62 nm.

21. Undecagold–Antibody Method

The electron dose necessary to see the gold cluster clearly is ~ 30 e/Å^2. A study of the visibility and stability of the undecagold in the electron beam has been reported (Wall *et al.*, 1982) and the fading of the clusters' spot intensity with beam dose has been well documented. The clusters do not move about as do single heavy atoms (Wall *et al.*, 1978) but remain stationary even at higher doses.

APPLICATIONS

The first undecagold–antibody probe constructed was with a rabbit polyclonal antiferritin antibody (Hainfeld, 1987). Ferritin has an iron core 8 nm in diameter and a protein shell 12.5 nm across consisting of 24 similar subunits. The Fab'-Au$_{11}$ was incubated with ferritin at 37°C for 1 hr. Unbound antibodies were separated by passing the mixture through a 0.66 × 50 cm TSK-HW55 (S) gel filtration column. A typical labeled ferritin molecule is shown in Fig. 21.6. The undecagold clusters are generally ≤5.0 nm from the surface of ferritin, which then also is the resolution limit of this label. Since ferritin has 24 subunits and the antibody used is polyclonal, 24 or more binding sites should be expected. This is why a number of Fab' fragments bind to each ferritin. Fab' fragments with more than one undecagold can be distinguished binding to ferritin in some images, and the multiple clusters therefore do not appear to affect immunoreactivity.

Further testing was done to verify the preservation of immunospecificity of the Fab'-Au$_{11}$. One question is, does the Fab'-Au$_{11}$ simply adhere nonspecifically to the ferritin or is it immunobiologically bound? The use of 0.2 M salt on the column that separates the excess and unbound Fab'-Au$_{11}$ from the ferritin helps to reduce nonspecific ionic binding. One experiment was carried out to prepare Fab'-Au$_{11}$ from an antibody to another protein, antidynein. The Fab'-Au$_{11}$ by itself looked identical in the electron microscope to the antiferritin Fab'-Au$_{11}$ (Fig. 21.5), but when incubated with ferritin and column purified, no Fab'-Au$_{11}$ was found on the ferritin. A further test was done to incubate antiferritin Fab'-Au$_{11}$ with ferritin, tobacco mosaic virus, and glutamine synthetase. The Fab'-Au$_{11}$ was found only on the ferritin. One further test was to prepare Fab' fragments, block the free sulfhydryl groups with N-ethylmaleimide, and incubate these Fab' fragments with ferritin. These fragments should bind to all of the antibody epitopes and coat the ferritin. Next, a second incubation was carried out with antiferritin Fab'-Au$_{11}$. Since all sites are already occupied, none of the gold-labeled Fab' should bind. The reaction mixture was column purified as described above. The result was as expected:

unlabeled Fab' coated the ferritin and no Fab'-Au$_{11}$ was bound. See Hainfeld (1987) for micrographs depicting these controls.

A number of other antibodies have been labeled with undecagold by this procedure with similar results. Column separations and electron microscopy are virtually indistinguishable for different antibodies. Other antibodies labeled include rabbit antibody to flagella, rabbit antibodies to 22S and 14S dynein, mouse monoclonal antibodies to bovine factor V, mouse monoclonal H$_2$4B5 to influenza, mouse monoclonal BR55-2 to breast cancer, and mouse monoclonal 17-1A to human colorectal carcinoma. Rabbit IgGs are well behaved, as are most mouse monoclonals. However, F(ab')$_2$ fragments must be formed in a reasonable yield.

Several diverse projects with the Fab'-Au$_{11}$ are currently in progress using the high-resolution STEM, where the increased resolution beyond that obtainable with colloidal gold labeling should permit meaningful mapping. The poor visibility of the gold clusters in the commercial TEM and the decreased contrast when they are used in sections or with negative stains, however, limit this technique to suitable instruments and samples at the present time.

I thank D. Safer for providing some of the undecagold reagents; J. Maracek and C. J. Foley for determining and providing synthetic details; N. I. Feng, K. D. Elmore, and G. G. Shiue for excellent technical assistance; M. N. Simon and F. E. Kito for microscope operation; and J. S. Wall and H. W. Siegelman for helpful discussions. This work was supported by the Office of Health and Environmental Research of the Department of Energy and by NIH Grant GM 31975.

REFERENCES

Bartlett, P. A., Bauer, B., and Singer, S. J. (1978). Synthesis of water-soluble undecagold cluster compounds of potential importance in electron microscopic and other studies of biological systems *J. Am. Chem. Soc.* **100,** 5085.
Bendayan, M. (1989). Protein A–gold and protein G–gold postembedding immunoelectron microscopy. In *Colloidal Gold: Principles, Methods, and Applications, Vol. 1* (M. A. Hayat, ed.), pp. 33–94. Academic Press, San Diego, California.
Bittner, J. W., and Wall, J. S. (1975). A new high resolution STEM for biological studies. *Proc.—33rd Annu. Meet. Electron Microsc. Soc. Am.* p. 114.
Hainfeld, J. F., and Wall, J. S. (1986). Mapping the domains of molecules and complexes by mass and heavy atom loading. *Ann. N.Y. Acad. Sci.* **483,** 181.
Hainfeld, J. F. (1987). A small gold-conjugated antibody label: Improved resolution for electron microscopy. *Science* **236,** 450.
Hainfeld, J. F. (1988). Gold cluster-labelled antibodies. *Nature (London)* **333,** 281.
Handley, D. A. (1989). Methods for synthesis of colloidal gold. In *Colloidal Gold: Principles, Methods, and Applications Vol. 1* (M. A. Hayat, ed.), pp. 13–32. Academic Press, San Diego, California.

Lipka, J. J., Hainfeld, J. F., and Wall, J. S. (1983). Undecagold labeling of a glycoprotein: STEM visualization of an undecagoldphosphine cluster labeling the carbohydrate sites of human haptoglobin–hemoglobin complex. *J. Ultrastruct. Res.* **84,** 120.

Lipka, J. J., Hainfeld, J. F., and Wall, J. S. (1984). Undecagold labelling of glutamine synthetase from *E. coli. Proc.—42nd Annu. Meet., Electron Microsc. Soc. Am.* p. 158.

McPartlin, M., Mason, R., and Malatesta, L. (1969). Novel cluster complexes of gold(0)–gold(I). *J. Chem. Soc., Chem. Commun.* p. 334.

Mosesson, M. W., Siebenlist, K. R., Diorio, J. P., Hainfeld, J. F., Wall, J. S., Soria, J., Soria, C., and Samama, M. (1986). Evidence that proximal NH_2-terminal portions of fibrinogen Metz (Aα 16 Arg → Cys)Aα chains are oriented in the same direction. *In Fibrinogen and Its Derivatives* (G. Müller-Berghaus *et al.*, eds.), p. 3. Elsevier, Amsterdam.

Parham, P. (1983). On the fragmentation of monoclonal IgG1, IgG2a, and IgG2b from BALB/c mice. *J. Immunol.* **131,** 2895.

Reardon, J. E., and Frey, P. A. (1984). Synthesis of undecagold cluster molecules as biochemical labeling reagents. I. Monoacyl and mono-*N*--hydroxysuccinimidosuccinyl–undecagold. *Biochemistry* **23,** 3849.

Safer, D., Hainfeld, J. F., Wall, J. S., and Riordan, J. (1982). Biospecific labeling with undecagold: Visualization of the biotin binding sites on avidin. *Science* **218,** 290.

Safer, D., Hainfeld, J., and Wall, J. S. (1985). The localization of cysteine-374 in F-actin determined by gold cluster labeling and scanning transmission electron microscopy. *Biophys. J.* **47,** 128a.

Safer, D., Bolinger, L., and Leigh, J. S. (1986). Undecagold clusters for site-specific labeling of biological macromolecules: Simplified preparation and model applications. *J. Inorg. Biochem.* **26,** 77.

Wall, J. S., Hainfeld, J. F., and Bittner, J. W. (1978). Preliminary measurements of uranium atom motion on carbon films at low temperatures. *Ultramicroscopy* **3,** 81.

Wall, J. S., and Hainfeld, J. F. (1986). Mass mapping with the scanning transmission electron microscope. *Annu. Rev. Biophys. Biophys. Chem.* **15,** 355.

Wall, J. S., Hainfeld, J. F., Bartlett, P. A., and Singer, S. J. (1982). Observation of an undecagold cluster compound in the scanning transmission electron microscope. *Ultramicroscopy* **8,** 397.

Yang, H., and Frey, P. A. (1984). Synthesis of undecagold cluster molecules as biochemical labeling reagents. III. A dimeric cluster with a single reactive amino group. *Biochemistry* **23,** 3836.

Yang, H., Reardon, J. E., and Frey, P. A. (1984). Synthesis of undecagold cluster molecules as biochemical labeling reagents. II. Bromoacetyl and maleimido undecagold clusters. *Biochemistry* **23,** 3857.

Yang, Y.-S., Furcinitti, P. S., Hainfeld, J. F., Wall, J. S., and Frey, P. A. (manuscript in preparation). Mapping undecagold labeled lipoyl groups in the pyruvate dehydrogenase complex by scanning transmission electron microscopy.

22

Immunogold Labeling for the Single-Laser FACS Analysis of Triple Antibody-Binding Cells

THOMAS H. TÖTTERMAN
and
ROGER FESTIN

Clinical Immunology Section
University Hospital
Uppsala, Sweden

INTRODUCTION
TECHNIQUES
  Cells
  Antibodies
  Staining of Cells
  Flow Cytometry (FCM)
APPLICATIONS
  Detection of Immunogold Stained Lymphocytes
  Accuracy of Gold versus Fluorochrome Staining
  Triple Antibody Staining
COMMENTS
  Advantages
  Limitations

Future Prospects
SUMMARY
REFERENCES

INTRODUCTION

The use of monoclonal antibodies (McAbs) to a range of different cell surface and cytoplasmic determinants has enabled improved analysis and understanding of cellular diversity among normal and malignant cells of the immune system (reviewed by Foon and Todd, 1986). Conjugation of McAbs to fluorochromes enables detection of McAb-binding cells in a UV-microscope or by flow cytometry (for a comprehensive review of flow cytometry, see, e.g., Shapiro, 1983). By combining McAbs conjugated to fluorochromes of different emission characteristics in a double-staining assay, it is possible to detect subpopulations of cells within a larger cell group. Dual staining with fluorescein isothiocyanate (FITC) together with either tetramethylrhodamine isothiocyanate (TRITC), phycoerythrin (PE) (Oi et al., 1982), or Texas Red (Titus et al., 1982) has already contributed considerably to the understanding of the structure of the normal or deranged immune system (Janossy et al., 1985).

In order to further dissect the phenotypic differentiation of cells, there is a need for more than two parameters. For example, the identification of subsubtype, activation state, or discrete steps in the differentiation stage of individual lymphoid cells may require three or more independent markers. The possibilities for multicolor immunofluorescence are, however, severely limited by overlap of emission spectra of available fluorochromes. Such analyses usually require two separate excitation sources of different spectral characteristics, which are found only in selected and expensive dual-laser flow cytometers (Dean and Pinkel, 1978).

The problem described above can be overcome by application of nonfluorescent probes such as colloidal gold. De Mey (1983) described the detection of McAb-stained cells binding colloidal gold-conjugated secondary layer antibodies in light microscopy by epi-illumination with polarized light. Van Dongen et al. 1985) applied this immunogold staining to multimarker analysis by combining the gold conjugates with FITC- and TRITC-labeled McAbs.

The multiparameter staining technique described is suitable for detection of small or very small cell populations, but the analysis is very cumbersome and time consuming with the microscope. Application of immunogold with a single antibody in flow cytometry was made by Böhmer and King (1984). We have developed this further and made use of the abilities of flow cytometry for rapid and accurate cell analysis. We extended the multiparameter immunogold technique to single-laser flow cytometry, and we demonstrate here the analysis of individual lymphoid cells binding three different McAbs detected with colloidal gold, FITC, and PE, respectively (Festin *et al.*, 1987).

TECHNIQUES

Cells

Peripheral blood lymphocytes from healthy blood donors were studied in single-antibody staining experiments. In triple-labeling experiments, lymphocytes from patients with rejecting allogeneic kidney grafts or from patients with regenerating T-cell populations following autologous bone marrow transplantation (ABMT) were obtained for analysis. Such cells included large populations of T suppressor/cytotoxic ($T_{s/cx}$, Leu-2a$^+$, CD8$^+$; see below) cells positive for both the Leu-15 (CD11) and HLA-DR markers (see below), as well as T helper/inducer ($T_{h/i}$, Leu-3a$^+$, CD4$^+$; see below) cells expressing both the Leu-18 (CD45R; see below) and HLA-DR markers. Lymphocytes were isolated from 20 ml of heparinized venous blood on routine density gradients (Lymphocyte Preparation Medium, Flow Labs, UK) and washed twice, and monocytes were carefully depleted by treatment with iron powder plus magnet.

Antibodies

Goat anti-mouse IgG and rabbit antigoat antibodies labeled with colloidal 40-nm gold particles (Auroprobe, GAM-IgG-G40, RAG-G40) were purchased from Janssen Pharmaceuticals (Belgium). GAM-G40 and RAG-G40 were used undiluted, following the procedures of Van Dongen *et al.* (1985) and Böhmer and King (1984) and because of lack of information concerning the concentration of the reagent. The following McAbs were obtained from Becton Dickinson (Mountain View, CA) as purified antibody or directly labeled with FITC or PE (all McAbs used as 1 : 10 dilution): anti-Leu-2a (CD8) identifying T suppressor/cytotoxic cells ($T_{s/cx}$) (Engleman *et al.*, 1981); anti-Leu-3a (CD4) identifying T helper/inducer cells ($T_{h/i}$) (Engleman *et al.*, 1981); anti-HLA-DR binding to a nonpolymorphic DR epitope (Lampson and Levy, 1980); anti-Leu-15 (CD11) binding to human complement receptor CR3 and identifying T suppressor (T_s)-like cells within the Leu-2a$^+$ population (Landay *et al.*, 1983); anti-Leu-18 (CD45R) identifying T suppressor/inducer ($T_{s/i}$) cells within the Leu-3a$^+$ population (Morimoto *et al.*, 1985). Pure cytotoxic-type T cells (T$_{cx}$) are Leu2$^+$Leu15$^-$, and pure T$_h$ cells are Leu-3a$^+$Leu-18$^-$.

Staining of Cells

For immunogold staining, 1×10^6 cells in 20 µl of medium (consisting of phosphate-buffered saline with 2% newborn calf serum and 0.1% sodium azide; hereafter referred to as medium A) were incubated with 50 µl of diluted (see above) unlabeled primary antibody for 10 min at room temperature (RT). After two washings and suspension in 20 µl of medium A, cells were incubated 10 min at RT with 50 µl of undiluted GAM-G40 and thereafter centrifuged (200 g for 5 min) and resuspended three times, followed by one wash in medium A. RAG-G40 (25 µl) was added to the pellet, followed by incubation and washings as above. The use of a tertiary gold-labeled antibody layer (RAG-G40) on top of the secondary GAM-G40 was shown to increase the signal of the gold-labeled cells further. In triple-labeling experiments, free binding sites on GAM- and RAG-G40 were blocked by incubation for 5 min at RT with normal mouse serum (1 : 100). After washing, the cells were incubated for 10 min at RT with FITC- and PE-conjugated McAbs. Following two further washings, cells were suspended in 0.5–1 ml of medium A and analyzed in flow cytometry.

Flow Cytometry (FCM)

FCM analysis was performed on a FACStar (Becton Dickinson) instrument equipped with a 5-W argon ion laser (Coherent Innova 90; Coherent, Palo Alto, CA) emitting 0.2 W at 488 nm. Green fluorescence (FITC) was collected through a 530/30-nm bandpass filter and red fluorescence (PE) through a 585/42-nm bandpass filter (both from Becton Dickinson). G40-positive cells were detected in the 90° light scatter (side scatter, SSC) photomultiplier tube. All these features are fully standard on the FACStar and on most commercially available flow cytometers. Instrument calibration was performed with 2.02-µm latex beads (Becton Dickinson) to obtain optimum resolution primarily in the SSC channel. Suitable amplification for fluorescence as well as electronic compensation for spectral overlap between FITC and PE channels was adjusted with CaliBRITE beads (Becton Dickinson). Data were collected in four-parameter list mode with linear amplification for forward light scatter (FSC) and logarithmic for SSC and green and red fluorescence, and were processed with the Consort 30 software in a Hewlett Packard 9000 series model 217 personal computer (Hewlett Packard, Fort Collins, CO).

APPLICATIONS

Detection of Immunogold Stained Lymphocytes

Normal peripheral blood lymphocytes were isolated and stained according to the described protocol with anti-Leu-2a or anti-Leu-3a McAbs followed by GAM-G40 and RAG-G40 and analyzed on the flow cytometer. A "secondary control," that is, cells stained with GAM- and RAG-G40 only, without primary antibody, produced no nonspecific background, as can be seen from Fig. 22.1a. However, incubation of the cells with a relevant primary antibody followed by G40 antibodies resulted in a population of G40$^+$ cells which is distinct and clearly separable from the negative cluster (Fig. 22.1b or c) and which can be gated on in three-color experiments. Figure 22.1c demonstrates that the use of the RAG-G40 as a tertiary layer to enhance the effect of the secondary GAM-G40 (Fig. 22.1b) does increase the signal strength of the G40$^+$ cells, a finding of importance when labeling weakly expressed antigens.

Note that the SSC channel (where G40$^+$ cells are detected) has logarithmic amplification in order to create the clear cluster appearance and avoid a widespread distribution of gold-labeled cells. It may also work in linear mode when labeling strongly and uniformly expressed antigens.

The necessity of depleting the sample of monocytes/macrophages is obvious from Fig. 22.1b. The presence of such cells would cause interference with the G40$^+$ cell population and make proper gating impossible.

Accuracy of Gold versus Fluorochrome Staining

To ensure that the gold label provides a correct estimate of the percentage of positive cells, we compared analyses of T lymphocytes stained with a T-cell antibody followed by GAM + RAG-G40 or conventional fluorochrome-conjugated McAbs. Figure 22.2a and b demonstrate that FITC- and PE-labeled anti-Leu-3a McAbs both detect virtually identical populations of Leu-3a$^+$ T cells. In Fig. 22.2c cells labeled with anti-Leu-3a-FITC and anti-Leu-3a-G40, respectively, were mixed together in a 1 : 1 ratio, indicating that there is no appreciable interference between the fluorochrome and the nonfluorescent gold label when these are on separate cells. Table 22.1 is a summary of five experiments made to verify the accuracy of the gold staining. It is evident that all three different labels, FITC, PE, and G40, detect the same population of Leu-2a as well as Leu-3a T-cell subsets. Gating on the G40$^+$ cell population is probably easiest to perform on the FSC-SSC two-parameter diagram (Fig. 22.1b), but the same results were obtained by gating on an SSC single-parameter histo-

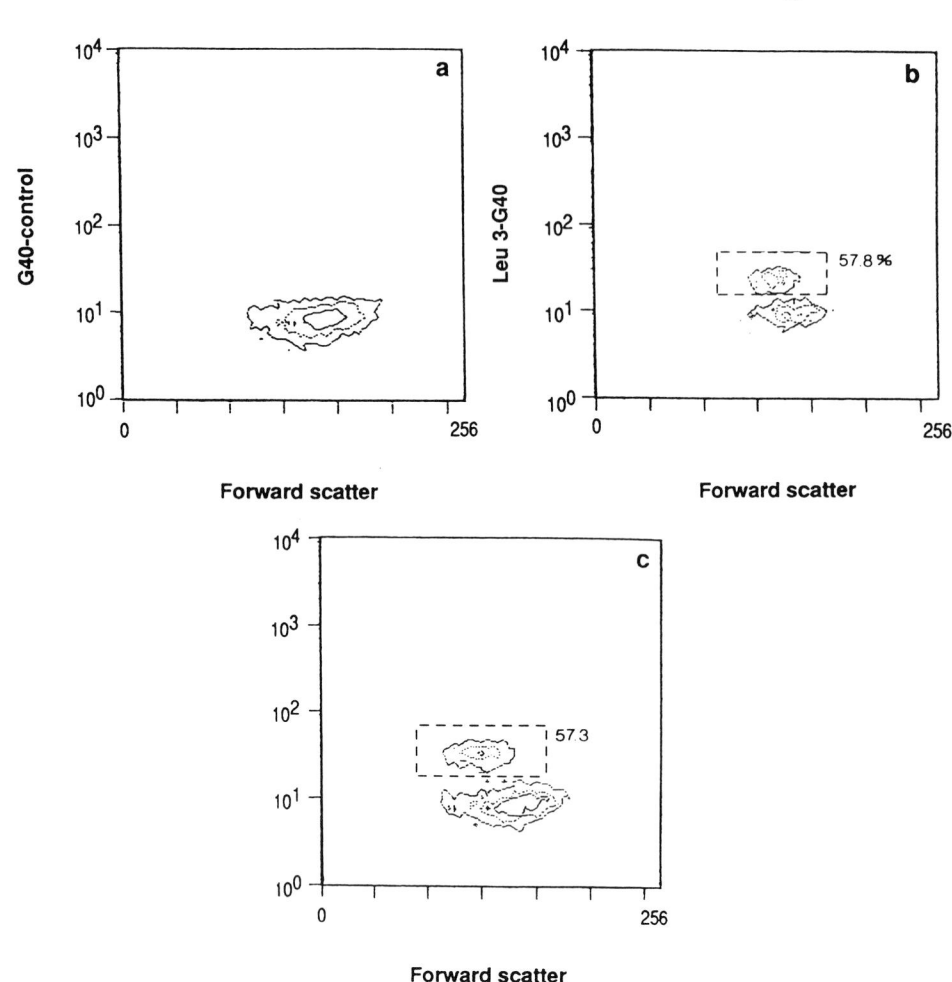

Fig. 22.1. Two-parameter diagram displaying the distribution of forward versus right-angle scatter of lymphocytes stained with (a) gold-labeled goat antimouse (GAM) and rabbit antigoat (RAG) antibodies only, (b) anti-Leu-3a followed by GAM-G40, and (c) anti-Leu-3a followed by GAM-G40 plus RAG-G40, illustrating the enhancing effect of a tertiary antibody layer.

gram (not shown) or a two-parameter diagram of G40 versus either FITC or PE (Fig. 22.2c).

Triple Antibody Staining

Peripheral blood lymphocytes were isolated from patients with rejecting kidney allografts (previously known to have substantial numbers of

22. Analysis of Triple Antibody-Binding Cells 437

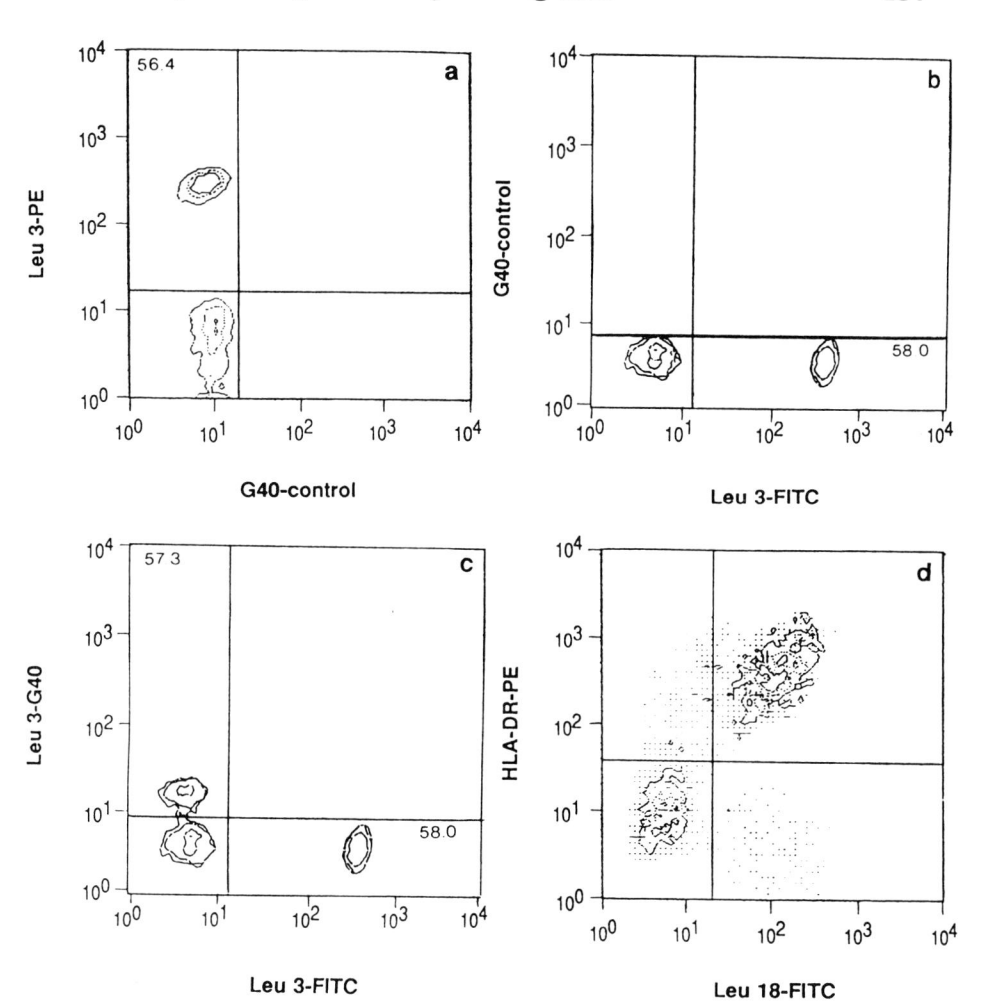

Fig. 22.2. Flow cytometric analysis (right-angle light scatter versus fluorescence distribution of lymphocytes stained with gold-labeled goat antimouse (GAM-G40) secondary antibodies and (a) anti-Leu-3a-PE or (b) anti-Leu-3a-FITC. (c) illustrates a 1 : 1 mixture of cells stained as in Figs. 22.1b and 22.2b, that is, stained with anti-Leu-3a-G40 or anti-Leu-3a-FITC, demonstrating lack of interference. (d) shows three-color staining of the $T_{h/i}$ cell subset. Lymphocytes were triple stained with anti-Leu-3a-G40, anti-Leu-18-FITC, and anti-DR-PE. Leu-3a-G40 positive cells were gated from the right-angle scatter signal and further analyzed for red and green fluorescence.

TABLE 22.1
Staining of T Lymphocytes with FITC-, PE-, or G40-Labeled Antibodies

	Percentage of Marker-Carrying Cells					
	Leu 2 ($T_{s/cx}$)[a]			Leu 3 (T_h)[b]		
Quantity	FITC	PE	G40	FITC	PE	G40
Mean	25.9	25.9	25.5	56.2	54.6	53.9
Range	25.2–33.0	23.1–32.5	23.8–31.0	55.0–63.0	52.0–58.6	49.8–59.8

[a] $n = 5$.
[b] $n = 3$.

activated suppressor/cytotoxic, i.e., Leu-2a$^+$DR$^+$, T cells) or from leukemia patients who had undergone ABMT within 3 months of time (by experience known to have a large fraction of activated helper/inducer, i.e., Leu-3a$^+$DR$^+$, T cells in the circulation). These cells were stained with anti-Leu-2a-GAM + RAG-G40, anti-Leu-15-PE, and anti-HLA-DR-FITC or anti-Leu-3a-GAM + RAG-G40, anti-Leu-18-FITC, and anti-HLA-DR-PE, respectively. Samples were run on the FCM; data were collected in four-parameter list mode and analyzed with the Consort 30 program. The Leu-2a-G40$^+$ or Leu-3a-G40$^+$ population was gated from an FSC versus SSC scattergram as in Fig. 22.1c and further analyzed for red and green fluorescence (Fig. 22.2d). Simultaneously, the same cells were also stained and analyzed by ordinary two-color immunofluorescence in order to compare the size of the DR$^+$ and Leu-15$^+$ populations within the Leu-2a$^+$ subset or the DR$^+$ and Leu-18$^+$ populations within the Leu-3a$^+$ subset with those obtained in the triple-labeling experiments.

By using these particular combinations of McAbs, which divide the T-cell subset into further subsubsets and determine the activation state (DR expression) of the cells, it is possible to distinguish phenotypically between activated and nonactivated T_s-like cells (Leu2$^+$Leu15$^+$DR$^+$ and Leu2$^+$Leu15$^+$DR$^-$, respectively) as well as activated and nonactivated T_{cx} cells (Leu2$^+$Leu15$^-$DR$^+$ and Leu2$^+$Leu15$^-$DR$^-$, respectively). Also, the Leu-3a$^+$ cells may similarly be divided into T_h (Leu3$^+$Leu18$^-$) and $T_{s/i}$ (Leu3$^+$Leu18$^+$) activated (DR$^+$) or resting (DR$^-$) cells.

COMMENTS

Advantages

The technique of using McAbs labeled with colloidal gold for identification of cell surface antigens originally described by De Mey (1983) is in many aspects different from as well as superior to conventional immunofluorescence. Since it is nonfluorescent, the problem of fading in the microscope is eliminated. Also, there is no risk of interference between gold and fluorescence, a feature that facilitates multiparameter analysis of cell populations and subsets thereof. Van Dongen *et al.* (1985) elegantly showed that this triple immunological staining technique works in an epi-illuminated microscope with polarized light. Böhmer and King (1984) applied one-parameter immunogold staining to FCM. We extended it even further and used FCM as a tool to detect triple antibody-binding T cells with colloidal gold as one label in addition to two fluorescence parameters (Festin *et al.*, 1987). Furthermore, we used a standard FCM system with only one laser, an Ar ion tube providing the 488-nm line used for optimal simultaneous excitation of FITC and PE.

The advantages of FCM over light microscopy are obvious and include sensitivity, accuracy, superior statistical significance, speed of analysis, and objectivity. If this is combined with the earlier-mentioned qualities of colloidal gold as a complement to fluorochromes for McAb conjugation, three-parameter analysis in single-laser flow cytometry is obtainable by any FCM user.

Limitations

Although triple immunological staining of this kind undoubtedly works in FCM, it has limitations and drawbacks that may prevent it from being applicable in all instances. The use of a 488-nm laser line provides suboptimal strength of the scattered light signal from gold-labeled cells. This can cause difficulties when trying to detect antigens which are weakly expressed (low cell surface density) or which display a broad and nonuniform distribution. Leu-3a and Leu-2a (as long as the dim Leu-2a$^+$ population is small or absent), as well as Leu-18, present no problems. Other antigens, such as Leu-15 or Leu-12 (CD19; weak) or the activation markers HLA-DR, IL2-receptor (CD25), or CD23 (broad), are too weak to display a separated population. Therefore, the gold-conjugated secondary antibody should be attached to the primary McAb detecting the most strongly expressed antigen and, if possible, also defining the largest subpopulation. Böhmer and King (1984) investigated the separation of negative and positive cells as a function of illumination wavelength and demonstrated that 488 nm was insufficient for the relatively broad antigen distribution they used. The conclusion was that 633-nm light (helium–neon laser) provided the best resolution.

Cell preparations must be devoid of granulocytes and monocytes/macrophages, which would otherwise interfere with the gold-positive cells in the scattergram by their inherent light-scattering properties and possibly also by nonspecific phagocytosis.

Another limitation of the colloidal gold, found by us as well as by

Böhmer and King (1984) is the necessity to use the gold-conjugated antibody undiluted, which of course is uneconomical. It is not clear from the work of Böhmer and King (1984) whether it was possible to obtain sufficient signal strength with a more diluted antibody using the 633-nm laser line. Surprisingly, even Van Dongen et al. (1985) shared our experience when working with the UV polarized-light microscope.

Future Prospects

We have demonstrated the feasibility of triple staining of lymphoid cells with three different McAbs detected with FITC, PE, and colloidal gold, respectively, and thereby determined relative sizes of subsubsets within the T-cell population. The technique could be improved in sensitivity by modifying several steps in the procedures, mainly concerning instrumentation. As mentioned earlier, the excitation wavelength of 488 nm used does not provide optimal and sometimes not even sufficient resolution of immunogold-labeled cells. It has been shown that illumination by an He-Ne laser emitting 633 nm light is superior (Böhmer and King, 1984), but multiparameter analysis would require a two-laser system. However, the Ar ion laser is capable of emitting monochromatic light at 514 nm as well, and we tried this wavelength to explore whether this would improve the gold signal strength. We also attempted using higher output power than the 200 mW routinely used. Neither of these approaches resulted in any appreciable improvement (data not shown). Raising the laser power to 1 W did not increase resolution but increased the background. The use of 514 nm for excitation did raise the gold signal marginally, but more important, it decreased the strength of the FITC signal because a higher-cutting emission filter had to be inserted in the FITC detection channel in order to block direct scattered laser light. Therefore, this is not an applicable solution in multimarker analysis.

Böhmer and King (1984) proposed that gold particles larger than 40 nm might be useful for obtaining stronger signals. This is a very promising alternative. However, to the best of our knowledge, no such particle-conjugated antibodies are commercially available, and attempts to manufacture particles of such sizes may result in nonuniform particles (Frens, 1973).

As for applications in biology and medicine, a triple-marker methodology implemented on a standard one-laser FCM system would certainly open new possibilities for improved analysis of, e.g., the immune system in research as well as in clinical fields. Two-color FCM is already an established technique, and it is our firm belief that the addition of a third parameter would add a new dimension to cell surface analysis.

Even if a third fluorochrome is found that can be combined with FITC and PE in single-laser excitation, colloidal gold can always be used as a complement to fluorescence and thereby provide four-color analysis.

SUMMARY

To further develop dual-parameter flow immunofluorescence to comprise a third independent cell surface marker, we have used antibodies conjugated to colloidal gold as a complement to the conventional fluorochromes FITC and PE. The gold label was available as 40-nm gold particles (G40) conjugated to secondary goat antimouse antibodies, and tertiary rabbit antigoat antibodies and produced no nonspecific background staining. Analysis of staining cells was performed on a fully standard flow cytometer equipped with a single laser emitting at 488 nm, and G40-labeled cells were detected in the side scatter (90°) channel in logarithmic amplification mode. Control experiments with single labeling of T cells with either FITC-, PE-, or G40-conjugated antibodies indicated that the gold label recognizes a population identical to that detected by the fluorochromes, and cell mixture experiments revealed a total lack of interference between the gold and fluorescence. Triple staining of lymphocytes with the three labels G40, FITC, and PE on different antibodies demonstrated that it is possible to detect individual triple antibody-binding cells, and identify subsets, subsubsets, and the degree of activation of these cells. The technology described is a further step in the refinement of multiparameter analysis of immunocompetent and other cells.

REFERENCES

Böhmer, R.-M., and King, N. J. C. (1984). Flow cytometric analysis of immunogold cell surface label. *Cytometry* **5**, 543.

Dean, P. N., and Pinkel, P. (1978). High resolution dual laser flow cytometry. *J. Histochem. Cytochem.* **26**, 622.

De Mey, J. (1983). The preparation and use of gold probes. In *Immunocytochemistry: Practical Applications in Pathology and Biology* (J. M. Polak and S. Van Noorden, (eds.), p. 82. Wright, Bristol.

Engleman, E. G., Benike, C. J., Glickman, E., and Evans, R. L. (1981). Antibodies to membrane structures that distinguish suppressor/cytotoxic and helper T lymphocyte subpopulations block the mixed leukocyte reaction in man. *J. Exp. Med.* **154**, 193.

Festin, R., Björklund, B., and Tötterman, T. H. (1987). Detection of triple antibody-binding lymphocytes in standard single laser flow cytometry using colloidal gold, fluorescein and phycoerythrin as labels. *J. Immunol. Methods* **101**, 23.

Foon, K. A., and Todd, R. F. III (1986). Immunologic classification of leukemia and lymphoma. *Blood* **68**, 1.

Frens, G. (1973). Controlled nucleation for the regulation of the particle size in monodisperse gold suspensions. *Nature (London), Phys. Sci.* **241,** 20.

Janossy, G., Campana, D. and Bollum, F. J. (1985). Immunofluorescence studies in leukaemia diagnosis. In *Monoclonal Antibodies* (P. Beverley, ed.), p. 97. Churchill-Livingstone, Edinburgh and London.

Lampson, L. A., and Levy, R. (1980). Two populations of Ia-like molecules on a human B cell line. *J. Immunol.* **125,** 393.

Landay, A., Gartland, G. L., and Clement, L. T. (1983). Characterization of a phenotypically distinct subpopulation of Leu-2a$^+$ cells which suppress T cell proliferative responses. *J. Immunol.* **131,** 2757.

Morimoto, C., Letvin, N. L., Distaso, J. A., Aldrich, W. R., and Schlossman, S. F. (1985). The isolation and characterization of the human suppressor inducer T cell subset. *J. Immunol.* **134,** 1508.

Oi, V. T., Glazer, A. N., and Stryer, L. (1982). Fluorescent phycobiliprotein conjugates for analyses of cells and molecules. *J. Cell Biol.* **93,** 981.

Shapiro, H. M. (1983). Multistation multiparameter flow cytometry: A critical review and rationale. *Cytometry* **3,** 227.

Titus, J. A., Haugland, R., Sharrow, S. O., and Segal, D. M. (1982). Texas Red, a hydrophilic, red-emitting fluorophore for use with fluorescein in dual parameter flow microfluorometry and fluorescence microscopic studies. *J. Immunol. Methods* **50,** 193.

Van Dongen, J. J. M., Hooijkaas, H., Comans-Bitter, W. M., Benne, K., Van Os, T. M., and De Josselin de Jong, J. (1985). Triple immunological staining with colloidal gold, fluorescein and rhodamine as labels. *J. Immunol. Methods* **80,** 1.

23

Silver-Enhanced Colloidal Gold for the Detection of Leukocyte Cell Surface Antigens in Dark-Field and Epipolarization Microscopy

M. DE WAELE

Laboratory of Hematology and Immunology
Academic Hospital, Free University Brussels
Brussels, Belgium

INTRODUCTION
IMMUNOGOLD–SILVER STAINING OF LEUKOCYTE CELL SURFACE ANTIGENS WITH MONOCLONAL ANTIBODIES
  Preparation of the Cell Suspensions
  Monoclonal Antibodies
  Colloidal Gold-Labeled Goat Antimouse Antibodies
  Physical Developer
  Immunogold–Silver Staining Procedure

EXAMINATION OF THE PREPARATIONS IN BRIGHT-
FIELD, DARK-FIELD, AND EPIPOLARIZATION
MICROSCOPY
  The Microscope
    Bright-Field Microscopy
    Dark-Field Microscopy
    Epipolarization Microscopy
  Appearance of the Preparations
STANDARDIZATION OF THE IMMUNOGOLD–SILVER
STAINING PROCEDURE
  Determination of the Silver Enhancement Time
  Titration of the Colloidal Gold-Labeled Goat Antimouse
    Antibodies
    Size of the Gold Probe
    Titration of the Gold Reagent
  Titration of the Monoclonal Antibody
  The Optimal Silver Enhancement Interval
  Reactivity of the Physical Developer: Influence of Light and
    Temperature
  Performance of Two Epipolarization Microscopes
CONCLUSION
REFERENCES

INTRODUCTION

Colloidal gold was originally introduced as an immunocytochemical marker for electron microscopy (Faulk and Taylor, 1971) but it is also visible in the light microscope (Geoghegan et al., 1978). The latter approach has been used for the detection of leukocyte cell surface antigens with monoclonal antibodies (De Waele et al., 1983; Rosenberg et al., 1984; Wybran et al., 1985). Unfixed cells were labeled in suspension and cytocentrifuge preparations were made. In bright-field light microscopy, positive cells had dark granules on their surface membrane. This immunogold staining was, however, rather weak so that sometimes methyl green was used as a counterstain, giving poor morphology. The cell identification was then improved by performing enzyme cytochemistry on the immunogold-stained cells.

The visibility of immunogold staining can be increased by silver enhancement (Holgate et al., 1983; Chapters 9 and 10 by Scopsi and Hacker, respectively, in Volume 1 of this series). In this technique immunogold-labeled specimens are treated with a physical developer, which deposits concentric layers of metallic silver around the gold particles. When silver enhancement was performed on the cytocentrifuge prepara-

23. Detection of Leukocyte Cell Surface Antigens

tions of the immunogold-stained leukocytes, dense black gold–silver granules on the surface membrane of the cells were obtained (Romasco et al., 1985; De Waele et al., 1986a,b). This staining allowed the use of dark Romanovsky counterstains, which gave a good cell morphology and permitted an accurate cell identification. Smaller gold particles than usually used for immunogold staining were applied and gave more efficient staining (De Waele et al., 1986a).

Immunogold and immunogold–silver staining can be visualized with dark-field and epipolarization microscopy (De Mey, 1983; De Waele et al., 1983, 1986b, 1988b). The staining appears as bright granules on a dark background. The efficiency of this detection is higher than that found with bright-field microscopy (De Waele et al., 1988b). Similar observations were made with reflection contrast microscopy (Hoefsmit et al., 1986). In this chapter we will describe the detection of leukocyte cell surface antigens with immunogold–silver staining in bright-field, dark-field, and epipolarization microscopy. The different steps of the labeling procedure, the equipment for microscopy, and the effect of different variables on the appearance of the staining will be discussed. In addition, we will describe how the procedure can be standardized.

IMMUNOGOLD–SILVER STAINING OF LEUKOCYTE CELL SURFACE ANTIGENS WITH MONOCLONAL ANTIBODIES

Preparation of the Cell Suspensions

Mononuclear cell suspensions of EDTA-anticoagulated venous blood are prepared by Ficoll-Hypaque density gradient centrifugation (Lymphoprep; Nycomed AS, Torshov, Norway). Suspensions of 3×10^{10} cells/liter are made in 0.01 M phosphate-buffered saline (PBS) (pH 7.2) containing 5% (w/v) bovine serum albumin (BSA) (Behringwerke AG, Marburg, FRG) and 5% heat-inactivated human AB serum (AB). The labeling procedure can also be performed on cells from other sources (bone marrow aspirate, lymph node cell suspension) or prepared with other methods (lysis of red blood cells, dextran sedimentation).

Monoclonal Antibodies

Mouse monoclonal antibodies raised against leukocyte cell surface antigens are used. The antigens identify the cell lineage, the stage of maturation, the state of activation, and the functional capacities of the cells (Pallesen and Plesner, 1987; Shaw, 1987).

TABLE 23.1
Monoclonal Antibodies Used in the Author's Laboratory to Enumerate Lymphocyte Populations in Peripheral Blood

Cluster Designation	Name	Isotype	Predominant Reactivity	Source[a]
CD2	OKT11	IgG_{2a}	E rosette receptor associated	1
CD3	OKT3	IgG_{2a}	T cells	1
CD4	OKT4	IgG_{2b}	T helper/inducer cells	1
CD8	OKT8	IgG_{2a}	T cytotoxic/suppressor cells	1
CD16	Leu11b	IgM	Fc IgG receptor on NK cells	2
CD19	Leu12	IgG_1	B cells	2
CD20	B1	IgG_{2a}	B cells	3
CD25	IL_2R_1	IgG_{2a}	IL_2 receptor on activated T cells	3
CD38	OKT10	IgG_1	Activated T cells	1
Anti-HLA-DR	OKIa	IgG_2	B cells, activated T cells	1

[a] 1, Ortho Diagnostic Systems Inc., Raritan, New Jersey; 2, Becton Dickinson, Sunnyvale, California; 3, Coulter Immunology, Hialeah, Florida.

Based on their reactivity with selected leukocyte suspensions and the molecular weight of the antigens, these antibodies have been classified in clusters of differentiation (CD groups). For each of these groups several monoclonal antibodies are commercially available. The antibodies used in the author's laboratory to identify major lymphocyte subpopulations in peripheral blood are listed in Table 23.1. These antibodies are diluted in PBS-BSA (5%). The optimal dilution is determined as described below.

Colloidal Gold-Labeled Goat Antimouse Antibodies

Goat anti-mouse IgG (GAMIgG-G5) and goat anti-mouse IgM (GAM-IgM-G5) antibodies coupled to 5-nm gold particles can be purchased from Janssen Biotech, Olen, Belgium (Auroprobe LM). The reagents are delivered at an optical density of 2.5 at 520 nm. They are diluted in 0.02 M Tris-buffered saline (TBS), pH 8.2, containing 1 mg/ml BSA and 0.02 M sodium azide. The optimal dilution is determined as described below. In some studies gold particles of large size were used.

Physical Developer

Different physical developers can be applied to enhance the immunogold staining. In our first reports on immunogold–silver staining (De

23. Detection of Leukocyte Cell Surface Antigens

Waele *et al.*, 1986a,b) a developer described by Danscher (1981) was used. It is prepared as follows:

1. 1 kg of crystalline gum arabic is dissolved in 2 liters of deionized water. After 5 days the solution is filtered and is ready for use.
2. Citrate buffer is obtained by dissolving 25.5 g of citric acid · $1H_2O$ and 23.5 g of sodium citrate · $2H_2O$ in 100 ml of deionized water.
3. 0.85 g of hydroquinone is dissolved in 15 ml of deionized water.
4. 0.11 g of silver lactate is dissolved in 15 ml of deionized water.

Exactly 60 ml of gum arabic solution and 10 ml of citrate buffer are mixed with 15 ml of hydroquinone solution, and then 15 ml of silver lactate solution is added immediately before use. All solutions containing silver lactate are protected from the light by working in a darkroom or by wrapping the staining jars in aluminum foil. In a modified version of this developer, the gum arabic is replaced by the same volume of deionized water (Moeremans *et al.*, 1984; De Waele *et al.*, 1986b).

Commercial physical developers are now available. In the Intense II kit (Janssen Biotech) an initiator and an enhancer solution are present. The developer is prepared by mixing equal volumes of these solutions immediately before use. The developer is less light sensitive than Danscher's developer, so that no protection against the light is needed during storage of the components and during the silver enhancement reaction. The cytocentrifuge preparations can be incubated in this developer in staining jars, or two drops of each component can be mixed directly on the cytocentrifuge preparations, which are incubated horizontally in a moist chamber.

The incubation time of the cytocentrifuge preparations in the physical developer can be determined as described later in this chapter. The incubation time depends on the reactivity of the physical developer, which is influenced by its composition and temperature. It is therefore recommended to standardize the reaction at a constant temperature. In our laboratory the staining jars with the developer are incubated in a water bath at 26°C. Working at a constant temperature is more difficult when a small volume of developer is applied directly on top of the horizontally incubated cytocentrifuge preparations.

Immunogold–Silver Staining Procedure

The immunogold–silver staining procedure is performed as follows:

1. Incubate 25 µl of the cell suspension in a test tube with 25 µl of the monoclonal antibody dilution for 30 min at room temperature. Agitate twice during this period.

2. Add 1 ml of PBS-BSA (1%) and cytocentrifuge the cells at 900 g for 3 min. Remove the supernatant without touching the pellet at the bottom of the tube. Repeat the washing procedure three times.

3. Resuspend the final pellet in 25 µl of PBS-BSA (5%)–AB (5%) and add 25 µl of an appropriate dilution of the goat antimouse–gold reagent for 60 min at room temperature.

4. Wash the cells three times as described above.

5. Resuspend the final pellet in 500 µl of PBS-BSA (5%) (final concentration of 2×10^6 cells/ml) and make cytocentrifuge preparations or smears. After air drying, these preparations can be stored at $-20°C$ or the silver enhancement can be performed immediately.

6. Fix the preparations with phosphate-buffered formol–acetone for 1 min at 4°C. This fixative is prepared by dissolving 20 mg of Na_2HPO_4 and 100 mg of KH_2PO_4 in 30 ml of distilled water, to which 45 ml of acetone and 25 ml of formaldehyde (37%) are added (Mason *et al.*, 1975).

7. Rinse the preparations for three times (3 min each) with deionized water. With Danscher's physical developer the last washing procedure is done with 1 : 10 diluted citrate buffer.

8. Incubate the preparations in the physical developer for the appropriate time. This can be done in plastic staining jars in a water bath at 26°C. With Danscher's developer these jars are wrapped in aluminum foil. The Intense II developer can also be put on the preparations incubated horizontally in a moist chamber.

9. Rinse the preparations three times (3 min each) in deinonized

10. After air drying, the preparations are counterstained with May–Grünwald–Giemsa.

11. After air drying, the preparations are mounted with a water-insoluble medium (e.g., DPX Mountant; BDH, Montreal, Canada).

To evaluate the nonspecific binding, adequate negative controls should be included. Therefore, a monoclonal antibody of the same isotype but with unrelated specificity can be used. To examine the nonspecific binding of the goat antimouse–gold reagent, the monoclonal antibody can be omitted from the procedure.

EXAMINATION OF THE PREPARATIONS IN BRIGHT-FIELD, DARK-FIELD, AND EPIPOLARIZATION MICROSCOPY

The Microscope

Bright-field microscopy can be done with any microscope of reasonably good quality. For dark-field and epipolarization microscopy, some special equipment is required. Microscopes with which the three types of microscopy can be carried out are now available from different companies. We use a Leitz Dialux 22EB microscope equipped for epi-illumination with an HBO 50-W mercury vapor lamp and a Ploemopak device.

Bright-Field Microscopy

The first position of the Ploemopak contains no filter combination and is used for bright-field and dark-field microscopy. Bright-field microscopy is done with a 12-V, 100-W lamp, a universal condenser, and an NPL Fluotar 50× (numerical aperture, NA, 1.00) or a PLAN APO 100× (variable NA, 1.32–0.60) oil immersion objective. With transmitted light the image is formed by differences in light absorption between the various parts of the specimen.

Dark-Field Microscopy

For dark-field microscopy, a dark-field front lens (D 1.19–1.44 oil S 11) with oil immersion is put on the condenser. A hollow cone of light then comes obliquely from the periphery to the specimen, glances off the edges of the structures, and misses the front lens of the objective. Only the light reflected or diffracted from the structures in the specimen is able to enter the objective. The edges and other structures of the cells which reflect the light will shine brightly against a dark background (Rose, 1977). It is necessary to place immersion oil between the condenser and the lower side of the microscopic slide and between the upper side of the slide and the objective to prevent reflection by the glass. The objective must have a lower numerical aperture than the condenser. The Leitz NPL Fluotar 50× objective can be used as such, but with the PLAN APO 100× objective, the aperture diaphragm has to be closed, while the maximum aperture is used for bright-field microscopy.

Epipolarization Microscopy

For epipolarization microscopy, the Ploemopak is switched to the second position. It contains a polarization block (Leitz 520509) consisting of two fixed crossed polarizers and a 50% dividing mirror. The first polarizer gives a linear polarization to the epi-illumination. This light is deflected by the 50% dividing mirror and reaches the specimen through the objective. Certain structures in the specimen (e.g., metal particles) reflect or scatter the light with a loss of the original polarization. The reflected light is now elliptically polarized and part of it is able to pass the second (crossed) polarizer, which serves as an analyzer (Cornelese-ten Velde *et al.*, 1988). In this way, the structures in the specimen which reflect the light appear bright against a dark background. Light reflected from the internal side of the objectives is unaffected in its polarization and is stopped by the analyzer.

An infrared-absorbing Calflex filter and a UV-absorbing filter K420 (Leitz 514170) are placed in front of the lamp housing to prevent heat damage to the polarizers and to protect the eyes from UV light, respectively. Between the lamp and the polarization block a V-diaphragm slide DF (Leitz 563345) is inserted in the Ploemopak. Closing this diaphragm darkens the background and increases the contrast in the image. Oil immersion objectives are used to eliminate reflections from the surface–air interface of the microscope slide. The NPL Fluotar 50× objective (NA 1.00) and the PLAN APO 100× objective at its highest numerical aperture are suitable for this purpose. The immunogold–silver staining is also visible with low-power objectives without oil immersion, but then more background light reflection occurs, which reduces the contrast in the image.

Reflection contrast microscopy is a similar microscopy system with image formation based on light reflection (Pera, 1979; Van der Ploeg and Van Duyn, 1979; Landegent *et al.*, 1985; Hoefsmit *et al.*, 1986; Cremers *et al.*, 1987; Cornelese-ten Velde *et al.*, 1988). It has also been used for the detection of immunogold and immunogold–silver-stained cell surface antigens (Hoefsmit *et al.*, 1986). In comparison with epipolarization, two additional contrast-enhancing devices are present. The first is a central stop inserted in front of the collecting lens of the lamp housing (Leitz 520508). It eliminates the light rays which otherwise would give the strongest reflections at the lens surfaces and would blur the image. If this device cannot be inserted, another diaphragm, as present in our epipolarization microscope, forms a good alternative. Second, special reflection contrast objectives, provided with a 1/4 lambda plate mounted on the front lens, are used. These objectives are available from different companies but are rather expensive.

The linearly polarized epi-illumination which comes through the objective now passes the 1/4 lambda plate and becomes circularly polarized at an angle of 45° to the axis of the plate. When this light beam is reflected by the specimen toward the objective, the sense of rotation is reversed. After the second passage through the 1/4 lambda plate, the net result is a polarization of 90° to the original plane, allowing the beam to pass the analyzer. The 1/4 lambda plate is rotatable to enable optimization of the reflection contrast image.

In the specimen, three types of light-reflecting objects can be distinguished: metal particles, dye crystals deposited in the cellular structure, and homogeneously stained cellular structures with a refractive index different from that of the surrounding medium (Van der Ploeg and Van Duyn, 1979). With metal particles, all wavelengths of the incident light will be reflected. Silver has a high reflecting power throughout the visible

23. Detection of Leukocyte Cell Surface Antigens

spectrum, while gold shows a gradual increase in reflecting power, which reaches a maximum at wavelengths of 600 nm and longer. Therefore these granules are respectively seen as white and yellow images. This approach has been applied to visualize silver grains in autoradiographs (Fujita *et al.*, 1974) or immunogold and immunogold–silver staining (De Mey, 1983; De Waele *et al.*, 1986b; Hoefsmit *et al.*, 1986; Cremers *et al.*, 1987).

The reflection of light by dye crystals has been used to visualize the reaction product of immunoperoxidase or immunoalkaline phosphatase in *in situ* hybridization procedures in reflection contrast microscopy (Landegent *et al.*, 1985; Cremers *et al.*, 1987). Feulgen- or Giemsa-stained cell nuclei or chromosomes are examples of the third type of light-reflecting objects. With dye crystals, nuclei, or chromosomes, only selective wavelengths will be reflected.

Appearance of the Preparations

In bright-field light microscopy the immunogold–silver-stained lymphocytes show numerous black granules on their surface membrane (Fig. 23.1A). As the cells are labeled at room temperature without prefixation, capping of the staining may occur but is not prominent. The May–Grünwald–Giemsa counterstain gives good morphology of the cells. Accurate differentiation of the different cell types is possible even in mixed cell suspensions (De Waele *et al.*, 1986b). The intracytoplasmic granules in neutrophils and large granular lymphocytes are less well stained by the May–Grünwald–Giemsa counterstain than those in unlabeled cells.

In dark-field microscopy the surface and nuclear membranes of the cells are visualized on a dark background (Fig. 23.1B). The gold–silver staining on the cells appears as bright white spots. Positive and negative cells can easily be differentiated. However, cell identification is difficult. In mononuclear cell suspensions, lymphocytes can be differentiated from monocytes by the cell size and the nuclear shape, but in more complex cell suspensions it may be impossible to identify accurately different cell types.

With epipolarization microscopy, the May–Grünwald–Giemsa-stained cell nuclei are visualized on a dark background (Fig. 23.1C). The gold–silver granules on the cells appear as bright silver-gray spots. Cell identification based on the size and shape of the cell nuclei is difficult. It can be facilitated by bright-field examination of the same field. This can be done by switching the Ploemopak to the bright-field position or by the simultaneous use of transmitted light (Fig. 23.2). The latter, however, reduces the brightness of the immunogold–silver staining and the contrast between staining and background.

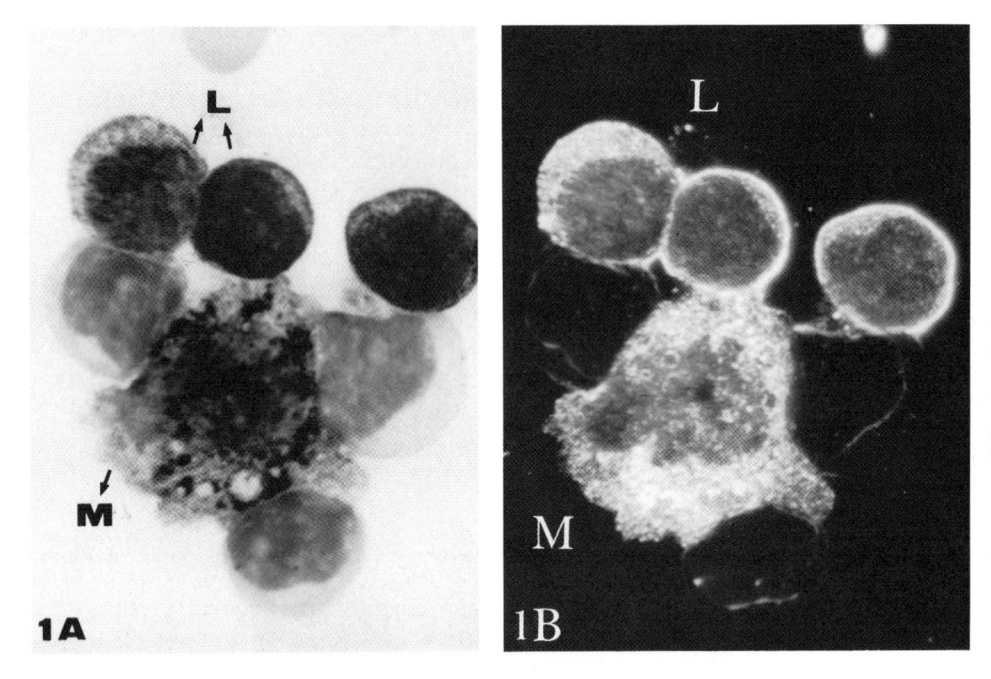

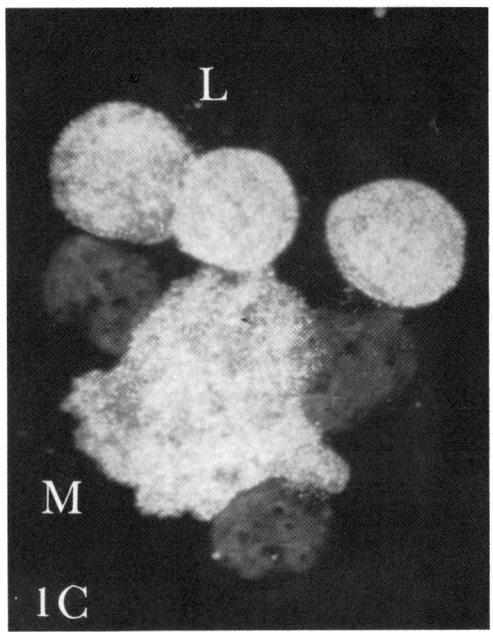

Fig. 23.1. A mononuclear cell suspension was labeled with anti-HLA-DR (OKIa) and immunogold–silver staining and was counterstained with May–Grünwald–Giemsa. With bright-field microscopy (A) the positive lymphocytes (L) and a monocyte (M) show dark granules on their surface membrane. With dark-field (B) and epipolarization (C) microscopy the granules appear as bright silver-gray dots on a dark background; ×1100.

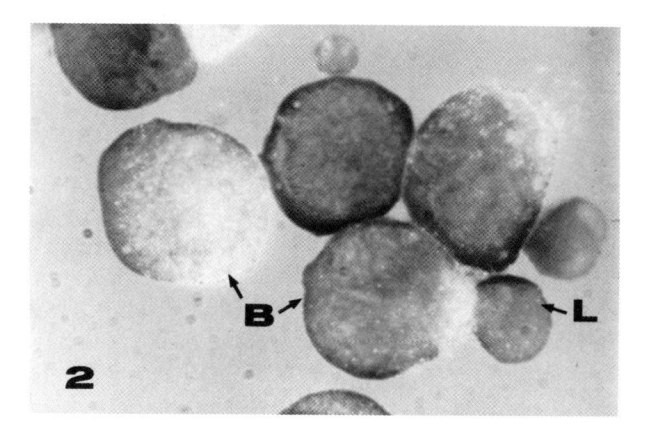

23. Detection of Leukocyte Cell Surface Antigens 453

Fig. 23.2. A suspension of bone marrow cells from a patient with an acute myeloid leukemia (M5A in the French-American-British classification) was labeled with VIM2 (BMA 0210; Behringwerke AG, Marburg, FRG) and immunogold–silver staining. The preparations were examined with epipolarization microscopy combined with low intensities of transmitted light. The staining appears as bright granules while the morphology of the cells is also visible. Most of the blastic cells (B) stain strongly with this antibody, whereas the lymphocytes (L) are negative; ×1100.

In general, the bright silver-gray granules on the dark background produce a higher contrast than the black granules on the purple May–Grünwald–Giemsa-stained cells. Therefore, the staining is more easily seen with dark-field and epipolarization microscopy than with bright-field microscopy. In mixed cell suspensions, however, the difficult cell identification is a serious disadvantage and the simultaneous use of transmitted light is necessary.

STANDARDIZATION OF THE IMMUNOGOLD–SILVER STAINING PROCEDURE

The final result of the immunogold staining procedure is influenced by variables concerning the monoclonal antibody, the goat antimouse–gold reagent, the physical developer, and the type of microscopy used to examine the preparations (De Waele *et al.*, 1986a, 1988b). The effect of different variables is illustrated below. In addition, we will describe how the method can be standardized. This is done by determining the silver enhancement time, followed by titrating the goat antimouse–gold reagent and the monoclonal antibody. The final result of the procedure must be compared with that of a reference method, e.g., immunofluorescence microscopy. When the optimal staining conditions for a particular set of reagents are already known, those for another batch, another concentration of the reagent, or another type of microscopy can be obtained by adapting the silver enhancement time.

Determination of the Silver Enhancement Time

Physical developers contain a silver salt and a reducing substance in a buffered medium. In the labeled specimens, the colloidal gold particles catalyze the reduction of the silver ions into metallic silver, which is deposited on the surface of the particles (Holgate et al., 1983). This increases their diameter as well as their light-microscopic visibility. However, unwanted metallic silver formation can also occur and leads to nonspecific staining of the specimens. Chemical groups in the tissue may reduce the silver ions so that silver granules appear (Gallyas, 1979). The silver ions and the reducing molecules may also interact autocatalytically, by direct collision, in the solution itself. This process is favored by contaminating chloride ions with the formation of silver chloride and by the illumination of the reagent. As a consequence of this autocatalytic reaction, the physical developer becomes black. This reaction may be so important that most of the silver in the developer is consumed before the desired level of specific labeling is obtained. Danscher (1981) recommended the use of a slow-acting physical developer in order to increase the specificity of the silver enhancement and decrease the autocatalytic reaction. To reduce the reaction rate of the physical developer, a colloid can be added. Alternatively, a silver salt with a low disintegration coefficient or a buffer at a nonoptimal pH for the reducing substance can be used; or the silver enhancement can be performed at a low temperature.

Some precautions can be taken to reduce this unwanted formation of metallic silver. The use of deionized water for the preparation of the developer and for rinsing the specimens before and after the silver enhancement minimizes the contamination with chloride ions (Danscher, 1981). All contact of the developer with metal should be prevented. The staining jars can be cleaned with chromic acid and rinsed with $0.1\ M$ HCl and distilled water to remove all traces of silver before being used again. If necessary, the developer should be protected from the light.

This unwanted metallic silver formation causes nonspecific staining in the specimens, which appears as small or large crystalline granules in between the cells. At least part of it is due to the deposition of metallic silver formed in solution by autocatalysis (Danscher, 1981). The intensity of this staining increases with longer incubation times in the physical developer. The incubation time at which this unwanted staining appears in the specimens depends on the reaction rate of the developer. It can be considered as the maximal silver enhancement time possible under these conditions. This time can be determined by incubating unlabeled cytocentrifuge preparations of mononuclear cells in the physical developer for increasing time intervals. In our experiments we found that at 26°C this time was 60 and 15 min for Danscher's developer with and without gum arabic, respectively. We arbitrarily decided to use for our routine work a silver enhancement time between 50 and 75% of the maximal time for that developer. These times were 40 and 10 min for Danscher's developer with and without gum arabic, respectively. We expected that in this way we would get a good intensification of the specific immunogold staining without background due to the developer. More should be known about the kinetics of the silver enhancement with different developers to make the choice of the silver enhancement time less arbitrary.

Titration of the Colloidal Gold-Labeled Goat Antimouse Antibodies

Size of the Gold Probe

With gold probes of 30- and 40-nm diameter or more, the immunogold staining is visible in bright-field, dark-field, and epipolarization microscopy without silver enhancement (De Waele *et al.*, 1983; Ellis *et al.*, 1988). Individual gold particles of 20-nm diameter can be demonstrated with reflection-contrast microscopy, while 5-nm particles require silver enhancement (Hoefsmit *et al.*, 1986). Staining with small gold probes gives more but smaller granules on the surface membrane of the cells than staining with large probes. The staining with the former is also more homogeneously distributed on the surface membrane. When used at the same optical density, small gold probes give more nonspecific staining on the cells in the negative control preparations than the large probes and must be diluted more to prevent this phenomenon. This could be due to the higher concentration of the gold particles in these reagents (De Mey, 1983) or to greater steric accessibility of the small probes to antigens in the cell surface membrane. At optimal dilutions, more efficient staining with small probes was found in several studies (De Waele *et al.*, 1986a; Ellis *et al.*, 1988; Yokota, 1988). The higher density of the staining and the possibility of diluting the probe further, resulting in lower costs, made us decide to use 5-nm probes for our routine work.

Titration of the Gold Reagent

In mononuclear cell suspensions, monocytes as well as some lymphocytes may fix the gold reagent, probably on Fc receptors. Protein solutions such as BSA or AB serum or gelatin may be added to the incubation medium to reduce this binding. The nonspecific staining of the lymphocytes may hamper the interpretation of the results, especially when quantitative data are required. When lymphocyte subsets are enumerated, the nonspecific staining of the monocytes is less important, provided that accurate cell identification is possible.

The nonspecific staining of the lymphocytes can be eliminated by diluting the gold reagent. This is demonstrated in Fig. 23.3. Different samples of a mononuclear cell suspension of normal peripheral blood were incubated with serial dilutions of the GAMIgG-G5 reagent varying between 1 : 1 and 1 : 10,240. A silver enhancement with Danscher's developer for 40 min at 26°C was performed. The number of positive lymphocytes was enumerated in each preparation. For a given GAMIgG-G5 dilution, the numbers found with dark-field and epipolarization microscopy were

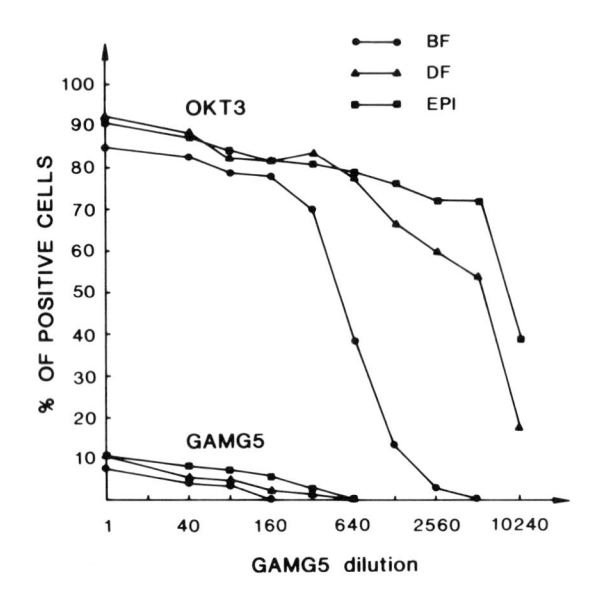

Fig. 23.3. A mononuclear cell suspension of normal peripheral blood was labeled with serial dilutions of goat antimouse–gold antibodies (GAMG5) from 1 : 1 to 1 : 10,240 (horizontal axis). Silver enhancement of 40 min at 26°C with Danscher's developer was performed. In each preparation, the percentage of positive lymphocytes was determined with the three microscopic techniques (vertical axis—GAMG5). With dark-field (DF) and epipolarization (EPI) microscopy higher dilutions of GAMG5 had to be used to eliminate the nonspecific staining on the lymphocytes than with bright-field microscopy (BF). In the second part of the experiment, the same cell suspension was labeled with OKT3 (CD3) and with the different dilutions of the GAMG5 reagent. The optimal dilutions of the GAMG5 reagents as determined above gave comparable numbers of OKT3-positive lymphocytes (vertical axis—OKT3) when examined with the respective microscopic techniques.

23. Detection of Leukocyte Cell Surface Antigens

slightly higher than with bright-field microscopy. The nonspecific staining decreased when the GAMIgG-G5 reagent was diluted and was no longer visible in bright-field microscopy from GAMIgG-G5 1 : 160 and in dark-field and epipolarization microscopy from GAMIgG-G5 1 : 640. In this experiment, these dilutions were considered as the optimal dilutions of this batch of GAMIgG-G5 reagent for these conditions of silver enhancement.

In the second part of this experiment the same cell suspension was first incubated with OKT3 (CD3) at 6.25 μg/ml and then with the same dilutions of the GAMIgG-G5 reagent as described above. The concentration of the monoclonal antibody had been determined before the experiment, as described further. The numbers of OKT3-positive cells found with dark-field and epipolarization microscopy were slightly higher than those seen with bright-field microscopy. These numbers decreased slowly when the GAMIgG-G5 was diluted from 1 : 1 to 1 : 160 for bright-field to 1 : 640 for dark-field and to 1 : 5120 for epipolarization microscopy. With higher dilutions, the positivity fell more rapidly. The dilutions 1 : 160 for bright-field and 1 : 640 for dark-field and epipolarization microscopy gave comparable numbers of OKT3-positive cells in the respective microscopic techniques. These numbers were similar to those found with an immunofluorescence method for this cell suspension (result not shown).

When the gold probe is diluted until all background staining is absent in the negative controls, it may be difficult to obtain an intense specific immunostaining with all antibodies. Therefore we often use higher concentrations of the gold probe and allow a nonspecific staining to occur on a maximum of 3% of the lymphocytes in the negative controls. With these concentrations the specific immunostaining is generally intense enough to permit rapid enumeration of the positive cells, while the clinical significance of the results is not affected by the small amount of background. In addition, the nonspecific staining is generally weaker than the specific staining with most monoclonal antibodies, so that the latter can be quantified separately. With a gold reagent of poor quality, producing high levels of background, it may be difficult to find adequate dilutions of the gold probe. Therefore, it is advisable to check the quality of each gold batch before using it for other experiments, by establishing a dilution curve as described above. In order to get comparable specific and nonspecific labeling, higher dilutions of the gold probes must be used for dark-field and epipolarization microscopy than for bright-field microscopy. This is in accordance with the higher sensitivity of the former methods for the detection of immunogold–silver staining (De Waele et al., 1988b).

It is clear that the dilutions determined in this way are optimal only for the silver enhancement time used. Shorter or longer silver enhancement times would require higher or lower concentrations, respectively, of the gold reagent for adequate staining. In our routine work, the same GAM-IgG-G5 dilution is used for all mouse monoclonal antibodies.

Titration of the Monoclonal Antibody

A primary antibody is generally used in a concentration which provides saturation of the antigen, gives a low background, and allows for biologic variations in the amount of antigen (Caldwell *et al.*, 1987). This concentration can be determined by labeling different samples of a normal mononuclear cell suspension with progressive dilutions of the monoclonal antibody. This is demonstrated for OKT3 in Fig. 23.4. Here we started from 6.25 µg/ml OKT3 (1 : 1), a concentration which is used in our laboratory for immunofluorescence procedures. The antibody was diluted to 1 : 32,678 (1.9 × 10^{-4} µg/ml). In a second step the cell suspension was incubated with GAMIgG-G5 1 : 160 for bright-field and 1 : 640 for dark-

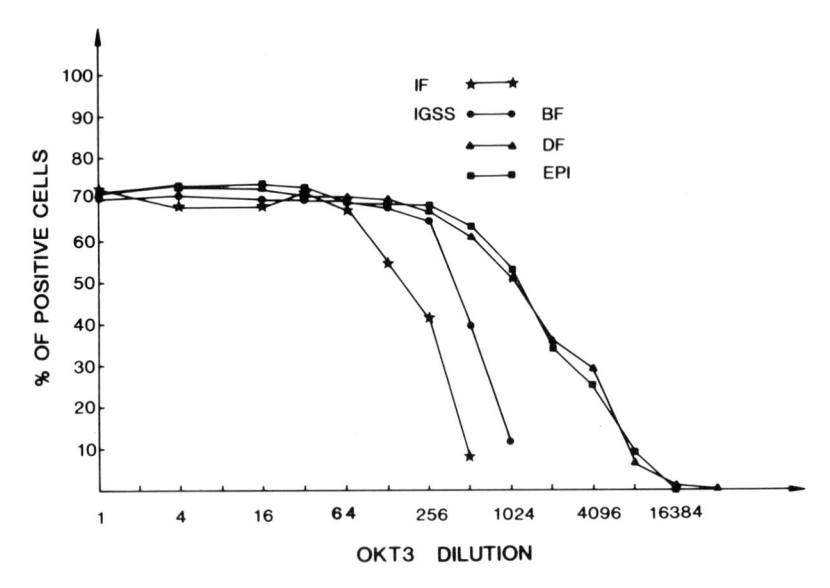

Fig. 23.4. A mononuclear cell suspension of normal peripheral blood was labeled with serial dilutions of OKT3 (CD3), starting from 6.25 µg/ml (dilution 1, horizontal axis). In the IGSS procedure, the optimal GAMIgG-G5 concentration for each microscopic technique was used. A silver enhancement of 40 min at 26°C with Danscher's developer was performed. The percentage of OKT3-positive cells was determined in each preparation. With dark-field (DF) and epipolarization (EPI) microscopy all OKT3-positive cells can be detected with slightly higher dilutions of the monoclonal antibody than with bright-field microscopy (BF) or immunofluorescence microscopy (IF). OKT3 dilutions 1 : 1 to 1 : 4 give a good staining intensity and can be used for routine analysis with all techniques.

field and epipolarization microscopy, the optimal concentrations as determined above for this batch of GAMIgG-G5. A silver enhancement of 40 min at 26°C was done with Danscher's developer. In this experiment, a similar OKT3 dilution curve was made with immunofluorescence microscopy. The fluorescein isothiocyanate (FITC)-conjugated goat antimouse antibodies were purchased from Tago, Inc. (Burlinghame, CA) and used in a 1 : 50 dilution. These preparations were examined with the same microscope, but with oculars 6.3× and an NPL fluotar 63× oil immersion objective (NA 1.30).

With bright-field microscopy the numbers of positive lymphocytes remained constant with OKT3 concentrations between 6.25 μg/ml (1 : 1) and 0.048 μg/ml (1 : 128) and then decreased rapidly. With dark-field and epipolarization microscopy comparable numbers of positive cells were found until an OKT3 concentration of 0.24 μg/ml (1 : 256) was reached. Nearly identical numbers of positive cells were obtained with immunofluorescence microscopy with OKT3 concentrations down to 0.97 μg/ml (1 : 64). In the plateau phase of the curve all cells expressing the antigen were stained, but the intensity of the staining decreased with decreasing concentrations of the monoclonal antibody. A good intensity of staining was found with 1 : 1 (6.25 μg/ml) and 1 : 4 (1.56 μg/ml) dilutions of the antibody. In the negative controls, using a monoclonal antibody of the same isotype but with unrelated specificity, less than 3% of the lymphocytes were positive with all techniques.

This type of antibody dilution curve should be established for each antibody used. In general, antibody concentrations optimized for immunofluorescence procedures are also suitable for immunogold–silver staining. In our laboratory, these concentrations vary between 1 and 10 μg/ml for the different antibodies used. Higher concentrations may also be applied, but they increase the cost of the procedure. With dark-field and epipolarization microscopy good results can be obtained with higher dilutions of the primary antibody than with bright-field microscopy and immunofluorescence. This illustrates the slightly higher efficiency of the former methods and is in accordance with previous findings described elsewhere (De Waele *et al.*, 1988b).

The Optimal Silver Enhancement Interval

An idea about the kinetics of the silver enhancement can be obtained by incubating different cytocentrifuge preparations of an immunogold-labeled cell suspension in the physical developer for increasing time intervals. This is illustrated in Fig. 23.5. A normal mononuclear cell suspension was incubated with OKT3 (6.25 μg/ml) and GAMIgG-G5 (1 : 160).

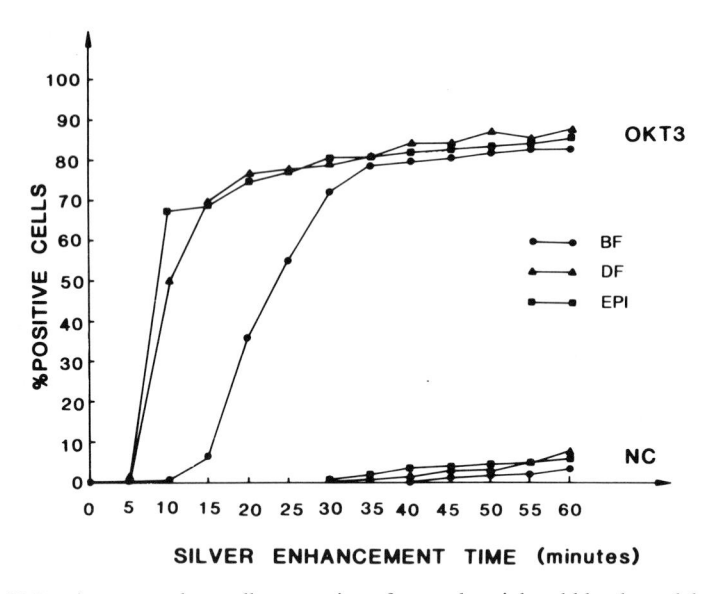

Fig. 23.5. A mononuclear cell suspension of normal peripheral blood was labeled with OKT3 (CD3) and immunogold–silver staining. Silver enhancement with Danscher's developer was varied from 10 to 60 min at 26°C (horizontal axis). In each preparation, the percentage of positive lymphocytes was determined with the different microscopic techniques (vertical axis). The nonspecific staining was evaluated in a negative control experiment (NC). The optimal silver enhancement interval was between 25 and 35 min for dark-field (DF) and epipolarization (EPI) microscopy and between 35 and 45 min with bright-field (BF) microscopy.

Silver enhancement with Danscher's developer was carried out for 10–60 min at 26°C. Without silver enhancement, the labeling was not visible with any of the microscopic techniques. With enhancement it became more rapidly visible with dark-field and epipolarization than with bright-field microscopy. With longer enhancement times, the intensity of the signal and the number of positive cells increased rapidly. The curve reached a plateau after 25 min of enhancement with dark-field and epipolarization and after 35 min with bright-field microscopy. With these and longer enhancement times all lymphocytes expressing the OKT3 antigen were detected. These percentages were nearly identical for the different microscopic techniques. In the negative control preparations, positive lymphocytes were first seen after silver enhancement times of 30 min with dark-field and epipolarization and after 40 min with bright-field microscopy. They increased slowly in number with increasing silver enhancement times. This paralleled a slow increase of the total number of positive cells in the OKT3-labeled preparations. The nonspecific staining was weaker than the OKT3 staining.

In the previous sections the optimal labeling conditions were determined by choosing a silver enhancement time and adapting the concentrations of the reagents to the microscopic method. In the experiment shown in Fig. 23.4, the optimal GAMIgG-G5 concentration for bright-field microscopy was used. The corresponding silver enhancement time of 40 min at 26°C with Danscher's developer gave a result in the plateau of the curve. With these conditions all OKT3-positive cells are detected and there is no nonspecific staining. With longer silver enhancement times the intensity of the staining increases and the number of positive cells rises slowly. The latter is mainly due to better visibility of the weak nonspecific staining on some of the lymphocytes. With a large number of preparations it may be difficult to control accurately the silver enhancement time.

For practical reasons, we consider that all enhancement times which detect all the antigen-expressing cells in the suspension and give acceptably low specific staining (e.g., a maximum of 3% of the lymphocytes in the negative isotype control preparations) can be used for routine work. In this way an optimal silver enhancement interval can be defined. In the previous experiment, this interval was between 35 and 45 min of silver enhancement for bright-field microscopy. For dark-field and epipolarization microscopy, however, enhancement times between 25 and 35 min were optimal. This illustrates again the higher sensitivity of the latter techniques. Gold–silver granules of smaller size were more visible than with bright-field microscopy but weaker enhancement conditions had to be used in order to get an acceptably low background.

When optimal labeling conditions for a particular set of reagents and a particular microscopic method are known, this silver enhancement curve can be established to determine the optimal silver enhancement conditions for other batches or concentrations of the reagents or another microscopic method. In this way, an immunogold-labeled cell suspension can be examined by any one of the three microscopic methods simply by adapting the length of the silver enhancement which is performed on the cytocentrifuge preparations.

Reactivity of the Physical Developer: Influence of Light and Temperature

The silver enhancement curve shown in Fig. 23.5 also gives an idea of the reactivity of the physical developer. This is also demonstrated in the experiment summarized in Table 23.2. We labeled a normal mononuclear cell suspension with OKT4 (CD4) at 10 µg/ml and GAMIgG-G5 1 : 40.

The silver enhancement was performed for increasing intervals with two developers in different conditions. The preparations were examined in bright-field microscopy. For each of the conditions, the optimal silver enhancement interval and the maximal silver enhancement time were determined as described above.

Alhough the results in Table 23.2 are not fully comparable to those in Fig. 23.5, since different concentrations of different gold batches were used, it is clear that the reactivity of Danscher's developer is considerably increased by omission of the gum arabic. This is in accordance with the findings in other studies (Moeremans *et al.*, 1984; Springall *et al.*, 1984; De Waele *et al.*, 1986a). The plateau of the curve was reached after shorter enhancement times. The optimal enhancement interval was smaller and the background staining in between the cells appeared more rapidly. The Intense II silver enhancement kit acted somewhat more slowly than Danscher's modified developer. The lengths of both optimal silver enhancement intervals are comparable, but the autocatalytic activity appears relatively later. This probably reflects a lower tendency for autocatalysis in the latter developer and suggests that with the Intense II kit higher degrees of silver enhancement are possible than with Danscher's modified developer.

The reactivity of the developers was not significantly influenced by protection against the light. When Danscher's modified developer was used under normal tube lighting rather than in the dark, the optimal silver enhancement interval and the maximal silver enhancement time did not change considerably (Table 23.2). This suggests that Danscher's modified developer is not so light sensitive as originally thought. This could per-

TABLE 23.2
Reactivity of the Physical Developers: Influence of Light and Temperature

Physical Developer	Temperature (°C)	Light Protection	Optimal Enhancement Interval (min)[a]	Maximal Silver Enhancement Time (min)[a]
Danscher's developer without gum arabic	26	+	8–12	15
	26	–	8–12	15
	4	–	15–30	40
Intense II kit[b]	26	+	12–17	35
	26	–	12–17	35
	4	–	30–50	90

[a]Time in minutes after the start of the silver enhancement. The optimal enhancement interval and the maximal silver enhancement time were determined as described in the text.
[b]Purchased from Janssen Biotech, Olen, Belgium.

haps be due to the use of cleaned plastic staining jars and of high-quality deionized water. Similar observations were made when the silver enhancement with the Intense II kit was carried out in the dark instead of the light.

Performing silver enhancement at a low temperature (4°C) reduced the reaction rate of both developers (Table 23.2). The appearance of the specific and the nonspecific staining was retarded. The optimal silver enhancement interval and the maximal silver enhancement time increased considerably. Danscher (1981) recommended the use of a slow-acting physical developer to increase the specificity of the silver enhancement. We obtained good results with rapid and slow-acting developers, but with the latter the silver enhancement time was less critical. With very rapid developers it may be difficult to control accurately the short enhancement times for all preparations. In contrast, however, the use of slow-acting developer is more time consuming.

Performance of Two Epipolarization Microscopes

The performance of the Leitz Dialux 22EB microscope and of the Nikon Labophot microscope, both equipped with the respective epipolarization blocks, was compared. The Nikon microscope had a 100-W mercury vapor lamp and a CF Plan Apochromat 100× oil immersion objective. The Leitz microscope was equipped with a 50-W merury vapor lamp and a PLAN APO 100× objective. A normal mononuclear cell suspension was labeled with serial dilutions of OKT3 starting from 6.25 µg/ml as described. The percentages of positive lymphocytes found with the two epipolarization microscopes were nearly identical. All cells expressing the antigen (plateau) were detected with OKT3 concentrations between 6.25 µg/ml (1 : 1) and 0.024 µg/ml (1 : 256). With higher dilutions the number of positive cells decreased. Although the intensity of the reflected light depends on the intensity of the incident light (Landegent *et al.*, 1985), the use of a 100-W lamp (Nikon microscope) instead of a 50-W lamp (Leitz microscope) for the epi-illumination did not significantly increase the sensitivity of detection of the immunogold–silver staining. Higher intensities of incident light lead to an increase in the intensity of the reflected light from the immunogold–silver staining but also give a clearer background with a loss of contrast. The background could be darkened by closing the diaphragm inserted in the pathway of the epi-illumination but this also reduced the intensity of the light reflected from the immunogold–silver staining. Therefore, the use of a 100-W or a 50-W lamp for the epi-illumination did not lead to a significant increase in the sensitivity of the detection.

With the Leitz microscope, the use of the NPL Fluotar 50× oil immersion objective instead of the PLAN APO 100× objective did not change the OKT3 dilution curve considerably. With both objectives, a plateau between OKT3 dilutions 1 : 1 (6.25 μg/ml) and 1 : 256 (0.024 μg/ml) was found. Low magnifications are especially interesting for scanning the preparations when looking for rare positive cells, but cell identification is somewhat more difficult. Similar observations were made in bright-field microscopy.

CONCLUSION

A few years ago, an immunogold–silver staining method for the detection of leukocyte cell surface antigens in bright-field microscopy was described (Romasco et al., 1985; De Waele et al., 1986a). Leukocytes were stained in suspension with monoclonal antibodies and colloidal gold-labeled secondary antibodies. Silver enhancement was performed on cytocentrifuge preparations of the immunogold-labeled cells. Positive cells had dark granules on their surface membrane. A Romanovsky counterstain permitted accurate cell identification.

In this chapter this method was applied to examine the potential of dark-field and epipolarization microscopy for the detection of immunogold–silver staining. Due to the light-reflecting properties of the metallic marker, the staining is visible as bright granules on a dark background. This signal is stable and is more easily visible than the dark granules on the purple Romanovsky-stained cells in bright-field microscopy. However, cell identification is more difficult. With epipolarization microscopy, this can be facilitated by the simultaneous or sequential use of transmitted light. A similar combination is not possible with dark-field microscopy and therefore the latter method is less suitable for the examination of immunogold–silver-stained mixed cell suspensions.

The detection of immunogold–silver staining with dark-field and epipolarization microscopy is more sensitive than that with bright-field microscopy (De Waele et al., 1988b). Gold–silver granules of smaller size can be seen. However, not only the specific but also the nonspecific staining is more easily detected. In order to get optimal staining, the background had to be removed by using weaker labeling or silver enhancement conditions. However, these conditions also reduce the intensity of the specific staining. Therefore, the efficiency of immunogold–silver staining detected with dark-field and epipolarization microscopy is only slightly higher than that found with bright-field microscopy (De Waele et al., 1988b). Only labeling conditions which could reduce the background staining without affecting the specific staining would allow higher degrees of silver enhancement and would lead to higher efficiency. The latter could perhaps also be obtained with cells from other sources showing a lower tendency for nonspecific staining than blood cells.

In our laboratory, immunogold–silver staining for bright-field microscopy is routinely applied for the enumeration of lymphocyte subsets and for leukemia and lymphoma cell typing (De Waele *et al.*, 1986a; 1988a). Epipolarization microscopy in combination with transmitted light is then sometimes used for the rapid detection of weak staining in these preparations or for scanning them at a lower magnification when looking for rare positive cells. Then the adequate negative controls are also examined in epipolarization microscopy to ensure the specificity of the staining. If epipolarization microscopy is used for routine work, weaker labeling conditions than for bright-field microscopy are often necessary. An advantage of the latter approach is that in transmitted light the morphology of the cells is less obscured by the staining, which improves the cell identification.

In this chapter the potential of dark-field and epipolarization microscopy for the detection of immunogold–silver staining was evaluated. The staining of leukocyte cell surface antigens was used as a model for this purpose. A scheme for the standardization of the method was proposed. This scheme can probably also be applied to optimize the detection of other surface membrane antigens on cells of other origin.

We thank A. Rosbach for typing this manuscript and F. Vanhoef for artwork. This work is supported by Grant No. 3.0077.87 from the Fonds voor Geneeskundig Wetenschappelijk Onderzoek, Brussels.

REFERENCES

Caldwell, C. W., Maggi, J., Henry, L. B., and Taylor, H. M. (1987). Fluorescence intensity as a quality control parameter in clinical flow cytometry. *Am. J. Clin. Pathol.* **88**, 447.

Cornelese-ten Velde, I., Bonnet, J., Tanke, H. J., and Ploem, J. S. (1988). Reflection contrast microscopy. Visualization of (peroxidase-generated) diaminobenzidine polymer products and its underlying optical phenomenon. *Histochemistry* **89**, 141.

Cremers, A. F. M., Jansen in de Wal, N., Wiegant, J., Dirks, R. W., Weisbeek, P., Van der Ploeg, M., and Landegent, J. E. (1987). Non-radioactive in situ hybridization. A comparison of several immunocytochemical detection systems using reflection-contrast and electron microscopy. *Histochemistry* **86**, 609.

Danscher, G. (1981). Histochemical demonstration of heavy metals. A revised version of the silver sulphide method suitable for both light and electron microscopy. *Histochemistry* **71**, 1.

De Mey, (1983). Colloidal gold probes in immunocytochemistry. In *Immunocytochemistry: Practical Applications in Pathology and Biology* (J. M. Polak and S. Van Noorden, eds.), p. 82. Wright, Bristol.

De Waele, M., De Mey, J., Moeremans, M., De Brabander, M., and Van Camp, B. (1983). Immunogold staining method for the light microscopic detection of leukocyte cell surface antigens with monoclonal antibodies: Its applications to the enumeration of lymphocyte subpopulations. *J. Histochem. Cytochem.* **31,** 976.

De Waele, M., De Mey, J., Renmans, W., Labeur, C., Reynaert, P., and Van Camp, B. (1986a). An immunogold–silver staining method for the detection of cell surface antigens in light microscopy. *J. Histochem. Cytochem.* **34,** 935.

De Waele, M., De Mey, J., Renmans, W., Labeur, C., Jochmans, K., and Van Camp, B. (1986b). Potential of immunogold–silver staining for the study of leukocyte subpopulations as defined by monoclonal antibodies. *J. Histochem. Cytochem.* **34,** 1257.

De Waele, M., Foulon, W., Renmans, W., Segers, E., Smet, L., Jochmans, K., and Van Camp, B. (1988a). Hematologic values and lymphocyte subsets in fetal blood. *Am. J. Clin. Pathol.* **89,** 742.

De Waele, M., Renmans, W., Segers, E., Jochmans, K., and Van Camp, B. (1988b). Sensitive detection of immunogold–silver staining with darkfield and epipolarization microscopy. *J. Histochem. Cytochem.* **36,** 679.

Ellis, I. O., Bell, J., and Bancroft, J. D. (1988). An investigation of optimal gold particle size for immunohistological immunogold and immunogold–silver staining to be viewed by polarized incident light (epipolarization) microscopy. *J. Histochem. Cytochem.* **36,** 121.

Faulk, W., and Taylor, G. (1971). An immunocolloid method for the electron microscope. *Immunochemistry* **8,** 1081.

Fujita, S., Ashimara, T., Fukuda, M., and Takeode, O. (1974). Deoxyribonucleic acid synthesis of regenerating hepatocytes studied by cytofluorometry in combination with tritiated thymidine autoradiography. *Acta Histochem. Cytochem.* **7,** 100.

Gallyas, F. (1979). Light insensitive physical developers. *Stain Technol.* **54,** 173.

Geoghegan, W. D., Scillian, J. J., and Ackerman, G. A. (1978). The detection of human B lymphocytes by both light and electron microscopy utilizing colloidal gold labeled anti-immunoglobulin. *Immunol. Commun.* **7,** 1.

Hoefsmit, E. C. M., Korn, C., Blijleven, N., and Ploem, J. S. (1986). Light microscopical detection of single 5 and 20 nm gold particles used for immunolabeling of plasma membrane antigens with silver enhancement and reflection contrast. *J. Microsc. (Oxford)* **143,** 161.

Holgate, C. S., Jackson, P., Cowen, P. N., and Bird, C. C. (1983). Immunogold–silver staining. New method of immunostaining with enhanced sensitivity. *J. Histochem. Cytochem.* **31,** 938.

Landegent, J. E., Jansen in de Wal, N., Ploem, J. S., and Van der Ploeg, M. (1985). Sensitive detection of hybridocytochemical results by means of reflection-contrast microscopy. *J. Histochem. Cytochem.* **33,** 1241.

Mason, D. Y., Farell, C., and Taylor, C. R. (1975). The detection of intracellular antigens in human leukocytes by immunoperoxidase staining. *Br. J. Haematol.* **31,** 361.

Moeremans, M., Daneels, G., Van Dijck, A., Langanger, G., and De Mey, J. (1984). Sensitive visualization of antigen–antibody reaction in dot and blot immune-overlay assays with immunogold and immunogold–silver staining. *J. Immunol. Methods* **74,** 353.

Pallesen, G., and Plesner, T. (1987). The Third International Workshop and Conference on Human Leukocyte Differentiation Antigens with an up-to-date overview of the CD nomenclature. *Leukemia* **1,** 231.

Pera, F. (1979). Effects of reflection-contrast microscopy in stained histological, hematological and chromosome preparations. *Mikroskopie* **35,** 93.
Romasco, F., Rosenberg, J., and Wybran, J. (1985). An immunogold–silver staining method for the light microscopic analysis of blood lymphocyte subsets with monoclonal antibodies. *Am. J. Clin. Pathol.* **84,** 307.
Rose, R. A. (1977). Light microscopy. In *Theory and Practice of Histological Techniques* (J. P. Bancroft and A. Stevens, eds.), P. L. Churchill-Livingstone, Edingburgh, and London.
Rosenberg, J. S., Weiss, E., and Wilding, P. (1984). Immunogold staining: Adaptation of a cell-labeling system for analysis of human leukocyte subsets. *Clin. Chem. (Winston-Salem, N. C.)* **30,** 1462.
Shaw, S. (1987). Characterization of human leukocyte differentiation antigens. *Immunol. Today* **8,** 1.
Springall, D. R., Hacker, G. W., Grimelius, L., and Polak, J. M. (1984). The potential of the immunogold–silver staining method for paraffin sections. *Histochemistry* **81,** 603.
Van der Ploeg, M., and Van Duyn, P. (1979). Reflection versus fluorescence. A note on the physical backgrounds of two types of light microscopy. *Histochemistry* **62,** 227.
Wybran, J., Rosenberg, J., and Romasco, F. (1985). Immunogold staining: An alternative method for lymphocyte subsets enumeration. Comparison with immunofluorescence microscopy and flow cytometry. *J. Immunol. Methods* **76,** 229.
Yokota, S. (1988). Effect of particle size on labeling density for catalase in protein-A–gold immunocytochemistry. *J. Histochem. Cytochem.* **36,** 107.

Index

A
Absorption spectrophotometry, gold concentration, 4
Accelerator, LR White and catalytic polymerization, 63
Acetone
  dehydration and LR White, 59
  PLT method, 81
Acrylic resins
  colloidal gold, 43
  immunocytochemistry and plant specimens, 304
  processing and embedding of plant cells, 305–306
Adenoviruses, immunonegative staining, 250
Adsorption
  albumin–gold complex, 166
  particle size, 7
  pH and, 9–10
  protein on particles, 5–7
  proteins and retrograde neuronal tracing, 205–206
Agarose, cell preservation, 267
Aggregation method, immunoelectron microscope, 244
Agitation, low temperature, 81
Air, LR White and polymerization, 63

Akabane virus, double labeling, 25
Albumin
  amount and stabilization, 166–167
  metabolic functions, 163–164
Albumin–gold complex
  preparation, 164–168
  research applications, 170, 172
  storage, 168–169
  working protocol, 169–170
Alcian blue, coating of plastic and glass surfaces, 260
Aldehyde cross-linking, embedding methods, 37
Aluminum block, polymerization, 88
Amylase–gold complex, cytochemistry, 136
Animal cells
  cleaving, 233–234
  fixation, 231
Anterograde tracers, WGA-apoHRP-Au, 221
Antibodies
  immunonegative staining, 244
  immunostaining for light microscopy, 362–363
  label-fracture, 181
  sources of and labeling, 313–314
  triple labeling, 433–34, 436, 438

Antigenicity
  deplasticization and preservation, 370–372
  fixation and glutaraldehyde concentration, 70
  fixatives and retention, 181
  glutaraldehyde and postosmication, 34
  LR White, 36
Antigens
  labeling and fixation, 34, 35–36
  low-concentration secondary sites, 78
  mapping on boar sperm cell surface and label-fracture, 192
Antisera, immunonegative staining, 244, 251
Apolar resins, background staining, 93
Asialofetuin binding, rat hepatocytes, 10
Aspergillus japonicus, pectinase–gold complexes, 142
Aspergillus niger, pectinase–gold complexes, 142
Atomic emission spectrometry, gold concentration, 4

B

Back-scattered electron imaging (BEI), label-fracture, 177, 194
Bacteria
  infiltration times and cryosubstitution, 103
  Lowicryl HM23, 107
Bacterial cell envelope, PLT embedding, 77
Bacterial DNA, cryosubstitution, 97
Bacterial pili, double labeling and specimen movement, 247
Bacteriology, immunonegative staining, 251–252
Bacteroides nodosus, immunonegative staining and vaccine, 251
Bee venom, phospholipase A_2, 142
Binding, nonspecific, error in protein A–gold labeling, 27, 29–30
Biochemistry, freeze-fracture technique, 194
Biology, triple-marker methodology, 440
Blood cells
  ferritin and scanning electron microscopy, 346
  preparation for immunogold labeling, 325
  triple labeling, 433, 435, 436, 438
Boar spermatozoa, label-fracture, 187, 192

Bovine serum albumin (BSA)
  adsorption parameters, 7
  molecular weight and adsorption, 8
Bright-field light microscopy
  immunogold staining, 444–445, 448–449, 451, 453, 464–465
  silver enhancement, 460, 461

C

Capillary endothelial cells, albumin–gold complex, 172
Carbon, coating and grids, 110
Cell biology, protein–gold complexes as tracers and markers, 10
Cell cultures, rapid infiltration and catalytic polymerization, 59, 61
Cells
  interaction with protein–gold complexes, 10–13, 16
  pretreatment and ultrastructural hybridization, 383–384
  study of virus-infected, 256
  surface and immunogold labeling, 335
  triple labeling, 433
Cell surface replica method, colloidal gold–lectin–ruthenium red, 350–351
Cellular components, immunogold labeling, 13, 16
Cellulase–gold complex, cytochemistry, 141
Central nervous system (CNS)
  earthworms, snails, and exocytosis, 293
  rat and methods of detecting exocytosis, 287, 288, 293, 297
Centrifugation, enzyme–gold complexes, 120
Chemical catalysts, polymerization, 87–88
Chitinase–gold complex, cytochemistry, 141
Chloroauric acid, preparation of stock solution, 165
Chloroplasts
  fixation, 305
  labeling of Lowicryl sections, 315
Chrome–alum gelatin, adhesion and glass slides, 67
Citrate reduction method, characterization of colloidal gold particle suspensions, 2
Citric acid, gold colloid preparation, 400

Index

Cleaving
  animal cells in culture, 233–234
  cell types, 239
  plant cells in tissues, 232
  protoplasts of plant cells, 232–233
Cold room, ultraviolet light polymerization, 87
Collagenase–gold complex, cytochemistry, 133
Colloidal gold
  as marker for electron microscopy, 118–119, 164
  characterization of particle suspensions, 2–5
  citric acid, 400
  detection of proteins, 408–409
  electron-dense marker for immunoelectron microscopy, 337–340
  immunocytochemical staining, 367–368
  immunostaining, 363–364
  modern acrylics, 43
  white phosphorus, 400
Colloidal gold–lectin–ruthenium red staining
  labeling procedure, 347–349
  sectioning and microscopic observation, 349–351
  whole cell mounts, 351–352
Colocalization, double labeling, 22, 24
Color, change and flocculation, 165–166
Colorimetric determinations, enzyme–gold complexes and biological activity, 121
Concanavalin A (Con A), label-fracture, 183, 187
Controls
  cytochemical labeling and enzyme–gold complexes, 124–125
  mutant plant lines, 319
  nonspecific or pseudospecific labeling, 317–319
  ultrastructural hybridization, 385–386
Conventional transmission electron microscope (CTEM)
  heavy metals and Lowicryls, 79
  undecagold–antibody method, 415
Copper, grids, 110, 230
Cotyledons, labeling of Lowicryl sections, 315
Counterstaining
  Lowicryl thin sections, 111
  semithin sections, 69
  thin sections, 69–70

Coverslips
  catalytic polymerization, 59, 61
  preparation of, 230
Critical-point drying, microtubules, 235
Cross-contamination
  errors in protein A–gold labeling, 26–27
  steps to reduce, 30
Cross-linking
  LR White and polymerization, 48, 49
  monomer and Lowicryls, 82–83
Cryoprotection, label-fracture specimens, 181
Cryosubstitution
  achieving substitution temperatures, 98
  embedding and Lowicryls, 84
  fixation, 95, 97–98
  freezing and tissue preparation, 99–100
  immunolabeling, 103, 105
  infiltration and embedding, 100–101
  Lowicryl HM23, 101, 102–103
  Lowicryl K11M, 101–102
  substitution media, 100
Cryoultramicrotomy, postosmication, 36
CS-Auto Substitution Apparatus, polymerization methods, 103
Cytochemical labeling, enzyme–gold complexes, 122–124
Cytochemistry
  enzyme–gold complexes and nucleic acids, 127
  extracellular matrix components, 133
  fungal and plant cell wall components, 141–142
  high-sensitivity research and LR White, 41
  phospholipids, 142, 144
  polysaccharides and glycoconjugates, 133, 136, 138, 141
  projection neurons and retrograde neuronal tracing, 221
Cytolytic viruses, cytopathic effect and preembedding labeling, 260
Cytomegalovirus (CMV), ultrastructural localization of DNA, 383, 384, 386
Cytoskeleton
  immunofluorescence microscopy, 228
  immunolabeling of cleaved cells and study, 237, 239
  nonspecific binding and colloidal gold labeling, 29

472 Index

D

Dark-field microscopy
  immunogold staining, 445, 448–449, 451, 453, 464
  silver enhancement, 460, 461
Dehydration
  antibody binding sites, 257
  choice of agent and PLT method, 81–82
  ethanol and Lowicryl embedding at subfreezing temperature, 307–309
  Lowicryl K4M, 37
  low temperature, 84
  microtubules, 235
  monolayers, 264
  N,N'-dimethylformamide at room temperature and embedding at -20°C, 309–310
Deplasticization
  antigenicity preservation, 370–372
  rate of resin removal, 369–370
  section thickness, 368–369
  thin sections, 366–367
  ultrastructure preservation, 372, 375
Detergents
  cytoskeleton and labeling, 228
  immunocolloidal gold reagents, 71
  preembedding peroxidase–antiperoxidase (PAP) staining procedures, 295
Diagnosis, rapid embedding and immunoelectron microscopy, 85
Dibenzoyl peroxide (DBP), chemical polymerization, 87–88
DNA
  cell pretreatment, 384
  CMV-infected fibroblasts and localization, 390–391
  fixation, 382
  in situ hybridization, 381, 386–387
  nucleic acid probes, 382
DNase, labeling and enzyme–gold complexes, 129, 132–133
Double glass-distilled water (DDW), preparation of gold markers, 165
Double labeling
  immunolabeling staining method, 247–248
  protein A–gold labeling, 22, 25
Double-replica method, label-fracture, 181, 182–183
Double-sided labeling method, immunonegative staining method, 247

Double staining
  monoclonal antibodies, 432
  protein–gold conjugates, 399
Drosophila melanogaster, Z contrast, 79
Dry ice, cryosubstitution temperatures, 98

E

Elastase–gold complex, cytochemistry, 133
Electron-dense markers, immunoelectron microscopy, 276–278, 337–340
Electron immunocytochemistry, colloidal gold probes, 244
Electron microscopy
  albumin and visualization, 164
  deplasticized adjacent thin sections, 365–368
  high voltage, 346
  immunolabeling, 228
  immunostained thick sections, 364–365
  lipoproteins and thin sections, 154
  Lowicryls K11M and HM23, 102
  microtubules, 236
  retrograde neuronal tracing, 209, 215
  undecagold–antibody method, 425–427
Embedding, *see also* Postembedding; Preembedding
  advantages of low-temperature, 77
  aldehyde cross-linking, 37
  cryosubstitution, 100–101
  cytochemistry and protocol, 122
  dehydration with N,N'-dimethylformamide at room temperature and, 309–310
  ethanol dehydration and Lowicryls at subfreezing temperatures, 307–309
  evaluation of microtest cultures, 264–265
  Lowicryls and heavy metals, 43
  Lowicryls and plant tissues, 306–307
  Lowicryl K11M, 102
  low-temperature and freeze-drying, 107, 109
  LR White and plant tissues, 310–313
  plant cells and acrylic resins, 305–306
  protocols at specific temperature, 84–85
  rapid methods and Lowicryls, 85
  reembedded sections, 360
Endoplasmic reticulum, immunoreactivity and labeling, 38
Endothelial cells, albumin–gold complex, 172

Index

"En face" observation, ligand and receptor interactions, 13, 16
Enterobacterium yersinia, immunonegative staining, 251
Enzyme digestion, advantages of LR White, 49
Enzyme–gold complexes
 assays for assessing biological activity, 121–122
 cytochemical controls, 124–125
 cytochemical labeling, 122–124
 cytochemistry of nucleic acids, 127
 DNase, 129, 132–133
 preparation, 119–121
 quantitative evaluations, 126
 research applications, 144–145
 RNase, 127, 129, 132
 tissue processing, 122
Eosin, counterstaining of semithin sections, 69
Epidermal cells, suspension preparation and immunogold labeling, 325
Epipolarization microscopy
 immunogold staining, 445, 448–451, 453, 464
 microscope performance, 463–464
 silver enhancement, 460, 461
Epon
 Lowicryl K4M and labeling, 91, 93
 polymerization, 264
 thin sections and LR White, 67
Epoxide embedding
 heavy metals, 43
 LR White and, 49
 reproducibility of results, 38
Esherichia coli
 bacteriology and immunonegative staining, 252
Etching
 antigens and osmication, 35
 LR White and epoxides, 49
Ethanol
 dehydration and Lowicryl embedding at subfreezing temperature, 307–309
 polymerization, 264
 rapid dehydration, 58
 tissue block processing, 53–54
Ethylene glycol, PLT method, 81
Eukaryotic DNA, cryosubstitution, 97

Exocytosis
 applications and limitations of methods, 292–293, 296–297, 299
 fixation methods, 289
 methods for detecting, 287–288
 protein A–gold labeling, 294
 study methods, 286
 tissue specimens, 287
Exoglucanase–gold complex, cytochemistry, 141
Extracellular matrix components, cytochemistry of, 133

F

Fab' fragments
 Au_{11} and purification, 424
 electron microscopy, 425–426
 ferritin, 427
 preparation, 423–424
 undecagold–antibody method, 417
Fab regions, IgG and cross-contamination, 26, 27
Fc regions, IgG and cross-contamination, 26, 27
Ferritin
 electron-dense markers, 276–277
 high-voltage electron microscopy, 346
 undecagold–antibody method, 427–428
Fibroblasts
 LDL–gold complex, 155
 ultrastructural hybridization, 383–384
Fixation, *see also* Glutaraldehyde
 aldehyde and antigens, 80
 animal cells, 231
 antigenicity and glutaraldehyde concentration, 70
 cell monolayers, 261–263
 chemical and immunoelectron microscopy, 280
 conventional and exocytosis, 289
 enzyme–gold complexes, 122
 formaldehyde, 231–232, 239
 immunocryoultramicrotomy, 267
 in situ hybridization, 381–382
 label-fracture, 181
 Lowicryls and embedding temperature, 79–80
 picric acid and glutaraldehyde, 51, 53
 plant cells and plasmolysis, 305
 plant cells in tissues, 230–231

Fixation (*continued*)
  protoplasts from plant cells, 231
  purity of reagents and reproducibility, 37–38, 41–42
  reembedded sections, 360
  retrograde neuronal tracing, 207
  sectioning and labeling, 33–36
  TAGO method and exocytosis, 289
  TARI method and exocytosis, 289
  ultrastructure preservation, 239
Flocculation
  albumin–gold complex, 165–166
  saturation and stabilization, 27, 29
  saturation curve, 7
Flow cytometry
  multiparameter immunogold technique, 432–433
  triple labeling, 434, 439
Fluorescent tracers
  four-color analysis, 441
  survival times, 222
Fluorochrome staining, gold and triple labeling, 435–436
Fluorogold, retrogradely labeled neurons, 211
Formaldehyde
  fixation, 231–232
  plant cell fixation, 305
  ultrastructure preservation, 239
Formalin, DNA fixation, 382
Formvar–carbon films, thin sections, 65, 366
Fracture-flip, label-fracture, 196, 198
Freeze-drying, low-temperature embedding, 107, 109
Freeze-fracture technique, electron microscopy, 176
Freezer
  cryosubstitution temperatures, 98
  ultraviolet light polymerization, 87
Freezing
  immunocryoultramicrotomy, 269–271
  label-fracture specimens, 181
  tissue preparation, 99–100
Fungus, cytochemistry of components, 141–142

G

Galactosidase–gold complex, cytochemistry, 136, 138
GAM-G40, triple labeling, 435

Gelatin capsules
  LR White processing, 54, 58
  ultraviolet light polymerization, 86–87
Gels
  gold colloids and detection of proteins, 399
  negative gold staining, 406
Glass
  cleanliness, 120, 165
  knives and cryosectioning, 271
  slide surface coating, 260
  slides, adhesion, and immunolabeling, 66–67
Globular proteins, molecular weight and adsorption, 8
Glucosidase–gold complex, cytochemistry, 136, 138
Glutaraldehyde
  concentration and antigenicity, 70
  cryosubstitution, 100, 103
  fixation of plant cells, 305
  picric acid and fixation, 51, 53
  processing methods and fixation, 34, 35
  purity and reproducibility of fixation, 37–38, 41–42
  RNA fixation, 382
  ultrastructural preservation, 89, 239
Glycoconjugates, cytochemistry, 133, 136
Glycogen, double fixation and retention, 136
Glycoproteins
  plant tissues and cross-reactivity, 320
  sodium metaperiodate, 36
Gold
  fluorochrome staining and, 435–436
  reflecting power, 451
  sol and treatment of blots before staining, 401
  titration of reagent, 455–458
Golgi apparatus, PLT embedding, 77
Green fluorescence (FITC), triple labeling and flow cytometry, 434
Grids
  preparation and handling, 230, 365–366
  thin sectioning and immunolabeling, 110–111

Index

H

Heat, polymerization, 88
Heavy metals
  epoxide embedding, 43
  Lowicryls, 79
Hematoxylin, counterstaining of semithin sections, 69
Hepatitis B, immunonegative staining, 251
Herpes simplex
  immunonegative staining, 251
  ultrastructural hybridization, 383, 384, 386
High-density lipoprotein (HDL)
  labeling, 152–153
  research applications of HDL–gold complex, 154
High-voltage electron microscopy (HVEM), research applications, 346
Histochemistry, immunological approach, 118
Human
  erythrocyte membrane and label-fracture, 183
  processing of skin for immunogold labeling, 333, 335
Hyaluronidase–gold complex, cytochemistry, 138, 141
Hybridization
  postembedding on thin sections, 392–393
  ultrastructural *in situ* nucleic acid, 385–387
Hydrogen peroxide, etching and antigens, 35–36

I

Ice
  cryosubstitution, 96–97
  low temperatures, 81
IgG antibodies
  F(ab') fragments, 422
  label-fracture and boar spermatozoa, 192
  protein A–gold and cross-contamination, 26–27
IgG–gold probes, electron-dense markers, 276–278
Immobilized substrate preparations, enzyme–gold complexes and biological activity, 121–122
Immunocryoultramicrotomy
  chemical prefixation, 280
  incubation and stabilization of thin sections, 273–275
  resin sections, 258
Immunocytochemical staining, colloidal gold, 367–368
Immunocytochemistry
  exocytosis and staining procedure, 293–294
  high-sensitivity research and LR White, 41
  plant sciences, 320
  plant specimens and acrylic resins, 304
  preparation of cells, 265, 267–271, 273
  research and resin interference, 358–360
  semithin cryosections, 275
  WGA-apoHRP-Au, 212, 215
Immunoelectron microscopy (IEM)
  chemical fixation, 280
  choice of markers and labeling techniques, 275–278
  colloidal gold as electron-dense marker, 337–340
  labeling and quality of primary antibodies, 179
  preembedding and postembedding of thin sections, 257–258
  preembedding and virus-infected cells, 258
  rapid embedding and diagnosis, 85
  virology, 280
  virus pathogenicity and morphology, 256
Immunoenzyme conjugates, electron-dense markers, 278
Immunofluorescence microscopy
  cytoskeletal elements, 228
  monoclonal antibodies, 438–439
Immunoglobulins, storage and deterioration, 66
Immunogold labeling
  blood cell preparation, 325
  cell surface, 335
  cellular components, 13, 16
  epidermal cell suspension preparation, 325
  human skin processing, 333, 335
  preembedding and cell surface-associated antigens, 339–340
  prefixation, 325
  rat kidney processing, 333
  silver enhancement, 444–445

Immunogold labeling (*continued*)
  single antigens, 326–327
  three distinct antigens on same cell surface, 329, 331
  two distinct antigens on same cell surface, 328–329
Immunogold negative staining, virology, 256
Immunogold–silver staining
  leukocyte cell surface antigens with monoclonal antibodies, 445–448, 464–465
  light microscopy and immunocytochemistry, 332
  physical developer, 446–447
  silver enhancement time, 454–455
  standardization, 453–454
Immunohistochemistry, WGA-apoHRP-Au, 212
Immunolabeling, *see also* Labeling
  advantages of protein A–gold, 20, 30
  cryosubstitution, 103, 105
  direct versus indirect techniques, 278
  Lowicryl K4M, 88–89
  Lowicryl HM23, 107
  Lowicryl thin sections, 110–111
  microtubules, 234
  semithin sections, 43, 68
  thin sections and on-grid labeling, 68–69
Immunonegative staining method
  bacteriology, 251–252
  double-labeling, 247–248
  immunolabeling schedule, 246
  methodology, 244–246
  quantitation, 248–249
  virology, 249–251
  viruses and variations of technique, 246–247
Immunoperoxidase method, whole-cell mount, 351
Immunoreactivity, osmium fixation, 317
Immunosorbent methods, immunoelectron microscopy, 244
Immunostaining, *see also* Staining
  antibodies and light microscopy, 362–363
  colloidal gold, 363–364
  electron microscopy and thick sections, 364–365
  microscopic techniques, 364
  peroxidase screening, 363

Incubation
  control experiments, 125
  immunocryoultramicrotomy and stabilization of thin sections, 273–275
India ink, double staining, 399
Infiltration
  catalytic polymerization, 61, 63, 65
  cryosubstitution, 100–101
  rapid and catalytic polymerization, 56, 58–59, 61
Influenza, immunonegative staining method, 250
Injection, retrograde neuronal tracing, 206
Ion exchange chromatography, undecagold cluster purification, 421

K

Kidney, processing of rat for immunogold labeling, 333

L

Label-fracture technique
  advantages and limitations, 192, 194–195
  choice of marker, 178, 181
  cryoprotection of specimens, 181
  fixation, 181
  fracture-flip, 196, 198
  fracturing and cleaning of replicas, 182–183
  freezing of specimens, 181–182
  images of replicas, 183
  mounting of specimens, 192
  preparation of cells, 178
  rationale and development, 177–178
  replica-staining, 196
  research applications, 176–177, 187, 192
Labeling, *see also* Immunolabeling
  advantages of protein A–gold, 30
  colloidal gold–lectin–ruthenium red staining, 347–349
  controls for nonspecific and pseudospecific labeling, 317–319
  DNase and enzyme–gold complexes, 129, 132
  fixation and sectioning, 33–36
  immunoelectron microscopy and choice of technique, 275–278
  lipoprotein–gold complex, 152–153

Index

Lowicryl K4M and Lowicryl HM20, 91, 93–94
Lowicryls and plant tissue sections, 315
osmicated tissue in LR White section without removal of osmium, 317
plant cell walls, 319
preembedding and postembedding of thin sections, 257–258
preembedding and preparation of cells, 258, 260
progressive lowering of temperature (PLT) method, 37
protein A–gold and double, 22, 25
protein A–gold and exocytotic release sites, 294
protein A–gold and single, 21–22
pseudospecific and plant tissues, 320
removal of osmium from LR White sections, 316–317
RNase and enzyme-gold complexes, 127, 129
sources of antibodies, 131–134
sources of error in protein A–gold, 25–27, 29–30
unosmicated LR White sections, 315–316
whole cell mounts, 351–352
Lead citrate, thin section counterstaining, 70
Leaves, ethanol dehydration and Lowicryl embedding at subfreezing temperatures, 307
Lectins
detection of glycoconjugates, 136
label-fracture, 183
Lectin–gold complex label-fracture, 181
Leukocytes
immunogold–silver staining and monoclonal antibodies, 445–448, 464–465
silver enhancement, 444–445
Ligand
adsorption to gold particles, 10–11
in section and "en face" observation, 12–13
Light, reactivity of physical developer, 462–463
Light microscopy
antibodies and immunostaining, 362–363
retrograde neuronal tracing, 208–209

Lipoproteins
labeling difficulties, 150
preparation, 150, 152
visualization, 154
Liquid nitrogen label-fracture specimens, 182
Listeria monocytogenes, bacteriology and immunonegative staining, 252
Locusta migratoria, exocytosis, 293
Low-density lipoprotein (LDL)
labeling, 152–153
molecular weight and adsorption, 8–9
research applications of LDL–gold complex, 155–158
Scatchard analysis, 5
topologically restricted receptors, 13
Lowicryl HM20
dehydrating agents, 81
freeze-drying and low-temperature embedding, 107, 109
Lowicryl K4M and labeling, 91, 93–94
progressive lowering of temperature (PLT) method, 43
thin sectioning, 110
Lowicryl HM23
cryosubstitution, 43, 101, 102–103
freeze-drying and low-temperature embedding, 109
future developments, 106–107
Lowicryl K4M
immunolabeling, 88–89
Lowicryl HM20 and labeling, 91, 93–94
low-temperature embedding, 109
rapid embedding, 85
thin sectioning, 110
tissue processing, 36–38, 41–43
Lowicryl K11M
cryosubstitution, 43, 101–102
freeze-drying and low-temperature embedding, 107, 109
future developments, 106–107
Lowicryls
advantages and versatility, 78–79
background staining, 125
composition and handling, 82–83
dehydration with N,N'-dimethylformamide at room temperature and embedding at -20°C, 309–310

478 Index

Lowicryls (*continued*)
  embedding of plant tissues, 306–307
  epoxide embedding and heavy metals, 43
  ethanol dehydration and embedding at subfreezing temperatures, 307–309
  fixation and embedding temperature, 79–80
  future prospects in the plant sciences, 310
  labeling plant tissue sections, 315
  storage and thin sectioning, 109–110
LR White
  chemical catalyst polymerization at subzero temperature, 313
  embedding of plant tissues, 310–311
  immunolabeling of sections, 68–70
  infiltration at 22°C and catalytic polymerization, 61, 63, 65
  labeling osmicated tissue without removal of osmium, 317
  labeling unosmicated sections, 315–316
  light and electron microscopy, 48
  osmicated tissue, 312–313
  physicochemical properties and advantages, 48–49
  polymerization and cross-linking, 70–71
  rapid embedding and Lowicryls, 85
  rapid infiltration and catalytic polymerization, 56, 58–59, 61
  removal of osmium and labeling, 316–317
  routine processing and polymerization by heat, 49, 51, 53–54
  sectioning tissue blocks, 66–68
  storage of tissue blocks, 65–66
  tissue processing, 36–38, 41–43
  unosmicated tissue, 311–312
Lymnaea stagnalis, methods of detecting exocytosis, 287–288, 292, 293
Lymphoblastoid cells, immunogold labeling *in situ*, 260
Lymphocytes
  bright-field light microscopy, 451
  triple labeling, 433, 435, 436, 438
"Lysis squirting" technique, labeling of cytoskeletal elements, 229

M

Macromolecules, *in situ* immunolabeling, 267
Maleimides, undecagold cluster, 422

Mannosidase–gold complex, cytochemistry, 138
Markers, immunoelectron microscopy and choice of labeling technique, 275–278
Medicine, triple-marker methodology, 440
Methanol, cryosubstitution, 100
Microchemistry, immunological approach in histochemistry, 118
Microtest cultures, embedding and evaluation, 264–265
Microtubules
  cleaving, 232–234
  dehydration and critical-point drying, 235
  electron microscopy, 236
  fixation, 230–232
  immunofluorescence, 236
  immunogold labeling, 234
  negative staining, 236
  postfixation, 235
Mitochondria, fixation, 305
Molecular weight
  proteins and adsorption parameters, 7–9
  undecagold clusters, 420
Monoclonal antibodies (McAb)
  advantages of triple labeling, 438–439
  immunogold labeling of single antigen, 326–327
  immunogold labeling of two distinct antigens on same cell surface, 328–329
  immunogold–silver staining of leukocyte cell surface antigens, 445–448
  labeling of plant proteins, 314
  localization of in rat kidney, 333
  research applications, 432–433
  titration, 458–459
Monocytes, bright-field light microscopy, 451
Monolayers
  processing, 261–263
  virus systems, 258, 260
Monomer, cross-linking and Lowicryls, 82–83
Mounting, label-fracture specimens, 181
Multiple retrograde labeling, WGA-apoHRP-Au, 220–221
Muscle cells, freeze-drying and low-temperature embedding, 109
Mutants, plant lines as controls, 319
Myoglobin, adsorption and pH, 9

Index

N

Negative charge, colloidal gold, 20
Negative gold staining, detection of proteins, 406–407
Negative staining
 lipoproteins and pH, 153, 154
 lipoproteins and visualization, 154
Neisseria gonorrhoeae, bacteriology and immunonegative staining, 251
Neuraminidase–gold complex, cytochemistry, 138, 141
Nickel, grids, 230
N,N'-dimethylformamide, dehydration at room temperature and embedding at −20°C, 309–310
N,N-dimethylparatoluidine (DMpT), chemical polymerization, 87–88
Nonspecific labeling
 microtubules, 234
 steps to reduce, 30
 trisodium citrate–tannic acid method, 29
Nucleic acids
 DNA and RNA probes, 382
 enzyme–gold complexes and cytochemistry, 127
Nucleoplasm, RNase and labeling, 129
Nucleus, RNase and labeling, 129

O

On-grid labeling, thin sections, 68–69
Osmicated tissue, LR White embedding, 312–313
Osmium tetroxide (OsO$_4$)
 antigenic sites and polypeptides, 80
 cryosubstitution, 98
 labeling osmicated tissue in LR White sections without removal, 317
 LR White and plant tissues, 312–313
 photochemical visualization, 49
 postfixation and enzyme–gold complexes, 122
 removal from LR White sections and labeling, 316–317
 survival of antigens, 35–36
Oxygen
 polymerization and acrylic plastic, 82
 polymerization and gelatin capsules, 54
Oxytocin
 exocytosis and posterior pituitary, 287
 rat CNS and exocytosis, 297

P

Pancreas, fixation and immunoreactivity of endoplasmic reticulum, 38
Pathogen, killing and examination, 246
Pectinase–gold complex, cytochemistry, 142
Perichromatin granules, RNase and labeling, 129
Peroxidase screening, immunostaining, 363
pH
 adsorption and, 9–10
 albumin–gold complex, 166
 colloidal gold particle size, 401
 DNase and labeling, 129
 enzyme–gold complexes, 120
 IgG and protein A, 276–277
 lipoprotein preparation, 150, 152
 negative staining and lipoproteins, 153, 154
 protein A–colloidal gold complexes, 20
Phaseolus vulgaris, enzymatically localized tracers, 212
Phosphate-buffered saline
 fracture-label fixation, 181
 microtubules and distortion, 231
Phosphines, undecagold cluster formation, 421
Phospholipase A$_2$–gold complex, cytochemistry, 142, 144
Phospholipids, cytochemistry, 142, 144
Physical developer
 immunogold–silver staining, 446–447, 454
 reactivity, light, and temperature, 461–463
Picric acid, fixation, 51, 53
Pigments, plant tissues and embedding, 306
Plant cells
 cleaving
  protoplasts, 232–233
  roots of seedlings, 232
 fixation
  plasmolysis, 305
  protoplasts, 231
  tissues, 230–231
 preparation of protoplast specimens, 229
 protein A–gold labeling and viruses, 250
Plants
 cell wall labeling, 319

Plants (*continued*)
cytochemistry of cell wall components, 141–142
glycoprotein cross-reactivity, 320
mutants as controls, 319
protein A–gold labeling and viruses, 250
pseudospecific labeling, 320
tissue embedding and Lowicryls, 306–307
tissue embedding and LR White, 310–311
Plant sciences
future prospects of Lowicryls, 310
immunocytochemistry and acrylic resins, 304, 320
Poliovirus, immunonegative staining and quantitation, 249
Polyclonal antiserum, labeling of plant proteins, 314
Polylysine, coating of plastic and glass surfaces, 260
Polymerization
catalytic and infiltration at 22°C, 61, 63, 65
catalytic and LR White embedding, 46, 54
chemical and LR White, 85
chemical catalyst at subzero temperature, 313
chemical catalysts and Lowicryls, 87–88
heat and Lowicryls, 78, 88
Lowicryls K11M and HM23, 102, 103
LR White and cross-linking, 48, 49, 70–71, 311
rapid infiltration and catalytic, 56, 58–59, 61
routine processing of LR White and heat, 49, 51, 53–54
ultraviolet light
cryosubstitution, 101
Lowicryls, 86–87
plant tissue pigments, 306
rapid embedding, 85
tissue block size, 80
Polysaccharides, cytochemistry, 133, 136
Postembedding
correlative immunocytochemistry, 359
hybridization on thin sections, 392–393
thin sections and preembedding, 257–258
Postfixation
microtubules, 235
OsO_4 and enzyme–gold complexes, 122
Postosmication
antigenicity and photochemical visualization, 34

cryoultramicrotomy, 36
membrane preservation and antigen damage, 49
omission and cross-linking, 35
Preembedding
correlative immunocytochemistry, 359
immunoelectron microscopy of virus-infected cells, 258, 260–265
immunogold labeling and cell surface-associated antigens, 339–340
thin sections and on-grid labeling, 69
thin sections and postembedding, 257–258
ultrastructural hybridization, 382–383
Prefixation, antigenicity, 257
Progressive lowering of temperature (PLT) method
acrylic resins and embedding methods, 37, 43
choice of dehydration agent, 81–82
embedding and Lowicryls, 77–78
methods of achieving low temperatures, 80–81
protocols, 83–84
ultrastructural preservation, 89
Projection neurons, cytochemistry, 221
Prokaryotic DNA, cryosubstitution, 97
Protein A, IgG antibodies and cross-contamination, 26–27
Protein A–colloidal gold complexes
advantages in immunolabeling, 30
double labeling, 22, 25
label-fracture, 181
labeling of exocytotic release sites, 294
preparation, 20–21
pseudospecific labeling and plant tissues, 320
single labeling, 21–22
sources of error in labeling, 25–27, 29–30
viruses, 250
Protein–gold complexes
broad reactivity and ease of preparation, 20
detection of proteins, 398–399
interaction with cells, 10–13, 16
Proteins
adsorption and retrograde neuronal tracing, 205–206
detection by gold colloids, 398–399, 408–409
detection of specific on blots, 401–403

Index

molecular weight and adsorption parameters, 7–9
nonspecific detection on blots, 399–400
quantitative determination, 407–408
staining performed sequentially on same blot, 404–406
Proteoglycans, cryosubstitution, 97
Protoplasmic faces, fracture-label technique, 177–178
Protoplasts
  cleaving, 232–233
  fixation, 231
  preparation of specimens, 229

Q

Quantitation
  enzyme–gold complexes, 126
  gold cluster labeling, 424–425
  immunonegative staining method, 248–249
  ultrastructural hybridization, 387, 389–391

R

Radioautography, freeze-fracture technique, 194
Radiolabeled substrates, enzyme–gold complexes and biological activity, 121
RAG-G40, triple labeling, 435
Rat
  hepatocytes and asialofetuin, 10
  kidney processing and immunogold labeling, 333
Receptors
  in section and "en face" observation, 12–13
  interaction of protein–gold complexes, 10–11
Red fluorescence (PE), triple labeling and flow cytometry, 434
Reembedded sections
  embedding, 360
  fixation, 360
  limitations of method, 375, 377
  removal of resin, 361–362
  sectioning, 360–361
Reflection contrast microscopy
  immunogold–silver stained cell surface antigens, 450
Replica-staining, label-fracture, 196

Replica technique, "en face" observation, 13
Resins, *see also* Lowicryls; LR White
  deplasticization and rate of removal, 369–370
  embedding and antibody binding capacity, 257–258
  ligand and receptor interactions, 13
  reembedded sections and removal, 361–362
Reticuloendothelial cells, plasma and ligand–gold complexes, 13
Retrograde labeling, WGA-apoHRP-Au, 210–211
Retrograde neuronal tracing
  adsorption to proteins, 205–206
  electron microscopy, 209
  injection, 206
  light microscopy, 208–209
  research applications, 209–212, 215, 220–221
  survival time, 207
  tissue fixation, 207
  tissue sectioning, 207–208
Ribosomal particles, RNase and labeling, 129
RNA
  cell pretreatment, 384
  fixation, 381–382
  in situ hybridization, 381
  nucleic acid probes, 382
RNase
  research applications of enzyme–gold complexes, 127, 129
Roots
  cleaving of plant cells, 232
  Lowicryl embedding, 307
Rotaviruses, immunonegative staining, 250
Ruthenium red, cell surface contrast, 348

S

Salts, flocculation and color change, 165–166
Saturation
  finite number of adsorption sites, 5
  flocculation and stabilization, 27, 29
Scanning electron microscope (SEM)
  fracture-label technique, 177
  transmission electron microscope and comparison of images, 353–354

Scanning transmission electron microscope (STEM), undecagold–antibody method, 415, 416
Scatchard analysis, adsorption of proteins to gold particles, 5–6
Sectioned, labeled-replica (SLR) method, postreplication labeling of E-leaflet molecules, 195
Sectioning
  colloidal gold–lectin–ruthenium red method, 349–351
  deplasticization and thickness, 368–369
  LR White blocks, 66–68
  reembedded sections, 360–361
  retrograde neuronal tracing, 207–208
Seed cotyledons, Lowicryl embedding, 307
Semithin sections
  counterstaining, 69
  immunocytochemistry and cryosections, 275
  immunolabeling, 68
  LR White blocks, 66–67
Serum, cross-contamination and labeling error, 26
Silver
  determination of enhancement time, 454–455, 457–458
  enhancement and light microscopy, 208, 209
  immunogold–silver staining, 332, 444–445
  light reflection, 450–451
  optimal enhancement interval, 459–461
  protein assays, 403
  retrogradely transportable dyes, 211
Single labeling, protein A–gold complexes, 21–22
Single retrograde labeling, WGA-apoHRP-Au, 215, 220
Sinusoidal endothelia, albumin–gold complex, 172
Size, colloidal gold particles
  adsorption parameters, 7
  citric acid, 400
  distribution analysis of suspensions, 3–4
  double labeling, 22, 24
  labeling intensity, 178, 181
  pH and, 401
  triple labeling, 440
Size exclusion chromatography, purification of undecagold clusters, 420

Skin, human and immunogold labeling, 333, 335
Slam-freezing, cryosubstitution, 99
Sodium metaperiodate
  cytochemical labeling, 124
  LR White, 49
  unmasking and antigens, 35–36
Somatostatin, labeling and heavy fixation, 34
Sophora japonica, labeling and removal of osmium, 317
Spectrophotometer, estimation of flocculation, 166
Spermatozoa, label-fracture and mammalian, 187, 192
Spinacia oleracea, label-fracture, 192
Stabilization
  albumin–gold complex, 166–167
  immunocryoultramicrotomy and stabilization of thin sections, 273–275
  protein A–colloidal gold complexes, 21
  saturation and flocculation, 27, 29
Staining, *see also* Immunostaining
  blots and gold sols, 401
  double, 398–399
  gold colloids and citric acid, 400
  gold colloids and white phosphorus, 400
  immunochemistry and exocytosis, 293–294
  immunogold–silver, 332
  lipoproteins and visualization, 154
  negative gold, 406–407
  quantitative determination of protein, 407–408
  sequential on same blot, 404–406
  triple labeling, 434
Staphylococcus aureus, protein A as stabilizer, 20
Storage
  albumin–gold complex, 168–169
  immunocryoultramicrotomy, 268–269
  Lowicryl blocks, 109–110
  LR White and tissue blocks, 65–66
Surface replica technique, visualization of cell surfaces, 194

T

T4 bacteriophage, vaccinia virus, immunonegative staining, 251

Index

Tannic acid
  contaminant in trisodium citrate–tannic acid procedure, 29
  tissue fixation, 286
Tannic acid–glutaraldehyde–OsO$_4$ (TAGO) method, fixation and exocytosis, 289, 293
Tannic acid–Ringer incubation (TARI) method, fixation and exocytosis, 289, 293
Temperature
  advantages of Lowicryl resins, 77, 78
  chemical catalyst polymerization at subzero, 313
  cryosubstitution, 97–98, 100
  dehydration agents, 81–82
  dehydration with N,N'-dimethylformamide and embedding, 309–310
  ethanol dehydration and Lowicryl embedding at subfreezing, 307–309
  fixation and Lowicryls, 79–80
  heat polymerization, 88
  infiltration at 22°C and catalytic polymerization, 61, 63, 65
  Lowicryl K4M and Lowicryl HM20, 93
  Lowicryls and cryosubstitution, 37, 101
  LR White and routine processing, 51
  methods of achieving low, 80–81
  progressive lowering technique (PLT), 83–85
  reactivity of physical developer, 463
Tetrachloroauric acid
  preparation of gold probe, 20
  reduction and size distribution, 3
Thermal denaturation temperature (T_m), *in situ* hybridization, 381
Thick sections
  electron microscopy and immunostaining, 364–365
  LR White blocks, 67
  resin removal and immunostaining, 370–371
Thin sections
  counterstaining and Lowicryls, 111
  counterstaining and LR White, 69–70
  electron microscopy, 365–368
  Formvar–carbon film, 65
  incubation and stabilization for immunocryoultramicrotomy, 273–275
  on-grid labeling, 68–69
  lipoproteins and visualization, 154
  Lowicryl and immunolabeling, 110–111
  Lowicryl blocks, 109–110
  LR White blocks, 67–68
  postembedding hybridization, 392–393
  preembedding and postembedding, 257–258
  resin removal and immunostaining, 371–372
Tissue blocks
  LR White
    catalytic polymerization, 63
    sectioning, 66–68
    storage, 65–66
    rapid infiltration and catalytic polymerization, 56, 58–59
    routine processing and LR White, 53
Tissue processing, enzyme–gold complexes, 122
Tobacco, labeling LR White sections, 315, 316
Tobacco mosaic virus, colloidal gold–protein A–antibody complexes, 250
Transmission electron microscopy
  scanning electron microscope and comparison of images, 353–354
  surface replicas and colloidal gold labels, 347
Triethylammonium bicarbonate, undecagold cluster purification, 420
Triple labeling
  advantages, 438–439
  antibodies, 433–434, 436, 438
  cells, 433
  flow cytometry, 434
  future developments, 440–441
  gold versus fluorochrome staining, 435–436
  immunogold stained lymphocytes, 435
  limitations, 439–40
  retrograde neuronal tracing, 215
  staining, 434
Trisodium citrate–tannic acid method, nonspecific labeling, 29
True Blue, silver intensification, 211
Trypanosoma brucei, topologically restricted receptors, 13

U

Ultracryomicrotomy, tissue fixation, 35
Ultrastructural *in situ* nucleic acid hybridization
 cell pretreatments, 383–384
 fixation, 381–382
 future developments, 392
 nucleic acid probes, 382
 preembedding, 382–383
 principles of technique, 381
 quantitation, 387, 389–391
 research applications, 380, 393
Ultrastructure, deplasticization and preservation, 372, 375
Ultraviolet light
 plant tissues and pigments, 306
 polymerization, 86–87
Undecagold–antibody method
 activation of gold cluster, 422–423
 advantages, 413, 415
 preparation of undecagold, 417–420
 purification of cluster, 420–422
 quantitation, 424–425
 research applications, 427–428
Unosmicated tissue
 labeling LR White, 315–316
 LR White embedding, 311–312
Uranyl acetate
 thin section counterstaining, 70
 Lowicryl thin sections, 111

V

Vaccine, *Bacteroides nodosus* and immunonegative staining, 251
Vasopressin, rat CNS and exocytosis, 297
Very low density lipoprotein (VLDL), labeling, 157
Virology
 immunoelectron microscopy, 280
 immunogold negative staining, 256
 immunonegative staining, 249–251
Viruses
 double labeling, 25
 immunonegative staining, 244, 249–251
 preembedding immunoelectron microscopy of infected cells, 258, 260–265
 variations of immunonegative staining method, 246–247

W

Wheat germ agglutinin (WGA)
 label-fracture, 183, 187
 receptor sites and label-fracture, 192
Wheat germ agglutinin-apoHRP (WGA-apoHRP-Au)
 enzymatically localized tracers, 211–212
 gold protein ratio, 224
 immunocytochemistry, 212, 215
 immunohistochemistry, 212
 projection neurons, 221
 retrograde labeling, 210–211
 retrograde tracing studies, 204
 WGA and gold–protein ratio, 224
Wheat germ agglutinin-horseradish peroxidase (WGA-HRP-Au)
 double labeling, 211
 enzymatically localized tracers, 212
 retrograde tracing studies, 204
White phosphorus method
 gold colloid preparation, 400
 nonspecific labeling, 29
Whole-cell mounts, colloidal gold–lectin–ruthenium red, 351–352
Woody tissues, Lowicryl embedding, 307

X

Xylanase–gold complex, cytochemistry, 141

Z

Z contrast, Lowicryl resins, 79